T0414113

Nobel Lectures

Chemistry
2016 – 2020

Nobel Lectures

Including Presentation Speeches
and Laureates' Biographies

Physics

Chemistry

Physiology or Medicine

Economic Sciences

Nobel Lectures

Including Presentation Speeches
and Laureates' Biographies

CHEMISTRY

2016 – 2020

Editor

Sven Lidin

Lund University, Sweden

World Scientific

NEW JERSEY • LONDON • SINGAPORE • BEIJING • SHANGHAI • HONG KONG • TAIPEI • CHENNAI • TOKYO

Published by

World Scientific Publishing Co. Pte. Ltd.
5 Toh Tuck Link, Singapore 596224
USA office: 27 Warren Street, Suite 401-402, Hackensack, NJ 07601
UK office: 57 Shelton Street, Covent Garden, London WC2H 9HE

NOBEL LECTURES IN CHEMISTRY (2016–2020)

ISBN 978-981-12-6057-5 (hardcover)
ISBN 978-981-12-6058-2 (ebook for institutions)
ISBN 978-981-12-6059-9 (ebook for individuals)

For any available supplementary material, please visit
https://www.worldscientific.com/worldscibooks/10.1142/12971#t=suppl

Printed in Singapore

PREFACE

Failure!

It is the most important feature of human endeavors. Failure is important not only because, as we all know, it is such a common feature of our existence, but because of all the good it brings. Failure is the best of teachers. Why else would we talk of trial and error rather than trial and success? Small children learn to walk and talk by trial and error and great thinkers achieve greatness through the same process. Failure also teaches us the hardest of human virtues, humbleness.

Most importantly failure is a prerequisite for success. Our Laureates know that better than anyone. Attempting to achieve what is unknown makes us rather more familiar with failure than with success. Success is a rare visitor. Failure is an old and trusted friend. To succeed you must go where no one went before, do what no one did before and think what no one thought before, but in doing so you are also almost certain to fail. Most often there is a reason why no one went, did or thought this way before, or at least why they never told anyone else that they did so. The fear of ridicule is a powerful driving force and the only thing that is more frightening than being ridiculed by others is being ridiculed by yourself. To withstand the onslaught of doubt takes great courage and a willingness to embrace failure. If you never fail, you haven't earned the right to succeed.

Why then do we need prizes for success? Is not success its own reward? The Nobel prizes that have so well deservedly been bestowed on our Laureates are not given because the Laureates need the prizes, but rather because the prizes need the Laureates. A prize is never more important than the people on whom it is bestowed.

When the Nobel prize was first awarded the timing was very fortunate. This was a period of great change and the Nobel prizes awarded at that time captured and reflected that change. Since then, the pace of science has been ever accelerating and to keep abreast with the development is one of many challenges for modern science.

The greatness of the Nobel prize comes from the greatness of the Laureates. But what is then the purpose of the prize? The purpose of the prize is to serve as a beacon of success in an ocean of failure. The importance of the prize is to show the importance of trying and failing. The greatness of the prize is to show the great value of courage. And what value the Nobel prizes in Chemistry from the years 2016 to 2020 has brought to mankind!

The 2016 Nobel Prize in Chemistry was awarded jointly to Jean-Pierre Sauvage, Sir J. Fraser Stoddart and Bernard L. Feringa *for the design and synthesis of molecular machines.*

Playfulness is at the core of good science and how else could we describe the work on molecular machines that earned this trio the 2016 Chemistry prize? In fact, in many other ways: Yes, there is a great deal of joyous, childish, what-would-happen-if-we-did-this-fullness in the Homo Ludens approach to synthesis, but there is also a great deal of seriously useful aspects of this work. While there are no technological applications of the discoveries to date and no nanocars to drive, the functionality-driven synthesis has driven development of synthetic methods in entirely new directions.

The 2017 Nobel Prize in Chemistry was awarded jointly to Jacques Dubochet, Joachim Frank and Richard Henderson *for developing cryo-electron microscopy for the high-resolution structure determination of biomolecules in solution.*

Electron microscopy was a singularly useful technique for structural elucidation, allowing for the study of minute crystals even before the work of the three Laureates of 2017. What cryo-electron microscopy brought onto the scene was the possibility to study molecules outside of a crystal. Apart from doing away with the rather difficult step of crystallization, this means that the molecules may be studied in conformations close to those they adopt in solution, and it even makes it possible to determine relative ratios of different conformations or different stages of a transformation.

The 2018 Nobel Prize in Chemistry was one half awarded to Frances H. Arnold *for the directed evolution of enzymes,* **the other half jointly to George P. Smith and Sir Gregory P. Winter** *for the phage display of peptides and antibodies.*

It was a long-held view that proteins could not be improved upon since they have been developed to perfection by Nature over eons and this is essentially true, but perfection is in the eye of the beholder. Proteins are certainly optimized for the, often quite general, tasks they perform in living organisms, but if we look to proteins for highly specialized function, there is a lot of room for improvement. The 2018 Laureates created the necessary tools for this development and made the first inroads into engineering proteins for specific functions never intended or attempted by Nature.

The 2019 Nobel Prize in Chemistry was awarded jointly to John B. Goodenough, M. Stanley Whittingham and Akira Yoshino *for the development of lithium-ion batteries.*

Batteries have been with us since the days of Volta and while they developed a lot over the years, it is telling that the lead-acid accumulator, invented in the mid-19th century is still the most common source for electricity in starting motors for internal combustion engines. A safe, efficient lithium-ion battery was long on the wish-list for powering mobile equipment beyond low consumption devices. The tremendous development in this area was driven by a number of ingenious discoveries made by the three 2019 Laureates over a period of time. To provide energy solutions for today is for the greatest benefit to mankind.

The 2020 Nobel Prize in Chemistry was awarded jointly to Emmanuelle Charpentier and Jennifer A. Doudna *for the development of a method for genome editing.*

It is a little ironic that bacteria have immune systems to help them defend themselves from infectious disease. The discovery of the ancient CRISPR/Cas system in *Streptococcus Pyogenes* was itself important, but this was just the starting point for the work of the two Laureates who realized that the search-and-destroy mechanism of the CRISPR/Cas system could be tailored to edit any DNA molecule at a precisely predetermined site. The importance of this development cannot be overestimated. It is indeed rare that a fundamental discovery permeates science and technology as quickly as this did, being used in research, academic and industrial, worldwide within a year.

The prizes 2016 to 2020 certainly show chemistry at its best. These discoveries in chemistry have given us tools for understanding and improving our world. The combination of playfulness and dedication, of hard work and brilliant insight, yields results that are diamonds of knowledge: they sparkle and glitter in their beauty and they are intensely useful and practical. There are no boundaries for hard facts.

Sven Lidin

CONTENTS

Chemistry 2016

Jean-Pierre Sauvage, Sir J. Fraser Stoddart and Bernard L. Feringa

"for the design and synthesis of molecular machines"

The Nobel Prize in Chemistry, 2016

Presentation speech by Professor Olof Ramström, Member of the Royal Swedish Academy of Sciences; Member of the Nobel Committee for Chemistry, 10 December 2016.

Your Majesties, Your Royal Highnesses, Esteemed Nobel Laureates, Ladies and Gentlemen,

Machines are an integral part of human development, helping us to perform tasks that often fall beyond our capacities. Over the millennia, our society has enjoyed an ever increasing plethora of useful machines for various purposes, in many ways leading to an enhanced quality of our lives. This progress has, in particular, accelerated since the industrial revolution, with its key discoveries resulting in a giant leap forward and dramatically changing the world.

Today, we are at the dawn of a new revolution that will bring us yet another giant leap forward. Humankind has always striven to push the limits of machine function and construction, for example regarding how small they can be made. The ultimate limit of this endeavour is to make molecular-sized machines, structures that are less than a thousandth of the width of a human hair. This fundamental challenge has been met by this year's Laureates, who have successfully demonstrated that the rational design and synthesis of molecular machines are indeed possible.

A groundbreaking step in this development was taken in the early 1980s, when Jean-Pierre Sauvage and his group discovered an efficient way to make molecules that are attached to each other using mechanical bonds. The group thus managed to link two molecular rings together in a so-called catenane, where the rings could move freely relative each without being separated. This marked a breakthrough towards molecular machinery, and Sauvage was able to show how such structures can undergo controlled motion.

By the turn of the 1990s, Fraser Stoddart and his group made other important advances towards molecular machinery. For example, the group used another type of mechanical bond and developed so-called rotaxanes in which a ring-shaped molecule was tied to move between set positions along an axle. Both Stoddart and Sauvage were also able to show how the movement in these structures could be controlled from external input, and subsequently developed a wide range of machine-like structures, such as molecular muscles, actuators, elevators, memories, motors and pumps.

The motor components are of central importance in machines, able to drive other parts of the constructions. By the late 1990s, Ben Feringa and his group made a significant breakthrough when they demonstrated a molecular rotary motor. The construction was driven by light and heat and was based on isomerisable bonds and molecular asymmetry, where the motor parts could rotate unidirectionally relative each other. The group was later able to improve the design to create motors that can rotate in either direction at very high speeds, and showed how the components can affect the rotation of much larger objects. In a more playful example, Feringa's group also constructed a four-wheel drive "nanocar" that can move over a surface.

Through the design and synthesis of very challenging structures, combined with the understanding and development of controlled motion and function, Sauvage, Stoddart and Feringa have created functional molecular machines. Their work has formed the basis for an entirely new field of research, for which the three Laureates have been groundbreaking pioneers and sources of inspiration.

Jean-Pierre Sauvage, Sir Fraser Stoddart, and Ben Feringa:

You are being awarded the Nobel Prize in Chemistry for the design and synthesis of molecular machines. On behalf of the Royal Swedish Academy of Sciences I wish to convey to you our warmest congratulations. May I now ask you to step forward and receive your Nobel Prizes from the hands of His Majesty the King.

Jean-Pierre Sauvage. © Nobel Prize Outreach AB. Photo: A. Mahmoud

Jean-Pierre Sauvage

Biography

I WAS BORN ON OCTOBER 21, 1944 IN PARIS, just before the end of the second world war and a few months after Paris had been liberated by the allies and the French army led by General Charles de Gaulle (August 19–25, 1944). My mother's name was Lydie Angèle Arcelin and her family came mostly from Normandy. She was born in 1920. My father was Camille André Sauvage and came from the northern part of France. Camille Sauvage was known as a successful jazz musician, both conductor and clarinettist on top of being a composer. Just after the war, he was a popular jazz player and, later on, he composed for French radio, television and movies. When I was a baby, my parents broke up. I stayed with my mother while my father departed to pursue an artist's life. While I was still a young child, my mother met an officer in the air force, Marcel Louis Grosse, and they founded a family. I was thus fortunate to have a stepfather who took care of me until I became an adult and whose role was truly that of my father. Thus, I always considered him as my real father and I still do. Since my stepfather was in the military and because my parents loved to travel from one place to another, I had a very mobile childhood. When I was three years old, we moved to North Africa, as Algeria and Tunisia were still French colonies at that time. I still have vivid memories of our time in Zarzouna, a small village close to Bizerte in Tunisia. I also spent some time near Oran, in Algeria. I went to school in Tunisia between the ages of 5 and 7. At that time, the French kids and the local Tunisian kids were together in the same classes and I do not believe there were any difficulties related to the fact that the French kids and the Tunisian ones were mingling.

My mother and my grandmother, Suzanne Arcelin, were very close to one another and my grandmother, who used to live in Paris, visited our family a few times. On one particular occasion, we went to southern Tunisia for sightseeing and the photo which is shown below is particularly representative. It was taken when I was about 4 years old.

Figure 1. In Tunisia with my mother and, on the left, my grandmother.

From 1951 to 1952, my family spent some time in the USA since my stepfather had to become a military engineer in the field of radar applications, a relatively new technology at the beginning of the 50s. We thus spent 6 months in Saint Louis, Missouri, followed by an additional period of 6 months in Denver, Colorado. I had no difficulty in adapting and, in Denver in particular, I used to play with the other children of our neighbourhood. In Saint Louis, we used to go to the movie theatre from time to time and it was of course a very enjoyable event for me. The picture shown below is interesting in the sense that one can see who were the most popular actors of the time.

When we returned to France, we started a long period of itinerancy, spending a few months in a given place before moving to a new city. I thus went to 4 or 5 different schools in the western part of France and in the Paris area when I was 8 to 10 years old. My mother became ill when I was about 10, and it was a difficult moment for my family. She had contracted tuberculosis, which was a very serious disease in the 50s. I was thus mostly with my grandmother for about a year, in the family village of Pacy-sur-Eure in Normandy. After this period, I was again with my parents, who moved to the eastern part of France in the Lorraine region. When I was 15 years old we moved to a small village in the north of Alsace, Drachenbronn, because my stepfather had been transferred to the radar station located on the Maginot Line named "Base Aérienne 901 Drachenbronn." This move coincided with the beginning of my high school studies and I thus went to the high school in Haguenau, a middle-size city not far from air base 901.

Figure 2. After a movie in Saint Louis (USA) in 1951.

I started to be very interested in chemistry when I was 15 or 16 years old and, in particular, I liked to play with natural molecules such as chlorophylls which I extracted from plants. I had a small and very primitive chemistry lab in the cellar where I was separating chlorophylls on paper or distilling various mixtures. At school, I was probably better in mathematics than in physics or chemistry but the interest of pupils for various topics is obviously very dependent on the personality of the teacher.

After my 'baccalauréat' (the French examination obtained at the age of 18 which enables one to begin university), I decided to enter a special and highly competitive structure named 'classes préparatoires' which was aimed at preparing the young people to compete for admission to engineering schools. Thus, after two years of a rigorous regime where leisure was limited to a strict minimum, I succeeded and I was admitted to the Chemical Engineering School of Strasbourg. This was exactly what I wanted since I could stay in my new but already beloved city.

I obtained my engineering diploma in 1967 and started my PhD thesis in 1967 under the guidance of Professor Jean-Marie Lehn. Jean-Marie had founded his own research team a few years before I started to work with him. He was only 5 years older than I and his research team was developing rapidly in terms of size and breadth of his research interests. Being more a physical chemist up until 1967, he had become interested in making new molecules with novel properties when I started with him. At the time Bernard Dietrich, formally a technician, decided to start his studies so as to become a graduate student, hoping also to do a PhD thesis. Bernard rapidly became my best friend and we worked together in the friendliest atmosphere imaginable between 1968 and 1971. Under the guidance of Jean-Marie we were highly successful, since we made the first macrobicyclic compounds able to encapsulate various ions, including alkali and alkaline earth metal cations (cryptands and cryptates for the metal-free compounds and their complexes respectively). In Jean-Marie Lehn's research group, I was able to acquire a solid background in organic and physical chemistry, thanks to my own work and to various seminars and group meetings as well as to the many hours I used to spend in the library. Discussions with Professor Lehn and with other members of the group were also very fruitful. Equally important was the influence that Jean-Marie had on me in terms of the relationship between researchers within a group. In particular, I enjoyed his way of managing a research team. He was very direct, placing basically no barrier between him and the PhD students or postdocs he was working with. In other words, hierarchy was reduced at its minimum and this is something which I tried to preserve later on in my own group. Inspired by Jean-Marie's passion for science, I also became enthusiastic and determined to devote my life to research.

I met my wife Carmen in 1967 and we got married in February 1971. Carmen was a student in History of Art and Archaeology. She was particularly interested in ancient ceramics and had participated in several excavation campaigns on the Anatolian plateau, in Turkey. Although we were not especially religious, we had a religious wedding mostly to respect family traditions (Figure 3). The wedding took place in Thierenbach, a village in the south of Alsace which used to be a place of pilgrimage and which is nowadays famous for its basilica.

We were very happy to become parents of a baby boy, Julien Clément Sauvage, on July 13, 1975. It was a great joy for us. Since Julien's early childhood we have been very close to him. In 2011 Julien married Diana, originally from Colombia, and since 2012 they settled down in San Francisco. They had a baby on April 9, 2016 so that Carmen and I became the happiest grandparents in the world.

At the end of my PhD thesis work, I obtained a CNRS position as 'chargé de recherche' (research assistant) in Jean-Marie Lehn's group, which corresponded exactly to what I was so eager to get.

Figure 3. The wedding ceremony in Thierenbach, February 8, 1971.

After my PhD thesis, I obtained a postdoctoral fellowship in Oxford (UK) where I spent a year in the research group of a very visible organometallic chemist, Dr Malcolm L. H. Green. Malcolm was considered as one of the most brilliant former students of Professor Geoffrey Wilkinson (1973 Nobel Laureate in chemistry with Professor Ernst Otto Fischer). He was a very influential person in expanding my interests to transition metal chemistry and organometallic chemistry. He was also a friendly person who used to do experimental work by himself from time to time. Life in Oxford was particularly pleasant and we used to enjoy the city, its colleges and its parks. We easily adopted the way of life of the other members of the Oxford community.

After my postdoctoral stay in Oxford, I came back to Strasbourg and more precisely to the Lehn laboratory as a permanent researcher. After some work in the field of chiral crown-ethers, Jean-Marie and I initiated a research project in a new field, at least in Strasbourg. It was related to photochemistry and solar energy. The first oil crisis took place in 1973 and it was a clear signal that alternative energies had to be found and that sustainable energies were crucially needed. Solar energy was an obvious and especially attractive option. A particularly appealing project was that of splitting the water molecule to H_2 and ½ O_2 in order to generate a non-polluting fuel, H_2, using photonic energy. Such a big project had already been explored for many years by several groups and there were already discussions and research works published on this general topic in the 50s and later on. More or less at the same time, the ruthenium complex $Ru(bipy)_3^{2+}$ (bipy : 2,2'-bipyridine) was shown to display promising electronic properties in its ground or excited states, in particular in relation to electron transfer and potentially photochemical water splitting. In 1977 we published one of the very first systems leading to photochemical water reduction to H_2 based on a combination of species such as in particular $Ru(bipy)_3^{2+}$ as a photoactive species and $Rh(bipy)_3^{3+}$ as an electron relay leading to H_2 formation. After two years of studies on this original system (with Jean-Marie Lehn and Michele Kirch) and related ones as well as on the development of a light-driven oxygen generating system from water (with Jean-Marie Lehn and Raymond Ziessel), I was lucky enough to be promoted to the position of CNRS Research Director, the equivalent of University Professor. I thus founded my own research group in 1980, at first with two highly motivated PhD students, Pascal Marnot and Romain Ruppert. After one or two years, Jean-Paul Collin, a CNRS fellow, and Marc Beley, an Associate Professor, joined our small team. Simultaneously a good friend of mine, Christiane Dietrich-Buchecker, also joined.

As it is often the case for young research teams, we tackled several research projects in parallel in relatively remote areas. Electrochemical reduction of CO_2 using $[Ni\ cyclam]^{2+}$ as an electrocatalyst led to remarkable data since CO_2 could be reduced very selectively to CO in water. This was somewhat surprising since H_2 was expected to be obtained as a major reduction product. We also did work in homogeneous catalysis and inorganic photochemistry. In this latter field, our projects were mostly triggered by a collaboration with David R. M. McMillin, who was on sabbatical leave in Strasbourg. David was an already well established photochemist and photophysicist. He was a professor at Purdue University (West Lafayette, Indiana, USA) and his main field of research was photochemistry of copper(I) complexes. This collaboration between our group, with its skill in organic synthesis, and David led to a series of particularly interesting photoactive copper complexes. Perhaps even more importantly, it led to a copper(I) complex containing two intertwined organic ligands which appeared to be the ideal precursors to a compound comprising two interlocking rings. It was thus very tempting to jump from inorganic photochemistry to interlocking ring compounds which, at the beginning of the 80s, seemed to be practically inaccessible molecules. This jump was made possible due to the expertise of Christiane Dietrich-Buchecker, who was a great organic chemist. After a few discussions within our team, we decided to take the risk and to embark on a totally new project concerned with the synthesis of catenanes (i.e. interlocking ring compounds). Within a few months, Christiane was able to develop an efficient preparative procedure for making our first [2]catenane (containing 2 interlocking rings). Respectable quantities could be obtained: batches of 0.5g could be prepared, in particular by Jean Weiss, a PhD student also supervised by Christiane, and the first compound was fully characterised using a variety of techniques. ^{1}NMR provided the first convincing evidence that a [2]catenane was indeed produced. These experiments were carried out by Jean-Pierre Kintzinger, a friend of mine and the brother-in-law of Christiane, who was at the same time an NMR expert. We published our first paper in this field in 1983 in an acceptable but not high-impact journal (*Tetrahedron Letter*). For us it was the beginning of a new era, mostly but not exclusively devoted to interlocking rings compounds and knotted molecules. The field has often been referred to as "Chemical Topology" due to the fact that the compounds have non-planar molecular graphs. In other words, contrary to almost all the molecules known, it is impossible to draw them in a plane (i.e. a sheet of paper) without crossings, regardless of the deformation the molecule can be subjected to.

Besides chemical topology and molecular machines, our group has been active in various relatively remote fields. The principal alternative research area has been that of artificial photosynthesis, with a particular emphasis on photo-induced charge separation, one of the key processes of natural or artificial photosynthesis. In order to elaborate efficient models of the natural photosynthetic systems, and in view of realising the complete water splitting cycle in the future, our group synthesised numerous multicomponent complexes, either incorporating metal-complexed porphyrins or second or third row transition metal complexes (Ru, Os, Rh and Ir) able to undergo light-induced charge separation. Following the synthetic work, the photochemical and photophysical properties of most of the compounds were investigated in various places by more physical chemistry-oriented research teams than ours. In particular a long-term collaboration with renowned photochemists located in Bologna turned out to be especially pleasant and fruitful (Balzani and his co-workers, University of Bologna, or Flamigni and Barigelletti, Consiglio Nazionale delle Ricerche, Bologna). Some of the charge separated states were shown to be remarkably long-lived, thus paving the way to real artificial photosynthetic devices reminiscent of the photosynthetic apparatus of green plants or photosynthetic bacteria.

I would like to stress that encounters with various people played a very important role in my professional life. Two teachers were particularly influential when I was a student: Raymond Weiss, who was a very rigorous physical and inorganic chemist, and Guy Ourisson, an exceptional organic chemist who was able to convince all the students he was teaching to that organic chemistry is exciting and can even be fun. Jean-Marie Lehn also had a great impact on my enthusiasm for science and to me he was the perfect model, although this model was totally out of reach. One of the most important encounters was that with Christiane Dietrich-Buchecker, a wonderful person and a great organic chemist whose contribution to the scientific production of our group turned out to be determinant. Finally, it may appear as surprising that Fraser Stoddart and I never looked at each other as competitors. Even more, we both tried to avoid any overlapping with the activities of the other research team. This is mostly because we became friends at the end of the 70s and this was the beginning of a faithful friendship which allowed us to work in a more serene atmosphere than if we had tried to overtake each other. Between 2010 and 2013, I was appointed as visiting professor at Northwestern University, where I collaborated primarily

Figure 4. One of the ceremonies held in Strasbourg (1988) to honour Jean-Marie Lehn's Nobel Prize (1987). On this special occasion, the chemist community gathered in a friendly atmosphere (as testified by the number of smiley people on the picture). From left to right, the four persons at the front row are Jean-Marie Lehn, Jean-François Biellman, myself and Guy Ourisson. My very good friend Bernard Dietrich (1940–2004), who was also one of the main contributors to the early work on cryptands and cryptates, is at the second row just behind J.-F. Biellman.

Figure 5. Picture taken on the stage after the Nobel Ceremony with the Royal Family. From left to right: Annelie Almkvist, our Attachée; Diana Sistiva, our daughter in law; Julien Sauvage, our son; Carmen Sauvage, my wife; myself and Jean-Marie Lehn.

with Fraser and his team. I also enjoyed interacting with other colleagues at this great university. I am particularly grateful to Fraser for arranging for me to get such a position.

I would like to conclude with two pictures separated by approximately 29 years. The first one, Figure 4, was taken at the ceremony to honour Jean-Marie Lehn's Nobel Prize in 1987. This ceremony took place in the Great Lecture Hall of the Chemistry Department of our university. The second picture, Figure 5, is much more recent since it was taken in Stockholm just after the December 10, 2016 ceremony.

From Chemical Topology to Molecular Machines

Nobel Lecture, December 8, 2016 by
Jean-Pierre Sauvage
University of Strasbourg, Strasbourg, France.

TO A LARGE EXTENT, THE FIELD OF "molecular machines" started after several groups were able to prepare reasonably easily interlocking ring compounds (named *catenanes* for compounds consisting of interlocking rings and *rotaxanes* for rings threaded by molecular laments or axes). Important families of molecular machines not belonging to the interlocking world were also designed, prepared and studied but, for most of them, their elaboration was more recent than that of catenanes or rotaxanes. Since the creation of interlocking ring molecules is so important in relation to the molecular machinery area, we will start with this aspect of our work. The second part will naturally be devoted to the dynamic properties of such systems and to the compounds for which motions can be directed in a controlled manner from the outside, i.e., molecular machines. We will restrict our discussion to a very limited number of examples which we consider particularly representative of the field.

CHEMICAL TOPOLOGY

Generally speaking, chemical topology refers to molecules whose graph (i.e., their representation based on atoms and bonds) is non-planar [1–2]. A planar graph cannot be represented in a plane or on a sheet of paper without crossing points. In topology, the object can be distorted as much as one likes but its topological properties are not modified as long as no cleavage occurs [3]. In other words, a circle and an ellipse are topologically identical. The most representative examples of topologically non-planar, and thus non-trivial, molecules are interlocking ring molecules or knotted cycles such as the trefoil knot. Interestingly, topology and chemistry were not at all connected till 1961 when two chemists working at Bell Telephone Laboratories, Frisch and Wasserman, published an important discussion simply entitled "Chemical Topology" [1]. For the first time, chemists were exposed to topology thanks to this seminal paper. It contains a very clear account of most of the ideas that constitute the background of chemical topology. For example, the idea of topological isomers (also named "stereoisomers") is introduced by comparing a [2]catenane (drawing III, two interlocked rings) to the set of the two separate cyclic molecules (object IV) as shown in Figure 1. Another example is provided by considering a single closed curve, which can be a normal cycle such as I (topologically trivial or planar) or the knotted cycle, the simplest example being the trefoil knot II. Hypothetic compounds I and II of Figure 1 are topological stereoisomers: although they may consist of exactly the same atoms in the same sequence, and the same chemical bonds between them, they cannot be interconverted by any type of continuous deformation in three-dimensional space.

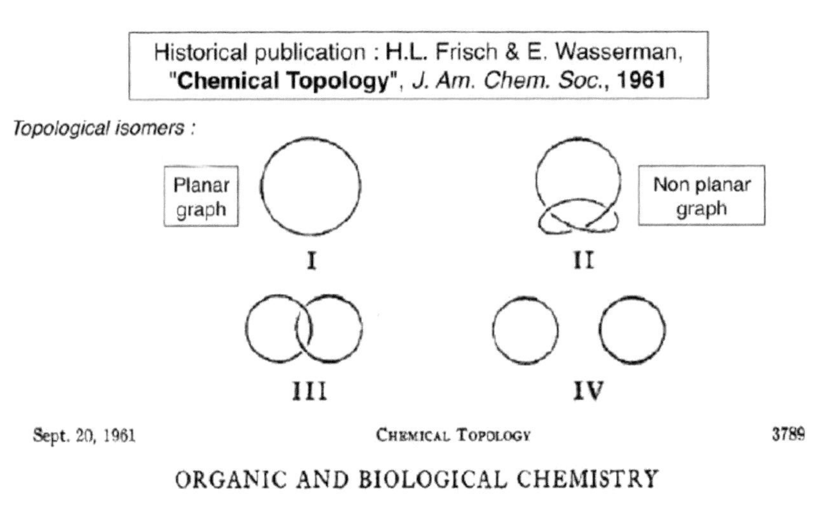

Sept. 20, 1961 CHEMICAL TOPOLOGY 3789

ORGANIC AND BIOLOGICAL CHEMISTRY

[CONTRIBUTION FROM THE BELL TELEPHONE LABORATORIES, INCORPORATED, MURRAY HILL, NEW JERSEY]

Chemical Topology[1]

BY H. L. FRISCH AND E. WASSERMAN

RECEIVED FEBRUARY 28, 1961

Figure 1. The historical publication introducing the notion of chemical topology in the chemical literature. Trefoil knot II and catenane III are topologically non-planar in the sense that their graphs, once projected on a plane, will always require crossings in order to be represented.

Schill's book *Catenanes, Rotaxanes and Knots*, written in 1971, is indispensable for the topologist [4]. In addition to interesting theoretical considerations regarding interlocked rings and knots, it contains much information on the experimental approaches used by Lüttringhaus, Schill and their co-workers to prepare such topologically novel systems. One of the first [2]catenanes to be reported in the literature was prepared by Schill and Lüttringhaus more than 50 years ago [5]. This remarkable piece of work was greeted by the scientific community due to the elegance of the procedure and also due to the impressive synthetic work it represented at the time. Unfortunately, the complete synthesis required a large number of steps and, as a consequence, the overall yield was limited. In spite of the novelty of the work, there were very few follow-up contributions from other organic chemists, mostly because of the extreme difficulty that a gram scale synthesis of such compounds would represent following a purely organic chemistry approach similar to that of Schill and Lüttringhaus.

Before our group started in the field of catenanes, we were interested in markedly different topics. We contributed to various research areas such as homogeneous catalysis of the water-gas shift reaction using iridium complexes [6], electrochemical reduction of carbon dioxide [7] and inorganic photochemistry [8] in relation to solar energy conversion to chemical energy. Like a relatively

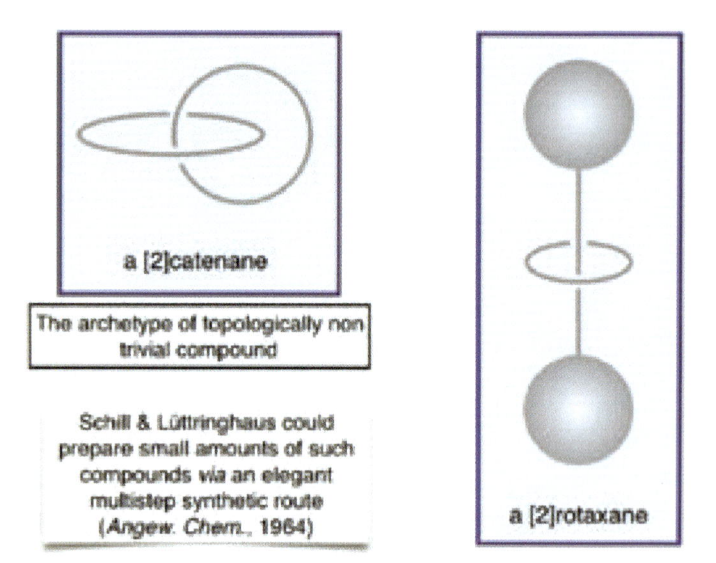

Figure 2. A [2]catenane and a [2]rotaxane. The number in square brackets (2 in this case) indicates the number of components incorporated in the interlocking structure. A [2]catenane is considered as the prototype of molecules having a non planar graph. In mathematics, the object consisting of two interlocking rings is called "Hopf link." Heinz Hopf was a German topologist.

large number of other research teams, we were fascinated by the "grand project" known by the community as "water splitting," i.e., photochemical splitting of the water molecule into oxygen and hydrogen, supposedly the ideal fuel for the future. Since the Nobel Foundation recommended that the laureates tell as honestly as possible "the story behind the discovery," I would like to explain how we moved from inorganic photosensitizers to catenanes and related species. At the end of the 70s, there was considerable interest for ruthenium complexes of the $Ru(bipy)_3^{2+}$ family (bipy : 2,2'-bipyridine) [9]. Several prestigious photochemistry groups had shown that this compound displays promising properties in terms of ability to transfer electrons, positive holes or electronic energy. Even more spectacular, several approaches to the "water splitting" reactions had already been reported [10]. A particularly interesting project was to replace the second raw ruthenium metal, notoriously know to be rare and expensive, by common first raw transition metals. Along this line of research, very few candidates appear as reasonable. The only one was copper(I) since this metal was known to form highly coloured complexes with aromatic diimine ligands of the bipy family. The absorption band responsible for the colour in the visible spectrum was shown long ago to be a Metal-to-Ligand Charge Transfer band. An American photochemist, David R. McMillin (Purdue University) had already published seminal work on the photochemical properties of copper(I) complexes in 1978 [11]. At the beginning of the 80s, our group prepared 2,9-diphenyl-1,10-phenanthroline (dpp) for various reasons, mostly in relation to sterically hindering ligands able to favour coordination unsaturation once complexed to rhodium or iridium centres [6]. David spent some time in Strasbourg when he was on sabbatical leave. He visited our group and, after several discussions, we embarked in a common project dealing with $Cu(dpp)_2^+$ and related complexes containing two intertwined ligands disposed more or less orthogonally to one another once coordinated to the central Cu(I). At that time $Cu(dpp)_2^+$ was unknown since dpp itself seemed to be a new compound. The beginning of this collaboration coincided with another major event which was the arrival of Christiane Dietrich-Buchecker in our team. Christiane had a superb organic chemistry background and we knew that she had all the necessary expertise for making novel ligands either by herself or associated to the PhD students she would co-supervise with me. Coming back to $Cu(dpp)_2^+$, by looking carefully at its three-dimensional representation [8] as shown on Figure 3, it was relatively easy to figure out that this compound, with its two intertwined ligands, was the ideal precursor to a [2]catenane.

There is perhaps an important factor in the fact that we recognised the link between $Cu(dpp)_2^+$ and catenanes: the molecules were drawn manually using Indian ink since this period corresponded to the "pre-ChemDraw era." We thus

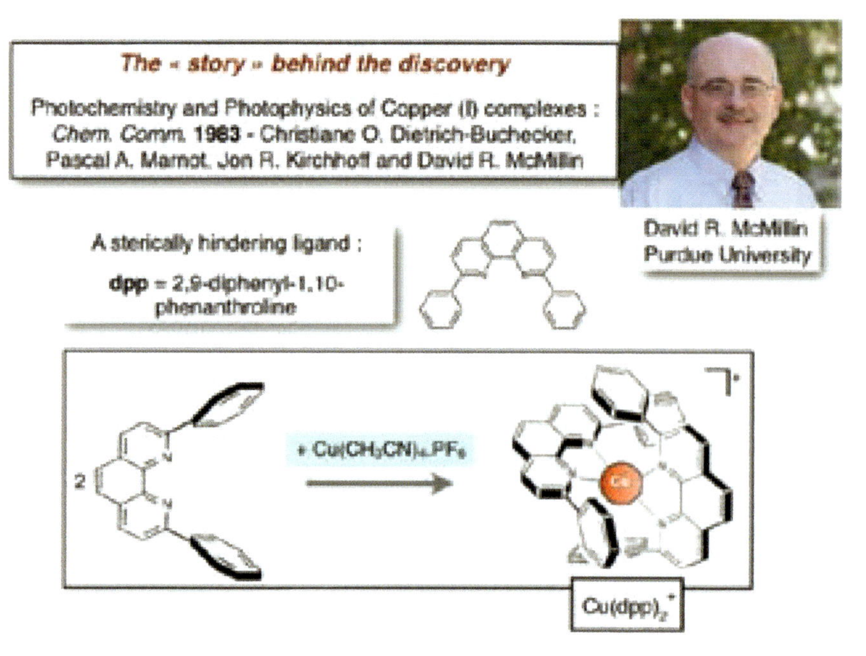

Figure 3. The simple coordination chemistry reaction leading to $Cu(dpp)_2^+$ and its 3-dimensional representation.

had time to completely "digest" the molecular structures of the compounds we were making and, first, that we had to draw. From the drawing of Figure 2, it is clear that by connecting the two para-positions of the phenyl nuclei attached to the same 1,10-phenanthroline group, one should obtain a ring. Furthermore, by doing this operation simultaneously on both dpp sub-units within the complex, formation of a [2]catenane appeared to be almost certain.

With Christiane, we thus embarked in the new field of catenanes. After a few months, we obtained the first positive results and about one year after the beginning of the project, the first communication on our new strategy for making interlocking rings was published [12] (Figure 4).

We described this new strategy as a *3-dimensional* template synthesis around a transition-metal centre by reference to classical template effects so efficiently used for making macrocycles by several groups such as Busch and his team in particular [13]. The work seemed to be the first practical synthesis of a [2]catenane in the sense that, for a good chemist, 500 milligrams could be obtained within a few weeks. After the preparative work performed by Christiane, what remained was to characterise the compound and convince ourselves that the molecule obtained was indeed a [2]catenane. ^1H NMR studies demonstrated

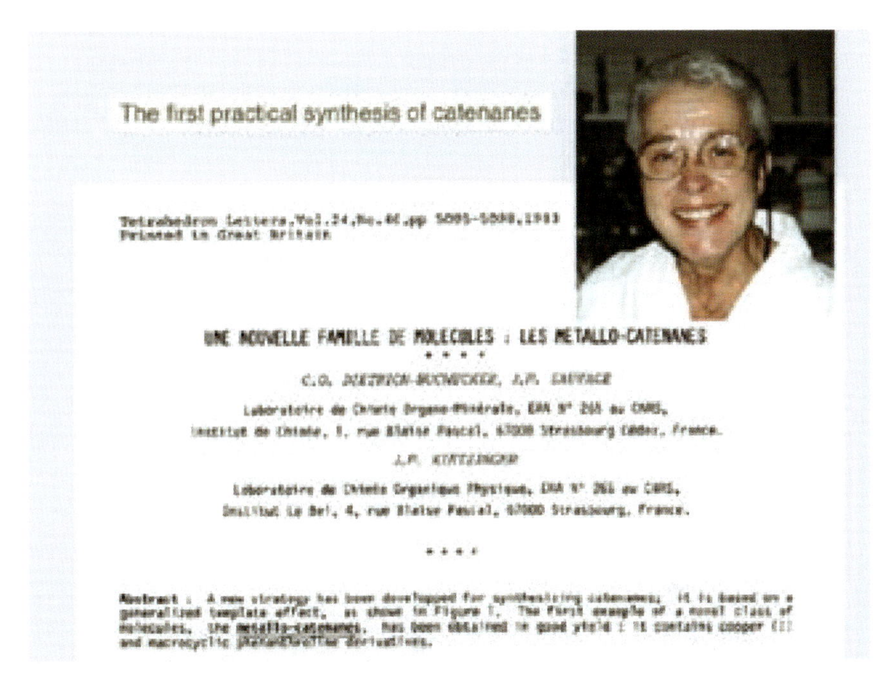

Figure 4. The title and the abstract of the first publication on catenanes from our group with the picture of the late Dr C.O. Dietrich-Buchecker (1942–2008).

that the two dpp coordinating fragments were intertwined, thus indicating that the rings were interlocking with one another. The general strategy is depicted in Figure 5.

Strategy **A** is both very general and simple. Both coordinating fragments have to bind to the metal centre so as to be mutually perpendicular. By an appropriate choice of metal, chelates, linkers (g-g), and functional groups (f and g), the system consisting of two intertwined organic fragments interacting with the central metal will react in the expected fashion with formation of two interlocked rings. The only apparent weakness of strategy **A** is that a total of eight reacting points have to find one another in the double-cyclisation reaction. The second approach, strategy B, reduces this problem since it involves only four reacting groups to be interconnected in a single cyclisation (twice g-f) leading to the catenane. The only requirement is that the starting macrocycle has to be pre-synthesised before the template reaction is performed. This ring contains a coordinating fragment (f-f) and a non-coordinating linker (g-g), and it is also likely to be formed as an intermediate in strategy **A**. It is noteworthy that, provided the threaded bis-chelate complex (bottom left) is stable enough, it must form *quantitatively* from a 1:1:1 mixture of the macrocycle, the metal ion, and the

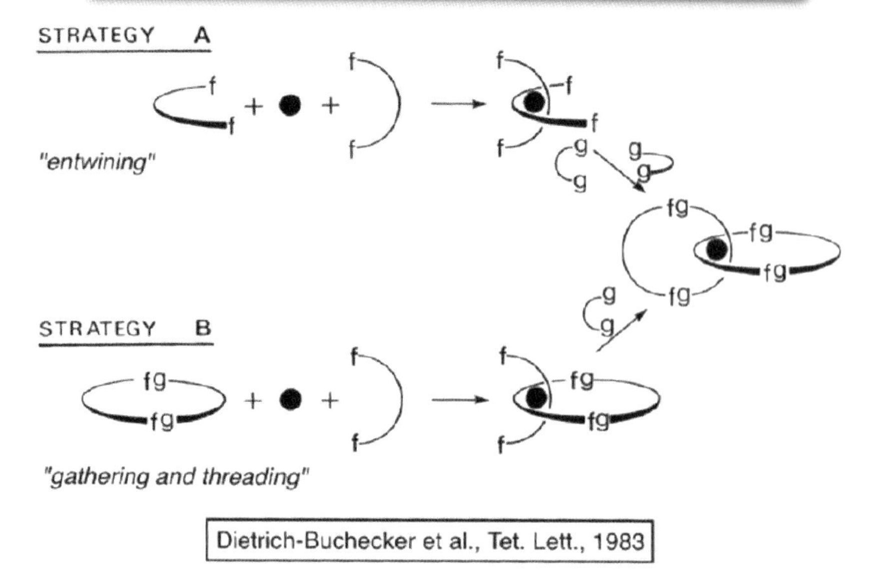

Figure 5. Synthesis of interlocking ring systems. The strategy is based on a 3D-template effect induced by a transition metal centre, copper(I) in the present case. The arcs of a circle and the rings contain coordinating fragments f-f able to interact with a transition metal centre (black disk). The f and g letters represent chemical functions able to react with one another and form an f-g chemical bond.

open-chain fragment. A statistical mixture of complexes, as would arise if two different open-chain ligands were reacted with a metal ion, cannot occur because it is impossible for two of the macrocyclic ligands to attach to the same metal ion.

Since it was less risky, we started with strategy B, which led to an interlocked system in good yield (1983) [12]. Later on, we used strategy A since it is slightly shorter than strategy B in terms of number of steps from commercially available compounds (1984) [14]. The actual reactions carried out are shown in Figures 6 and 7. It is noteworthy that strategy B is more general than strategy A since it opens the way to asymmetric [2]catenanes, i.e., compounds containing two different rings. In addition, the threaded intermediate is close to a [2]rotaxane: it suffices to end-functionalise the threaded fragment of the compound using bulky groups to generate such a [2]rotaxane. We have used this synthesis route extensively for making porphyrin-containing catenanes and rotaxanes as well as variously substituted compounds of the same family.

The copper-complexed catenane of Figure 7 could be quantitatively demetalated using a cyanide salt. The back reaction leading to the starting copper(I)

"entwining" two ligands around a copper(I) centre
OH
N
N
2
OH
Cu(CH3CN)4+
HO
OH
N
N
Cu
N
N
HO
OH
(100%)
Dietrich-Buchecker et al., 1983-1984

Figure 6. The starting compound represented on the left is formed from commercially available 1,10-phenanthroline in two steps. It can be made at the 10-gram scale within one or two weeks. In the presence of copper(I), the entwined compound represented on the right is formed quantitatively.

Double cyclisation reaction leading to the [2]catenane
HO
OH
HO
OH
Cs2CO3
I-CH2(CH2OCH2)4CH2-I
(27%)
Dietrich-Buchecker et al., 1983-1984

Figure 7. The double cyclisation reaction leading to the catenane is relatively low-yielding. Nevertheless, it was possible to prepare batches of 0.5 g within a few weeks. By slightly modifying the structure of the starting molecules and by applying the ring-closing methodology brilliantly proposed by Grubbs, we substantially improved the yield (up to 92%) [15].

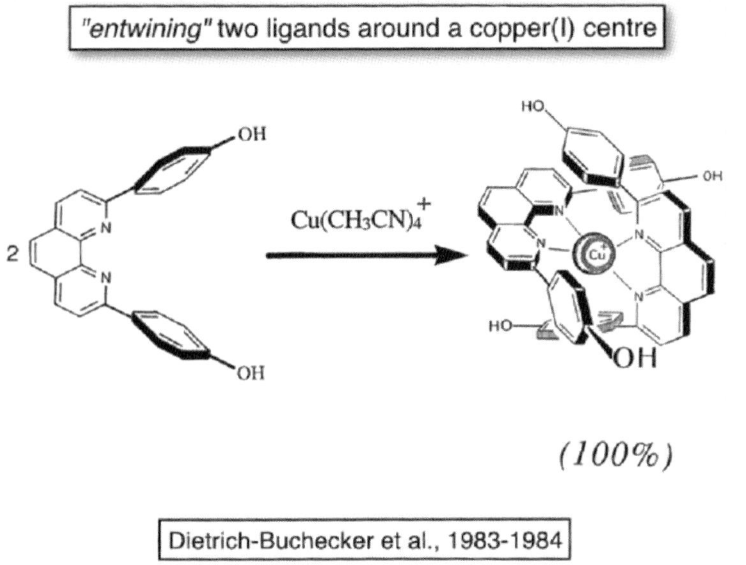

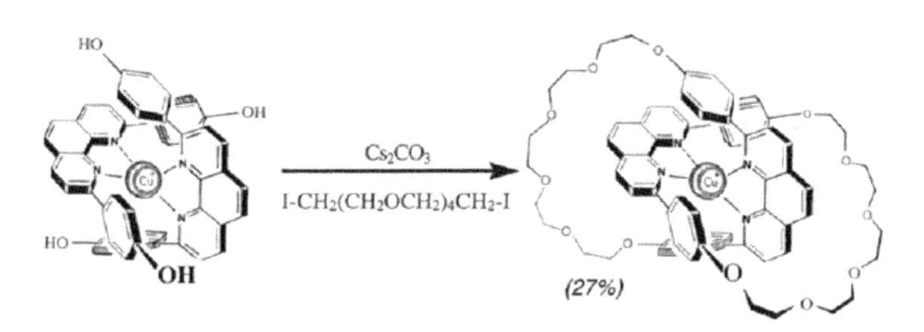

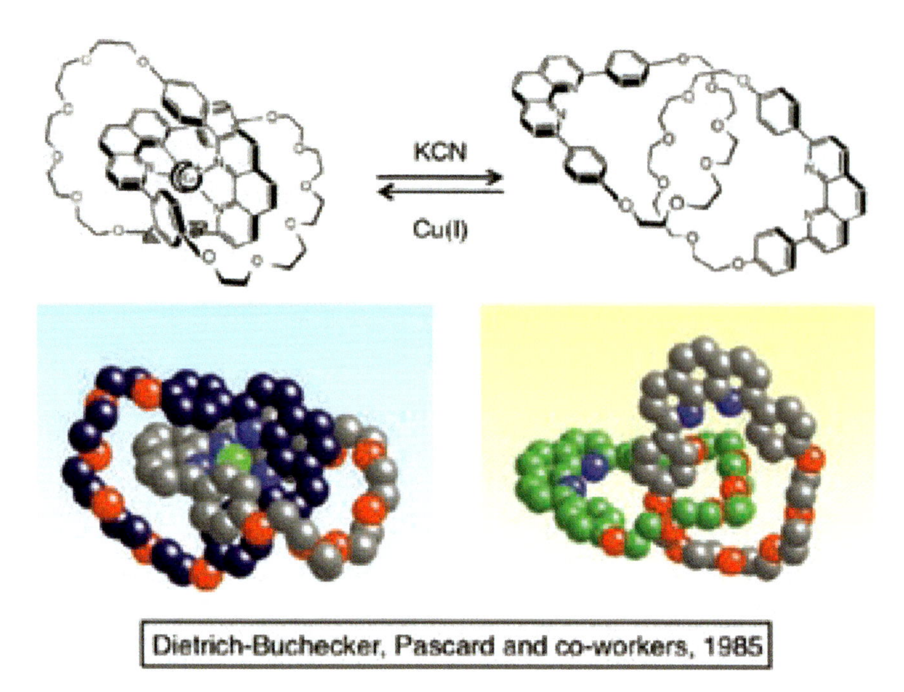

Figure 8. The demetalation-remetalation reaction and the X-rays structures of the copper(I) catenate and the catenand.

complex from the metal-free compound was perfectly reversible as shown on Figure 8. X-ray quality crystals of both forms could be grown and their crystallographic structures were obtained by Pascard and co-workers [16]. We coined the terms "catenates" and "catenands" for complexes whose ligands are interlocking rings and for the corresponding metal free-compounds, respectively.

The X-ray structures are represented below the corresponding chemical drawings of the copper-complexed and metal-free [2]catenanes. As far as the copper complex is concerned, the similarity between the drawing and the X-ray structure is striking. By contrast, the shape of the solid state free catenane is significantly different from the drawing. In any case, demetalation of the complex or complexation of the free form results in a complete rearrangement of the organic backbone. This metamorphosis can clearly be connected to the molecular machinery field although it is still relatively far from movements taking place in real molecular machines.

The Strasbourg approach to catenanes represented significant progress in the catenane and rotaxane field. All of a sudden, the transition metal templated approach developed by our team opened the way to novel topologies and to rotaxanes. In fact, the potential of transition metal in relation to catenanes, rotaxanes and even knotted rings appeared to be limitless. Although the contribution of Lüttringhaus and Schill was conceptually important, the template approach completely modified the way molecular chemists looked at catenanes. Within a few years, they changed status from very exotic and almost impossible-to-make species, at least at a macroscopic scale, to normal and reasonably accessible compounds. A few pioneering contributions based on organic templates were successfully explored by other groups. In 1989 Stoddart and his group reported a particularly efficient and attractive synthesis of a [2]catenane [17]. They took advantage of acceptor-donor interactions between an electron deficient 4,4'-bipyridinium derivative and an aromatic electron donor, combined to favourable hydrogen bonds, to predispose the various groups implied in a very favourable situation before a cyclisation reaction leading to the desired [2]catenane. This seminal piece of work was followed by two reports based on hydrogen bonding only, published in 1994 by Hunter and his group [18] and Voegtle and co-workers [19]. Both teams reported almost simultaneously very similar data although they were working far away from one another. These two pioneering hydrogen bonding-based approaches were followed by remarkable work performed by Fujita and co-workers [20]. It was based on kinetically labile Pd-N bonds and hydrophobic interactions. Fujita introduced the concept of "magic rings" since the [2]catenane could be reversibly dissociated to the set of two non-interlocking rings by just changing the nature of the medium in which the catenane was dissolved or the concentration of its constitutive components.

As far as rotaxanes are concerned, cyclodextrins turned out to be particularly interesting. These natural rings were used as the cyclic components of a large variety of rotaxanes. In the presence of appropriate string-like organic fragments, threading through the cyclodextrin was shown to take place thanks to hydrophobic forces. The first convincing example of a cyclodextrin-based [2]rotaxane was reported by Ogino in 1981 [21]. This pioneering publication was followed by remarkable work from various groups and, in particular, from Harada and co-workers [22].

Figure 9 gathers a few remarkable topologies which could be obtained at the molecular level by our group using a generalised transition metal-based strategy, i.e., by assembling two or more transition metal centres and molecular threads containing two or more coordinating fragments: a [3]catenane [23], the trefoil knot [24] and a doubly interlocking [2]catenane also known as the Solomon rings [25]. Other groups have also been interested in chemical topology. Impressive topologies at the molecular level have been reported by the research teams of Stoddart, Leigh and Siegel, just to cite a few [26, 27, 28].

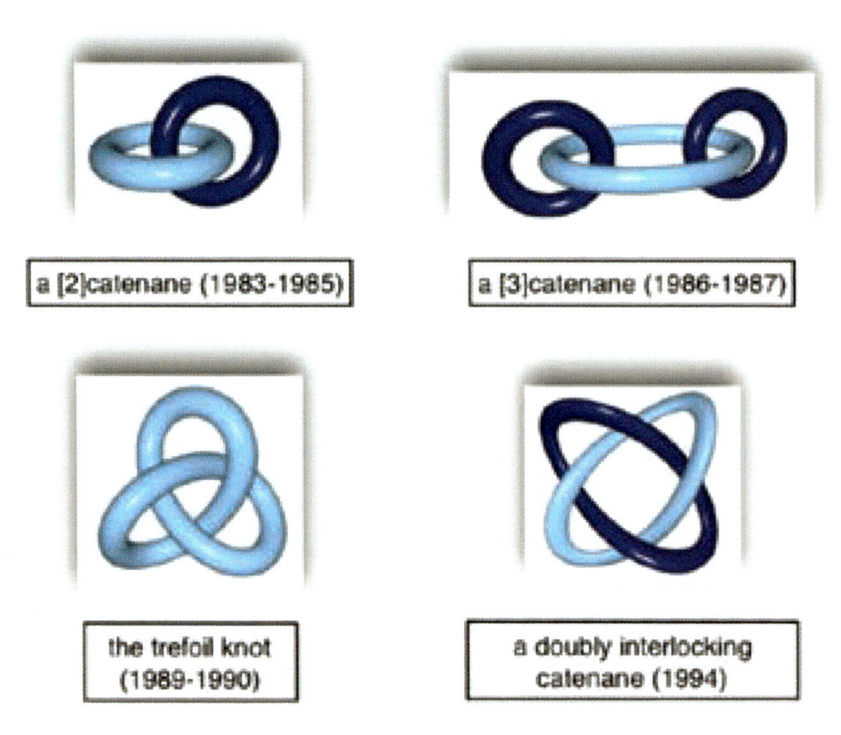

Figure 9. From a simple [2]catenane to more complex topologies. The examples represented on Figure 9 are those made in Strasbourg.

MOLECULAR MACHINES

The field of molecular machines, mostly initiated by Stoddart, Balzani and their co-workers [29–30], was to a large extent an extension of that of interlocking ring molecules, at least at the beginning. The structures of catenanes and rotaxanes seem to be ideally suited to the making of molecular switches and machines, i.e., compounds which can undergo large amplitude motions in a controlled and reversible way.

Electrochemically-driven swinging motion of a copper-complexed catenane

The very first system made by our group for which a large amplitude motion can deliberately be triggered by an external signal was reported in 1994 [31]. A ring could be forced to rotate inside the other ring by playing with the copper(I)/copper(II) redox couple which led us to name the compound a "swinging catenane." Since that time, molecular machinery has blown up in a very impressive

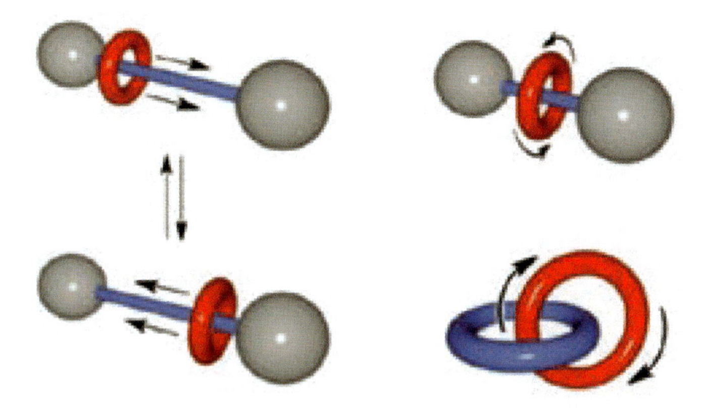

Figure 10. Catenanes and rotaxanes in motion: towards molecular machines.

way thanks to the great contributions of various groups, either working with catenanes or rotaxanes, or using non-interlocking compounds. Stoddart, Kaifer and co-workers reported their first molecular "shuttle" in the same year [30]. The compound made and studied in Strasbourg is a copper-complexed [2]catenane, displaying two very distinct coordination modes corresponding to two extremely different geometries. Having taken advantage of the roughly tetrahedral arrangement of the copper(I) centre in bis-diimide complexes to construct catenanes and molecular knots, the electrochemical properties of related copper complexes have later been used in our group to design and prepare a large variety of catenanes and rotaxanes and for investigating their controlled dynamic properties, playing on the two classical oxidation states of copper, +1 and +2. As far as the first "swinging catenane" is concerned, the interconversion between both forms of the complex is electrochemically triggered and corresponds to the sliding motion of one ring within the other. It leads to a profound rearrangement of the compound and can thus be regarded as a complete metamorphosis of the molecule. The principle of the process is explained in Figure 11.

The key feature of the transformation is the difference in preferred coordination number (*CN*) for the two different redox states of the metal: *CN* = 4 for copper(I) and *CN* = 5 (or 6) for copper(II). The organic backbone of the

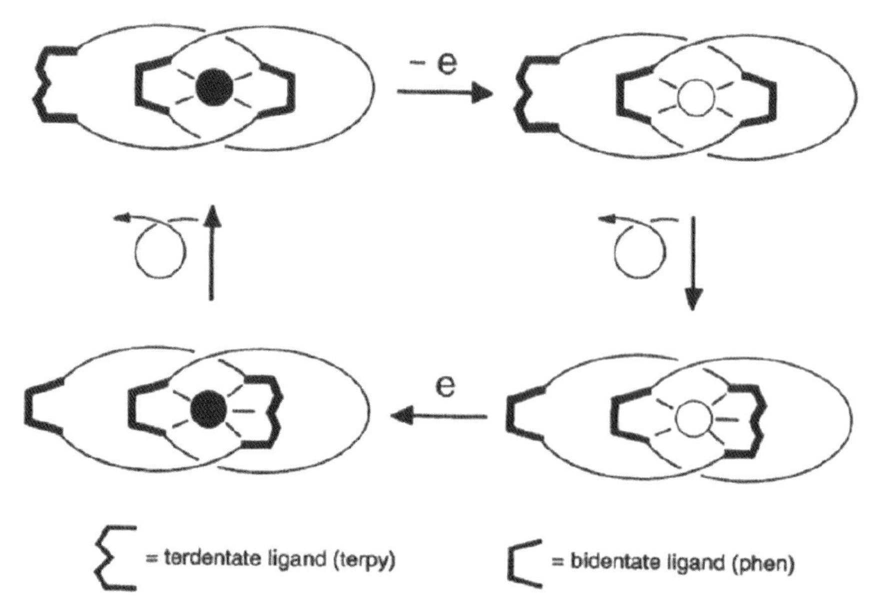

Figure 11. Principle of electrochemically induced molecular motions in a copper-complexed [2]catenane. The stable 4-coordinate monovalent complex (*top left*, the black circle represents Cu(I)) is oxidised to an intermediate tetrahedral divalent species (*top right*, the white circle represents Cu(II)). This compound undergoes a complete reorganisation process to afford the stable 5-coordinate Cu(II) complex (*bottom right*). Upon reduction, the 5-coordinate monovalent state is formed as a transient (*bottom left*). Finally, the latter undergoes the conformational change that regenerates the starting complex. The bidentate coordinating fragment is phen (1,10-phenanthroline), part of the dpp ligand, and the tridentate ligand is terpy (2,2':6',2"-terpyridine).

asymmetric catenate consists of a dpp bidentate chelate included in one cycle and, interlocked to it, a ring containing two different subunits: a dpp moiety and a terpy ligand. Depending upon the mutual arrangement of both interlocking rings, the central metal copper can be tetrahedrally complexed (two dpp units) or 5-coordinate (dpp + terpy). Interconversion between these two complexing modes results from a complete pirouetting of the two-site ring in one given direction or the other. It can easily be induced electrochemically, by means of a chemical reductant or oxidant or even photochemically [32]. From the stable tetrahedral monovalent complex, oxidation leads to a 4-coordinate Cu(II) state, which rearranges to the more stable 5-coordinate compound. The process can be reversed by reducing the divalent state to the 5-coordinate Cu(I) complex, obtained as a transient species before a changeover process takes place to afford back the starting tetrahedral monovalent state.

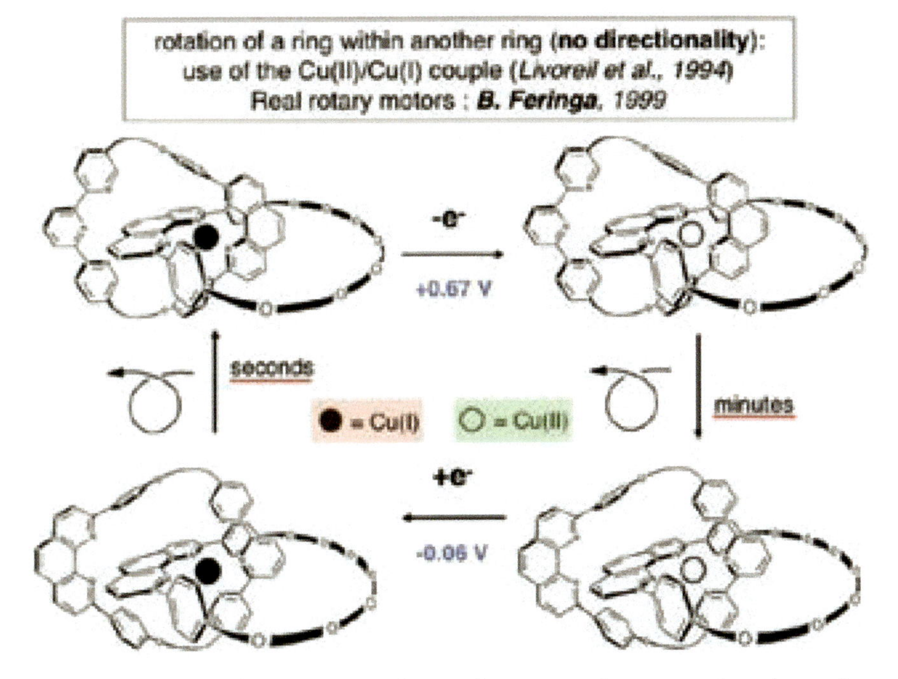

Figure 12. Square scheme corresponding to the process whose principle is depicted in Figure 11, with the chemical structures of the compounds. The dpp-containing cycle is a 30-membered ring whereas the dpp+terpy incorporating one is a 31-membered ring. It could be shown that the 5-coordinate Cu(II) complex has a square pyramidal geometry and not a trigonal bipyramid one. [33]

The real molecules are represented in Figure 12 as well as the square scheme interconverting the 4-and the 5-coordinate species.

Interestingly, the transformations of Figure 11 and Figure 12 is accompanied by a change in the electrochemical properties of the complex as well as spectroscopic changes. As expected, the tetrahedral copper complex has a relatively high redox potential: $Eo = +0.63$ V vs. SCE in CH_3CN, whereas the 5-coordinate species has a slightly negative potential, pointing to greater stabilisation of the divalent copper than in the 4-coordinate species: $Eo = -0.07$ V.

The obvious weak point of the swinging catenane of Figures 11 and 12 is its kinetic inertness. It should be noted that this first system was dramatically improved by modifying structural parameters of the compounds and, in particular in replacing the [2]catenane organic backbone by a [2]rotaxane, either acting as a pirouetting species or as a two- or three-station molecular shuttle. The rearrangement process was also markedly sped up by modifying the structure of the bidentate chelate incorporated in the rings. By replacing the highly sterically hindering dpp fragment by an endocyclic but non-sterically hindering chelating group of the 8,8'-diaryl-3,3'-biisoquinoline family, fast moving rotaxanes were obtained [34–35]. Finally, it should be noted that the motion taking place in the swinging [2]catenane is by no means a rotation motion. The ring that incorporates both a bidentate and a tridentate chelating group undergoes a "pirouetting" motion corresponding to a 180° rotation in a given direction or in the other. Real rotary machines were reported much later. Feringa's contribution to the field of rotary motions was indeed crucial [36].

TOWARDS SYNTHETIC MOLECULAR MUSCLES: CONTRACTION AND STRETCHING OF A LINEAR ROTAXANE DIMER

The topic of this paragraph is dealing with a special type of molecular machine related to contraction and extension of switchable species reminiscent of muscles [37–38]. Rotaxane dimers as represented in Figure 13, can be regarded as synthetic analogues of natural muscles. Since real muscles mostly consist of filaments able to glide along one another, a molecular assembly in which two string-like molecular fragments can glide along one another was designed. This is the process taking place in the sarcomere, in which the thick filament (containing

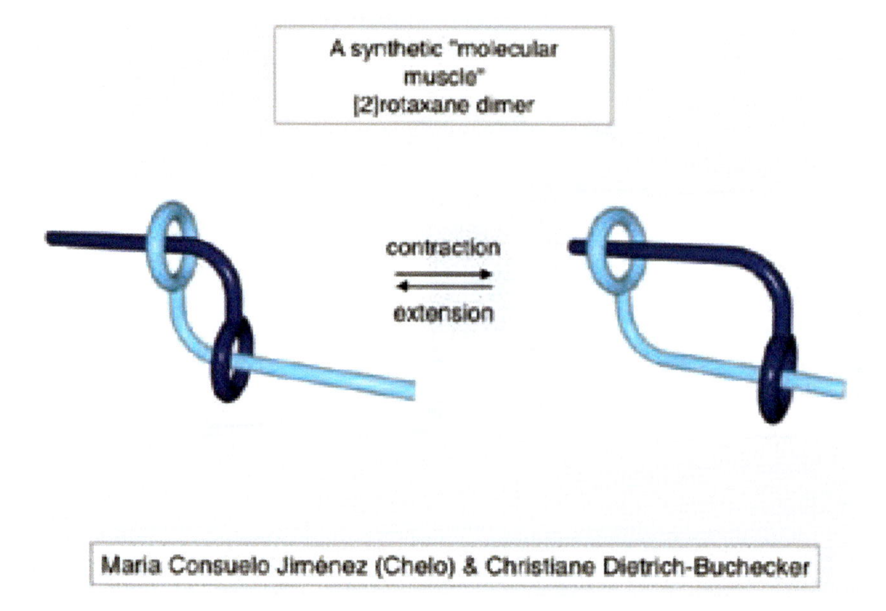

Figure 13. The molecular structure of a rotaxane dimer or pseudo-[2]rotaxane dimer (i.e., without bulky stoppers at the ends of the threaded filaments) is adapted to the contraction/extension motion mimicking the dynamic properties of a muscle.

myosin) moves along the thin filament (actin polymer) in one direction or the other so as to induce contraction or elongation. The actin-myosin complex as well as other motor proteins are fascinating biological systems which represent an inexhaustible source of inspiration for synthetic chemists.

After careful design mostly based on space-filling model, we started the synthesis work. It turned out to be particularly delicate and time consuming. In the present discussion, we will skip the preparation of the ring-and-lateral thread conjugate although the synthesis of this compound was far from trivial.

A 31-membered ring incorporating a dap fragment (dap: 2,9-diaryl-1,10-phenanthroline), prepared in several steps from commercially available compounds, was covalently linked to a linear fragment incorporating a 2,9-dimethyl-3,8-bis-(p-hydroxyphenyl)-1,10-phenanthroline so as to afford the bis-chelating target consisting of a coordinating ring attached to a lateral arm containing another bidentate chelating group. The reaction of the ring-and-string conjugate with a stoichiometric amount of $[Cu(CH_3CN)_4]PF_6$ in CH_3CN/CH_2Cl_2 at room temperature led to the doubly threaded species whose structure is represented in Figure 15. The success of this latter reaction was certainly far from being certain, considering the number of Cu(I) complexes which could be formed by coordinating the two different chelating fragments (endocyclic and

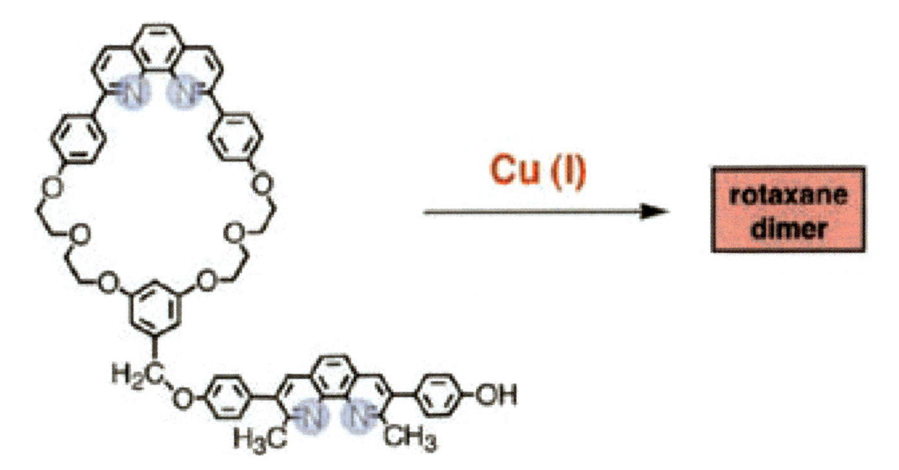

Figure 14. Cu(I)-induced double threading process leading to a dicopper(I) rotaxane dimer. The reaction is quantitative. The structure of the doubly threaded compound obtained is shown in Figure 15.

lateral coordination sites) of the ligand to metal centres. Nevertheless, we were happy to observe that formation of the doubly threaded complex was quantitative. In fact, the complexity of the reaction leading to the doubly threaded species is reflected by the slowness of the reaction leading to the desired product (several days at room temperature). Many coordination/de-coordination steps have to occur before the system finds its thermodynamic well. The desired [2] pseudorotaxane was formed under thermodynamic equilibrium and thus it was obtained pure as a deep red crystalline solid without further purification. X-ray quality crystals were grown from acetone/diethyl ether by diffusion. The crystallographic structure of the dicopper(I) [2]rotaxane dimer is represented in Figure 15. The most striking feature of the structure is its linear extended arrangement, which results in a distance between the two terminal phenolic oxygen atoms of 36.3 Å. The two copper atoms are 18.3 Å apart and occupy similar environments. The coordination tetrahedron around both copper(I) atom are strongly distorted.

In order to complete the synthesis of the switchable dimer rotaxane we had to introduce additional function. In particular, the project involved two potential

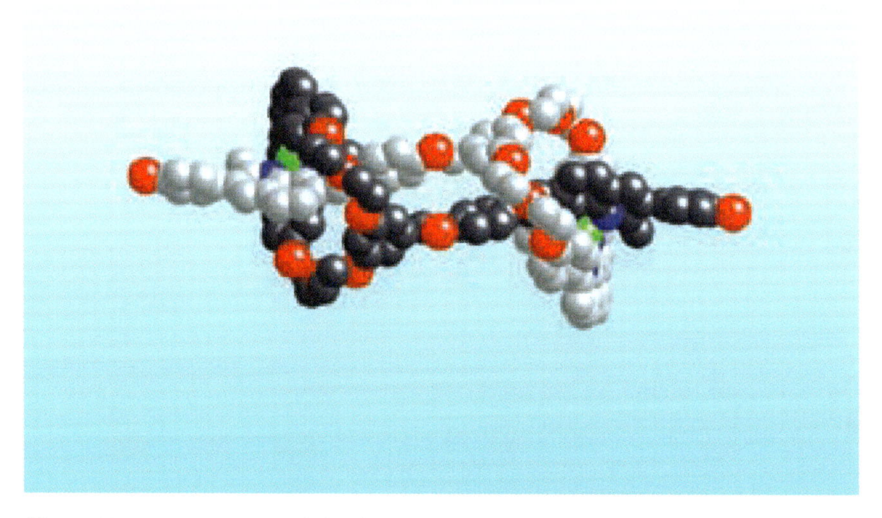

Figure 15. X-ray structure of the dicopper(I) rotaxane dimer. The pale grey ring-and-string component and the black subunit are identical. The carbon atoms are pale grey or black. The two copper(I) centres are green. The nitrogen and oxygen atoms are deep blue and red, respectively.

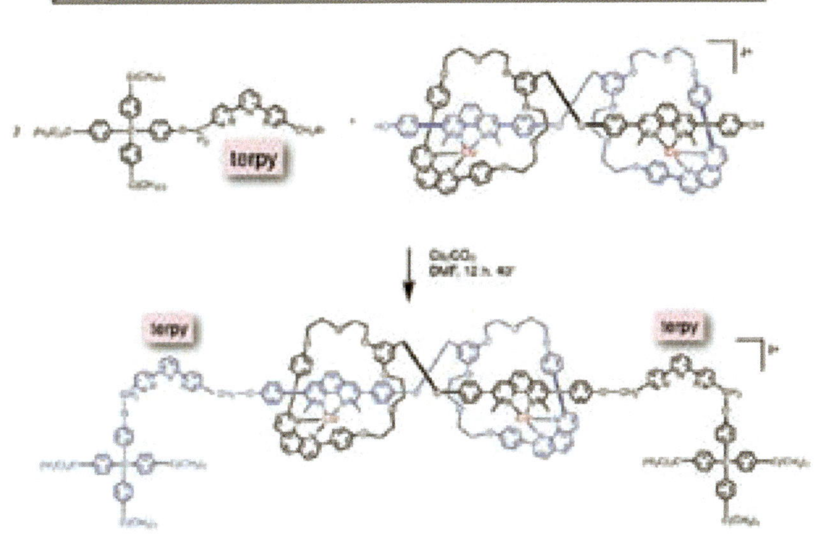

Figure 16. Functionalisation of the [2]pseudorotaxane dimer, precursor to the "muscle," by attaching a terpy-bulky stopper conjugate to its both ends. The "muscle" represented at the bottom of the figure is in its extended situation.

binding modes to a transition metal centre for the organic backbone: 4-coordinate or 5-coordinate. Tridentate ligands of the terpy family were thus introduced simultaneously to the stoppers, whose function was simply to prevent unthreading of the molecular filament from the rings they were threaded through. The last organic reaction of the synthesis sequence is shown in Figure 16. The functionalised terpy represented on the upper part of Figure 16 incorporates a tridentate chelating group and is attached to a bulky stopper. It was prepared separately and subsequently it was attached to both ends of the dicopper(I) precursor of Figure 15. This last reaction afforded the muscle-like compound depicted on Figure 16 (bottom) as a dark red solid in 60% yield. The last coupling step not only allowed the introduction of terdentate sites on both sides of the molecular assembly but also afforded a real rotaxane which cannot undergo any unthreading reaction.

As usual in our group, Cu(I) was used as a gathering and templating metal. The movement was induced by a chemical reaction, namely a metal exchange. As shown in Figure 17 in a very schematic fashion, the doubly threaded compound can interact simultaneously with two metal centres, either in a four- or in a five-coordinate geometry each.

Due to the synthesis procedure, each Cu(I) centre was coordinated to two 1,10-phenanthroline units, resulting in a four-coordinate situation. This binding

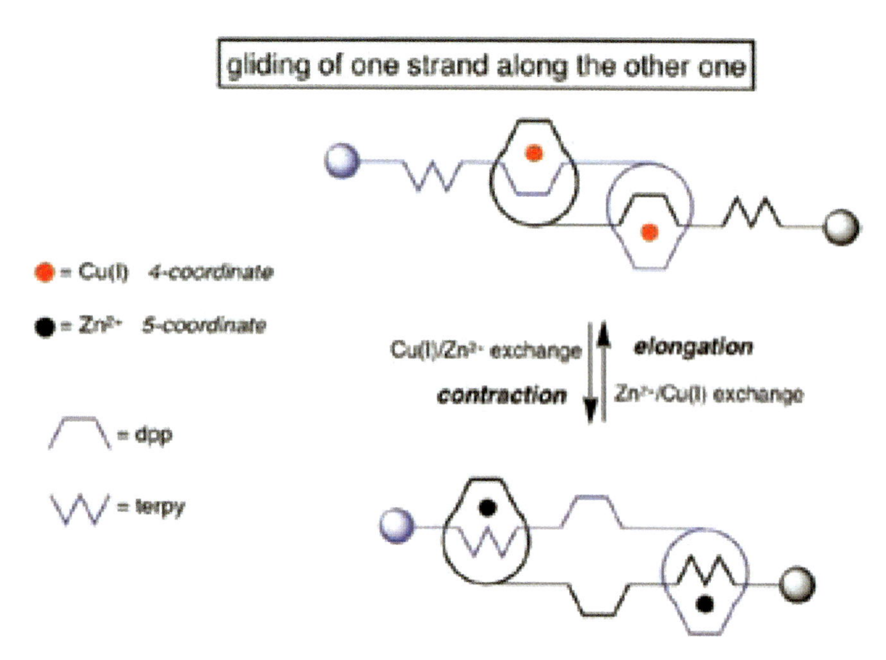

Figure 17. Principle—Contraction/extension of a [2]rotaxane dimer by gliding one strand along the other one. The W-shaped fragment represents a terdentate chelating group such as terpy whereas the U-shaped species is a bidentate ligands of the dpp type.

mode corresponds to an extended geometry (upper part of Figure 17), whereas coordination of a divalent metal ion such as Zn^{2+} affords a five coordinate situation corresponding to interaction of each metal centre with one 1,10-phenanthroline or dpp unit and one terpy. It corresponds to a contracted geometry (bottom of Figure 17).

As shown on Figure 18, by treating the dicopper(I) pseudorotaxane dimer with KCN, the metal-free ligand was obtained in quantitative yield (not shown on the figure). After remetalation using $Zn(NO_3)_2$, the colourless di-zinc complex (bottom) was obtained quantitatively. It is now in the contracted form. The extended dicopper(I) dimer could be regenerated by the reaction of the di-zinc complex with an excess of $[Cu(CH_3CN)_4]PF_6$ at room temperature. Using a metal exchange reaction, the muscle-like compound is thus set in motion. The Cu(I) complex corresponds to the extended form and the Zn(II)-complexed species is the contracted state. A metal exchange reaction was utilised to set the muscle in motion due to the fact that the electrochemical method turned out to be inefficient. The copper(I) complex was oxidised with the purpose of contracting the system to a dicopper(II) complex with two 5-coordinate Cu(II) centres but the motion was very slow which led us to use a chemical signal. The kinetic barrier to the rearrangement is very large due to the fact that the gliding motion of a given thread is strongly coupled to that of the other thread. The tight connection between the two subunits makes it impossible for the [2]rotaxane dimer to move in a stepwise manner.

The two rotaxane dimers of Figure 18 represent one of the first examples of a unimolecular linear array able to elongate and contract at will under the action of a given stimulus. From Corey–Pauling–Koltun (CPK) model, one can estimate that the length of the compound changes from 83 to 65 Å between both situations (Figure 18). This corresponds roughly to the same relative amount as what is found in natural muscles (~27 %).

It is noteworthy that several muscle-like compounds have been proposed in recent years. In most cases, they were set in motion using principles which were somewhat different from ours. It is not the place in this article to review in detail

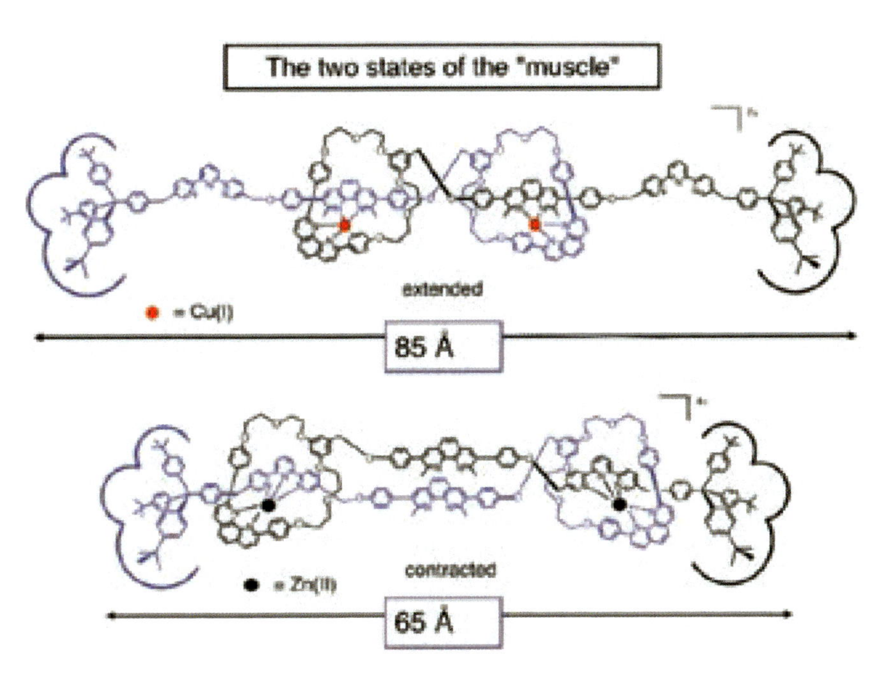

Figure 18. The two forms of the muscle-like [2]rotaxane dimer. Whereas the interconversion between both forms is easily induced by metal exchange, oxidation of copper(I) to copper(II) (upper form of the figure) was unfortunately not followed by rearrangement of the complex to the extended situation, at least on a reasonable timescale. Due to the rigid structure of the molecule, the gliding motions of both threads along one another have to take place simultaneously and in a concerted manner. The kinetic barrier to the global motion is thus very high which explains why the motion is so slow when no demetalation step is involved.

the work done by other groups in this area but we would like to mention some of the most significant contributions. Several groups prepared dynamic [2]rotaxane dimers based on cyclodextrins which, under the action of a given signal, could be contracted or elongated [39–40]. Reversible molecular motions mimicking natural muscles proceed through hydrophobic forces, size constraints, solvents polarity changes or photochemical isomerization of azobenzene or stilbene derivatives.

Another family of muscle-like compounds was developed based on other types of interlocking architectures. These systems contain two types of recognitions sites, one of them being based on hydrogen bonding (and thus pH sensitive) and the other one operating via acceptor-donor interactions. A few years ago, Stoddart and co-workers showed that a contraction and expansion motion was triggered by changing the pH of the solution containing the rotaxane [41]. They described a new bistable rotaxane dimer architecture in 2008, namely a two-component [c2]daisy chain topology. This system consists of two self-complementary interlocking molecular fragments in which a secondary ammonium ion-containing lateral arm is attached to a crown ether. The end of each arm is connected to a bulky stopper. Each crown ether glides between the two specific binding sites incorporated in the thread, the motion being triggered by modifying the pH. The gliding motions result in contracting or elongating the rotaxane dimer.

Simultaneously with the former piece of work, Coutrot et al. published another example of pH-switchable molecular muscle [42] based on a related principle. The system incorporates a cyclic rotaxane dimer consisting of two interlocking monomers. Again, each monomer incorporates a crown ether attached to a linear fragment. This fragment contains two different stations and the system can be switched from a contracted state to an elongated situation by modifying the pH. A remarkable extension of Coutrot's work was published in 2012 [43]. Giuseppone and co-workers designed and synthetised a pH-driven muscle-like system similar to Coutrot's "muscle" in which [2]rotaxane dimers are linked together linearly by taking advantage of a very efficient coordination chemistry-based polymerisation process. A remarkably high degree of polymerisation was obtained (~3,000 units) for both the contracted and extended forms. As in Coutrot's work, at low pH, a doubly-threaded dimer adopts an extended geometry. At basic pH, the compound adopts a contracted form. These two states were very convincingly characterized by dynamic light-scattering and small angle neutron scattering experiments. This remarkable piece of work is probably one the first examples of molecular muscle operating at a microscopic scale and thus functioning as a real device.

CONCLUSION

The main field of research of our group since the beginning of the 80s has clearly been that of transition metal chemistry. In relation to the molecular machine area, it should be stressed that the metals had a dual function:

1. Interlocking and knotted ring compounds were obtained by using transition metals such as Cu(I) in particular for gathering and disposing in a precisely defined geometry various organic fragments before they were incorporated in the target molecule.
2. The presence of transition metals provided the molecular systems with *electrochemical* or *photochemical* properties which allowed to set given fragments of the molecule in motion and thus to obtain molecular machines.

Generally speaking, motivation for doing research in the field of controlled dynamic systems of the muscle-like family is manifold. The most important one is related to the scientific challenge that the synthesis and the study of complex and functional molecular systems that can be set in motion in a controlled way represents. Making contractile and extensible molecules of the rotaxane dimer type was probably not even envisioned a few decades ago due to expected synthetic and analytical difficulties. Preparing and studying linear or rotary motors or artificial muscles is related to the fascination that biology exerts on chemists and engineers. It is very challenging for a synthetic chemist to make a molecule or molecular system whose function and mode of action are reminiscent of those of biological systems. The few examples discussed in the present chapter are certainly very remote from natural muscles and they are undoubtedly very primitive compared to their natural analogues. Nevertheless, it was a great challenge to elaborate these compounds and to show that they can be set in motion in a well-controlled fashion.

The field of molecular machines in general, still a basic science area of research, has been acknowledged by the Royal Swedish Academy of Science and the Nobel Committee. I would like to thank these prestigious organisations for this recognition. This field is nowadays more than 20 years old and it is certainly time to address the question of applications, either in a general fashion for machine-like compounds or more precisely for contractile and extensible molecular species. In which field of application will the area of molecular muscles be the most important in the future? It is today very risky to give an answer to this question. Presently, one of the most impressive extension of molecular machines towards applications is concerned with "molecular computing" and the molecular approach to electronic memory devices. The Stoddart and Heath groups reported promising work in this area [44].

Many types of nanodevices and nanomachines can be envisaged in relation to chemical applications (for example, sorting and transport of molecules in solution or through membranes) but also to purely mechanical applications. In the future, nano- and microrobots able to fulfil a large variety of functions (from medicine to everyday life) will have to be articulated. The use of molecular components to control their movements is certainly a promising possibility. In a relatively long term prospective, artificial muscles of various length (from microns to millimetres or even centimetres) might be needed for various applications such as humanoid robots, actuators for microfluidic science and technology or prosthetic organs. However, in spite of all the possibilities offered by molecular machines in terms of potential applications for the future, it should be stressed that basic research is still of utmost importance and is or has been at the origin of the many technologies which are nowadays part of our daily life.

Finally, I would like to stress that the recent book written by Bruns and Stoddart represents a monumental piece of work which, today, gathers all the important data obtained by hundreds if not thousands of researchers since the beginning of catenanes, rotaxanes and molecular machines. Any interested reader must of course have a copy of this book.

ACKNOWLEDGMENTS

The work on molecular machines performed in Strasbourg was financially supported by the University of Strasbourg (formally Université Louis Pasteur), the CNRS (Centre National de la Recherche Scientifique), the European Commission and the International Center for Frontier Research in Chemistry (ICFRC). We would like to thank these organisations. I also acknowledge the important contribution of Dr Christiane O. Dietrich-Buchecker (deceased in 2008). Her contribution to the early work performed in Strasbourg on molecular catenanes and molecular machine prototypes has really been crucial.

I would also like to thank my mentor, Professor Jean-Marie Lehn. Jean-Marie has represented a model for me since I worked with him as a PhD student. I am grateful to two teachers who had a strong influence on me: Professor Guy Ourisson (organic chemistry) and Raymond Weiss (inorganic and physical chemistry). I would also like to express my gratitude to Professor Malcolm L.H. Green. As a postdoctoral fellow in his team, I learnt transition metal and organometallic chemistry, which also had a strong impact on the future projects of my group.

Laboratoire de Chimie Organo-Minérale

1. Members of the University. of Strasbourg or of CNRS
Marc Beley, Christiane Dietrich-Buchecker †, Jean-Claude Chambron, Jean-Paul Collin, Valerie Heitz, Jean-Marc Kern †, Stéphanie Durot, Angélique Sour, Valérie Sartor

2. Ph D students
Pascal Marnot, Romain Ruppert, Jean Weiss, Jean-Claude Chambron, André Edel, Abdelaziz Jouaiti, Sylvie Chardon-Noblat, Catherine Hemmert, Stéphane Guillerez, Christophe Coudret, Valerie Heitz, Abdelhakim Bailal, Angélique Sour, Jean-François Nierengarten, Fabrice Odobel, Sandrine Chodorowski, Aude Livoreil, Nathalie Solladié, Jean-Luc Weidmann, Gwénaël Rapenne, Myriam Linke, Laurence Raehm, Isabelle Dixon, Anne Chantal Laemmel-Gouget, Didier Pommeranc, Christine Hamann-Schaffner, Etienne Baranoff, Pierre Mobian, Benoit Colasson, Damien Jouvenot, Nicolas Belfrekh, Sylvestre Bonnet, Benoit Champin, Christian Tock,Julien Frey, Fabien Durola, Jacques Lux, Julie Voignier, Yann Trolez, Maryline Beyler, Cécile Roche

Figure 19. 1. Permanent researchers of the "Laboratoire de Chimie Organo-Minérale" (LCOM): University members and CNRS researchers including the late Dr Christiane Dietrich-Buchecker (1942–2008) and the late Prof. Jean-Marc Kern (1944–2004). 2. PhD students listed by chronological order. The two first students, Pascal Marnot and Romain Ruppert, started their PhD work in 1980.

Many researchers have been associated to the work of our team, "Laboratoire de Chimie Organo-Minérale." In spite of the expected and unexpected difficulties they frequently had to face in the course of their work, these researchers have always been very enthusiastic and efficient. I owe them the success of our research team. The names of our CNRS researchers or University members (from assistant to full professors), of our PhD students and post-doctoral researchers as well as those of other visitors who contributed to our research are listed in Figure 19 and 20.

Laboratoire de Chimie Organo-Minérale

Post-doctoral researchers and other visitors

David Amabilino, Audrey Auffrant, Christine Beemelmanns, Martial Billon, Maria-Jesus Blanco-Pillado, James I. Bruce, Diego J. Cardenas, Ricardo Carina, John D. Crane, Christian S. Diercks, Vincent Duplan, Jonathan A. Faiz, Yoshio Furusho, Pablo Gaviña, Christine Goze, David Hanss, Neri Geum Hwang, Akiko Hori, Elisabetta Iengo, Fumiaki Ibukuro, Stuart James, Maria Consuelo Jimenez, Antoine Joosten, Robert Kayhanian, Jérome Kieffer, Abdel Khemiss, Masatoshi Koizumi, Tomáš Kraus, Ulla Létinois-Halbes, Jacques Lux, Alexander G. Martynov, Dennis Mitchell, Bernhard Mohr, Cécile Moucheron, Frédéric Niess, Laure-Emmanuelle Perret-Aebi, Ingo Poleschak, Alexander I. Prikhod'ko, Felipe Reviriego, Zeinab Saad, Efstathia G. Sakellariou, Xavier J. Salom-Roig, Emma Schofield, Hideki Sugihara, Pierre Louis Vidal, Michael D. Ward, Noémie Weber, Oliver Wenger, J. A. Gareth Williams

Figure 20. Post-doctoral researchers and other visitors.

Finally, I thank my parents, my wife, Carmen, and my son, Julien, for their continuous support and for their forgiveness regarding my pace of work and travel.

REFERENCES

1. Frisch, H.L., Wasserman, E. (1961) Chemical Topology. *Journal of the American Chemical Society*, **83**, 3789–3795.
2. Walba, D.M. (1985) Topological Stereochemistry. *Tetrahedron*, **41**, 3161–3212.
3. Adams, C.C. (1994) *The Knot Book*. W.H. Freeman and Company, New York.
4. Schill, G. (1971) *Catenanes, Rotaxanes and Knots*. Academic Press, New York/London.
5. Schill, G., Lüttringhaus, A. (1964) The Preparation of Catena Compounds by Directed Synthesis. *Angewandte Chemie International Edition*, **3**, 546–547.
6. Marnot, P.A., Ruppert, R., Sauvage, J.-P. (1981) Catalysis of the water gas shift reaction by complexes of rhodium and iridium with 2,2'-bipyridine and similar ligands. *Nouveau Journal de Chimie*, **5**, 543–548.
7. Beley, M., Collin, J.-P., Ruppert, R. (1984) Nickel(II)-cyclam: An extremely selective electrocatalyst for reduction of CO_2 in water. *Chemical Communications*, 1315–51316.
8. Dietrich-Buchecker, C.O., Marnot, P.A., Sauvage, J.-P., Kirchhoff, J.R., McMillin, D.R. (1983) Bis(2,9-diphenyl-1,10-phenanthroline) copper(I): a copper complex with a long-lived charge-transfer excited state, *Chemical Communications*, 513–514.
9. Juris, A., Balzani, V., Barigelletti, F., Campagna, S., Belser, P., Von Zelewsky, A. (1988) Ru(II) polypyridine complexes: photophysics, photochemistry, electrochemistry and chemiluminescence, *Coordination Chemistry Reviews*, **84**, 85–277 and references.
10. Lehn, J.-M., Sauvage, J.-P. (1977) Chemical storage of light energy: Catalytic generation of hydrogen by visible light or sunlight irradiation of neutral aqueous solutions. *Nouveau Journal de Chimie*, **1**, 449–457. Kirch, M., Lehn, J.-M., Sauvage, J.-P. (1979) Hydrogen generation by visible light irradiation of aqueous solutions of metal complexes. An approach to the photochemical conversion and storage of solar energy. *Helvetica Chimica Acta*, **62**, 1345–1384 and references.
11. Buckner, M.T., McMillin, D.R. (1978) Photoluminescence from copper(I) complexes with low-lying metal-to-ligand charge transfer excited states, *Chemical Communications*, 759–761.
12. Dietrich-Buchecker, C.O., Sauvage, J.-P., Kintzinger, J.-P. (1983) Une nouvelle famille de molécules: les métallo-caténanes, *Tetrahedron Letters*, **24**, 5095.
13. Thompson, M.C., Busch, D.H. (1964) Reactions of coordinated ligands. IX. Utilization of the template hypothesis to synthesize macrocyclic ligands *in Situ*, *Journal of the American Chemical Society*, **86**, 3651–3656
14. Dietrich-Buchecker, C.O., Sauvage, J.-P., Kern, J.-M. (1984) Templated synthesis of interlocked macrocyclic ligands: The catenands, *Journal of the American Chemical Society*, **106**, 3043–3045.
15. Mohr, B., Sauvage, J.-P., Grubbs, R. H., Weck, M. (1997) High-Yield Synthesis of [2] catenanes by intramolecular ring-closing metathesis, *Angewandte Chemie International Edition*, **36**, 1308–1310.

16. Cesario, M., Dietrich-Buchecker, C.O., Guilhem, C., Pascard, C., Sauvage, J.-P. (1985) Molecular structure of a catenand and its copper(I) catenate: complete rearrangement of the interlocked macrocyclic ligands by complexation, *Chemical Communications*, 244–247.
17. Ashton, P.R., Goodnow, T.T., Kaifer, A.E., Reddington, M.V., Slawin, A.M.Z., Spencer, N., Stoddart, J.F., Vicent, C., Williams, D.J. (1989) A [2]Catenane made to order, *Angewandte Chemie International Edition*, **28**, 1396–1399.
18. Hunter, C.A. (1992) Synthesis and structure elucidation of a new [2]catenane. *Journal of the American Chemical Society*, **114**, 5303–5311.
19. Vögtle, F.; Meier, S.; Hoss, R. (1992) One-step synthesis of a fourfold functionalized catenane, *Angewandte Chemie International Edition*, **31**, 1619–1622.
20. Fujita, M., Ibukuro, F., Hagihara, H., Ogura, K. (1994) Quantitative Self-Assembly of a [2]Catenane from Two Preformed Molecular Rings. *Nature*, **367**, 720.
21. Ogino, H. (1981) Relatively high-yield syntheses of rotaxanes. Syntheses and properties of compounds consisting of cyclodextrins threaded by α,ω-diaminoalkanes coordinated to cobalt(III) complexes. *Journal of the American Chemical Society*, **103**, 1303–1304.
22. Harada, A. (2001) cyclodextrin-based molecular machines. *Accounts of Chemical Research*, **34**, 456 and references therein.
23. Dietrich-Buchecker, C.O., Sauvage, J.-P., Khemiss, A. (1986) High-yield synthesis of multiring copper(I) catenates by acetylenic oxidative coupling. *Chemical Communications*, 1376–1377.
24. Dietrich-Buchecker, C.O., Sauvage, J.-P. (1989) A synthetic molecular trefoil knot. *Angewandte Chemie International Edition*, **28**, 189–192.
25. Nierengarten, J.-F., Dietrich-Buchecker, C.O., Sauvage, J.-P. (1994) Synthesis of a doubly interlocked [2]-catenane. *Journal of the American Chemical Society*, **116**, 375–376.
26. Forgan, R.S., Sauvage, J.-P., Stoddart, J.F. (2001) Chemical topology: complex molecular knots, links, and entanglements. *Chemical Reviews*, **111**, 5434–5464 and references.
27. Gil-Ramirez, G., Leigh, D., A., Stephens, A. J. (2015) Catenanes: fifty years of molecular links. *Angewandte Chemie International Edition*, **54**, 6110–6150.
28. Arias, K. I., Zysman-Colman, E., Loren, J.C., Linden, A., Siegel, J. S. (2011) Synthesis of a D_3-symmetric "trefoil" knotted cyclophane. *Chemical Communications*, **47**, 9588–9590.
29. Ballardini, R., Balzani, V., Gandolfi, M. T., Prodi, L., Venturi, M., Philp, D., Ricketts, H. G., Stoddart, J. F. (1993) A photochemically driven molecular machine. *Angewandte Chemie International Edition*, **32**, 1301–1303.
30. Bissel, R. A., Cordova, E., Kaifer, A.E., Stoddart, J.F. (1994) A chemically and electrochemically switchable molecular shuttle. *Nature*, **369**, 133–137.
31. Livoreil, A., Dietrich-Buchecker, C.O., Sauvage, J.-P. (1994) Electrochemically triggered swinging of a [2]-catenate. *Journal of the American Chemical Society*, **116**, 9399.
32. Livoreil, A., Sauvage, J.-P., Armaroli, N., Balzani, V., Flamigni, L. Ventura, B. (1997) Electrochemically and photochemically driven ring motions in a dissymmetrical copper [2]catenate. *Journal of the American Chemical Society*, **119**, 12114–12124.

33. Baumann, F., Livoreil, A., Kaim, W., Sauvage, J.-P. (1997) Changeover in a multimodal copper(II) catenate as monitored by EPR spectroscopy. *Chemical Communications*, 35–36.
34. Durola, F., Sauvage, J.-P. (2007) Fast Electrochemically Induced Translation of the Ring in a Copper-Complexed [2]Rotaxane: The Biisoquinoline Effect. *Angewandte Chemie International Edition*, **46**, 3537–3540.
35. Collin, J.-P., Durola, F., Lux, J., Sauvage, J.-P. (2009) A Rapidly Shuttling Copper-Complexed [2]Rotaxane with Three Different Chelating Groups in Its Axis. *Angewandte Chemie International Edition*, **48**, 8532–8535.
36. Koumura, N., Zijistra, R.W.J., van Delden, R.A., Harada, N., Feringa, B.L. (1999) Light-driven monodirectional molecular rotor. *Nature*, **401**, 152–155.
37. Jiménez-Molero, M.C., Dietrich-Buchecker, C., Sauvage, J.-P. (2000) Towards synthetic molecular muscles: contraction and stretching of a linear rotaxane dimer. *Angewandte Chemie International Edition*, **39**, 3284–3287.
38. Jiménez-Molero, M.C., Dietrich-Buchecker, C., Sauvage, J.-P. (2002) Chemically induced contraction and stretching of a linear rotaxane dimer. Chemistry: A European Journal, **8**, 1456–1466.
39. Dawson, R. E., Lincoln, S. F., Easton, C. J. (2008) The foundation of a light driven molecular muscle based on stilbene and α-cyclodextrin. *Chemical Communications*, **34**, 3980–3983.
40. Tsukagoshi, S., Miyawaki, A., Takashima, Y., Yamaguchi, H., Harada, A. (2007) Contraction of Supramolecular Double-Threaded Dimer Formed by α-Cyclodextrin with a Long Alkyl Chain. *Organic Letters*, **9**, 1053–1055.
41. Wu, J., Leung, K.-F., Benítez, D., Han, J.-H., Cantrill, S.J., Fang, L., Stoddart, J.F. *et al* (2008) An Acid–Base-Controllable [c2]Daisy Chain. *Angewandte Chemie International Edition*, **47**, 7470–7474.
42. Coutrot, F., Romuald, C., Busseron, E. (2008) A New pH-Switchable Dimannosyl[c2] Daisy Chain Molecular Machine. *Organic Letters*, **10**, 3741–3744.
43. Du, G., Moulin, E., Jouault, N. *et al* (2012) Muscle-like Supramolecular Polymers: Integrated Motion from Thousands of Molecular Machines. *Angewandte Chemie International Edition*, **51**, 12504–12508.
44. Bruns, C.J., Stoddart, J.F. *The nature of the mechanical bond: from molecules to machines*, Wiley Ed., 2016.
45. Green, J.E., Choi, J.W., Boukai, A., Bunimovich, Y., Johnston-Halperin, E., Delonno, E., Luo, Y., Sheriff, B.A., Xu, K., Shin, Y.S., Tseng, H.-R., Stoddart, J.F., Heath, J.R. (2007) A 160-kilobit molecular electronic memory patterned at 10^{11} bits per square centimetre. *Nature*, **445**, 414–417.

Sir J. Fraser Stoddart. © Nobel Prize Outreach AB. Photo: A. Mahmoud

Sir J. Fraser Stoddart

Biography

I WAS BORN IN THE CAPITAL OF SCOTLAND on Victoria Day in the middle of World War II. The nursing home in Edinburgh where this cliff-hanger of an event took place, during the early evening of 24th May 1942, was located at 57 Manor Place. It was not anticipated that a little boy weighing in at just under two kilograms would live until the next morning. I have the doctor's bill, dated 7th October 1942, confirming that I defied the odds he gave against my survival. The bill reads, "*Dr. Douglas Miller presents his compliments to Mrs. Stoddart and begs to intimate that his fees for professional attendance amount to £4:4/-.*" It is not the realization that my parents were obliged to pay the princely sum of Four Guineas—equivalent to my father's monthly salary at the time—to have me brought into this world that resonates with me most of all; rather it is the four and a half months the doctor was prepared to wait before sending out his bill to my mother. How times have changed.

My mother, christened Jane Spalding Hislop Fortune, but known as Jean to the family, had made her own way into the world on 23rd May 1911 at Seggarsdean Farm, in the vicinity of Haddington, a small town about 20 miles east of Edinburgh, in East Lothian. One claim to fame for this town is that it is the birthplace of John Knox, the Scottish minister who was the leader of the Reformation in Scotland and the founder of the Presbyterian Church of Scotland. While still a toddler, Jean Fortune made the move with her parents and two elder brothers, Jim and Tom, to Colstoun Mains which is located three miles south of Haddington. This farm on prime agricultural land was to become the seat of the Fortunes for most of the 20th century. From all reports, my mother was quite a sickly child and did not achieve as much as she might have done at school, leaving the Knox Academy in Haddington when she was 14. Her health improved during her teenage years on the farm and eventually she attended the Edinburgh College of Domestic Science on Atholl Crescent, graduating in

February 1935 with a First Class Institutional Management Diploma. Following brief experiences as the manageress of private boarding schools in Yorkshire and Devonshire she became, with financial support from her father, the proud owner and proprietor of the Edenholm Private Hotel in Dunbar, a seaside resort on the North Sea, some eight miles east of Haddington. Old photographs indicate that it was around this time in 1937 that my mother and father met, became engaged and were married in St Cuthbert's Church in Edinburgh on 16th October 1938, just before the onset of World War II in September of 1939. My maternal grandmother rented a holiday home in Dunbar every summer in the late 1940s so that she could bring all her grandchildren under one roof for a few weeks. I have vivid memories, while in the company of my cousins, of watching pigs swim in a paddock on the outskirts of the town during the Great Floods of 1947 that hit the United Kingdom. These holidays by the seaside bring back happy memories, aside from when my grandmother found the urge to have us all visit the unheated swimming pool, open to the chill waters of the North Sea. How we all dreaded the experience that she informed us was good for our constitutions, yet apparently not hers!

My father, Thomas Fraser Stoddart, always referred to as Tom by the rest of the family, was born on 20th January 1910 in Irvine, Ayrshire on the West Coast of Scotland. His father was a golf professional who, together with his wife, ran the Bogside Golf Course until he retired in 1945. As summer is the high season

Figure 1. Seated between my father, Tom, and my mother, Jean c. 1946.

for golf, Tom Stoddart, together with his two younger sisters Anna and Clem (short for Clementine), were packed off each summer to the farms of cousins in East Lothian. It was at Howden that my father struck up a close, life-long relationship with his cousin, Tom Scott, while also being bitten by the farming bug. After attending Irvine Academy, my father continued his education at the West of Scotland Agricultural College in Glasgow where, after three years' training, he gained First Class Certificates in most of his classes and won the McAlpine Memorial Prize as the best student of his year in agricultural botany. I can vouch for the fact that my father knew all that there was to know about grasses to be found in the Lowlands of Scotland. When he graduated from the college in December 1932, the then Principal and Professor of Agriculture was to comment in a testimonial that "*He is a young man of energetic and painstaking habits, is methodical in his work, and is possessed of more than the average endowment in grit and determination. These attributes, combined with his sound theoretical and practical knowledge, mark him out as one well fitted for a responsible position in agriculture and dairying.*" That position turned out to be the manager of the University of Edinburgh's farms, one of them being Shothead, in the neighborhood of Balerno on the west side of the city, within sight of the Pentland Hills.

EDGELAW

When I was only six months old, my father decided to forsake the comparative comfort and relative security of being a farm manager to take on the tenancy of Edgelaw Farm about a dozen miles south of Edinburgh. Part of the Rosebury Estate, it was the middle of three farms on a dead-end road, which defined its remoteness and lack of electricity until I was almost 18. These circumstances, coupled with the fact that I was an only child, were to define much of my early life's experiences in what I was later to refer to as the 'University of Life'.

I grew up during the 1940s and 1950s in a post-World War II society coping with the rationing of food, clothes, and petrol (gas), and without access to modern-day conveniences in the home and workplace that we take for granted these days. The consequences for me were that I had to live out a very simple lifestyle, and I also had to find ways of amusing myself in a home where only a few rooms through the winter months were habitable. The need for warmth meant that we often lived as a small and close-knit family huddled together in the kitchen, which was fired by a Rayburn cooker that not only provided localized heat, but also hot water for the scullery, wash-house and single bathroom, in addition to some limited cooking space. It was augmented by another gas cooker that was fueled from a large cylinder of rural (liquid) gas. Other rooms in the farmhouse

Figure 2. Top: Edgelaw Farm House c. 1963. Middle: Kneeling in the middle of the middle row with 11 of my Melville College classmates on my 11th birthday. Bottom: Tethered to a young Ayrshire bull c. 1959.

had to be heated by open coal- and wood-burning fires that were often influenced in an unpredictable manner by the wind and rain outside. Up would go the cry that the fire in the drawing room was 'smoking', which meant that the room was filling up rapidly with smoke and would soon have to be evacuated. The one and only telephone was located in the hall, which was rarely, if ever, warm, and so conversations tended to be short during the winter months. Light through the long, dark winter months was provided by a vast array of Tilley and oil lamps.

I remember, as if it was only yesterday, the day 'the electricity,' as it was called, came to the farmhouse for the first time. It was Christmas Eve 1959 when we received word that the meter man would not be coming to install the meter until the New Year. The disappointment in the household was palpable. We had been so looking forward to celebrating Christmas and the New Year and the long-awaited (17 years!) arrival of 'the electricity' with family and friends. I had helped the electrician—a character if ever there was one by the name of Phil MacKay—during the preceding months wire the farmhouse, steading and cottages, and so was pretty knowledgeable when it came to wiring. Unbeknown to my parents, but egged on by Phil, I waited until the cows had been milked and the assembled company were all getting ready to sit down and have a Christmas Eve supper. At that point, I fetched a pair of stepladders, climbed up to a point near the ceiling where the meter would eventually be installed, and joined up the first pair of wires between the house and the grid with a pair of pliers. As I expected, nothing of significance happened. On bringing the second pair of wires together, however, there was blinding flash and much of the house was ablaze with light for the first time. There was a lot of noise. My mother was beside herself. She was convinced that someone would report us to the police and we would all end up in jail! Reason prevailed. My mother was soon convinced that by closing the curtains (drapes) in all the rooms we could harbor our secret and have 'the electricity' after all. And so, it was that for more than a week we had 'the electricity' for free and we used it to full advantage. In later years, we became much more conscious of switching off lights, for that practice had some bearing on the size on 'the electricity' bill. I reckon we all read more and I had no excuse not to do my homework. A television set arrived not so long afterwards and life was never quite the same ever again.

The whole episode brought out the daredevil side of my character. I discovered on the farm that defying regulations and breaking rules was a way to achieve distant goals on a shorter time-scale and, while there might be a price to pay, there would always be supporters, even secret admirers, and after the deed had been done there was no going back.

From a young age, I was addicted to solving jigsaw puzzles and would stack them up when completed between sheets of newspaper. I ascribe my early fascination in stereochemistry and topology to this addiction, which was to give way gradually to one of the more sophisticated of toys in Britain in the 1950s, namely 'Meccano'. The opportunity to construct a gadget I had designed myself and then put it to work after a fashion was to find expression later on when my passion for the chemical synthesis of unnatural products began to develop. There is also little doubt that my 'Meccano' set whetted my appetite many years later for constructing artificial molecular machinery from the bottom up. My interest in tinkering with machines and motors was increased considerably during those times on the farm, when I would take car and tractor engines apart, decoke them, replace the spark plugs and put them back together again, with the prospect that I would be going through exactly the same routine a few months later. The early internal combustion engines were not all that efficient or reliable: they demanded a lot of care and attention.

The late 1940s through the 1950s into the mid 1960s were times of rapid development in agriculture. My parents had no choice but to embrace change like there was no tomorrow. The horse and cart gave way rapidly to the tractor and trailer. The binder and all the labor-intensive and time-consuming paraphernalia that followed in its wake yielded more gradually to the combine harvester and the baler. Our 32 cows, distributed between three byres, were some of the first in the district to be milked by machine. Not all change was seen to be desirable: right up to the last days of the farm in 1968, my mother remained a strong advocate of producing eggs from free-range hens. With the onset of mechanization, collaboration between farmers was commonplace. For all the 26 years that my father ran a flock of 160 lambing ewes, the sheep-shearing was completed in one day (weather permitting) by the shepherds from Colstoun Mains, who would arrive in the early morning with all their motorized clippers. It was an occasion when my mother captured their hearts and souls with a wholesome dinner in the middle of the day that was surely second to none.

The farm had two cottages—one for the byreman and the other for the ploughman as well as a bothy (a single-room cottage) that was home to an Irish laborer for part of the year. The cottages experienced a fairly regular turnover of families usually with quite a number of children who were my playmates. We were left free to run wild around the farm and also to roam the countryside at will on our homemade buggies and old bicycles. Creativity and risk-taking came into our play on the grandest of scales in a playground we fashioned to changing circumstances. We invented our own games and learned the hard way about the dangers of climbing on roofs, burrowing through passages between bales of hay in the hay shed, and speeding down hillsides on carts adorned with a variety of wheels in the summer and on homemade sledges in the winter. The concept of playdates had still to be invented.

My formal education began when I was four years old with mornings only attendance at the local village school in Carrington, around three miles from the farm. My mother recalled that when she collected me from the school at noon on the first day and inquired as to how I had got on, my answer was to ask if I could go the next day for the whole day. At first there were only four other children, all girls, including one, Muriel Logan, a very bright girl from Aikendean Farm. As a consequence of the gender imbalance, I learned to knit, particularly stockings, rather well. By the time I left the village school in 1950, the number of pupils had risen sharply to 28. In this rapidly changing educational environment, where the older pupils helped to look after the younger ones, I discovered very quickly that Miss Morrison did not hesitate to use the tawse—a leather strap having one end cut into thongs that was used by schoolteachers in Scotland in the 1950s as an instrument of punishment—for poor performance, let alone bad behavior.

At the age of eight, my mother decided that I should go to one of the many fee-paying boys' day schools in Edinburgh. She chose Melville College—formerly the Edinburgh Institution and now, as a result of a merger, Stewart's Melville College—because she was attracted to its predominantly red and black uniform. I was obliged to take an entrance examination which, apparently, I passed with flying colors as a consequence of all that I had learned in the village school from Miss Morrison. I was blessed with some really outstanding primary school teachers—Miss Christie and Miss Pratt come to mind, both of them now in their nineties and still going strong today. They recall a very shy little boy, shyness being a trait that was to take me more than three decades to overcome.

Before I reached the age of 16, when I could negotiate the journey to school on a Lambretta scooter, my mother would drive me in the family's 1938 Hillman Minx—purchased from James Ross and Sons for £155—the three miles to the nearest bus terminal in Rosewell, at that time a small coal-mining village, to catch the 7:40 a.m. bus to Edinburgh. The popularity of cigarette smoking amongst the office workers and shop assistants meant that you could cut the atmosphere with a knife on the top deck of the bus towards the end of the 45-minute journey to St Andrew's Square. The bus journey was followed by a mile-long walk down George Street to the school on Melville Street. I was to realize many years later that the education I received at Melville was second to none, maybe because the 1950s were less than two centuries removed from the period of the Scottish Enlightenment that was graced by eminent scholars, such as philosopher David Hume, economist Adam Smith, poet Robert Burns and chemist Joseph Black. I

Figure 3. Top: Seated to the right of Mr Richardson, the Headmaster with the other nine school prefects in 1960. Bottom: Third from the left in the back row of the 1960 First Hockey Eleven.

was taught by teachers, most of whom could have been university professors, in Latin, English Language, English Literature, French, History, Geography, Mathematics, Physics and Chemistry. The school's music master, W. O. (Bill) Minay, was the organist at St. Cuthbert's Church—where my parents were married on 19th October 1938—at the west end of Princess Street. It was from Bill Minay that I took piano lessons for many years. With a huge amount of practice and no little encouragement from him, I was able to play the first two movements of Beethoven's First Piano Concerto. Although enjoyable, that experience told me I was not cut out to be a concert pianist.

Sport was also a major part of the school curriculum. Rugby, cricket and field hockey were compulsory, along with swimming all the year round. During our last three years in school we found ourselves in army-style uniforms as part of a Cadet Force, in which I rose to become the Signals Sergeant. In my final year, the Headmaster appointed me to be the Second Prefect of the School, a position that gave me adequate opportunities to develop leadership skills. This experience was to prove invaluable when I became Head of the School of Chemistry at Birmingham—and later, the Director of the California NanoSystems Institute.

My mother was a terrific cook and an awesome baker, knowing instinctively when to add and mix ingredients, and rarely, if ever, measuring or weighing anything out. She knew just the right moment to stop whisking, heating and beating mixtures. She went about all these activities and more, including dress-making and patching up clothes, while feeding hens, mucking out henhouses, rearing chickens, gathering eggs and selling them in the neighboring villages and townships. My early successes at practical work in chemical laboratories owed much to watching this remarkable time and motion machine in action. My father, by far the best educated and most well-read farmer in the district, set very high standards for himself, expressed most intensely when a heifer was being groomed to perfection, prior to being sold at the Lanark Stock Market, or a flock of lambs, suitably washed in the dipper and individually manicured to perfection, were on their way to the auctioneer at the St. Boswells Sheep Sales.

I was to witness, from a very young age, the essential mating activities that were an integral part of maintaining a herd of dairy cows and orchestrating in October and November the running of around 150 ewes with tups (rams), at an approximate ratio of 50:1 to ensure the arrival of around 250 lambs during a frantic three-week window in March. While tending to cows calving throughout the year on a fortnightly basis, often in the middle of the night, was more or less routine, the lambing season never failed to reduce myself and my parents to states of utter physical and mental exhaustion, from a combination of lack of sleep and very long working days, for spring was also the time to be in the fields from morning to night sowing wheat, barley and oats, to be followed immediately thereafter by potato planting and the sowing of kale and turnips (swedes). In the summer months, I enjoyed nothing more than walking round the 365-acre (one for every day of the year) farm with my father in the evenings of long light. He knew all there was to know about the flora and fauna of the countryside. He was also a walking dictionary—a kind of Google before its time—that was useful for me in building up a vocabulary, and when we were engaged in the evenings in solving crossword puzzles in *The Scotsman*. My vocabulary was also broadened through my friendship with the farmhands, who taught me to swear from a young age. Later in life, my mother reflected that, much to her chagrin, I could swear like a trooper well before I could talk.

EDINBURGH

During my four years as an undergraduate student at Edinburgh (1960–1964), I managed to hold my own in Mathematics, Physics, Chemistry and Biochemistry classes, in the face of stiff competition from many very bright students drawn, in large part, from the east coast of Scotland, many coming from the elite Edinburgh schools. A cohort of English students—who entered the Scottish higher educational system having covered much of the first-year science curriculum at A-level in England—got off to a flying start in their first year, but in subsequent years we Scots started to pull ahead of most of them. The chemistry teaching at Edinburgh in the early 1960s was not particularly taxing or stimulating, apart from some excellent lectures given by Tom Cottrell, John Knox, Peter Schwartz and Dai Rees. Organic chemistry, under the leadership of Professor Sir Edmund Hirst, was heavily skewed towards carbohydrate chemistry. A transformation occurred in my third year during a laboratory course in quantitative analytical chemistry. During his introduction, the somewhat abrasive Dougie Anderson announced to more than 100 of us that we would be pipetting by mouth enough cyanide to kill the whole of Edinburgh! After having made this spine-chilling remark, he went on to state that he had been running the 10-week course for more than a decade and in that time no student had ever completed it. Here was my opportunity, I thought, to apply the multitasking skills I had acquired from working on a mixed-arable farm for a couple of decades. I used this experience and finished the course inside seven weeks, gaining a mark close to 100%. This achievement earned me my first visit to the office of Sir Edmund who told me that Dr. Anderson would like to offer me a paid position in his research group during the following summer. I jumped at the opportunity. I felt much more at home in this new research environment, where I was given the opportunity to

Figure 4. Top: Standing on the right with Douglas Anderson and my fellow postgraduate students at Kings Buildings, University of Edinburgh c. 1965. Bottom: Second on the right in the third row back with Walter Szarek and Ken Jones at the front taken at Queen's University in Canada c. 1968.

unravel the structural complexities of plant gums of the *Acacia* genus. There was little doubt from what was already published in the literature that these acidic polysaccharides—accompanied mysteriously by a small amount of protein—were high molecular weight polyelectrolytes constituted around a branched carbohydrate backbone. I was to continue researching these biomacromolecules well beyond a fourth-year research project into the pursuit of a PhD degree as a postgraduate student. My main contribution to the field was to challenge the "main-chain" hypothesis, implying a brush-polymer constitution, and replacing it with a much more highly branched constitution without having the foresight to describe it as a dendrimer before its time. My postgraduate research was to leave me with one lasting impression—namely, that the many gum trees in the Sudan, from whence the nodules I studied came, had never managed to produce between all of them through all of time, two gum molecules which were identical in size and constitution. After this period of handling highly heterogeneous mixtures, I longed to grow acquainted with a molecular world where homogeneity ruled the roost, at least for a time.

Between continuing to work on the farm, and becoming bitten by the research bug, I had to settle for graduating with a BSc Honours Degree in Chemistry and being the top Upper Second, in fifth place overall, in the 1964 Class of 45 students. By contrast, my postgraduate research was a resounding success and I was able to graduate with a PhD degree in just over two years in November 1966, having met the love of my life, Norma Scholan, who had joined the Anderson group as a fourth-year undergraduate research student. Norma made up for my lackluster performance in my Finals by coming top of her class of over 80 Final Year Chemistry students in 1966. In the years to come our two daughters, Fiona and Alison, were to graduate in Chemistry—from Imperial College London and the University of Cambridge, respectively—with First Class Honours degrees just like their mother before them, leaving me the dunce of the family!

KINGSTON

During the first 25 years of my life I had travelled very little and I yearned to go to North America, with enthusiastic support from my parents and somewhat less so from Norma, who had transferred her allegiance to the Biochemistry Department in the Medical School to begin her postgraduate work in steroid biosynthesis, under the tutelage of George Boyd. For my part, Sir Edmund sprung into action and did not take long to arrange for me to go to Queen's University in Kingston, Ontario as a National Research Council of Canada Research Fellow. Here I would join the Chemistry group, headed up by Ken Jones, one of his own postgraduate students from his Bristol days. The 1960s witnessed the end of an era in UK chemistry departments, arranging for the department's best students to go overseas to pursue postdoctoral fellowships in research. How times have changed for the better.

I left Prestwick for Montreal aboard a British Overseas Airways Corporation (BOAC) plane, taking to the air for the first time in my life, on 1st March 1967, with Sir Edmund's words ringing in my ears, "*Whatever you do in research, Stoddart, make sure you work on a big problem.*" I was not at all sure what he meant by a 'big problem' but I was determined to heed his advice to the best of my limited ability. I suspect he anticipated that I would remain a carbohydrate chemist for the remainder of my professional life but that did not turn out to be the case. In the event, as soon as I set foot in the Jones laboratory, Ken confided in me that come 1st April he would be leaving for Curitiba in Brazil to spend one whole year there on sabbatical leave. This totally unexpected piece of breaking news, although quite a shock for me at the time, was to work to my advantage in the long run. I found myself assisting Walter Szarek, a former graduate student in Ken's group, who had returned to Queen's from Rutgers University to take over its supervision. It was good early experience for me in helping Walter run and mentor a medium-sized research group.

Communications between Canada and Brazil were dependent on the back and forth delivery of airmail letters, with a complete turnaround of information taking about three weeks, by which time the news was often obsolete. We were quickly relieved of this frustration when the Canadian postal service was brought to a halt by strikes for months on end. These circumstances left me with enough time on my hands to go in search of Sir Edmund's 'big problem'. I stumbled upon it in the chemistry department library under the guise of a short communication by Charles Pedersen in the *Journal of the American Society* (*JACS*) in the Spring of 1967, describing the efficient template-directed synthesis of dibenzo[18]crown-6 in 48% yield. This breaking news, coming out of the Dupont Laboratories in Delaware, flew in the face of all the teaching I had experienced as an undergraduate student at Edinburgh, where I had been led to believe that, while making five-, six-, and seven-membered rings was commonplace, large-sized rings were a totally different kettle of fish. I also realized that these macrocyclic polyethers—or crown ethers as Pedersen had called them—shared some of the constitutional features (OCCO repeating units) with the sugars. So, I set off on a mission to pursue what I referred to as 'lock-and-key chemistry' by simply marrying conceptually Pedersen's crown ethers with Emil Fisher's carbohydrates. Of course, it was easier said than done, for I was only one pair of hands with many other things on my mind. One of them was to return to Edinburgh in the Fall of 1968 to say goodbye to the farm—for my parents had decided that after my leaving for Canada it was simply too much for them to handle on their own—and the other was to get married in Glasgow in the presence of close family members to Norma on 8th October 1968. We returned to Canada the next day via Montreal, my newlywed wife occupying her time during the flight by completing mountains of immigration paperwork. At the airport, we were greeted by a customs officer who took one look at us, summed up the situation, crumpled the papers into a ball, threw them into a waste-paper bin (trash can) with the words "*we grow trees in Canada and far too many of them get turned into paper*" and waved us through to begin our married life in a foreign country with a welcome we were never to forget. Perish the thought that such a welcome would occur at an international border in today's world.

Our remaining 15 months at Queen's were blissful ones. We lived at 432 Alfred Street after I had negotiated to rent the house from the owners, Thelma and Dave Buchan, who had more or less become my Canadian 'aunt and uncle' during my first 18 months as a boarder in their home. Norma had completed research for her PhD degree, like myself in just over two years, but not without a never-to-be-forgotten incident following the decision that I would type the manuscript on my portable Olivetti typewriter. It was approaching midnight and I was typing the last few pages of her thesis. Norma decided I needed a cup of coffee and duly set the cup down on the table next me. The next time I triggered the carriage return it hit the cup fair and square on its side and propelled most of the contents right over the stack of 150 typed pages. Norma retired to a corner of the room sobbing her heart out. After a kiss and a cuddle, I sent her off to bed and then stayed up all night, retyping much of the thesis by the following morning. When disaster strikes, it is best to waste no time in putting the experience to rest.

During my stay at Queen's I found it easy to interact with the faculty. Saul Wolfe, in particular, took me under his wing and transmitted to me the importance of being on top of the current literature. He brought to my attention the teachings of Kurt Mislow at Princeton on the importance of applying molecular symmetry to stereochemistry. Mislow had just introduced the concept of topism for analyzing the topic relationships between atoms and ligands in molecules. Amongst other attributes, it rendered the interpretation of NMR spectra a much easier task and helped to save me the embarrassment of coming to a wrong conclusion more than once. I had the opportunity to travel down to Princeton with Saul to meet this sage of stereochemistry. There were other opportunities to listen to lectures by the intellectual leaders of their time in organic chemistry, among them the famous Harvard professor and synthetic chemist par excellence, R.B. Woodward, whom I recall holding an audience in the palm of his hand in Ottawa for more than three hours. Then the Queen's chemistry department invited Saul Winstein from the University of California at Los Angeles (UCLA) to give the MacCrae lectures in the Spring of 1969. Winstein was considered by many to be the intellectual leader in physical organic chemistry at that time and would almost certainly have been the recipient of a Nobel Prize in Chemistry had he not died very suddenly of a heart attack, at age 57, in November of that same year. What I recall most vividly about the MacCrae lectures was the manner in which Winstein launched into a 20-minute diatribe against H. C. Brown, reflecting the bitter controversy that raged between them for years over classical (HCB) versus non-classical (SW) carbocations. I could not have known in 1969 that almost 30 years later I would be making my way to UCLA to become the second holder of the Winstein Chair, following Donald Cram who shared the 1987 Nobel Prize in Chemistry with Charles Pedersen and Jean-Marie Lehn from the University of Strasbourg.

Saul Wolfe was a pupil of the highly influential and renowned carbohydrate chemist, Ray Lemieux, for whom I had acquired an enormous respect after hearing him give a series of remarkable named lectures (Purves, if I recall correctly) at McGill University, which ultimately led me to write a monograph on the *Stereochemistry of Carbohydrates*. I set out on this mission with the support of Ken Jones, who had returned from Brazil, without realizing the responsibility one assumes when writing a book! My attendance at a symposium hosted by the US Army Laboratories at Natick led to my meeting Ernest Eliel, the author of *The Stereochemistry of Carbon Compounds*, a classic published by McGraw-Hill in 1965. It had been my bible from my Edinburgh days and so I decided that I would approach Dr. Eliel at the end of his inspirational talk and ask him if he would be kind enough to look over and comment on my manuscript. I sent him the manuscript and within a very short space of time it came back plastered in red ink. This experience taught me that having my manuscripts scrutinized by experts wherever possible would save me no end of embarrassment in the fullness of time. On this occasion, no doubt, Ernest saved my bacon: he and his wife Eva were to become close friends of myself and Norma for the rest of their lives.

SHEFFIELD IN THE SEVENTIES

As the 1960s came to a close, Norma convinced me that it was time to return to Old Blighty, where we would give some thought to raising a family. Sometime in the summer of 1969 Ken Jones came back from a conference in the Caribbean with the news that David Ollis from Sheffield had given a lecture (with demonstrations) on the conformational behavior of a 12-membered ring compound known as tri-*o*-thymotide, or TOT for short, that had captured everyone's imagination. I decided to apply for an ICI Fellowship to go to Sheffield but initially failed to make the cut. Three months later I heard the good news that I had, after all, landed this prestigious fellowship, as one of the successful candidates had decided not to accept the offer. We decided it would be practical to ship our goods and chattels across The Pond and enjoy an ocean liner experience onboard the West German flagship *Bremen* during the week before Christmas. Five days after leaving New York we arrived in Southampton to be greeted by thick fog, which made the drive north to Edinburgh, stopping off in Sheffield on the way, all the more challenging.

There were several reasons for going to Sheffield. One was to attend the Annual Sheffield Stereochemistry Meeting, where I had the opportunity to hear Jean-Marie Lehn speak for the first time. It was such a pleasure to listen to this young French chemist with a research agenda in the making that was destined to chart new territory for the subject beyond the molecule or, as Lehn named it subsequently, supramolecular chemistry. Another reason for being in Sheffield was to introduce myself to David Ollis. When the subject of my start date came up he insisted I should be present in the department for the 1st of January 1970. This edict infuriated Norma, and I was not best pleased either, given the fact that we were heading to Scotland where New Year's Day is a national holiday. The crossing of swords with Ollis would go on for the best part of two decades.

On my return to the chemistry department on 1st January it became apparent that I was not going to be allowed the independence to carry out the kind of research that was the fellowship's official remit. In addition, when Ollis learned that I would be spending some of my time writing the final chapter of the book, he immediately expressed his displeasure, stating quite emphatically that "*people at my stage should not be writing books*." Norma, who was illustrating the manuscript with India ink and stencils, convinced me to ignore his decree and the monograph was published in 1971 by Wiley. If the welcome to Sheffield was muted from on high, Norma and I were made to feel very welcome by the postgraduate community, particularly by David (Dave) Brickwood (whom I was delegated to supervise), Stephen (Steve) Potter and Richard (Dick) Taylor. Little do they really know how much they helped us through those difficult times.

I was working in my laboratory (E19) on Good Friday in 1970 when Ollis walked in to tell me that at a meeting of the Organic Staff the day before, it had been decided that I should be offered a Lectureship in Chemistry—a position that had unexpectedly fallen vacant with the resignation of the youngest member of the staff—from 1st October. I was, of course, happy to have some long-term job security, although it in no way earned me my independence. It was 1973 before Andrew Coxon became my first independently supervised postgraduate student. For my first lecturing assignment, I was handed a poisoned chalice in the shape of teaching the first-year medical students (all 180 of them) organic chemistry, in the knowledge that the course would soon be discontinued. The refrain from the students was very much along the lines of "*Why are we having to take this course when it's about to be withdrawn?*" It was a tall order to hold their attention in lectures and laboratory classes, but I did my very best to engender their enthusiasm by introducing all sorts of innovations into my teaching. Nonetheless, when brought before a group of medical staff in the presence of their dean, I was informed by him that "*we might as well be teaching our students biblical studies.*" It was a crushing put-down but I reasoned that I should not have been the person from the chemistry department finding himself in this particular lion's den!

Despite the fact that my progress in research was being forestalled at every turn by the antics of the professors in the department, Andrew Coxon, and later Dale Laidler, made some notable advances in their research with carbohydrate precursors to crown ethers, to the extent that when I was invited to speak at international conferences and symposia I had some interesting results to talk about under the banner of 'lock-and-key chemistry'. A major turning point in my fortunes came in 1976, when I was invited to give no less than 17 lectures and seminars, nine in the UK, including Oxford, Imperial College London, Edinburgh and Glasgow, four in the US, including Columbia, Princeton and Dupont, and four in Canada, including McGill and Queen's. I was also fortunate in being invited to give a talk at the Centennial American Chemical Society Meeting in New York in early April. This invitation afforded me the opportunity to listen to Donald Cram speak and to meet with him one-on-one—along with his shopping bag full of CPK space-filling models—for the first time. Once again, I found myself in the company of an eminent American chemist, who not only enthused about his own research, but also about mine, an experience for me that was uplifting beyond my wildest dreams. Don was also the RSC Centenary Lecturer in May 1976. He insisted that I would be one of the supporting speakers in Manchester and, two days later, in London, at University College. When the powers that be at the RSC questioned my double act, Don swept aside their protestations with the comment that "*apart from Fraser and myself, those in the audiences in the two places will be different*" and, of course, no one could argue with him, for he was right! David Ollis was livid, but Don was drawing considerable satisfaction from the situation because he knew how I was being treated on home turf. Don also presented his Centenary Lecture in Sheffield and went out of his way to say he had come because of my presence in the department. He went on to lavish praise on my research group, leaving Ollis red with rage!

During these meetings, Don encouraged me to apply to the Science Research Council (SRC) for a Senior Research Fellowship to spend the first three months of 1978 on sabbatical leave at UCLA. This short stay in the UCLA Department of Chemistry and Biochemistry was a real breath of fresh air and served to increase my yearning to move to the US one day in the future. Interest in hiring me had been mooted in a number of different US universities, but then something else happened in the UK that I could live with very comfortably, and that left Norma happy that our two girls, who had arrived on the scene in 1973 (Fiona) and 1976 (Alison), could continue their education in the UK. That development involved the SRC, who were ready, willing and able to support my secondment to the ICI Corporate Laboratory in Runcorn, under the auspices of a brand new Cooperative Research Scheme for three years, from 1978 to 1981. It also received the backing of a number of ICI's senior management, including Tom McKillop and Bernard Langley. I was over the moon. I was free at last to carry out my own research in a highly supportive and amazingly well-equipped environment, staffed with research scientists who were second to none. We sold our home on Derriman Avenue in Sheffield and moved across the Pennines to a brand-new house in Curzon Park in Chester, with a six-month layover in a small rented property in Little Sutton, on the Wirral. The next three years were amongst the happiest that we spent as a family in England.

RUNCORN

I joined Warren Hewertson's Catalysis Group at ICI's Corporate Laboratory and supervised a couple of postgraduate students in Runcorn, plus half a dozen who remained in Sheffield, where I spent minimally one day a week. The Corporate Laboratory, situated on The Heath at Runcorn, was probably the closest one could get to a Bell Laboratories experience in the UK. It was in this setting that I quickly struck up a highly productive collaboration with a brilliant young chemist, Howard Colquhoun, who had only recently joined the laboratory. Following some discussions about what different kinds of complexes could be formed with crown ethers, we came to the conclusion that, as far as we knew, transition metal ammines had not been put to the test. We were fortunate insofar as there was a treasure trove of these ammines down in the basement of the laboratory that had been prepared by Joseph Chatt when he was an employee of ICI during the 1950s. I could not believe our luck. Before long we had lots of crystals of adducts of transition metal ammines with crown ethers, whose solid-state superstructures were solved at the drop of a hat by David Williams, X-ray crystallographer extraordinaire, down in London at Imperial College.

Amidst all these many superstructures, one caught our attention. It was the 1:1 adduct in which dibenzo[30]crown-10 (DB30C10) wraps itself round a dicationic platinum complex, carrying a 2,2'-bipyridyl ligand in addition to a couple of *cis*-diammine ligands, in such a manner that the ammine ligands form hydrogen bonds with the polyether loops of the crown ether, while the two π-electron rich catechol units sandwich the π-electron deficient bipyridyl ligand in a stacking manner. The structural similarities between this bipyridyl ligand and the bipyridinium herbicide Diquat (DQT) was pointed out to us by former ICI research scientist Eric Goodings. Sure enough, when the transition metal complex was replaced by DQT we obtained deep orange crystals of a 1:1 complex with DB30C10, as revealed yet again by its solid-state superstructure. Both the adduct and the complex, when associated with soft counterions, are reasonably stable in acetonitrile solution, as indicated by the presence of diagnostic charge-transfer bands that render the solutions light yellow and bright orange, respectively.

We had injected new life into Alfred Werner's concept of second-sphere coordination in the process of establishing donor-acceptor interactions as a force to be reckoned with in molecular recognition processes. They would ultimately serve as the sources of templation in the making of molecules with mechanical bonds. Although we had still to address the need to form complexes between crown ethers and Paraquat (PQT)—the other component of the wipe-out weed-killer

Figure 5. Left: With Norma outside our third Sheffield home in Bradway c. 1982. Right: After graduating from Edinburgh in 1980 with a DSc degree.

that ICI marketed worldwide for many years—we had given the search for the 'big problem' an enormous fillip from an unlikely starting point. If I had not spent those years at ICI's Corporate Laboratory, my role in the development of mechanically interlocked molecules, that has led to designing and synthesizing molecular machines, would either not have happened or would have taken a very different course. All of what I was subsequently to achieve in research can be traced back to these three years.

I left Runcorn in the late summer of 1981 with a heavy heart, but there was no option. My three-year secondment was coming to a close and, more disturbingly, the writing was on the wall for the Corporate Laboratory. Norma and the girls had come to enjoy life in Chester and it was going to be challenging for all of us to return to Sheffield. Once again, the transfer was staged by my acquiring a small semi-detached home in Bradway, from which we were able to purchase the ideal family home in the shape of an Edwardian house on Dore Road.

SHEFFIELD IN THE EIGHTIES

My situation at Sheffield had been strengthened by my industrial experience and I was promoted to a Readership in Chemistry in 1982. Although many of the same issues still existed in the chemistry department at Sheffield, I was much more able to handle the slings and arrows of outrageous fortune. With growing confidence, I became quite vocal at the national level about the weaknesses, as I saw them, in the British academic system. My pronouncements and my writings—often to the British national newspapers—did not win me many friends, but at the same time they served to define where I stood on a wide range of issues. Eventually those in influential positions started to notice and take note.

At this time, I struck up another important relationship that not only turned out to be of immense value in the promotion of my research as it developed during the 1980s, but also helped me launch some university-wide initiatives, such as the Sheffield Industrial Forum in 1986. That relationship was with Roger Allum, the Press Officer for the University. He was extremely supportive and would always seek to make our research intelligible to the wider public. If I ever felt a little depressed from working in a department that was brim full of politics, I could take a walk up to the Edgar Allen Building and have a reassuring chat with Roger. He always had time for me, no matter how busy he was tending to other university business. I would leave his office with my spirits lifted and ready to take on the world.

As I moved from one university to another, the importance of maintaining good and close relationships with the talented individuals in media relations remained with me. Martin Hicks at Birmingham continued in the footsteps of Roger and once I reached the University of California, Los Angeles (UCLA), I was to learn a lot from Stuart Wolpert on how to handle live and recorded interviews for radio and television. At Northwestern University (NU) I have been blessed many times over to have Megan Fellman working closely with myself and members of my research group in getting story after story out into the public domain. More recently, I have discovered a soulmate in Stephanie Russell, Editor of the *Northwestern* magazine, who has gone to considerable lengths, and well beyond the call of duty, in presenting me and my research to the alumni and friends of NU, following my award of the Nobel Prize in Chemistry.

On the scientific front, after some wasted effort and unproductive years, we were able to demonstrate quite simply that a constitutional isomer of DB30C10, namely bis-*para*-phenylene[34]crown-10 (BPP34C10), forms a strong 1:1 complex with PQT. The fact that the solid-state superstructure of this complex was 'rotaxane-like' in its appearance led me to suggest that it be called a [2]pseudo-rotaxane, a name which eventually transmogrified into meaning a template that could subsequently be converted into a catenane as well as a rotaxane. We had established that we could thread a *p*-acceptor through a ring containing two laterally disposed π-donor units. Our next challenge was to reverse this recognition motif by making a cyclophane in the form of cyclobis(paraquat-*p*-phenylene) and containing a couple of parallely disposed bipyridinium units held rigidly apart at a plane-to-plane separation of approximately 7 Å by two *para*-xylylene units, through which π-donors of many different persuasions could thread. In the first instance, Mark Reddington was able to prepare this cyclophane starting from 4,4'-bipyridine and xylylene dibromide in a 12% yield.

During our efforts to publish the synthesis and full characterization of this cyclophane in *Angewandte Chemie*, I received a curt letter from Siegfried Hünig at the University of Würzburg, explaining that one of his students had synthesized a whole range of very similar cyclophanes and studied their ability to complex aromatic hydrocarbons, a piece of information that was available, but overlooked by me, in *Dissertation Abstracts*. I wrote back to Professor Hünig, who had clearly been one of the reviewers of our communications, and suggested that he write up a communication on his work while we delayed the publication of our communications so that all three could appear in the journal in a row. Thereafter, Siegfried and I became close friends, to the extent that he and his wife invited Norma and myself to Würzburg to help celebrate his 80th birthday in 2001. The publication of our two communications coincided with the beginnings of my use of color—red for π-donors and blue for π-acceptors—so that the cyclophane soon became known in the literature as the 'little blue box' and was to gain considerable notoriety as a promiscuous host for a wide range of π-donors, including benzidine and tetrathiafulvalene. Subsequently, employing both templates and catalysts—and some other tricks—we have been able to prepare the little blue box in all but quantitative yield.

The stage was now set to carry out the template-directed synthesis of the first donor-acceptor [2]catenane in a remarkable 70% yield, by very simply employing the ingredients used in the preparation of the little blue box in acetonitrile at room temperature in the presence of three molar equivalent of BPP34C10. This experiment, which was carried out by Cristina Vicent and Neil Spencer, was one of the most memorable as we all gathered to watch the reaction mixture turn orange and crystals start growing on the side of the reaction flask within 10 minutes. I realized there and then that we were sitting at the entrance of a gold mine as we prepared the manuscript for publication in *Angewandte Chemie* in October of 1989. While the manuscript was out for review, I received a phone call from Jean-Pierre Sauvage in Strasbourg saying how impressed he was by the contents and offering me his congratulations. He was obviously one of the reviewers.

The 1980s represented a sea change for my group, as I began to realize that our level of research performance could be raised out of all recognition by welcoming postgraduate students and postdoctoral fellows from overseas. The arrival of Franz Kohnke from the University of Messina, not to mention the short visit of Cristina Vicent from Madrid, had a profound effect on the group culture as we became increasingly international in our composition. The cultural change also encouraged home-grown PhD students to raise their sights. Following graduation with their PhD degrees, David Leigh went to Ottawa in search of postdoctoral experience with David Bundle, while John Mathias, equipped with a postdoctoral fellowship, was invited by George Whitesides to go to Harvard.

Pier Lucio Anelli, who came to Sheffield as a postdoctoral researcher from the University of Milan, was another of a growing number of makers and shakers. Employing a pre-prepared dumbbell-shaped molecule as a template, he synthesized by templation a degenerate [2]rotaxane with two π-donating, hydroquinone-based, recognition sites for encirclement by one little blue box, which could be shown by dynamic NMR spectroscopy to be darting back and forth between the recognition sites at around 2000 times per second. I called it a molecular shuttle and concluded in a 1991 *JACS* communication that it was "*the prototype for the construction of more intricate molecular assemblies where the components will be designed to record, store, transfer and transmit information in a highly controllable manner following their spontaneous self-assembly at the supramolecular level.*" The development of this next step in the research program had to wait until a move to the University of Birmingham had been planned and executed.

Figure 6. With Don and Jane Cram at the 16th International Symposium on Macrocyclic Chemistry held at Sheffield University in September 1991.

BIRMINGHAM

I had been approached in 1991 by the then Vice-Chancellor of the University of Birmingham, Sir Michael Thompson, to consider moving to Birmingham as the Professor of Organic Chemistry. He had been attracted by my refusal to join the large group of whingers in British academia at that time. The Department of Chemistry was in a badly run-down state and morale was low to say the least. After much discussion and an undertaking by the Vice-Chancellor to implement a staged refurbishment of the Haworth Building and invest in some key state-of-the-art equipment, including NMR and mass spectrometers, I accepted the chair and started a phased move of my research group, now growing in size, from Sheffield to Birmingham. Norma remained in Sheffield to look after the everyday needs of the group members there, while I oversaw the revamping of the top (seventh) floor of the building and prepared for the new spectrometers to arrive on the scene. While Neil Spencer accepted the challenge of establishing the new NMR facility, I managed to persuade the highly gifted senior technician, Peter Ashton, to also make the move from Sheffield to Birmingham and establish a mass spectrometry facility that was second to none in the country. I commuted between Birmingham and Sheffield for more than a year, given the added responsibility of being one of the organizers, along with Norma and David Fenton, of the 1991 International Symposium on Macrocyclic Chemistry, at which both Donald Cram and Jean-Marie Lehn received Honorary Degrees in Science from Sheffield University. It was also the occasion when the first International Izatt-Christensen Award was presented to Jean-Pierre Sauvage.

A life-changing event was to occur in February 1992 when I took an early morning phone call in my Birmingham office from Alison, her first words being, "*Something terrible has happened, Daddy.*" She went on to explain that her mother was in hospital, having suffered a brain hemorrhage overnight. I wasted no time in jumping into my car and driving up to Sheffield, only to be told by the surgeon in charge of her case that he was going to have to operate and that there was no better than a 50% chance that Norma would survive the surgery. It was a long day that was to take a turn for the better when the surgeon informed me in the early evening that the artery in Norma's brain had self-healed and he would not need to operate. Relief all round! We moved from our Edwardian home in Sheffield to a 1930s home in Edgbaston, close to the campus of the University of Birmingham, on 1st April. With Norma still very much in a convalescent state, I was approached by Ken Houk at UCLA, who asked me if I would consider moving to UCLA to assume occupancy of the Winstein Chair on the impending retirement of Don Cram. My reply—with mixed emotions—was an easy one. Norma was too ill for me even to share this news with her and I was in the throes of a complicated relocation. I assumed that my message to Ken declining his offer would be the last I would hear of the Winstein Chair at UCLA and that this tantalizing prospect had slipped out of my grasp. This assumption proved to be incorrect.

As far as research was concerned, my seven years at Birmingham were to exceed my wildest dreams. Our first bistable [2]rotaxane, that could be switched both chemically and electrochemically, reached the literature in 1994, following a sojourn by Richard Bissell at the University of Miami with Angel Kaifer. Olympiadane was self-assembed by David Amabilino, while Gunter Mattersteig synthesized the first bistable [2]catenane, in which the two π-donating hydroquinone recognition sites in the degenerate [2]catenane were replaced with tetrathiafulvalene and dioxynaphthalene recognition sites. A highly fruitful collaboration, in which this catenane and many other bistable MIMs were switched chemically, electrochemically and photochemically, was struck with Vincenzo Balzani and Alberto Credi at the University of Bologna. Jon Preece spent time in the laboratory of Helmut Ringsdorf at the University of Mainz learning how to produce Langmuir monolayers and films of both degenerate and bistable [2]catenanes and preparing the way for device fabrication when we reached UCLA in 1997.

Douglas Philp was a major intellectual driving force in the group during its early days in Birmingham. Aside from his high level of productivity that matched his creativity every inch of the way, he left a considerable legacy by writing a much-cited review on "Self-Assembly in Natural and Unnatural Systems" that was published in *Angewandte Chemie* in 1996. While Peter Glink established hydrogen bond templation (known within the group as 'ammonium binding') as a means of templating the synthesis of MIMs, Narayanaswamy Jayaraman and Sergey Nepogodiev launched ambitious programs of research into glycodendrimers and the synthesis of cyclic oligosaccharides related to the cyclodextrins. Steven Langford and Matthew Fyfe took over where Douglas Philp left off by bringing their keen intellects and dedicated commitment to the development of MIMs to a highly sophisticated level in relation to their physical organic chemistry.

Unwelcome news kept breaking in 1992. In August, Norma was diagnosed with breast cancer and underwent surgery in the form of a lumpectomy, followed by radiation and chemotherapy. The cancer recurred two years later in 1994, resulting in a mastectomy and yet more of the inevitable back-up treatment. This did not halt the progress of the disease, which was diagnosed as having become metastatic in 1996. During a visit to UCLA in 1994 to participate in a symposium to mark Don Cram's 75th birthday, Ken Houk raised once again the availability of the Winstein Chair, reiterating the interest of the Department of Chemistry

Figure 7. The research group at Birmingham c. 1995.

and Biochemistry in my coming to occupy it at UCLA. My feeling that Norma was not receiving the best of medical care in Birmingham was accepted by her in early 1997 and so we decided to go on a trip to the US, visiting the M. D. Anderson Clinic in Houston and the Jonsson Comprehensive Cancer Center at UCLA, where Norma was told by the oncologists we met that, while she had a chronic disease, they had 50 different ways of treating it. At this point it was decided that I would step down from being the Head of the School of Chemistry at the end of June and formally move to Los Angeles to take up the Winstein Chair on 1st July 1997. Some 15 members—including first-year graduate students Stuart Cantrill, David Fulton, Sarah Hickingbottom, James Lowe and Anthony (Ant) Pease—of my research group made the transition from the middle of England to the West Coast of America, with postdoctoral fellow Françisco Raymo acting out the role of the scout. Coming to grips with the very different way American academia operates compared with that in the UK, together with getting my mind round the funding system from the federal agencies and beyond, constituted a baptism of fire for a 55-year-old. We simply rolled up our sleeves and got on with it. Norma, for the first time gainfully employed as a research assistant to my group by UCLA, helped in all this.

UNIVERSITY OF CALIFORNIA LOS ANGELES (UCLA)

In just over a decade at UCLA, from 1997 to 2008, we broadened the scope of our template-directed approaches to mechanically interlocked molecules (MIMs) by appealing to both hydrogen-bond and metal templation, as well as developing donor-acceptor templation to cover the production of a wide range of molecular switches. Stuart Rowan joined my research group in 1998, with the intention of establishing his own independent academic career in the United States, after having played a major role in the furtherance of dynamic covalent chemistry (DCC) at the University of Cambridge with Jeremy Sanders. This thermodynamically controlled approach to the template-directed synthesis of MIMs can be extraordinarily powerful. It eventually led to high-yielding syntheses of molecular Borromean rings and Solomon knots. Template-directed approaches under kinetic control to MIMs began to rely more and more on the use of 'click chemistry', as popularized by Barry Sharpless. Amongst the ring leaders during this period—in addition to Stuart Rowan (University of Chicago)—were Ivan Aprahamian (Dartmouth College), Adam Braunschweig (Hunter College), Sheng-Hsien Chin (National Taiwan University), William Dichtel (Northwestern University), Amar Flood (Indiana University), David Fulton (University of Newcastle), Jan Jeppesen (University of Southern Denmark), Steve Joiner

(Moorpark College), Ken Leung (Hong Kong Baptist University), Cari Meyer (Pierce College), Ognjen Miljanić (University of Houston), Al Nelson (University of Washington), Brian Northrop (Wesleyan University), Hsian-Rong Tseng (University of California, Los Angeles), Bruce Turnbull (University of Leeds), Sebastian Vidal (University of Lyon), Scott Vignon (Washington DC) and Jishan Wu (National University of Singapore).

The UCLA era was characterized by numerous efforts to uncover applications for molecular switches, both the non-degenerate catenated and rotaxanated varieties. One of the most rewarding and fulfilling collaborations was with Jim Heath in the field of molecular electronics. The marriage between molecular switches and electrodes is far from being an easy one, and I have to say that Jim picked his way through what turned out to be a bit of a minefield with the greatest of

Figure 8. Norma and I with David Leigh, Stuart Rowan and Stuart Cantrill after the International Symposium on Macrocyclic Chemistry held at St Andrews University in July 2000.

ease. By employing crossbar devices, he and his highly skilled team of graduate students and postdoctoral fellows were able, using the LB technique established during the Birmingham days, to lay down monolayers of switchable catenanes and rotaxanes between parallel wires of polysilicon (bottom electrodes) and orthogonally disposed parallel wires of titanium capped with aluminum. By 2007, using an amphiphilic bistable [2]rotaxane, a 160,000-bit molecular electronic memory circuit had been fabricated at a density of 100,000,000,000 bits per square centimeter. The entire 160-kbit crossbar device was smaller than the cross-section of a white blood cell. It transpired that there is one fatal weakness with the crossbar devices, and that is their lack of robustness. When Omar Yaghi arrived at UCLA in 2006 we started a joint program of research, whereby bistable MIMs are being incorporated inside metal-organic frameworks—and it continues today at Northwestern University (NU) in collaboration with Joe Hupp and Omar Farha.

For a time Norma's oncologists kept her cancer at bay, chiefly by moving in the face of resistance to treatment from one anticancer drug to another, and subsequently to a cocktail of two or three or more of them. She and I were able to travel the world together for a while, visiting many cities, including Paris, Stockholm and Vienna in Europe and Kyoto and Nara in Japan. Slowly and perceptibly, Norma's state of health started to wane as the side effects of the drugs began to sap her energy, causing her to seek refuge in our small Santa Monica townhouse, assisted by a kind and marvelous caregiver, Sylvia Mena, and no end of material and psychological support from Alice Jung, wife of my colleague Mike Jung, who was a dab hand at making Norma laugh and in so doing lifting my spirits. She referred to Mike's other half as 'Alice the Angel'. By late November 2003, the 25th to be precise, Norma's head oncologist, John Glaspy, told me what I had already guessed: it was that the disease had reached her brain and that it was only a matter of time, a few weeks at most, before a battle that had occupied a fifth of her life and demanded our attention for a third of our married lives was about to end. Norma always insisted that her brain was her last refuge: if and when it was invaded by the "little buggers," she would throw in the towel. Her final foray into the outside world was a sight to behold. It was an excursion to Gap in Santa Monica to purchase a large selection of garments for her grandson, only a few weeks away from being born to Fiona and Quentin McCubbin, yet she was not going to set eyes upon James Fraser (the Second!). Norma's shopping sprees were legendary, but this one stole the show. For the first time since 1966, she was oblivious to the spirit and trappings of Christmas as she prepared to make a dignified exit, simply commenting that she had drawn the short straw. During the final days of her life she communicated with me using a pencil and writing pad, being too weak to speak. Her last comment, written the night before she passed away on 12th January 2004, was, "Am I dead yet?" She sank into oblivion as I was struggling to decipher her question and so I was not able to provide her with an answer. In the last few weeks of her life she was insistent that her main legacy were 'her girls' and there is no arguing with that statement to this day. She had every right to feel proud of Fiona and Alison.

The departure in 2003 of Jim Heath to the California Institute of Technology (CALTECH) signaled two changes in my professional life. One was the taking over of the Directorship of the California NanoSystems Institute (CNSI) from Jim, the founding director, first of all in an acting capacity and then subsequently for real. Despite these developments, Jim and I maintained our collaboration in the realm of molecular electronics, aided and abetted by Bill Goddard's entry into the program. Through his impressive computational investigations, he did much to vindicate our proposed switching mechanism exhibited by monolayers of bistable rotaxanes in crossbar devices. I became a great admirer of Bill's command of his science and the fearless manner in which he tackles large and complicated problems, a trait that continues to this day. No one knows and understands our donor-acceptor catenanes and rotaxanes all the way from bistable molecules through to devices better than Bill: this belief is supported by arguments presented in more than 30 joint publications. Another positive change that occurred around 2003 was the forging of a close and equally fruitful collaboration at UCLA with Jeff Zink, whose knowledge and practical expertise in relation to the preparation of mesoporous silica nanoparticles led to the covering of the surfaces of these 100–200 nanometer diameter particles with both bistable/switchable rotaxanes (nanovalves) and their supramolecular counterparts, which we called snap-tops, as a sophisticated means of controlling the release of small molecules, such as anticancer drugs. My foray into drug delivery systems was undoubtedly influenced by my day-to-day experiences of living for 12 years with a cancer patient. I have to admit, however, that I am coming to the opinion, after having co-authored more than 30 articles with Jeff, that, while more and more sophisticated ways of delivering drugs to patients suffering from degenerative diseases can prolong their lives, they are probably never going to provide the cures that many people would like to think are just around the corner.

After graduating with his PhD from UCLA in 2001, Stuart Cantrill spent a couple of years at CALTECH as a postdoctoral scholar with Bob Grubbs, one of the 2005 Nobel Laureates in Chemistry. The outcome of this association was yet another collaboration, in which Grubbs-catalyzed olefin metathesis in many of its different manifestations was introduced into the thermodynamically controlled syntheses of MIMs using hydrogen bonding as the source of templation.

Figure 9. Left: Three peas in a pod. With Alison and Fiona at Fiona's wedding in June 2000. Right: Outside Buckingham Palace with Alison and Fiona in June 2007 after being knighted by HM Queen Elizabeth.

Stuart returned to UCLA in 2003 to take on my undergraduate teaching responsibilities and to assist me in the running of my research group while I was CNSI Director. During what turned out to be a three-year sojourn, he also became the *de facto* Associate Editor of *Organic Letters*. It was during this time that he not only made sure that the research he had initiated in relation to the dynamic synthesis of the molecular Borromean rings reached the light of day in the literature, but he was also to discover that his own future lay in scientific publishing. On his return to the UK in 2005 he found employment with the Nature Publishing Group, first of all as an Associate/Senior Editor with *Nature Nanotechnology*, before being given the responsibility in 2008 to launch *Nature Chemistry* as its Founding Chief Editor. It is this kind of career progression by my students that I look back upon with immense pride.

Two bolts appeared out of the blue in late 2006 and early 2007 that could be considered as serious game-changers. The first one arrived in the context of a phone call on 13th November 2006 from Bob Pierce, the British Consul General in Los Angeles. To my consternation, Bob asked me if I would be prepared to accept an appointment to Her Majesty the Queen as a Knight Bachelor. My acceptance, under a cloak of secrecy, became public knowledge in the 2007 New Year Honours List. It led to my attending an investiture in June 2007, accompanied by David Leigh, a 1987 PhD graduate from my Sheffield days, along with Fiona and Alison. The second great surprise came in the form of a call to my cell phone from my assistant, Christina Oliver, while I was attending the Third Annual FENA Review Meeting at the LUXE Hotel in Los Angeles. On this occasion, the message, which encroached upon a conversation I was having with Youssry Botros (Intel), who was appointed as a consultant from industry to the

Center for Functional Engineered Architectonics (FENA) directed by Kang Wang at UCLA, was from the King Faisal Foundation in Riyadh to say that I had been selected to receive the 2007 King Faisal International Prize (KFIP) in Science. Youssry, who was born and brought up in Egypt, was immediately raised to a highly excited state on learning the news from me. In the event, I invited Youssry, a fluent Arabic speaker, to accompany Alison, her then fiancé Mikey Ho, and myself on our first trip to Saudi Arabia to receive the Prize from the King in the middle of April. During a week-long visit, Youssry and I had our first meeting with Prince Turki Al-Saud who was then Vice-President of the King Abdulaziz City for Science and Technology (KACST), and he is now the President of KACST. Following this meeting, KACST has generously funded six projects at Northwestern University related to energy storage, energy harvesting, molecular electronics, porous materials, membrane technology and drug delivery, under the auspices of a Joint Center of Integrated Nanosystems (JCIN). Managing JCIN along with my highly supportive co-Director Majed Nassar, has

Figure 10. Meeting President Obama in the Oval Office along with Fiona and Alison on 30 November 2016.

been aided and abetted in a big way by Alyssa Avestro, Ashish Basuray, Tracy Chen and Mark Lipke.

It became clear towards the end of 2006 that the fortunes of the CNSI were set to suffer as a result of a change in the State of California Administration in Sacramento. Although buildings were nearing completion at both UCLA and the University of California, Santa Barbara (UCSB), it was apparent that there would be next to no funds made available from the State to equip the two buildings. I had little desire to find myself in charge of two white elephants.

I had tried to interest Chad Mirkin at Northwestern University in moving to UCLA to take over the Directorship of the CNSI from me. He was not interested but then turned the tables on me by inviting me to move to NU. Negotiations with the then President Henry Bienen at NU began in February 2007 and proceeded at such a pace that an announcement of my move could be made in August of that same year, allowing a few members of my group to start relocating a month later into newly refurbished laboratories in the Technological Institute. Although it was to take until August 2014 for a brand-new building, sanctioned and supported on my advice by the President, to house some of the major items of departmental equipment (spectrometers and diffractometers) to materialize, it was well worth the wait. I moved up to Evanston on 1st January 2008 amidst a spate of gong-collecting. It seems that awards and prizes feed off one another to a considerable extent. In January 2010 my research group, under the guidance of Doug Friedman, moved with military-like precision from the Tech building into the newly opened Silverman Hall to occupy research space second to none in our previous 40-year history. It did not take long for it to be called the Research Palace—or RP for short.

NORTHWESTERN UNIVERSITY

At Northwestern, a decade of broadly based activity in research relating to supramolecular chemistry, as well as mechanostereochemistry, has relied heavily on simply allowing a team of extremely creative graduate students and postdoctoral scholars free rein within the remit of the grants that support their research. This approach to invention and innovation in research has been highly successful, leading to a host of serendipitous events. One of these accidental discoveries, by Ron Smaldone—who was joined on its realization by Jeremiah Gassensmith and Ross Forgan—relates to the unexpected ability of γ-cyclodextrin to form highly porous extended structures with Group IA metal cations, particularly potassium, rubidium and cesium ions. This discovery in turn has led to the establishment

Figure 11. Signage. Top: The research group beside the Northwestern Arch in October 2016. Bottom: The front of the American Chemical Society Building in Washington, DC. In an e-mail received on 3 November 2016, Stu Borman of C&E News comments that 'all I can see out my window now is "RAS" and "DDA."

of a start-up company, PanaceaNano, in 2010 with Youssry Botros as its Chairman and Chief Executive Officer (CEO). The company has developed several Organic Nano-Cube (ONC) based materials that are completely safe for use in many industries, such as cosmetics, home and personal care, health and medicine, chemical, environment, food and beverage, and agriculture. In less than one and a half years, the company has developed and shipped many prototypes for testing by collaborators and distributors in the cosmetics, fragrances and drugs areas. The other chance discovery, by Zhichang Liu, relates to a remarkable lock-and-key fit between α-cyclodextrin and potassium tetrabromoaurate in a 2:1 ratio within a linear and rigid supramolecular polymer, which aggregates in its thousands—like drinking straws in a box—to form needle-like crystals within minutes in aqueous solutions. A start-up company, Cycladex, was launched in 2014 with Roger Pettman, one of my early postgraduate students from my Sheffield days, as its CEO. The company is offering the opportunity to the gold-mining industry to abandon the use of cyanide and mercury in the isolation of gold from ore and to adopt a much less expensive environmentally friendly way of achieving a better outcome.

In the realm of supramolecular chemistry, the trio of Michal Juríček, Jonathan Barnes and Edward Dale devised much more user-friendly and efficient approaches to the synthesis of the little blue box, before going on to expand the dimensions of this tetracationic cyclophane to yield much larger receptors they called ExBox and ExCage, which turned out to be ideal for complexing polycyclic aromatic hydrocarbons. An intellectually satisfying piece of research carried out in collaboration with Jay Siegel at Tianjin University was the induced-fit catalysis of corannulene bowl-to-bowl inversion, which illustrates very nicely the principles of enzyme catalysis. In what is a simple textbook example, catalysis of the inversion process in corannulene, induced by its stereoelectronic binding inside ExBox, can be followed along a single 'reaction' coordinate, where the reactant and product are the same. A full paper published in the *Journal of the American Chemical Society*, "ExCage"—one of the shortest titles ever for an article on chemistry—amounts to a tour de force in contemporary physical organic chemistry enacted by the ExBox/ExCage trio.

The reason that the laissez-faire approach to supervising graduate students and postdoctoral scholars in the Northwestern chemistry department thrives so well is because it is an approach that is endorsed to the full by the vast majority of the faculty. The experimentally and computationally active members in different research groups interact with each other so well – more often than not from the bottom up – that it has led to the comment that we hunt in packs in

Figure 12. Top: Family gathered together at the Nordic Museum in Stockholm, 2016. Bottom: In the midst of some young budding scientists at a party in the Grand Hotel in Stockholm, 2016.

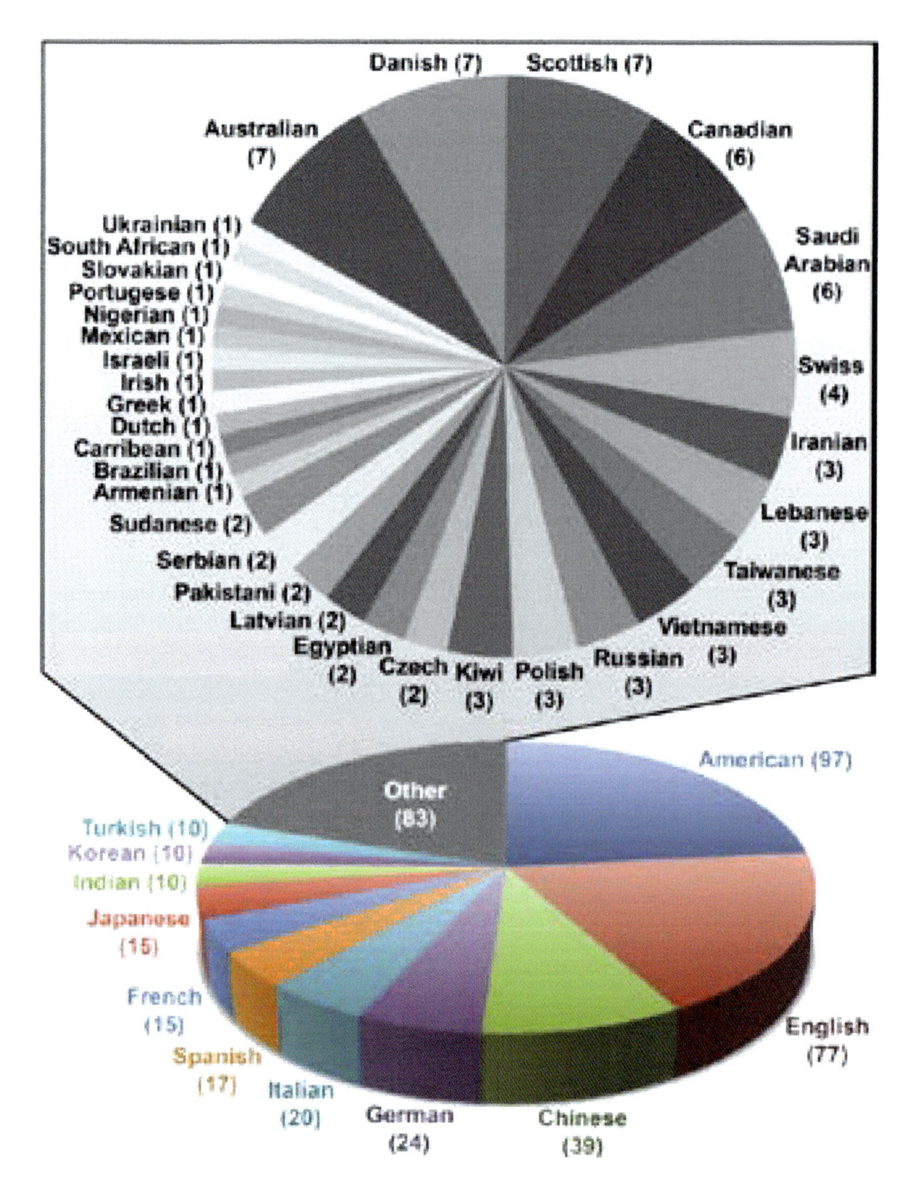

Figure 13. Breakdown of Stoddart group members during the past 45 years from 43 different countries. Those group members (a few) who have changed their nationality are counted twice, reflecting both their original and present nationalities, leading to a total number of entries of 417 on the pie charts. I thank Carson Bruns for producing this illustration at the drop of a hat!

the Department of Chemistry at Northwestern. Allowing these interactions to take their own course without meddling or interference is an extremely effective dynamic when it comes to the attainment of high-quality research. It is a dynamic that, is by and large, lost on university administrators and the regulatory authorities, who are much attracted and enamored by the concept of research being performed in silos. The fact that the laissez-faire approach is the dominant practice in the Northwestern Chemistry Department despite the presence of rules and regulations that would dictate otherwise, has rendered it possible for my research group personnel—Gokhan Barin, Ali Coskun, Marco Frasconi, Sergio Grunder, Chenfeng Ke, Severin Schneebeli, Cory Valente and many others, to collaborate with their counterparts in the groups led by Mike Wasielewski, Joe Hupp, Omar Farha, Bartosz Grzybowski, Emily Weiss, Chad Mirkin, Mark Ratner, George Schatz and Lin Chen, while maintaining active collaborations with Bill Goddard's group at CALTECH and Omar Yaghi's group at UC Berkeley. Add to this list the name of my fellow Nobel Laureate Jean-Pierre Sauvage, who spent a couple of years (2010–2012) coming from Strasbourg to Evanston from time to time as a visiting professor, and you have yet another source of intellectual stimulation *par excellence*. His overarching presence encouraged us all to spend a considerable amount of time thinking deeply, practicing painstakingly and writing wisely about chemical topology, a subject area that will surely come into its own right in years to come.

Two additional developments at Northwestern deserve special recognition. One was the demonstration in 2010 by Ali Trabolsi, and followed through by Albert Fahrenbach, of the strong 1:1 complex formed between viologen radical cations and the bisradical dicationic cyclophane, obtained on reduction of the little blue box in the presence of methyl viologen or its derivatives. It was somewhat counterintuitive that three 'free' electrons would hold a complex together in the face of substantial Coulombic repulsion, but it is a fact! This discovery led to the template-directed synthesis of catenanes and rotaxanes. While Jonathan Barnes set about making the homo[2]catenane of the little blue box—an achievement which Diego Benítez described as being intellectually disruptive—Hao Li employed radical templation to make rotaxanes that have been introduced subsequently into the design and synthesis of a rapidly growing range of artificial molecular pumps by Chuyang Cheng, Paul McGonigal and Cristian Pezzato. This story is featured in my Nobel Lecture—produced, as in the case of hundreds of other presentations, with the help and expertise of graphic artist Alex Bosoy.

The other development worthy of special mention was the writing, along with Carson Bruns, of a major treatise on MIMs. In every respect—both words

Figure 14. Meeting Chinese Premier Li Keqiang in the Great Hall of the People on 20 January 2017.

and pictures—the heavy lifting was done by Carson during a 30-month period that spilled over into some of his time as a Miller Research Fellow at Berkeley. It was quite fortuitous that Wiley ended up introducing the work to the world at large in the early part of November last year, halfway between the announcement of the 2016 Nobel Prize in Chemistry on 5th October and the prize-giving in Stockholm on 10th December. The manuscript was reviewed critically by many colleagues and the production of its six chapters, along with all the necessary components that go into the making of any book, were orchestrated by two people in particular. One was Xirui Gong, who read the proofs sentence-by-sentence, word-by-word, letter-by-letter, and number-by-number. The other was Margaret (Peggy) Schott, who assisted me in the demanding task of quality control as well as assuming responsibility for the production of the index in a highly efficient manner. Over the past 10 years at Northwestern University, Peggy has helped me, day-in and day-out, to guide a team of highly talented, yet often quite demanding, young researchers, from the day of their arrival to that of their departure and beyond into their own independent careers. These activities reflect only the tip of the iceberg when it comes to hailing the support Peggy—a PhD chemist and Northwestern alumna—provides to so many in the chemical community—locally, nationally and internationally.

EPILOGUE

In reflecting upon my peripatetic journey, which started with my valuable early experiences at the 'University of Life' on the farm, I can look back with feelings of pleasure interspersed with times of personal and professional hardships. There have been occasions marked by joy and others by sorrow. There have been periods that were characterized by success and others by failure, which I did my best to mitigate. Through all my life's experiences, the aim has always been the same: to emerge from life's roller coaster better informed and more knowledgeable about the ways of the world.

Putting all the ups and downs aside, I have been immensely privileged to be able to practice my hobby almost every day of my life in the presence of highly intelligent and outstandingly gifted young people, roughly aged between 18 and 32 drawn from nearly all quarters of the globe—and to do the things I love doing with them as a result of the generosity of those institutions and people, often without my being able to put a label or face to them, who have lent their support to my vision and mission from the Athens of the North (Edinburgh University) to the Windy City beside Lake Michigan (Northwestern University) with interludes on the edge of the Canadian Shield beside Lake Ontario (Queen's University), in the Socialist Republic of Yorkshire (University of Sheffield), on the Plains of Cheshire beside the Wirral (ICI Corporate Laboratory), in the Heartland of Albion (University of Birmingham and in the City of Angels alongside the Peaceful Sea (University of California, Los Angeles). My journey is far from over: it will continue as long as family and friends fail to raise a red card telling me that I have reached my sell-by date.

Science is global and there's no going back: scientists the world over live in a global village. There are no better words to catch this sentiment than those of the Scottish poet Robert Burns. In an epic poem, he emphasizes that "we're all the same under the skin." It is a statement of egalitarian sentiments. The poem reads—

> Then let us pray that come it may
> (As come it will for all that)
> That Sense and Worth over all the earth
> Shall have pre-eminence and all that
> For all that, and all that,
> It's coming yet for all that
> That man to man the world over
> Shall be brothers for all that.

Oh, that people who exercise power and influence in the world over we ordinary mortals might be guided by these sentiments. What a wonderful world it would be for all humankind if there were no borders—and rejoice at the thought, no countries.

Mechanically Interlocked Molecules (MIMs)—Molecular Shuttles, Switches, and Machines

Nobel Lecture, December 8, 2016 by J. Fraser Stoddart
Northwestern University, Evanston, IL, USA.

PREAMBLE

One of the most influential books ever to have been written in the field of chemistry is *The Nature of the Chemical Bond* by Linus Pauling [1], the first edition of which was published in 1939. In this classic work, Pauling distinguishes between electrostatic, covalent and metallic bonds, while recognizing that chemical bonds between two atoms or groups of atoms exist when the forces acting between them lead to the formation of aggregates we call molecules. Three decades later came the realization that there is another field of chemistry that exists beyond the molecule, which Donald Cram [2, 3] referred to as host-guest chemistry, and Jean-Marie Lehn [4, 5] as supramolecular chemistry. Chemistry beyond the molecule relates to organized entities that can be neutral molecules, or even cations, anions or radicals, which come together to form higher-order aggregates—call them adducts or complexes—under the influence of stabilizing intermolecular forces that are considerably weaker than are the covalent bonds which define the entities themselves.

The process involving the coming together by organized entities is often referred to [6] as self-assembly and the noncovalent bonding that accompanies it as molecular recognition. The noncovalent bonds include hydrogen and halogen bonds amidst a gamut of weak interactions which have been exploited to considerable effect during the past half century. On this time scale, another type of bonding in chemistry had been lurking in the background. That bond is the *mechanical bond* [7], which was the subject of an extensive treatise [8] entitled, *The Nature of the Mechanical Bond: From Molecules to Machines*, published as recently as 2016. Just as the chemical bond is associated in our minds with *attractive* forces, such as those associated with the sharing of electrons between atoms or the electrostatic forces that exist between ions of opposite charges, the mechanical bond is first and foremost a physical bond which is governed in the final analysis by *repulsive* forces that prevent chemical bonds from intersecting. Whereas chemical bonds are shared between atoms or groups of atoms, mechanical bonds are shared between molecular entities called *component parts*. It follows that a mechanical bond can be defined as an entanglement in space between two or more component parts such that they cannot be separated without breaking or distorting the chemical bonds between atoms.

MECHANICALLY INTERLOCKED MOLECULES

We have referred to molecules that possess mechanical bonds as mechanically interlocked molecules or *MIMs* for short. The two archetypal examples (Figure 1) of MIMs are the *catenanes* and the *rotaxanes*, which were the subject of a monograph [9] published in 1971 entitled *Catenanes, Rotaxanes, and Knots* by Gottfried Schill. A catenane, whose name is derived from the Latin word *catena*, meaning chain, is a molecule with two or more topologically linked macrocyclic component parts, while a rotaxane, whose name is derived from the Latin words *rota* for wheel and *axis* for axle, is a molecule comprising at least one macrocyclic component part, i.e., ring(s) with at least one linear component part, i.e., axle(s), threaded through the ring(s) and terminated by bulky end-groups (stoppers) large enough to prevent dethreading of the dumbbell(s) resulting from the existence/formation of chemical (covalent and/or coordinative) bonds between the stoppers and the axle(s). Two points are worthy of mention at this juncture. Catenanes and rotaxanes are molecules: they are *not* supramolecular entities or supermolecules despite the fact that they most likely will harbor intramolecular noncovalent bonds. While catenanes assume the trivial topologies of links, rotaxanes are topologically non-trivial for the simple reason that their component parts may be separated by continuous deformation, e.g., expanding the diameter of a ring or shrinking the cross-section of a stopper, both of which have been demonstrated [8] chemically. Related is the physical process of making rotaxanes known as slippage [10, 11].

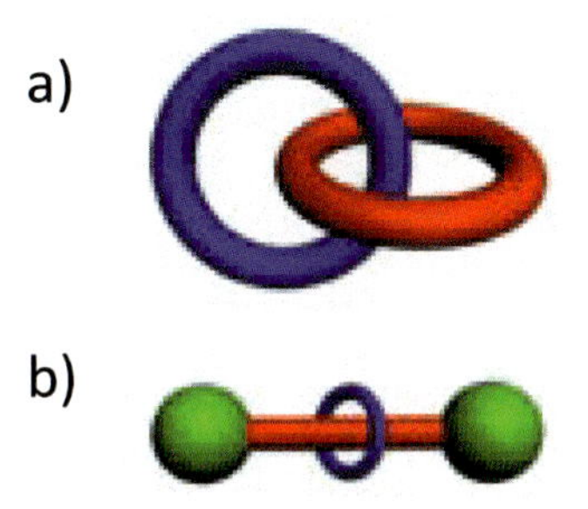

Figure 1. Graphical representations of a) a catenane and b) a rotaxane. A catenane is a mechanically interlocked molecule consisting of two or more interlocked rings. The word catenane is derived from the Latin *catena* meaning chain. A rotaxane is a mechanically interlocked molecule consisting of a dumbbell component threaded by one or more rings. The word rotaxane is derived from the Latin *rota* for wheel and *axis* for axle.

OLYMPIADANE

In the early 1990s the community was somewhat skeptical about the existence and worth of MIMs. Did they really exist and, if they did, how easy were they to make? If they were easy to make, in what context would they become useful? These were not unreasonable questions and they had to be addressed. In the event, it might well take, as of today, a decade or more to start providing answers to the second question. As far as the first question was concerned, it could be addressed rather easily, given the ability of X-ray crystallographer and structural chemist David Williams at Imperial College London to provide a fast turnaround on the solid-state structures of catenanes, and occasionally rotaxanes as well. In the case of donor-acceptor catenanes [12], we were intent on demonstrating that higher-order analogues could be produced using a template-directed protocol [13]. In this manner, both [3]- and [4]catenanes were synthesized in good yields overall and fully characterized. The envelope was pushed by two postdoctoral researchers, Anatoly Reder and David Amabilino, who, one after the other, took up the formidable challenge of making a [5]catenane (Olympiadane) whose constitution was confirmed [14] in 1996 by a solid-state structure (Figure 2), courtesy of David Williams. Ju-Young Lee gilded the lily by synthesizing a branched [7]catenane [15] in one step from Olympiadane. Both the collection of the crystallographic data, which took a couple of weeks, and the solving of the solid-state structure of the branched [7]catenane after a period of five months constituted a tour de force at the time by David Williams. The presence of 20 disordered PF_6^- counterions plus disordered solvent molecules on top of a crystallographic symmetry highlights the complexity of the task from both an experimental point of view and a computational one. The solid-state structures of these two higher-order catenanes revealed a veritable array of noncovalent bonding interactions in the form of face-to-face [π. . .π] stacking interactions (cf. DNA) and [C–H. . .π] edge-to-face alignments of aromatic rings which are commonplace in proteins rich in aromatic amino acids, along with multiple [C–H. . .O] hydrogen bonds. It is the installing of these (ultimately) intramolecular interactions that aids and abets the templation that makes it possible to synthesize these higher-order donor-acceptor catenanes. The point one learns from all this information is that it is the same weak interactions that are present in naturally occurring compounds that show up time and time again in exotic unnatural products. At the level of noncovalent bonding interactions, their presence is ubiquitous throughout biology, chemistry and materials science.

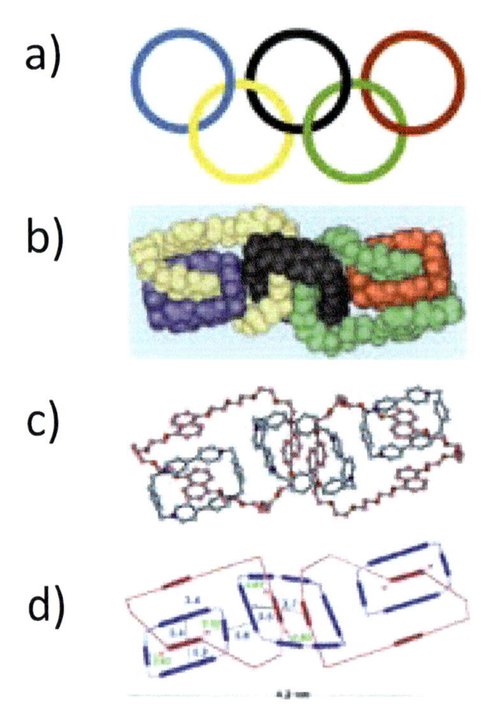

Figure 2. a) The Olympics logo consisting of five interlocked rings picked out from left to right in blue, yellow, black, green and red. b) A space-filling representation of the solid-state structure of a [5]catenane called olympiadane in which the five mechanically interlocked rings are colored according to the Olympics logo. c) A ball-and-stick structure of olympiadane which reveals that two outer blue rings are the same while the blue ring in the middle is a larger homologue of the two terminal rings. The other two red rings enjoy the same constitution in which three 1,5-disubstituted naphthalene rings are linked in a circular fashion, by tetraethylene glycol links. d) A graphical representation of Olympiadane where the (red) π-electron rich rings are mechanically interlocked with three (blue) π-electron poor rings.

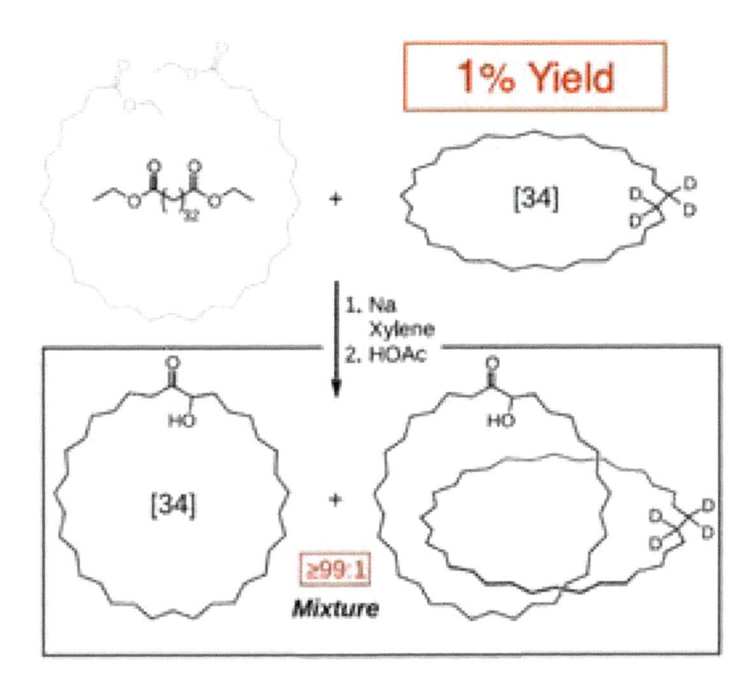

Figure 3. Wasserman's statistical synthesis reported in 1960 of arguably the first wholly synthetic [2]catenane employing an acyloin condensation in order to clip an acyclic diester around a deuterated cyclohexane derivative. The ring sizes, expressed as the number of carbon atoms in the rings, are denoted in square brackets. Deuterium labeling was employed as evidence in support of catenation following cleavage of the a-hydroxyketone functions with alkaline hydrogen peroxide.

Figure 4. Photograph of Ed Wasserman.

HISTORY

It was an absence of appreciation by chemists, in the era before there was a realization of the importance of noncovalent bonds and a recognition regime beyond the molecule, i.e., supramolecular chemistry, that led to a lack of success in the making of catenanes and rotaxanes. Wasserman's synthesis (Figure 3) of a [2]catenane [16] in no more than a 1% yield in 1960 bears witness to the fact that, without any appreciable noncovalent bonding interactions between the precursors to the component parts, the possibility of achieving mechanical interlocking to afford a [2]catenane was largely down to chance, hence the use of the term statistical synthesis to describe much of the research carried out in the 1960s. The achievement of Ed Wasserman (Figure 4) was to serve notice on the chemical community that, although catenation which relies on a chance event is, most likely not going to be efficient, it can be demonstrated. While Wasserman was carrying out his research at Bell Laboratories in Murray Hill, New Jersey, Gottfried Schill and Arthur Lüttringhaus at the University of Freiburg in Germany were devising ways [17] by which the component parts of a [2]catenane could be brought together using a covalent bond that could be cleaved in the final steps of the synthesis. The key compound shown at the top in Figure 5 was obtained crystalline after 14 initially linear steps towards the directed synthesis of the [2]catenane shown at the bottom in Figure 5. The authors described the key compound as "*one which is linked intra-annularly and in which the chains of the double ansa-system are situated on opposite sides of the benzene ring.*" They also drew attention to the fact that the chain in the precursor to the key compound, which is attached by means of a cyclic acetal to the benzene ring, is held at right angles to the plane of the ring on account of the tetrahedral configuration about the aliphatic (acetal) carbon atom, thus ruling out the formation of the isomer with the extra-annular attachment of the ring. The final four steps of the 18-step synthesis all went in nearly quantitative yields to afford the [2] catenane. Aside from it being a long and difficult synthesis, the fact that once all the covalent bonds holding the component parts together have been cleaved, the two mechanically interlocked rings are essentially devoid of any 'cross-talk' between them. Nonetheless, Schill continued to be active in the field of directed synthesis of MIMs, including rotaxanes, until the early '90s. I have commented elsewhere [18] that Gottfried Schill (Figure 6) can be looked up as the father of the mechanical bond: he is a chemist who was decades ahead of his time!

A seminal publication [19] by Jean-Pierre Sauvage (Figure 7) in 1983 describing (Figure 8) the use of a copper(I) ion to template the formation of

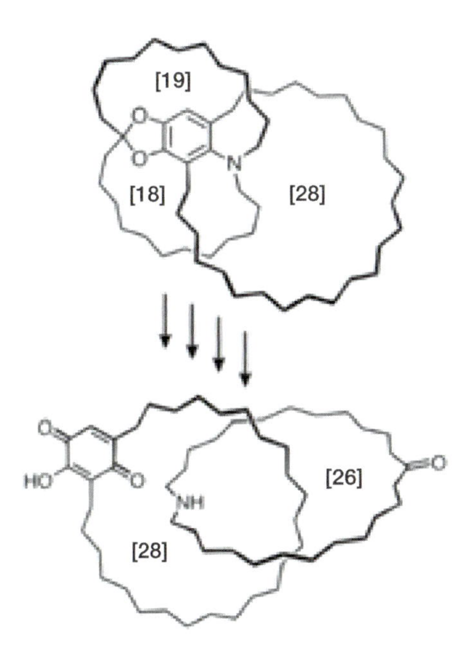

Figure 5. The telescoped final four steps in a 22-step covalent-directed synthesis of a [2] catenane reported by Schill and Lüttringhaus in 1964. The ring sizes, given in terms of the number of carbon atoms present in the rings, are denoted in square brackets.

Figure 6. Photograph of Gottfried Schill.

Figure 7. Photograph of Jean-Pierre Sauvage.

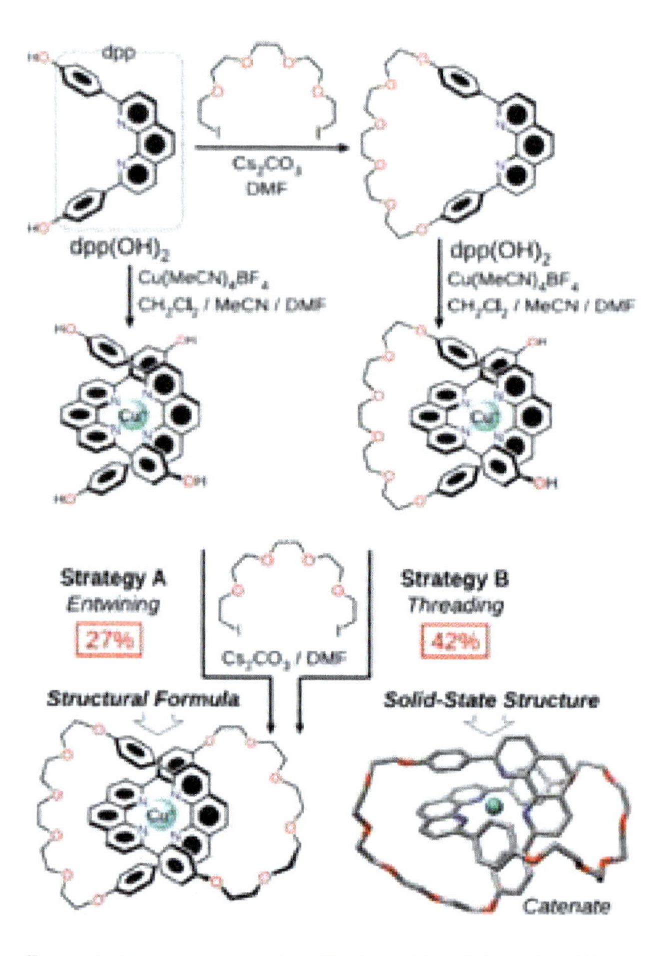

Figure 8. Sauvage's entwining (Strategy A) and threading (Strategy B) for the copper(I)-templated synthesis of the first catenate in 1983. In the bottom right-hand corner is a tubular representation of the solid-state structure of the catenate.

catenates—from whence, on demetallation, catenanes can be obtained—was a game-changer. His introduction of transition metal templation into the field of MIMs was transformative: it demonstrated that catenanes and rotaxanes were readily accessible and set the stage for the subsequent emergence and rapid growth of the field.

COMMERCIAL BUILDING BLOCKS

During a three-year secondment to the ICI Corporate Laboratory between 1978 and 1981, I joined forces with Howard Colquhoun in the investigation of the second-sphere coordination of transition metal ammines by crown ethers [20]. This program of research was sustained in its speed and efficiency by our striking up a highly fruitful collaboration with David Williams. One of the solid-state superstructures he obtained in 1981 was to set us on a road to discovery and invention. The superstructure (Figure 9a,b) in question was that of a 1:1 crystalline adduct that is formed [21] when a dicationic platinum ligand in addition to two *cis*-diammine ligands is crystallized (CH_2Cl_2/Et_2O) in the presence of a molar equivalent of dibenzo[30]crown-10 (**DB30C10**). While the two ammine ligands from [N–H. . .O] hydrogen bonds with the two polyether loops of **DB30C10**, the bipyridyl ligand finds itself slotted (Figure 9c) in between the two catechol rings of the crown ether. The structural similarity between the platinum

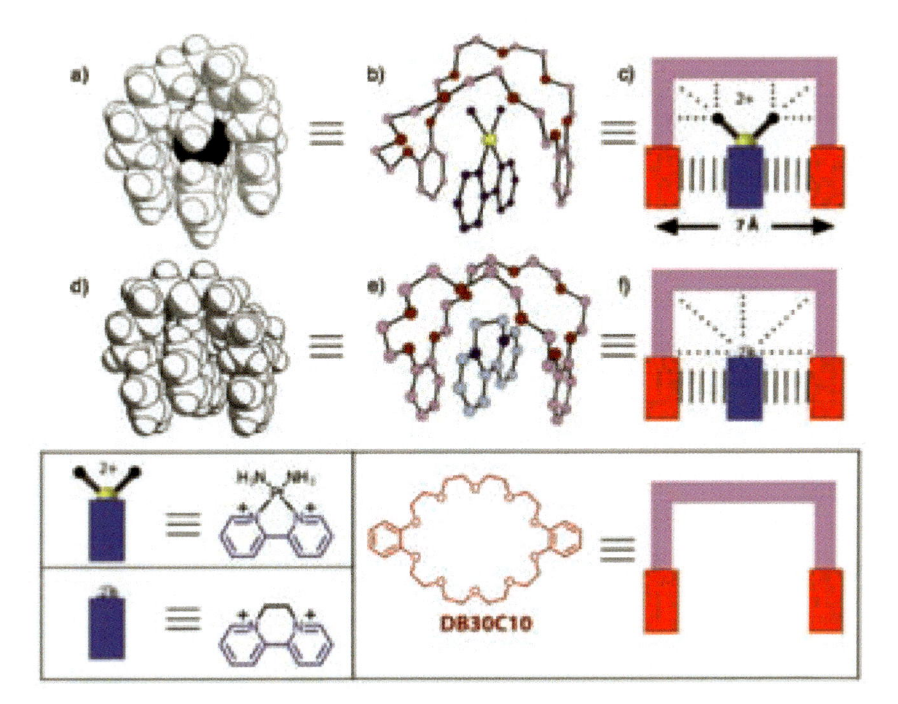

Figure 9. a) Space-filling and b) ball-and-stick representations of the 1:1 adduct formed between $[Pt(bipy)(NH_3)_2]^{2+}$ and DB30C10 in the solid state. c) A graphical representation of the 1:1 adduct showing the [π. . .π] stacking interactions (sets of vertical lines) and the [N–H. . .O] hydrogen bonds and pole-dipole interactions between the dicationic transition metal diammine and some of the oxygen atoms in the polyether loops of DB30C10. d) Space-filling and e) ball-and-stick representations of the 1:1 complex formed between the diquat dication and DB30C10. f) A graphical representation showing the [π. . .π] stacking interactions (sets of vertical lines) and the [C–H. . .O] hydrogen bonds between the bismethylene bridge in the diquat cation and some of the oxygen atoms in the polyether loops of DB30C10.

complex and diquat—a compound marketed by ICI at the time in admixture with paraquat as a 'wipe-out' weedkiller—led us to show that it forms [22] a 1:1 crystalline complex with **DB30C10** as well. Its solid-state superstructure (Figure 9d,e) mirrors that of the 1:1 adduct **DB30C10** forms with the platinum complex. The diquat dication finds itself enjoying (Figure 9f) charge transfer and π–π stacking interactions with the two catechol rings in the crown ether, aided and abetted by [C–H. . .O] hydrogen bonds.

BLUE INSIDE RED AND RED INSIDE BLUE

The next step, on my return to Sheffield in 1981, was to uncover a good crown ether receptor for paraquat. We had observed that **DB30C10** was able to form a weak 1:1 complex with diquat and so we argued that a small change to the constitution of the crown ether might be all that is required in order to pinpoint a good crown ether receptor for paraquat. This strategy proved to be successful. We found that a constitutional isomer of **DB30C10**, namely bisparaphenylene[34] crown-10 (**BPP34C10**), forms [23] a strong 1:1 complex with paraquat. Moreover, the solid-state superstructure (Figure 10a) of the crystalline 1:1 complex, when compared with the solid-state structure of **BPP34C10**, carried two very

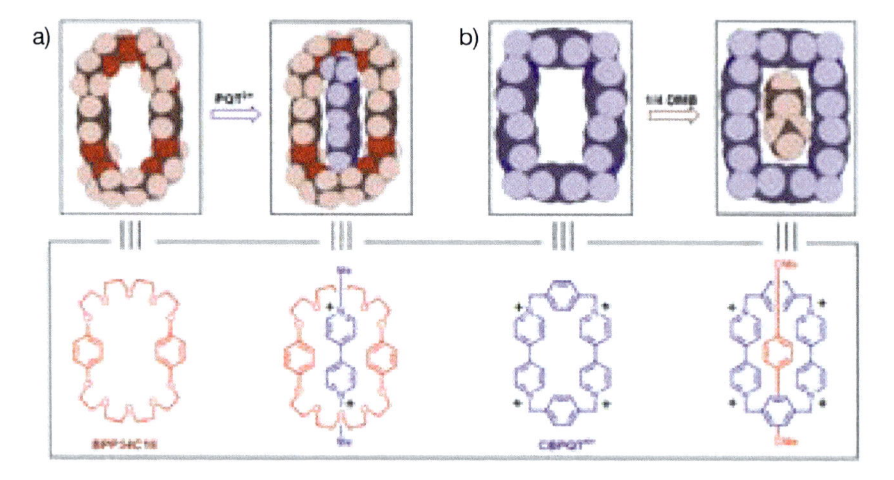

Figure 10. Comparison of the space-filling representations of the solid-state (super) structures of the two receptors a) BPP34C10 and b) $CBPQT^{4+}$ and their 1:1 complexes obtained, respectively, with paraquat (PQT^{2+}) and 1,4-dimethoxybenzene (1/4DMB). The box at the bottom relates the solid-state (super)structures to the structural formulas for the two receptors, BPP34C10 and $CBPQT^{4+}$ and their 1:1 complexes with PQT^{2+} and 1/4DMB, respectively.

strong messages. One was the fact that the crown ether, at least as portrayed in the solid state, is preorganized to complex with the paraquat (**PQT^{2+}**) dication. The other was that the manner in which the **PQT^{2+}** dication threads through the **BPP34C10** ring is highly suggestive of it being a MIM precursor. In the context of molecular recognition, we had established how to bind a π-electron rich host—or using our color scheme, we could put blue inside red. The next question was—could we reverse the recognition motif and put red inside blue? Or expressed another way, could we bind a π-electron rich guest, e.g., 1,4-dimethoxybenzene (**1/4DMB**), inside a π-electron deficient host? With his synthesis [24] of cyclobis(paraquat-*p*-phenylene) (**CBPQT^{4+}**), Mark Reddington was able to show that **1/4DMB** does indeed bind, albeit somewhat weakly, with **CBPQT^{4+}**. Once again the solid-state superstructure (Figure 10b) of the crystalline 1:1 complex, when compared [25] with the solid-state structure of **CBPQT^{4+}**, reveals a highly preorganized match between the guest and the rigid host.

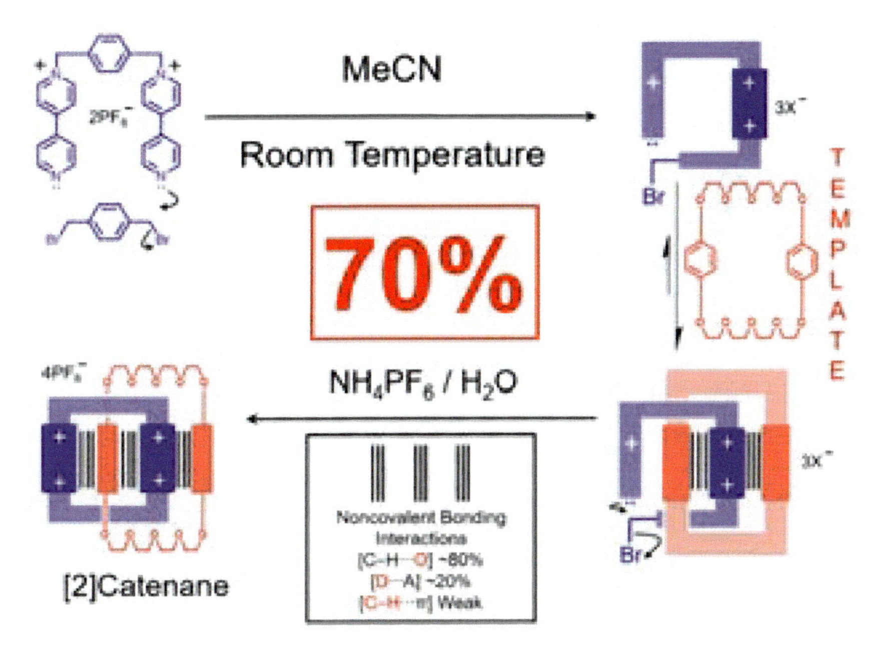

Figure 11. The template-directed synthesis of the first donor-acceptor [2]catenane under kinetic control at room temperature in acetonitrile. BPP34C10, which was present in 3 molar equiv, acts as a template in the stepwise formation of the CBPQT^{4+} ring in a process where covalent bond formation leads to the production of a bipyridinium unit which is recognized noncovalently by the BPP34C10 when threading occurs: thereafter a second covalent bond is formed resulting in the formation of the [2]catenane where the noncovalent bonding interactions 'live on' inside the MIM. They are indicated by the parallel vertical lines.

FIRST DONOR–ACCEPTOR CATENA

We were now poised to perform the all-important experiment which would answer the question—what would happen if we carried out the synthesis of **CBPQT^{4+}** in the presence of **BPP34C10**? Cristina Vicent and Neil Spencer provided the answer by carrying out the reaction (Figure 11) in acetonitrile at room temperature in the presence of **BPP34C10** as a template. The outcome from the first reaction exceeded our wildest dreams: we were able to isolate our first donor-acceptor [2]catenane in a remarkable 70% yield. Our excitement did not stop there: a space-filling representation of the solid-state structure (Figure 12), which graced the front cover of the October 1989 issue of *Angewandte Chemie* [26], was a sight to behold. Moreover, both ^{1}H NMR spectroscopy and electrochemical experiments, performed in Miami by Angel Kaifer, conveyed a very strong message. The weak noncovalent bonding interactions (Figure 13a) that are accrued during the template-directed synthesis 'live on' in the [2]catenane afterwards. The implications of this observation were to be profound when it

Figure 12. A space-filling representation of the solid-state structure of the first donor-acceptor [2]catenane adorning of the front cover of the October issue of *Angewandte Chemie* in 1989.

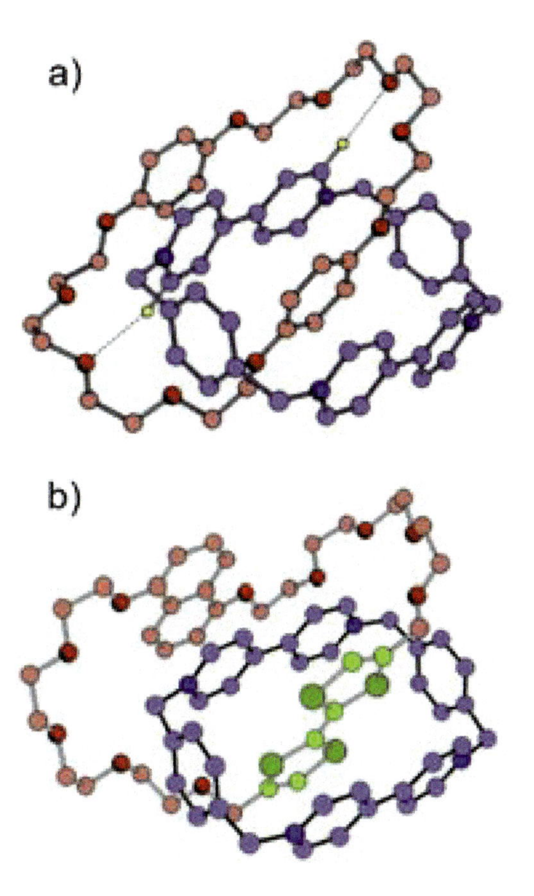

Figure 13. The solid-state structures, depicted in ball-and-stick representations, of a) the first donor-acceptor [2]catenane and b) a switchable donor-acceptor [2]catenane based on a stronger tetrathiafulvalene (green) and a weaker 1,5-dioxynaphthalene donor (pink). Note the presence in both [2]catenanes of π-π stacking interactions (3.5 Å) between the aromatic donors and acceptors. In the degenerate [2]catenane, note the presence of strong [C–H. . .O] interactions.

came to designing and synthesizing, first of all molecular switches, and then ultimately molecular machines.

MOLECULAR SHUTTLE

As we took our leave of Sheffield in 1990 to move to Birmingham, Pier Lucio Anelli had completed the template-directed synthesis (Figure 14) of a degenerate [2]rotaxane [27] modeled on that of the degenerate [2]catenane. The lower yield of 32% reflects the fact that the dumbbell template is much less preorganized

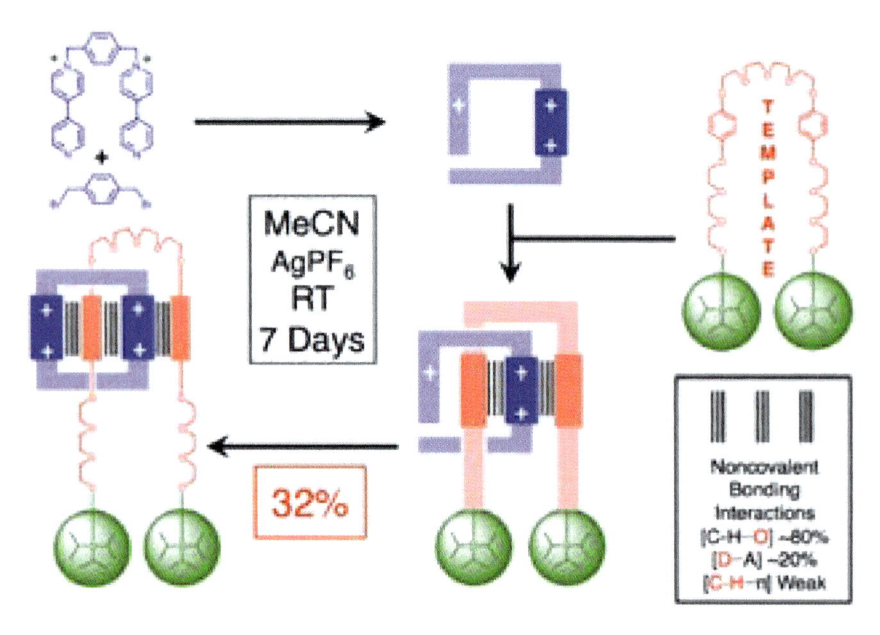

Figure 14. The template-directed synthesis of a degenerate donor-acceptor [2]rotaxane, also known as a 'molecular shuttle' under kinetic control at room temperature in acetonitrile. The dumbbell acts as the template for the formation of the $CBPQT^{4+}$ ring in a stepwise manner. The noncovalent bonding interactions which 'live on' inside the [2] rotaxane are indicated by parallel vertical lines.

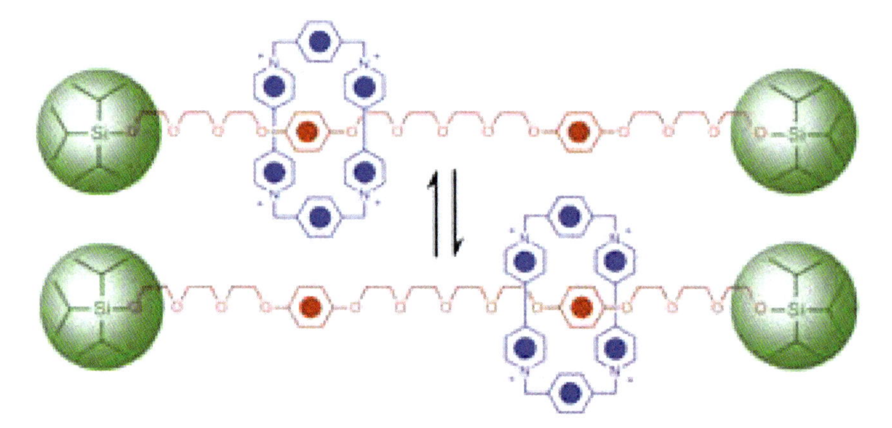

Figure 15. A degenerate donor-acceptor [2]rotaxane for which the phrase 'molecular shuttle' was introduced in 1991. The $CBPQT^{4+}$ ring darts back and forth between the two hydroquinone recognition sites about 1000 times a second in acetone at room temperature. The molecular shuttle was seen as a prototype for the construction of molecular switches and machines based on molecules containing mechanical bonds.

than **BPP34C10**. Dynamic ^{1}H NMR spectroscopy, performed on the rotaxane in hexadeutero acetone, revealed that the $CBPQT^{4+}$ ring shuttles (Figure 15) back and forth between the two hydroquinone units ('stations') around 1000 times a second. For obvious reasons, I found myself referring to the compound as a molecular shuttle and was prompted to draw the following conclusion in the communication [27] published with this title in the *Journal of the American Chemical Society* in 1991—

> *The opportunity now exists to desymmetrize the molecular shuttle by inserting nonidentical 'stations' along the polyether 'thread' in such a manner that these different 'stations' can be addressed selectively by chemical, electrochemical, or photochemical means and so provide a mechanism to drive the 'bead' to and fro between 'stations' along the 'thread'. Insofar as it becomes possible to control the movement of one molecular component with respect to the other in a [2]rotaxane, the technology for building 'molecular machines' will emerge.*
>
> *The molecular shuttle described in this communication is the prototype for the construction of more intricate molecular assemblies where the components will be designed to receive, store, transfer and transmit information in a highly controllable manner, following their spontaneous self-assembly at the supramolecular level. Increasingly, we can look forward to a 'bottom up' approach to nanotechnology, which is targeted toward the development of molecular-scale information processing systems.*

MOLECULAR SWITCHES

The next challenge we faced was to desymmetrize the degenerate [2]rotaxane in an attempt to obtain a rotaxane with two recognition units on the dumbbell component, one of which is more attractive to being encircled by the $CBPQT^{4+}$ ring than the other. After much initial experimentation we settled on a design where one of the hydroquinone units in the molecular shuttle is replaced by a benzidine unit and the other by a biphenol residue. Since we were prevented from using benzidine in the United Kingdom, postdoctoral scholar Richard Bissell made the journey across the Pond in order to complete the template-directed synthesis of the non-degenerate [2]rotaxane in collaboration with Angel Kaifer. On this occasion, dynamic ^{1}H NMR spectroscopy revealed [28] that the $CBPQT^{4+}$ ring spends 84% of its time on the benzidine unit and 16% on the biphenol unit at

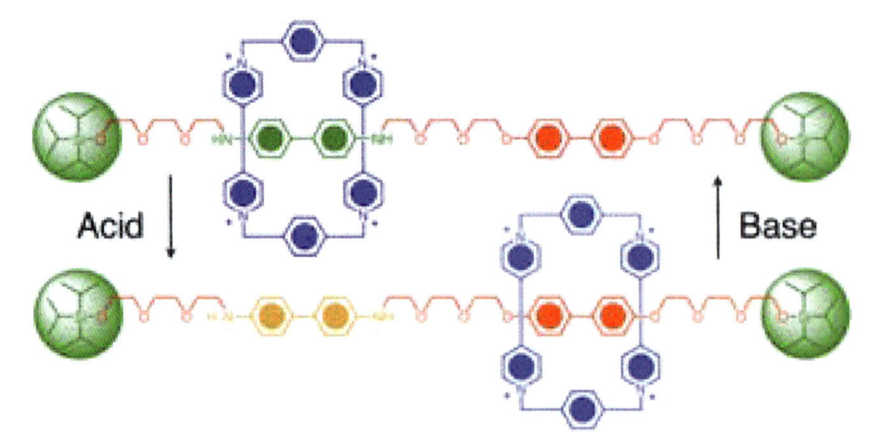

Figure 16. The first switchable donor-acceptor [2]rotaxane, where switching can be implemented by adding a drop of acid to protonate the benzidine recognition unit (the more preferred site for the location of the ring) whereupon the $CBPQT^{4+}$ ring migrates to the biphenol unit, the less preferred site for the location of the ring on the neutral dumb-bell component. The switch can be reset by adding base.

equilibrium at room temperature in CD_3CN solution. The unequal distribution of the ring between the two recognition units was sufficient to allow us to intervene in the equilibrium between the two translational isomers. It was possible, by adding acid to protonate the nitrogen atoms on the benzidine unit, to induce (Figure 16), as a result of Coulombic forces, the $CBPQT^{4+}$ ring to move to the neutral biphenol unit rather than reside on the deprotonated benzidene unit carrying its two positive charges. While this bistable [2]rotaxane was to rank as

Figure 17. Photograph of Jim Heath.

our first switchable donor–acceptor MIM, a bistable [2]catenane, [29] in which one of the two hydroquinone rings in the BPP34C10 component of the original [2]catenane (Figure 13a) was replaced by a tetrathiafulvalene (TTF) unit and the other with a 1,5-dioxynaphthalene (DNP) unit was designed and synthesized by Gunter Mattersteig. The solid-state structure (Figure 13b) of this bistable [2] catenane reveals that the TTF unit resides inside the cavity of the $CBPQT^{4+}$ ring, while the DNP unit is located outside and alongside one of the two bipyridinium ($BIPY^{2+}$) units in the $CBPQT^{4+}$ ring.

MOLECULAR ELECTRONICS

This bistable [2]catenane was used [30] to construct a solid-state electronically reconfigurable switch in a simple crossbar device by Jim Heath (Figure 17) when we entered into a highly fruitful collaboration, following my move from Birmingham to the University of California, Los Angeles (UCLA). Prior to introducing

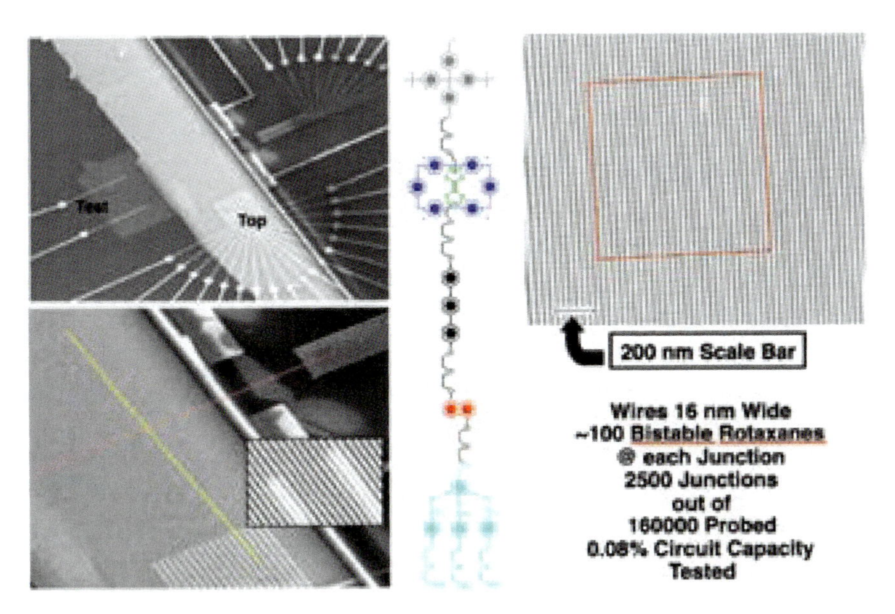

Figure 18. Scanning electron microscope image (top left) of the nanowire crossbar memory. The array of 400 Si bottom nanowires is portrayed as the light grey rectangular patch extending diagonally up from the left. Scanning electron microscope image (middle) showing the cross-point of the top (red) and bottom (yellow) nanowire electrodes. Each cross-point corresponds to an ebit in memory testing. High resolution scanning electron microscope image (top right) of approximately 2500 junctions out of a 160,000-junction nanowire crossbar circuit. The red square highlights an area of the memory that is equivalent to the number of bits tested. Positioned in the center of the illustration is the structural formula of the bistable [2]rotaxane used in the memory.

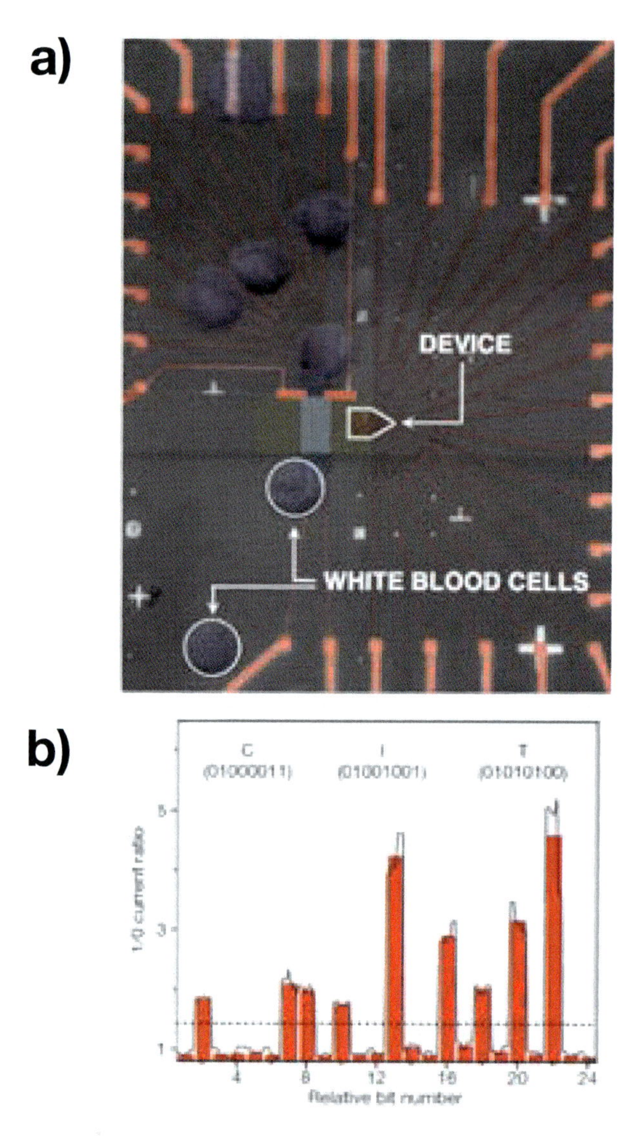

Figure 19. a) A 160-kilobit crossbar molecular memory device (orange in the middle of the illustration) is smaller than the cross-sectional area of a white blood cell (purple). b) A demonstration of point-addressability within the crossbar. Good ebits were selected from the defect mapping of the tested portion of the crossbar. A string of 0s and 1s corresponding to the ASCII characters for 'CIT' (California Institute of Technology) were stored and read out sequentially. The dotted line indicates the separation between the 0 and 1 states of the individual ebits. The black trace is raw data showing 10 sequential readings of each bit while the red bars represent the average of these 10 readings.

this bistable [2]catenane into the crossbar device we had ascertained [31] how to effect its transfer onto the device using the Langmuir-Blodgett (LB) technique from a Langmuir monolayer where the counterions are dimyristoylphosphatidyl anions. The ON/OFF ratios for the molecular switch tunnel junctions (MSTJs) in crossbar devices, incorporating the bistable [2]catenane, were only around 2. The desire to increase this ratio took us inexorably in the direction of bistable [2] rotaxanes (Figure 18) which, it transpired, can be switched with an order of magnitude higher ON/OFF ratios. Reversible, electronically driven switching was not only observed in MSTJs incorporating a molecular monolayer of this bistable [2] rotaxane sandwiched between a bottom polysilicon electrode and a top titanium/aluminum electrode, but a 160,000-bit molecular electronic memory circuit has also been fabricated (Figure 19a) at a density of 100,000,000,000 bits per square centimeter [31, 32]. The entire 160-kbit crossbar assembly was smaller than the cross-section of a white blood cell and could be switched (Figure 19b) up to around 100 times. The search for a more robust setting [34] in metal-organic frameworks for these bistable MIMs continues [35].

DRUG DELIVERY SYSTEMS

In a collaboration with Jeff Zink (Figure 20) at UCLA, we have functionalized the surfaces of mesoporous silica nanoparticles (MSNPs) with both rotaxane-based nanovalves and snap-tops (Figure 21) for the controlled release of drugs. The nanovalves which adorn the surfaces of the MSNPs can be bistable donor–acceptor [2]rotaxanes [36] where redox chemistry can be used to open and close the

Figure 20. Photograph of Jeff Zink.

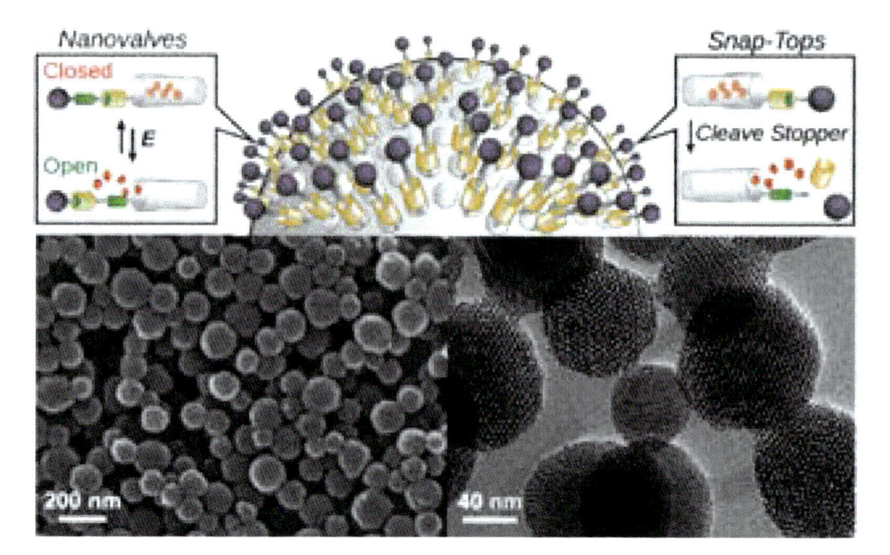

Figure 21. Examples of mechanized mesoporous silica nanoparticles (MSNPs). Illustrations of the mode of operation of rotaxane-based nanovalves and snap-tops for the controlled release of drugs. Scanning electron microscopic (bottom left) image of MSNPs. Transmission electron microscopic (bottom right) image of MSNPs.

nanovalves. This type of integrated device holds considerable promise for drug delivery systems.

MOLECULAR MACHINES

A fundamental property of biological molecular machines, e.g., the motor proteins [37–44], is that they consume energy and drive systems away from equilibrium by controlling kinetic barriers. The emergence of the science of artificial molecular machines (AMMs) in the hands of chemists, for the most part, presents us with a steep learning curve to negotiate. Nonetheless, it is a challenge that the chemistry community has embraced increasingly during the past two decades with much of the teaching of the fundamental theory coming from the physics community, and, in particular, one very prominent physicist from the University of Maine, namely Dean Astumian [45–48] with whom we have collaborated as well as published several reviews [49–51] that stand alongside a crop [52–69] that can be traced back for at least two decades from the present day. In my Nobel Lecture, I chose to highlight the progress [50, 51] we have made in recent times on the design and synthesis of artificial molecular pumps, based on the building blocks that found their origins almost 40 years ago in the ICI

Corporate Laboratory and were used by us [21–27] to put the mechanical bond on a firm footing before introducing [28] them into molecular switches in the first instance.

UNIDIRECTIONAL TRANSPORT

At the outset, we designed and synthesized a constitutionally unsymmetrical dumbbell [34] and demonstrated (Figure 22) that it transports the **CBPQT**$^{4+}$ ring in a unidirectional manner. In its midriff, the dumbbell houses an electron rich 1,5-dioxynaphthalene (DNP) unit which is capable of recognizing the electron poor **CBPQT**$^{4+}$ ring. Kinetic control of the threading and dethreading of the dumbbell by the ring is managed by arranging to have a neutral isopropylphenyl (IPP) unit at one end of the dumbbell and a positively charged 3,5-dimethylpyridinium (PY^{+}) unit at the other end. The **CBPQT**$^{4+}$ ring finds its most thermodynamically stable location on the DNP unit by passing over the IPP unit for the simple reason that the positively charged ring is repelled by the PY^{+} unit. When the **CBPQT**$^{4+}$ ring is reduced to **CBPQT**$^{2(\bullet+)}$, the relative heights of the two barriers change on account of the significantly decreased Coulombic barrier confronted by the ring which also, most likely, undergoes a reduction in size, resulting in increasing steric interactions with the PY^{+} unit. At the same time, the donor–acceptor interactions between the **CBPQT**$^{2(\bullet+)}$ ring and the DNP recognition unit are nullified, resulting in a preference for the rings to dethread into solution over the charged end of the dumbbell. In what is essentially a supramolecular system, no work is done: rings are extracted out of solution onto the dumbbell momentarily before being released back into solution.

RADICAL INTERACTIONS AND TEMPLATION

Our initial approach [70] to introducing unidirectionality into a supramolecular system employing donor–acceptor interactions demonstrated that we can control the relative motion between a dumbbell and a ring. The system, however, lacks the potential for it to be developed into a more sophisticated molecular machine where useful work can be done since the free energy change during the redox cycle is simply too small to be useful in a practical context. A discovery (Figure 23) made in my research laboratory in 2010 by Ali Trabolsi and Albert Fahrenbach was to come to the rescue. They showed [71] that a relatively strong tricationic trisradical complex ($K_a = 10^4$–10^5 M^{-1} in MeCN) is formed between **CBPQT**$^{2(\bullet+)}$ and bipyridinium (**BIPY**$^{\bullet+}$) radical cations under reducing conditions. The strong binding affinity (attraction) can be switched back to being

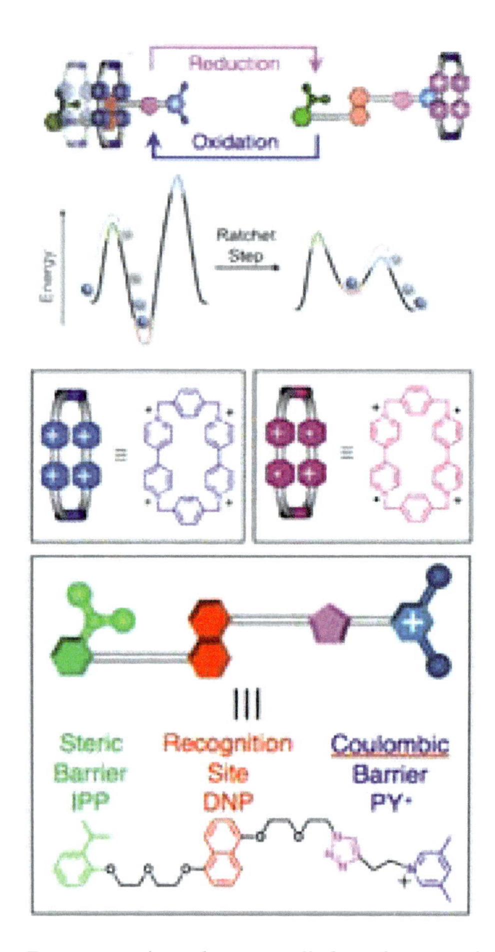

Figure 22. The redox-controlled unidirectional movement of a $CBPQT^{4+}$ ring along a dumbbell with a neutral steric barrier (green) on the left, a recognition site (red) for the ring under oxidative conditions in the middle and a positively charged Coulombic barrier (blue) on the right. Underneath are energy profiles illustrating the operation of an energy ratchet. The cartoons are defined in the boxes below.

highly repulsive, following oxidation of the radical cations, in both instances, to give $\mathbf{CBPQT^{4+}}$ and $\mathbf{BIPY^{2+}}$, resulting in a massive enthalpy change, something which struck us as being highly promising when it comes to designing and synthesizing AMMs to perform useful work.

We reasoned that we could achieve much better and more efficient unidirectional transport of $\mathbf{CBPQT^{4+}}$ rings if we were to replace (Figure 24) the DNP unit in the constitutionally unsymmetrical dumbbell with a $\mathbf{BIPY^{2+}}$ unit and employ radical chemistry to form a stable [2]rotaxane-like tricationic trisradical on reduction of both the ring and the $\mathbf{BIPY^{2+}}$ unit. We had already employed

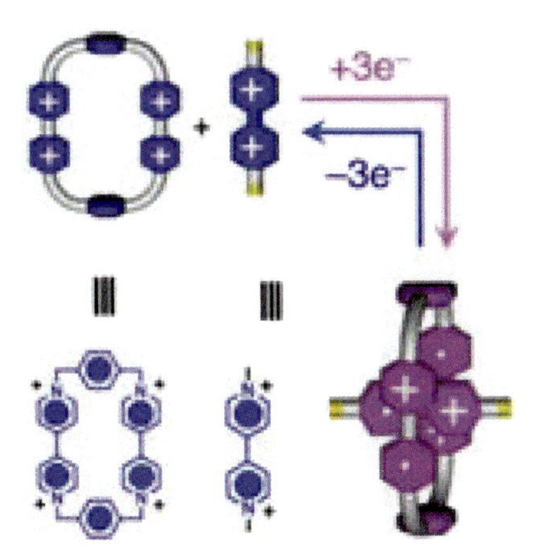

Figure 23. The formation of a $CBPQT^{2(\bullet+)}$ ring and $PQT^{\bullet+}$ on reduction. The dissociation of a $CBPQT^{4+}$ ring and PQT^{2+} on oxidation of the 1:1 trisradical tricationic complex.

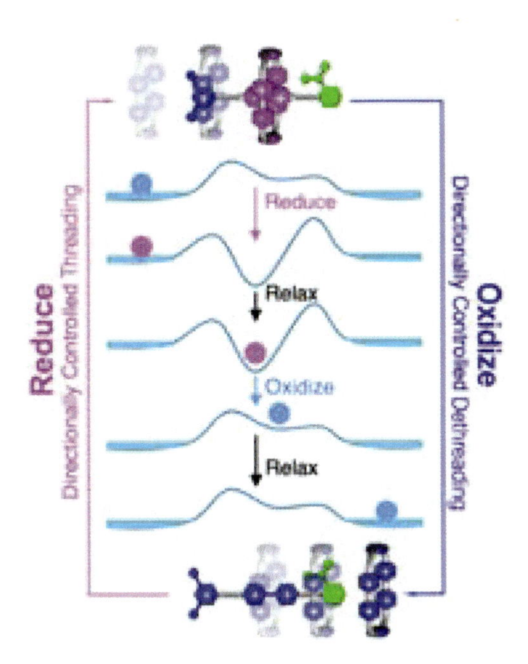

Figure 24. Graphical representations and energy profiles for an artificial molecular pump prototype based on radical-radical stabilizing interactions. In this prototype a bipyridinium radical cation ($BIPY^{\bullet+}$) centered on a dumbbell with the charged end on the left and the neutral one on the right. The $CBPQT^{2(\bullet+)}$ bisradical dication threads onto the dumbbell over the charged end on the left. On oxidation, a push-button molecular switch comes into action as a result of the generation of six positive charges in place of three that were suppressed previously by radical-radical interactions. The net result is the generation of a lot of potential energy which obliges the $CBPQT^{4+}$ ring to depart from the neutral right-hand end of the tricationic dumbbell. No work has been done.

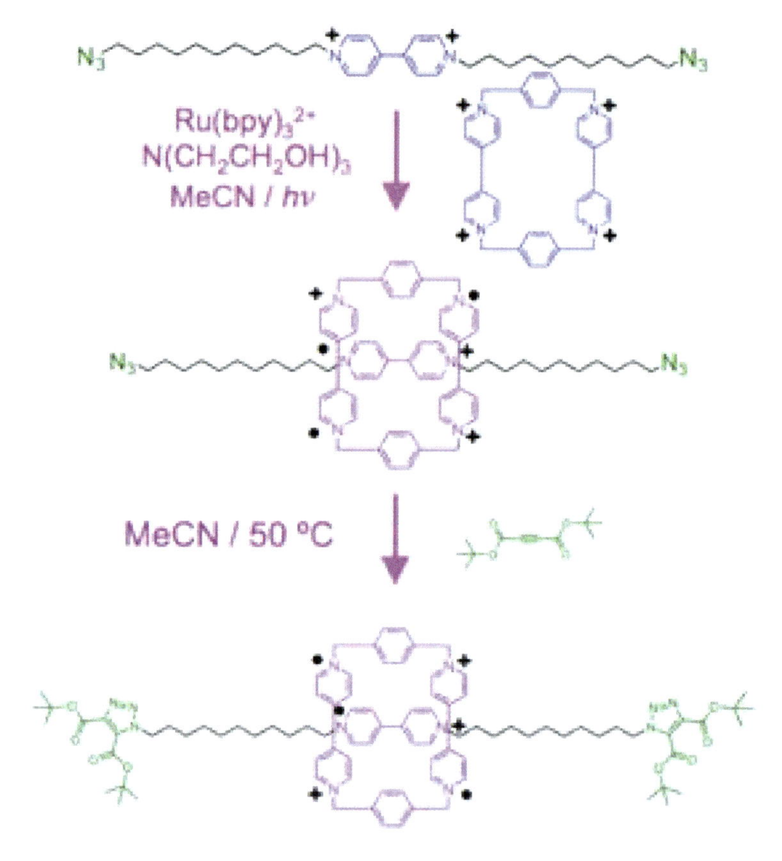

Figure 25. The use of radical templation, promoted by light in the presence of a photosensitizer and sacrificial electron donor, to obtain, following click chemistry, a trisradical tricationic [2]rotaxane.

radical templation [72] to synthesize (Figure 25) a [2]rotaxane—as well as a homo[2]catenane [73]—and demonstrated (Figure 26) the strong repulsion that comes into play when the ring and recognition unit are oxidized back to $CBPQT^{4+}$ and $BIPY^{2+}$, respectively.

ARTIFICIAL MOLECULAR PUMP

Based on all these previous observations, and a large number of exploratory experiments conducted painstakingly by Chuyang Cheng, he settled on the design (Figure 27) of an artificial molecular pump [74, 75]—call it Mark I—where an oligomethylene chain terminated by a stopper is attached at its other end to an IPP speed bump so that this portion of the dumbbell can act as a

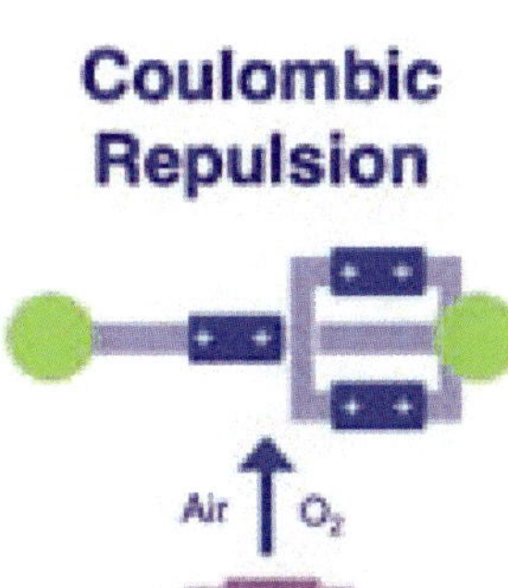

Figure 26. Exposing the trisradical tricationic [2]rotaxane to air produces a hexacationic [2]rotaxane which behaves like a molecular shuttle with an electrostatic barrier to shuttling.

collecting chain for the rings transported from solution. In the oxidized state, the **CBPQT**$^{2(\bullet+)}$ ring and the molecular pump portion of the dumbbell repel each other. Upon reduction, the **CBPQT**$^{2(\bullet+)}$ ring passes quickly over the PY$^+$ unit in search of the radical recognition site, namely a BIPY$^{\bullet+}$ unit, to form the thermodynamically favored trisradical tricationic complex. On oxidation, this complex becomes a highly unstable species carrying six positive charges. Under these circumstances, the CBPQT^{4+} ring would like to relax to a more favored location. Although its returning to the bulk solution is thermodynamically favored, this pathway is blocked kinetically by the PY$^+$ unit, which acts as a Coulombic barrier to deslipping of the ring. Hence, with the aid of thermal energy the ring passes over the IPP speed bump onto the collecting chain. When a second reduction is performed, the IPP unit prevents the association of the trapped ring with the **BIPY**$^{\bullet+}$ recognition site, enabling this site to attract a second **CBPQT**$^{2(\bullet+)}$ ring from solution. In this AMM two positively charged rings are collected by a positively charged dumbbell, reflecting a process that is neither entropically nor enthapically favored. Work is done! The artificial molecular pump operates away from equilibrium by relying on the consumption of redox chemical energy. Our findings have shown that the kinetics associated with two cycles are the same to all intents and purposes (Figure 28), suggesting that that the first ring to be trapped does not have a significant influence on the threading of the second ring. These results are promising since they indicate that it is possible to extend the two-cycle system into a multi-cycle one, opening up the possibility

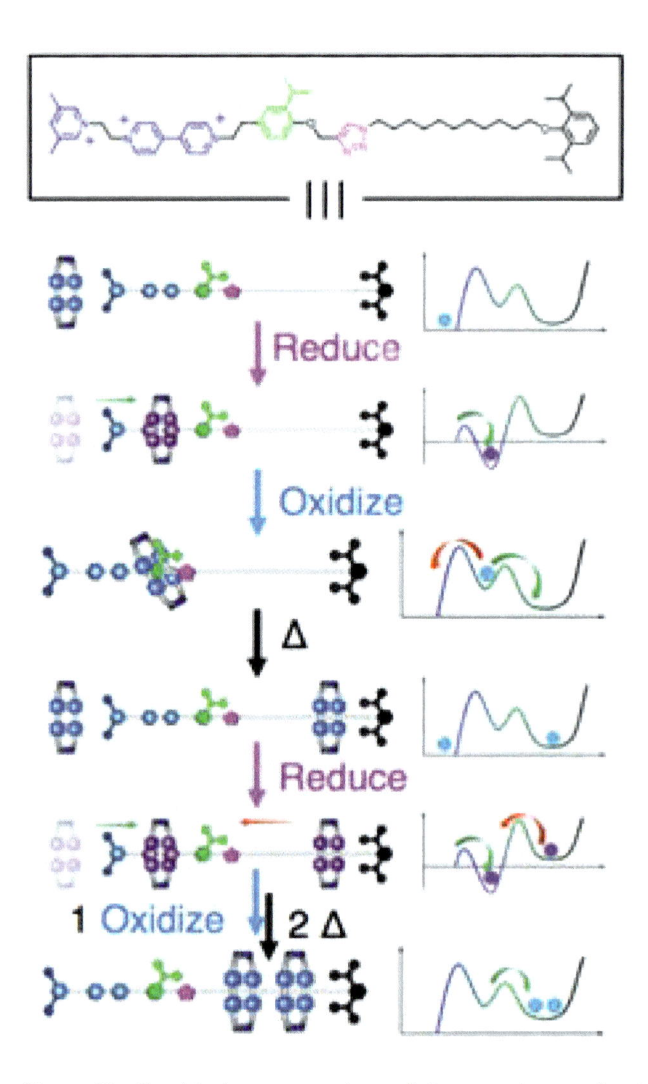

Figure 27. Graphical representations of the pumping mechanism which operates in the case of the artificial molecular pump, portrayed as its structural formula in the box atop the illustration. The energy profiles illustrate two strokes of the pump whereby the $CBPQT^{4+}$ ring is plucked out of solution in its reduced $CBPQT^{2(\bullet+)}$ form and then forced over a steric barrier (green) onto a long collecting chain by thermal energy following oxidation. Work is done.

of synthesizing slide-ring materials [76]. A couple of design modifications to the artificial molecular pump are being explored currently: one is to speed up [77] the action of the pump in a Mark II version at least four-fold from two hours per redox cycle to 30 minutes, and the other is to double the addition of rings to polymer chains (Figure 29) by locating pumps at both ends of the chains.

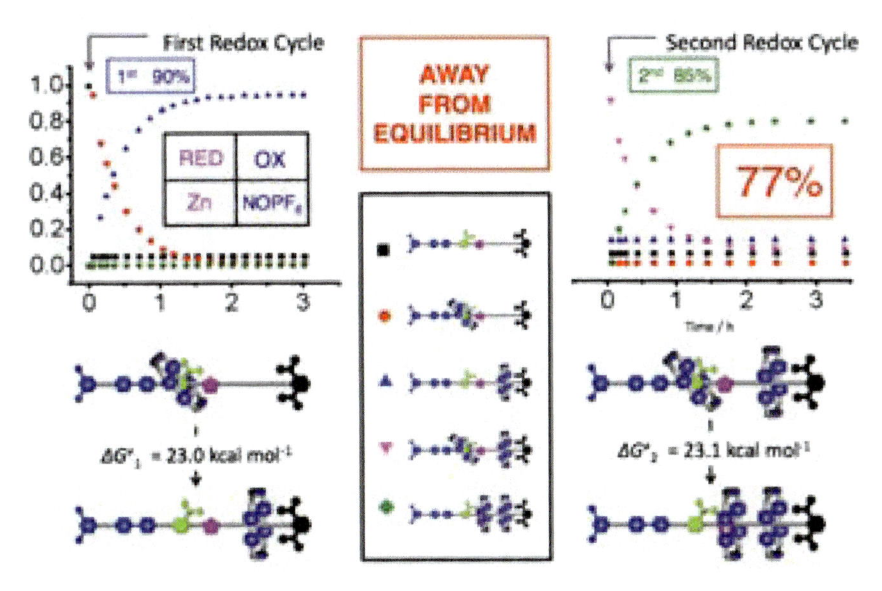

Figure 28. The artificial pump working away from equilibrium. Plots in the change of mole fractions of co-conformations over time in CD_3CN at 42 °C during the co-conformational rearrangements that occur after the first cycle to give the [2]rotaxane and after the second cycle to give the [3]rotaxane.

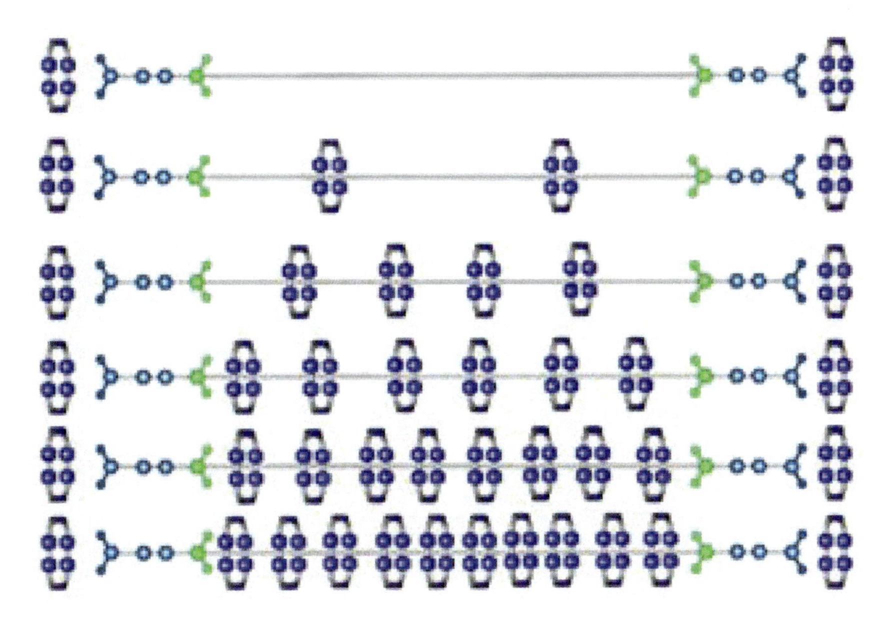

Figure 29. Pumping $CBPQT^{4+}$ rings in pairs onto polymer chains terminated at both ends with artificial molecular pumps.

EPILOGUE

In summary, our interest in AMMs was given a considerable fillip in 1991 with the advent [27] of the molecular shuttle, and the demonstration [28] of the first donor–acceptor bistable switch in 1994. During the past couple of decades, there have been many investigations [52, 54, 57] carried out in collaboration with Vincenzo Balzani at the University of Bologna. Our entry into radical chemistry [71] and radical templation [72] in 2010 set the stage for going forward from entropically driven unidirectional translation with donor–acceptor systems under thermodynamic control in 2013 to enthalpically driven unidirectional transport in 2015 under kinetic control and continuing up to the present time.

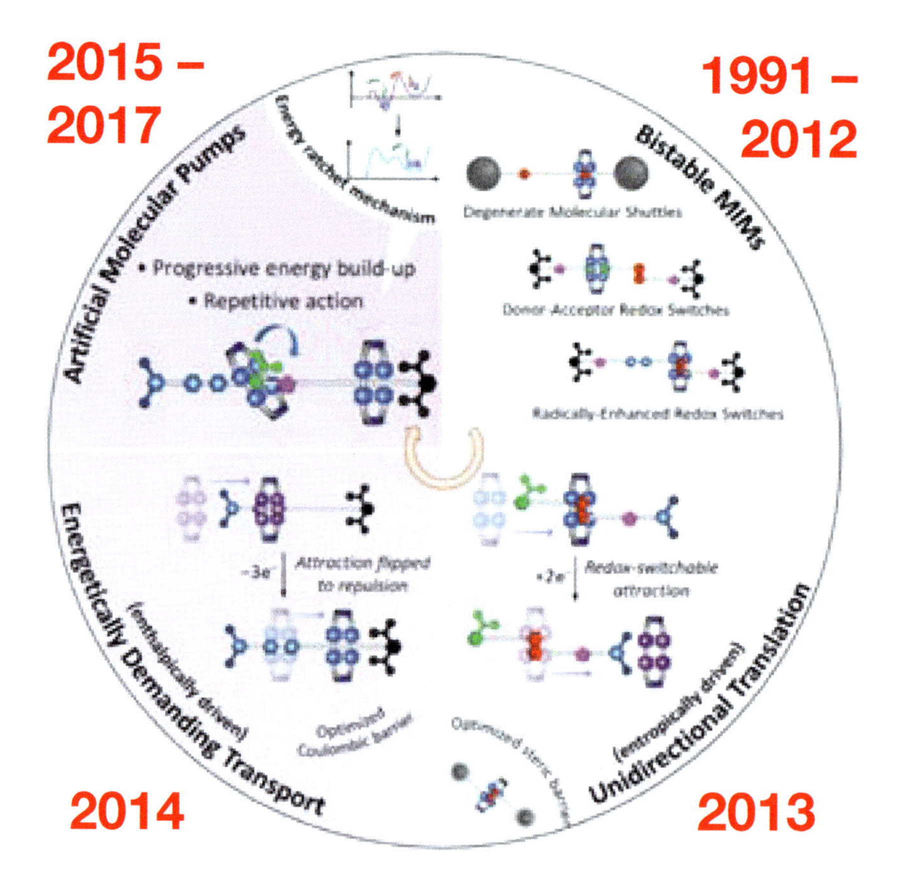

Figure 30. A timeline for the design and synthesis of bistable mechanically interlocked molecules (1991–2012), leading to unidirectional transport (2013) and energetically demanding transport (2014), and artificial molecular pumps (2015–2017).

In a recent Tutorial Review [51] in *Chemical Society Reviews* on *Mastering the Non-Equilibrium Assembly and Operation of Molecular Machines*, written in collaboration with Dean Astumian, we focus on the thermodynamics and kinetics associated with the operation of AMMs and discuss how theory can influence the design principles for constructing molecular machines going forward. A lot of

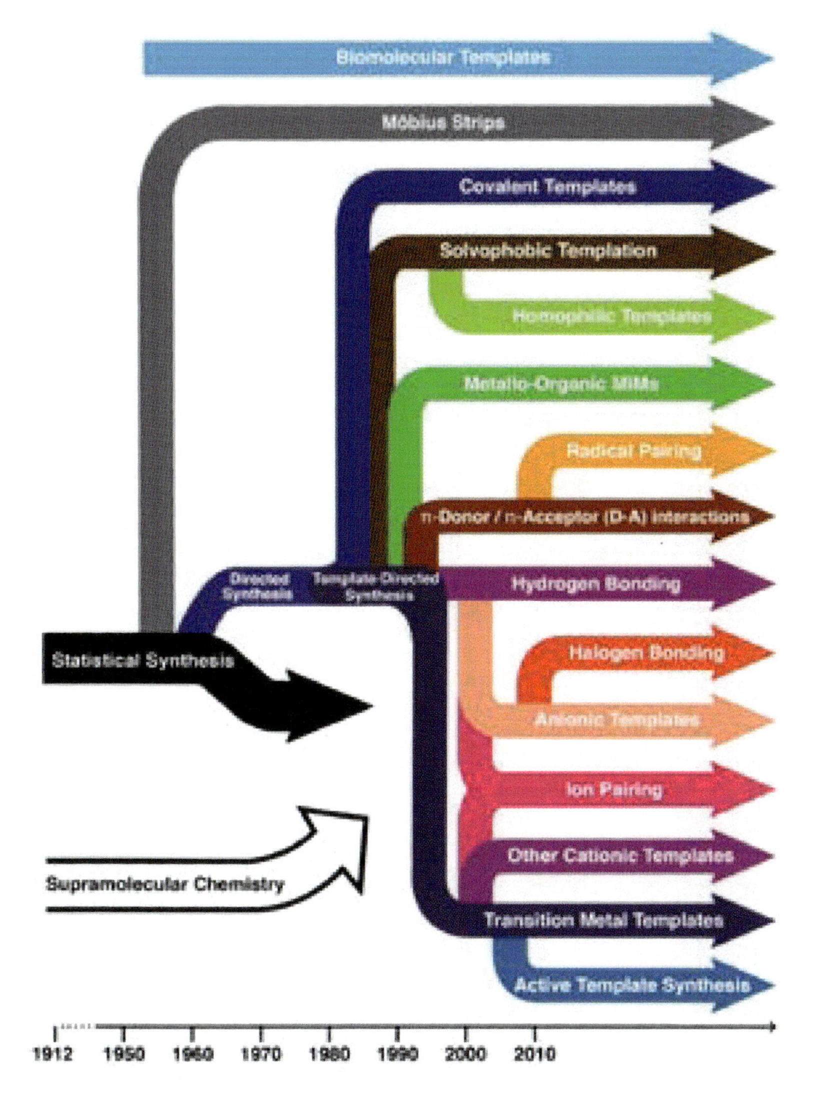

Figure 31. The evolution of the mechanical bond. A time-resolved evolutionary tree highlighting milestones and differentiation events in the template-directed synthesis of mechanical bonds. See ref [8].

fundamental work remains to be done. It is too early to speak in an authoritative and informed manner about what will be the killer applications of AMMs. We are at the very early stages of knowing how to build them, let alone use them. Let me use the analogy of manned flight, particularly in relation to where aviation had reached in 1927, the year in which Charles Lindbergh crossed the Atlantic Ocean in the Spirit of St Louis. An account of the practice of manned flight in 1927 is told [78] in all its amazing glory and gory detail in *One Summer* by Bill Bryson. Compare and contrast the situation for aviation in 1927 with where it stands today as a form of mass transport that has not only opened up country-wide travel, but has also brought Continents together on the grandest of scales. As far as MIMs and AMMs are concerned, they await the engagement of the next generation of chemists eager to exploit the nature of the mechanical bond in chemistry, while taking up the task of designing and synthesizing MIMs and putting them to good use. Figure 31 presents a time-resolved evolutionary tree which highlights the milestones and differentiation events in the synthesis of MIMs. Insofar as the template-directed protocols have established themselves as the most efficient ways to make 'intelligent' MIMs, a breakdown [8] of the recent literature, based on a random selection of 500 articles on catenanes and rotaxanes, has led to the visualization of the data in a pie chart (Figure 32). Presently, solvophobic, hydrogen bonding, donor–acceptor, and metal templation account for a good three-quarters of the chemical literature on the mechanical bond. The histogram illustrated in Figure 33a tracks the number of publications on catenanes (red) and rotaxanes (blue), and those containing both MIMs (purple), on a year-by-year basis. Publications relating to the mechanical bond originate (Figure 33b) from all over the world with the United States, Japan, and the United Kingdom at the top of the list and China not all that far behind. There is every reason to believe that the rapidly accelerating, widespread interest in the chemistry of the mechanical bond, which only started to grow in a linear fashion year-by-year in 1990, is all set now to experience an era of exponential growth.

ACKNOWLEDGEMENTS

I wish to put on record my most sincere thanks to the over 400 students, from close on 50 different countries during the past 45 years, who have contributed quite magnificently to the intellectual life as well as the research achievements and impressive productivity coming out of laboratories in four universities and one international company, situated in two different countries, namely the United States and the United Kingdom. I acknowledge the constant stimulation

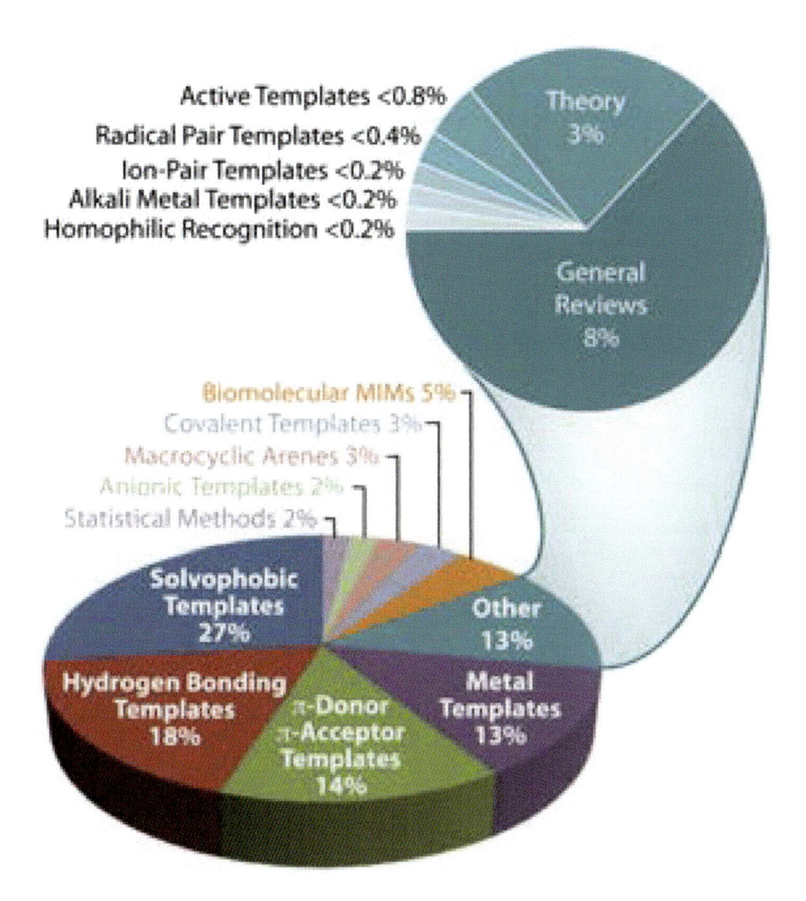

Figure 32. Breakdown of the literature on mechanical bonds according to the mechanisms of their formation / templation. See ref [8].

and invaluable support in research provided by more than 60 collaborators, some of them colleagues in the five institutions in which I have carried out research and others drawn from research centers of one sort or another, dotted all around the world. I thank all those universities, foundations, research councils, funding agencies and industries who have supported my research enthusiastically and generously—in particular, the Universities of Sheffield and Birmingham in the United Kingdom and at the University of California, Los Angeles (UCLA) and Northwestern University in the United States, Imperial Chemical Industries, the Engineering and Physical Sciences Research Council (and its predecessors), the Biotechnology and Biological Sciences Research Council (and its predecessors) in the United Kingdom, the California NanoSystems Institute, the International

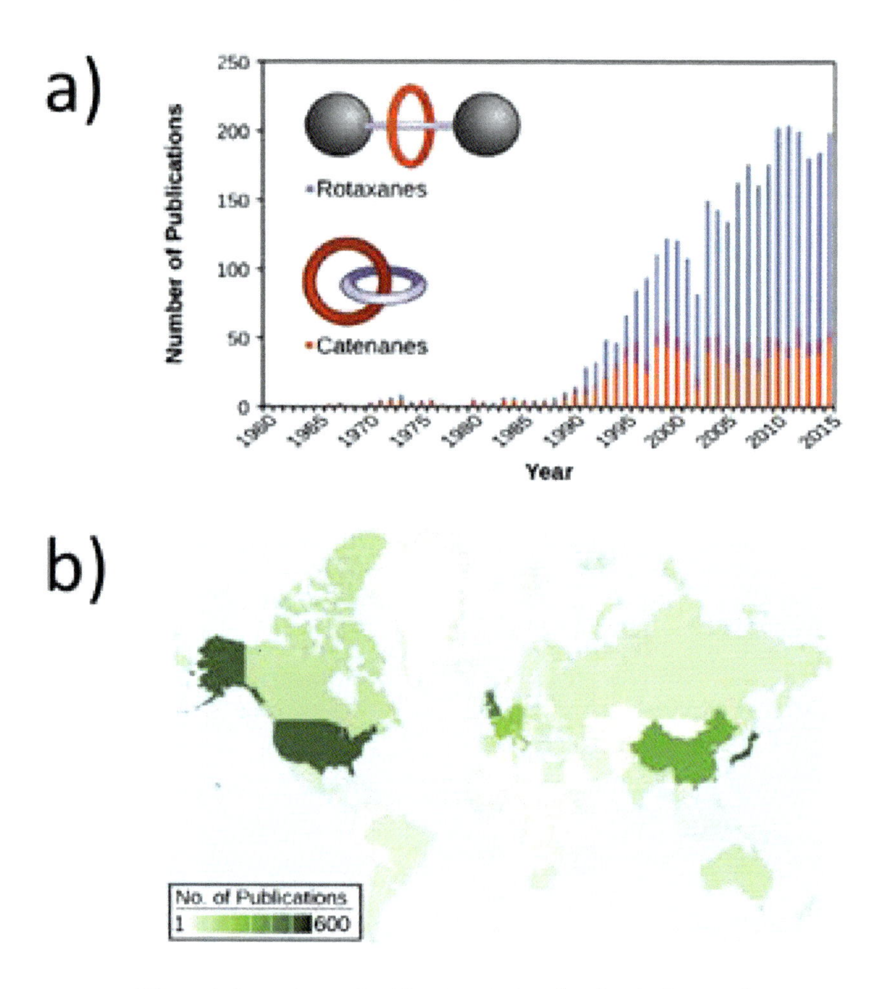

Figure 33. The statistics on journal publications indexed under the keywords, catenanes and rotaxanes. a) The histogram indicates the approximate number of publications (total >3000) from 1960 to 2015. Purple relates to articles indexed under both catenanes and rotaxanes. b) The distribution of publications, indexed under catenane or rotaxane as keywords, color-scaled in green on a map of the world according to their country of origin. See ref [8].

Institute for Nanotechnology, the National Science Foundation, the National Institutes of Health, the American Chemical Society, the Department of Energy and the Department of Defense in the United States, and finally, but by no means least, the King Abdulaziz City for Science and Technology in the Kingdom of Saudi Arabia.

REFERENCES

1. L. Pauling, *The Nature of the Chemical Bond: An Introduction to Modern Structural Chemistry*, Cornell University Press, New York, 1939.
2. "The design of molecular hosts, guests, and their complexes (Nobel Lecture)," D. J. Cram, *Angew. Chem. Int. Ed. Engl.* 1988, **27**, 1009–1020.
3. D. J. Cram, J. M. Cram, *Container Molecules and Their Guests*, The Royal Society of Chemistry, Cambridge, UK, 1994.
4. "Supramolecular chemistry—Scope and perspectives: Molecules, supermolecules, and molecular devices (Nobel Lecture)," J.-M. Lehn, *Angew. Chem. Int. Ed. Engl.* 1988, **27**, 89–112.
5. J.-M. Lehn, *Supramolecular Chemistry: Concepts and Perspectives*, Wiley-VCH, Weinheim, 1995.
6. J. Rebek Jr., *Hydrogen-bonded Capsules Molecular Behavior in Small Spaces*, World Scientific Publishing, Singapore, 2015.
7. "Zur Struktur der Polysiloxene. I," H. Frisch, I. Martin, H. Mark, *Monatsh. Chem.* 1953, **84**, 250–256.
8. C. J. Bruns, J. F. Stoddart, *The Nature of the Mechanical Bond: From Molecules to Machines*, Wiley, New Jersey, 2016.
9. G. Schill, *Catenanes, Rotaxanes, and Knots*, Academic Press, New York, 1971.
10. "The effect of ring size on threading reactions of macrocycles," I. T. Harrison, *J. Chem. Soc., Chem. Commun.* 1972, 231–232.
11. "Synthetic supramolecular chemistry," M. C. T. Fyfe, J. F. Stoddart, *Acc. Chem. Res.* 1997, **30**, 393–401.
12. "Molecular meccano. 1. [2]Rotaxanes and a [2]catenane made to order," P. L. Anelli, P. R. Ashton, R. Ballardini, V. Balzani, M. Delgado, M. T. Gandolfi, T. T. Goodnow, A. E. Kaifer, D. Philp, M. Pietraszkiewicz, L. Prodi, M. V. Reddington, A. M. Z. Slawin, N. Spencer, J. F. Stoddart, C. Vicent, D. J. Williams, *J. Am. Chem. Soc.* 1992, **114**, 193–218.
13. "Big and little meccano," J. F. Stoddart, H. M. Colquhoun, *Tetrahedron* 2008, **64**, 8231–8263.
14. "Olympiadane," D. B. Amabilino, P. R. Ashton, A. S. Reder, N. Spencer, J. F. Stoddart, *Angew. Chem. Int. Ed. Engl.* 1994, **33**, 1286–1290.
15. "The five-stage self-assembly of a branched heptacatenane," D. B. Amabilino, P. R. Ashton, S. E. Boyd, J. Y. Lee, S. Menzer, J. F. Stoddart, D. J. Williams, *Angew. Chem. Int. Ed. Engl.* 1997, **36**, 2070–2072.
16. "The preparation of interlocking rings: A catenane," E. Wasserman, *J. Am. Chem. Soc.* 1960, **82**, 4433–4434.
17. "The preparation of catena compounds by directed synthesis," G. Schill, A. Lüttringhaus, *Angew. Chem. Int. Ed. Engl.* 1964, **3**, 546–547.
18. "Putting mechanically interlocked molecules (MIMs) to work in tomorrow's world," J. F. Stoddart, *Angew. Chem. Int. Ed.* 2014, **53**, 11102–11104.
19. "Une nouvelle famille de molecules: Les metallo-catenanes," C. O. Dietrich-Buchecker, J.-P. Sauvage, *Tetrahedron Lett.* 1983, **24**, 5095–5098.

20. "Second-sphere coordination—A novel role for molecular receptors," H. M. Colquhoun, J. F. Stoddart, D. J. Williams, *Angew. Chem. Int. Ed. Engl.* 1986, **25**, 487–507.
21. "Second sphere coordination of cationic platinum complexes by crown ethers—The X-Ray crystal structure of [Pt(bpy)$(NH_3)_2$ dibenzo[30]crown-10]$^{2+}$$[PF_6]^{2-}$ xH_2O (x≈0.6)," H. M. Colquhoun, J. F. Stoddart, J. B. Wolstenholme, D. J. Williams, R. Zarzycki, *Angew. Chem. Int. Ed. Engl.* 1981, **20**, 1051–1053.
22. "Complex formation between dibenzo-3*n*-crown-*n* ethers and the diquat dication," H. M. Colquhoun, E. P. Goodings, J. M. Maud, J. F. Stoddart, D. J. Williams, J. B. Wolstenholme, *J. Chem. Soc., Chem. Commun.* 1983, 1140–1142.
23. "Complexation of paraquat by a bisparaphenylene-34-crown-10 derivative," B. L. Allwood, N. Spencer, H. Shahriari-Zavareh, J. F. Stoddart, D. J. Williams, *J. Chem. Soc., Chem. Commun.* 1987, 1064–1066.
24. "Cyclobis(paraquat-*p*-phenylene). A tetracationic multipurpose receptor," B. Odell, M. V. Reddington, A. M. Z. Slawin, N. Spencer, J. F. Stoddart, D. J. Williams, *Angew. Chem. Int. Ed. Engl.* 1988, **27**, 1547–1550.
25. "Isostructural alternately-charged receptor stacks. The inclusion complexes of hydroquinol and catechol dimethyl ethers with bisparaquat (1,4)cyclophane," P. R. Ashton, B. Odell, M. V. Reddington, A. M. Z. Slawin, J. F. Stoddart, D. J. Williams, *Angew. Chem. Int. Ed. Engl.* 1988, **27**, 1550–1553.
26. "A [2]catenane made to order," P. R. Ashton, T. T. Goodnow, A. E. Kaifer, M. V. Reddington, A. M. Z. Slawin, N. Spencer, J. F. Stoddart, C. Vicent, D. J. Williams, *Angew. Chem. Int. Ed. Engl.* 1989, **28**, 1396–1399.
27. "A molecular shuttle," P. L. Anelli, N. Spencer, J. F. Stoddart, *J. Am. Chem. Soc.* 1991, **113**, 5131–5133.
28. "A chemically and electrochemically switchable molecular device," R. A. Bissell, E. Cordova, A. E. Kaifer, J. F. Stoddart, *Nature* 1994, **369**, 133–137.
29. "A chemically and electrochemically switchable [2]catenane incorporating a tetrathiafulvalene unit," M. Asakawa, P. R. Ashton, V. Balzani, A. Credi, C. Hamers, G. Mattersteig, M. Montalti, A. N. Shipway, N. Spencer, J. F. Stoddart, M. S. Tolley, M. Venturi, A. J. P. White, D. J. Williams, *Angew. Chem. Int. Ed.* 1998, **37**, 333–337.
30. "A [2]catenane-based solid-state electronically reconfigurable switch," C. P. Collier, G. Mattersteig, E. W. Wong, Y. Luo, K. Beverly, J. Sampaio, F. M. Raymo, J. F. Stoddart, J. R. Heath, *Science* 2000, **289**, 1172–1175.
31. "Current/voltage characteristics of monolayers of redox-switchable [2]catenanes on gold," M. Asakawa, M. Higuchi, G. Mattersteig, T. Nakamura, A. R. Pease, F. M. Raymo, T. Shimizu, J. F. Stoddart, *Adv. Mater.* 2000, **12**, 1099–1102.
32. "A 160-kilobit molecular electronic memory patterned at 10^{11} bits per square centimeter," J. E. Green, J. W. Choi, A. Boukai, Y. Bunimovich, E. Johnston-Halprin, E. DeIonno, Y. Luo, B. A. Sheriff, K. Xu, Y. S. Shin, H.-R. Tseng, J. F. Stoddart, J. R. Heath, *Nature* 2007, **445**, 414–417.
33. "High hopes: Can molecular electronics realise its potential?," A. Coskun, J. M. Spruell, G. Barin, W. R. Dichtel, A. H. Flood, Y. Y. Botros, J. F. Stoddart, *Chem. Soc. Rev.* 2012, **41**, 4827–4859.

34. "Robust dynamics," H. Deng, M. A. Olson, J. F. Stoddart, O. M. Yaghi, *Nature Chem.* 2010, **2**, 439–443.
35. "A redox-active bistable molecular switch mounted inside a metal–organic framework," Q. Chen, J. Sun, P. Li, I. Hod, P. Z. Moghadem, Z. Kean, R. Q. Snurr, J. T. Hupp, O. K. Farha, J. F. Stoddart, *J. Am. Chem. Soc.* 2016, **138**, 14242–14245.
36. "Mesoporous silica nanoparticles in biomedical applications," Z. Li, J. C. Barnes, A. Bosoy, J. F. Stoddart, J. I. Zink, *Chem. Soc. Rev.* 2012, **41**, 2590–2605.
37. "Energy, life, and ATP (Nobel Lecture)," P. D. Boyer, *Angew. Chem. Int. Ed.* 1998, **37**, 2296–2307.
38. "ATP synthesis by rotary catalysis (Nobel lecture)," J. E. Walker, *Angew. Chem. Int. Ed.* 1998, **37**, 2308–2319.
39. "ATP synthase," W. Junge, N. Nelson, *Annu. Rev. Biochem.* 2015, **84**, 631–657.
40. "Walking to work: Roles for class V myosins as cargo transporters," J. A. Hammer, J. R. Sellers, *Nat. Rev. Mol. Cell Biol.* 2012, **13**, 13–26.
41. "Hibernating bears, antibiotics, and the evolving ribosome (Nobel Lecture)," A. Yonath, *Angew. Chem. Int. Ed.* 2010, **49**, 4340–4354.
42. "Unraveling the structure of the ribosome (Nobel Lecture)," V. Ramakrishnan, *Angew. Chem. Int. Ed.* 2010, **49**, 4355–4380.
43. "From the structure and function of the ribosome to new antibiotics (Nobel Lecture)," T. A. Steitz, *Angew. Chem. Int. Ed.* 2010, **49**, 4381–4398.
44. "Ribosome profiling reveals the what, when, where and how of protein synthesis," G. A. Brar, J. S. Weissman, *Nat. Rev. Mol. Cell Biol.* 2015, **16**, 651–664.
45. "Fluctuation driven ratchets: Molecular motors," R. D. Astumian, M. Bier, *Phys. Rev. Lett.* 1994, **72**, 1766–1769.
46. "Thermodynamics and kinetics of a Brownian motor," R. D. Astumian, *Science* 1997, **276**, 917–922.
47. "Design principles for Brownian molecular machines: How to swim in molasses and walk in a hurricane," R. D. Astumian, *Phys. Chem. Chem. Phys.* 2007, **9**, 5067–5083.
48. "Microscopic reversibility as the organizing principle of molecular machines," R. D. Astumian, *Nat. Nanotechnol.* 2012, **7**, 684–688.
49. "Great expectations: Can artificial molecular machines deliver on their promise?," A. Coskun, M. Banaszak, R. D. Astumian, J. F. Stoddart, B. A. Grzybowski, *Chem. Soc. Rev.* 2012, **41**, 19–30.
50. "Design and synthesis of nonequilibrium systems," C. Cheng, P. R. McGonigal, J. F. Stoddart, R. D. Astumian, *ACS Nano* 2015, **9**, 8672–8688.
51. "Mastering the non-equilibrium assembly and operation of molecular machines," C. Pezzato, C. Cheng, J. F. Stoddart, R. D. Astumian, *Chem. Soc. Rev.* 2017, *Advance Article*, DOI: 10.1039/C7CS00068E
52. "Molecular machines," V. Balzani, M. Gómez-López, J. F. Stoddart, *Acc. Chem. Res.* 1998, **31**, 405–414.
53. "Transition metal-containing rotaxanes and catenanes in motion: Toward molecular machines and motors," J.-P. Sauvage, *Acc. Chem. Res.* 1998, **31**, 611–619.
54. "Artificial molecular machines," V. Balzani, A. Credi, F. M. Raymo, J. F. Stoddart, *Angew. Chem. Int. Ed.* 2000, **39**, 3349–3391.

55. "Molecular machines," J. F. Stoddart, *Acc. Chem. Res.* 2001, **34**, 410–411.
56. "Synthetic molecular motors and mechanical machines," E. R. Kay, D. A. Leigh, F. Zerbetto, *Angew. Chem. Int. Ed.* 2007, **46**, 72–191.
57. V. Balzani, A. Credi, M. Venturi, *Molecular Devices and Machines—Concepts and Perspectives for the Nanoworld*, Wiley-VCH, Weinheim, 2008.
58. "Molecular rotors and motors: Recent advances and future challenges," J. Michl, E. C. H. Sykes, *ACS Nano.* 2009, **3**, 1042–1048.
59. "Crystalline molecular machines: Function, phase order, dimensionality, and composition," C. S. Vogelsberg, M. A. Garcia-Garibay, *Chem. Soc. Rev.* 2012, **41**, 1892–1910.
60. "Controlling motion at the nanoscale: Rise of the molecular machines," J. M. Abendroth, O. S. Bushuyev, P. S. Weiss, C. J. Barrett, *ACS Nano.* 2015, **9**, 7746–7768.
61. "Rise of the molecular machines," E. R. Kay, D. A. Leigh, *Angew. Chem. Int. Ed.* 2015, **54**, 10080–10088.
62. "Artificial molecular machines," S. Erbas-Cakmak, D. A. Leigh, C. T. McTernan, A. L. Nussbaumer, *Chem. Rev.* 2015, **115**, 10081–10206.
63. "Wholly synthetic molecular machines," C. Cheng, J. F. Stoddart, *ChemPhysChem.* 2016, **17**, 1780–1793.
64. "Progress toward a rationally designed molecular motor," T. R. Kelly, *Acc. Chem. Res.* 2001, **34**, 514–522.
65. "The art of building small: From molecular switches to molecular motors," B. L. Feringa, *J. Org. Chem.* 2007, **72**, 6635–6652.
66. "Artificial molecular motors," S. Kassem, T. Van Leeuwen, A. S. Lubbe, M. R. Wilson, B. L. Feringa, D. A. Leigh, *Chem. Soc. Rev.* 2017, **46**, 2592–2621.
67. "Designing dynamic functional molecular systems," A. S. Lubbe, T. van Leeuwen, S. J. Wezenberg, B. L. Feringa, *Tetrahedron* 2017, **73**, 4837–4848.
68. "The art of building small: From molecular switches to motors (Nobel Lecture)," B. L. Feringa, *Angew. Chem. Int. Ed.* 2017, **56**, 11060–11078.
69. "From chemical topology to molecular machines (Nobel Lecture)," J.-P. Sauvage, *Angew. Chem. Int. Ed.* 2017, **56**, 11080–11093.
70. "Relative unidirectional translation in an artificial molecular assembly fueled by light," H. Li, C. Cheng, P. R. McGonigal, A. C. Fahrenbach, M. Frasconi, W. G. Liu, Z. X. Zhu, Y. L. Zhao, C. F. Ke, J. Y. Lei, R. M. Young, S. M. Dyar, D. T. Co, Y. W. Yang, Y. Y. Botros, W. A. Goddard III, M. R. Wasielewski, R. D. Astumian, J. F. Stoddart, *J. Am. Chem. Soc.* 2013, **135**, 18609–18620.
71. "Radically enhanced molecular recognition," A. Trabolsi, N. Khashab, A. C. Fahrenbach, D. C. Friedman, M. T. Colvin, K. K. Cotí, D. Benítez, E. Tkatchouk, J. C. Olsen, M. E. Belowich, R. Carmielli, H. A. Khatib, W. A. Goddard III, M. R. Wasielewski, J. F. Stoddart, *Nat. Chem.* 2010, **2**, 42–49.
72. "Mechanical bond formation by radical templation," H. Li, A. C. Fahrenbach, S. V. Dey, S. Basu, A. Trabolsi, Z. Zhu, Y. Y. Botros, J. F. Stoddart, *Angew. Chem. Int. Ed.* 2010, **49**, 8260–8265.
73. "A radically configurable six-state compound," J. C. Barnes, A. C. Fahrenbach, D. Cao, S. M. Dyar, M. Frasconi, M. A. Giesener, D. Benítez, E. Tkatchouk, O. Chernyashevskyy, W. H. Shin, H. Li, S. Sampath, C. L. Stern, A. A. Sarjeant, K. J.

Hartlieb, Z. Liu, R. Carmieli, Y. Y. Botros, J. W. Choi, A. M. Z. Slawin, J. B. Ketterson, M. R. Wasielewski, W. A. Goddard III, J. F. Stoddart, *Science* 2013, **339**, 429–433.

74. "Energetically demanding transport in a supramolecular assembly," C. Cheng, P. R. McGonigal, W.-G. Liu, N. A. Vermeulen, C. Ke, M. Frasconi, C. L. Stern, W. A. Goddard III, J. F. Stoddart, *J. Am. Chem. Soc.* 2014, **136**, 14702–14705.
75. "An artificial molecular pump," C. Cheng, P. R. McGonigal, S. T. Schneebeli, H. Li, N. A. Vermeulen, C. Ke, J. F. Stoddart, *Nature Nanotech.* 2015, **10**, 547–553.
76. K. Ito, K. Kato, K. Mayumi, *Polyrotaxane and Sliding-Ring Materials*, The Royal Society of Chemistry, England, 2016.
77. "An efficient artificial molecular pump," C. Pezzato, M. T. Nguyen, C. Cheng, M. T. Otley, D. J. Kim, J. F. Stoddart, *Tetrahedron* 2017, **73**, 4849–4857.
78. B. Bryson, *One Summer*, Anchor Books, First Edition, New York, 2013.

Bernard L. Feringa.

Bernard L. Feringa

Biography

IT IS A GREAT PRIVILEGE TO BE ABLE TO STAND ON the shoulders of the giants of chemistry and in doing so experience the marvels of the molecular world and provide "challenges for our youth, dreams for the people, and opportunities for industry." For me being a scientist engaged in designing new molecules and chemical systems is a life-long "adventure into the unknown," entering an uncharted territory of astonishing beauty, surprises and amazing perspectives. Over the past decades on many occasions we have lost track on our intended journeys, reaching places in chemical space we could never have imagined. On these occasions, one of my heroes, Abel Tasman, comes to mind. Several hundred years ago, Tasman, an adventurer, departed from a small village close to where we live, sailed in a primitive wooden ship to the edge of the known world, lost his bearings and as a consequence made the serendipitous discovery of what we now call Tasmania and New Zealand. From the outset of my academic studies as a young adult I ventured on an unexpected odyssey into chiral space, however my fascination for the unknown, for "exploring beyond the border," began in my childhood.

THE EARLY DAYS

In 1866, my grandfather, then 3 years old, moved with his family, poor Roman Catholic buckwheat farmers from Emsland, a few miles across the German-Dutch border, to settle in the great Bourtanger moor; a vast, largely uninhabited and remote area in the northeastern part of the Netherlands. The two main reasons for these "Siedler" to build a living in this desolate area were a lack of fertile soil and the threat of conscription into the Prussian army. It was in that same year that the Kingdom of Hanover was dissolved. They were among the founding families of the village of Barger-Compascuum. Starting in primitive turf houses, they slowly established themselves by farming and digging peat. The rather harsh living conditions imbued the family with a strong work ethic, being independent and self-supportive and with a strong desire for knowledge, which we also experienced in our childhood. My father Geert Feringa, who was the youngest of the family of ten, ran the farm while being involved in village community organizations including the local bank, school and church councils. The family of my mother Elizabeth Hake has a similar background, also originating from the border region. Facing poverty, the whole family of her ancestors decided to emigrate to the USA in the 1800s except for the youngest son, who became the first headmaster of the elementary school in Hebelemeer, a German village close to where we lived. Her parents also moved across the border, reclaiming land, and my mother grew up at their farm as the eldest of a family of ten.

My parents married in 1949 and I was born in 1951 as the second of ten children. I cannot remember that I ever left the village during my early youth; most of the first 10 years I spent within 800 meters of the border (except while attending school). The farm and the vast wilderness just behind our fields being my world and that of my brothers and sisters as well as the dozens of nephews and nieces that formed our community. This playground definitively stimulated my imagination, sense of teamwork and desire to explore. Crossing the border behind our farm was always a hard-to-resist adventure and the wilderness on the other side provided many unexpected engagements and findings. Our family

Figure 1. The farmhouse of my grandparents around 1900, the farm I was raised and my parents.

was largely self-supporting with animals for milk, eggs and meat, peat for heating, a water well, and a large garden for vegetables and fruit, the latter being my mother's pride and joy. There were no luxuries but we were comfortable and to this day, I am amazed at how she managed to feed all ten of us with an abundance of healthy food even throughout the winter. From an early age, each of us had our own tasks, and as I grew, I tended the chickens, helped in the garden and later would cut peat for the stove. Observing the behavior of animals, growing three-meter-tall sunflowers and questioning the origin of peat without doubt stimulated greatly my inexorable desire for knowledge.

BASIC AND HIGH SCHOOL EDUCATION

I am extremely grateful to my elementary school teachers, who provided us with a solid primary education. My life long appreciation of history and geography started with their accounts while covering these topics, which was further stimulated by the fascinating stories told by my father and uncles during long-winter evening gatherings at the farm. Being asked frequently why "playing with molecules is so much fun," the proper answer is perhaps that I am striving to fill the gap in my early education left by fact that I did not attend a kindergarten. Both of my parents had little more than elementary school education but they were nevertheless top of their classes and on the occasions that I failed to deliver the proper answer to the headmaster, he would remind me that my mother would have known the answer. Our parents were certainly role models for learning and encouraged us to seize opportunities absent to them in a remote farmer's community in the pre-war period. It may be hard to imagine today but we should remember that there were no TVs, PCs or smartphones; but there were certainly books at home or in the local church library that we could reach for in our search for knowledge.

The next step in my education was to attend the Katholiek Drents College, a secondary education called the HBS, which was held in high esteem in the Netherlands. I had the good fortune to attend a rather small school with a team of excellent academically trained young teachers covering a wide range of topics. Confronted with biology, mathematics, physics, and chemistry, a new world opened for me fueling my thirst to know how and why. I remember vividly that most of our teachers could address topics beyond the textbook and put the material we had to learn in a broader context.

Our chemistry teacher, Op de Weegh, was an exceptional inspiration, always eager to challenge us. In the later part of my high school education, when the next step in academic education was approaching, he was particularly influential

Figure 2. The family in the 1970s.

in my decision to do chemistry. Although mathematics was my most successful subject, the fact that in chemistry you could experience color, odor or beautiful crystals and see practicality ranging from fertilizer to drugs were decisive factors. At a recent reunion of my high school, talking to my chemistry teacher reminded me of one of his sayings: "I wish every child in his or her life at least one excellent teacher." I had the good fortune to have several! Cycling 15 km every day to school with my friends—there was no public transport—also gave room for intense debates, sharpening our minds. This was also the time that I started to play for the local soccer team, and although I was a player of modest talent, and digressed for a few years playing handball, I have enjoyed playing soccer for a long period extending well into my academic career. Perhaps the best gift of my high school education was that I learned to appreciate many disciplines.

A perhaps unexpected influence during my late high school and early university studies, that wild period of the student revolts and social upheaval, were the endless discussions at home among my brothers and sisters. Our Sunday debates on topics ranging from world politics to inventions, religion, and human behavior are still vividly remembered by all of us. Let me not conclude describing this period without mentioning perhaps the single most influential person. I always bore the desire to become a farmer but had the good sense to follow my father's wise advice to study first and only later, perhaps, reconsider my options.

Figure 3. Me with my chemistry teacher G. Op de Weegh at a recent reunion of our high school.

As a consequence, I spent most of the long summer holidays during high school and university working alongside my father on the farm. He shared with me the fascination and admiration for the natural world, the wonder of ears of wheat growing from a tiny seed, the beautiful colors of the flowers in the fields, and cows giving birth to their offspring. Such wonder alleviated the muscle ache that followed the solid day's work and while we were puzzled by the shape of clouds or the flow of water, and as we struggled with the nature of gravity, it invariably guided us back to our work with the soil.

UNIVERSITY EDUCATION

I entered the University of Groningen as a major in chemistry in 1969 and I quickly learned to appreciate the academic environment, the various aspects of student life and the many hours of demanding courses and lab work. Two factors I consider of major importance for this period of my undergraduate education. First, we were the first cohort of students to work in our then brand-new laboratories; we take pride in being a part of that community. Second, several of our professors were either US citizens or trained in the USA and they challenged us—we felt their sense of expectation. They had modelled the chemistry department after top US institutes and their rather unique spirit did not go unnoticed. My real love for synthetic chemistry started in my third year when I had my first opportunity to work on a short research topic. I hold fondly the memory of the exhilaration that I felt making my first new compound—a compound never prepared anywhere in the world. My next experience of research was a period in the inorganic department, where I learned to handle the most air and moisture sensitive early transition organometallic reagents, in particular organotitanium compounds. Every time I see a nice painted wall the vivid memory of a leaking seal of the Schlenk flask, with oxygen slowly creeping in, springs to mind.

My decision to carry out my Masters research, I think, says a lot about my character then. I had declined a project proposal from a chemistry professor who had indicated that prior to working on that topic I should do a lot of routine measurements, as "the problem was too difficult for me." I was eager to be challenged and was fortunate that another professor, Hans Wijnberg, struck the right cord by providing a topic that had no prior art whatsoever. Asymmetric coupling of phenols; how to couple two radicals generating axial chirality, as in BINOL? I started exploring Fe-analogs of chiral camphor-based -diketonate ligands, reported in 1974 by George Whitesides for his chiral europium NMR shift reagents. Although during my Masters research I failed to accomplish the asymmetric coupling of 2-naphthol, it was rewarding that ultimately during my PhD studies I was able to realize BINOL formation with 16% optical purity using a chiral copper amine complex as oxidant. These were the years that I became fascinated by stereochemistry, not least by the excitement that arose in the field as a result of many amazing discoveries in asymmetric catalysis. The general interest in the group on fundamental aspects of stereochemistry ranging from ORD and CD spectroscopy, absolute configuration and absolute asymmetric synthesis to enantiomers lacking optical activity and the pioneering work on asymmetric organocatalysis using cinchona alkaloids was a fertile learning environment. It was also important that numerous prominent (stereo-) chemists—among them Sharpless, Eliel, Barton, Turro and Kagan—visited Groningen during that period and we were strongly encouraged to discuss with these great scientists. I continued my PhD studies in the Wijnberg group and discovered among others small differences in selectivity between a racemic mixture and pure enantiomers in stoichiometric reactions. We named this phenomenon the antipodal effect and, although our initial submission met with disbelief from the referees, ultimately our work was published. Much to our delight, 10 years later, Henri Kagan

Figure 4. My mentor and PhD supervisor Professor Hans Wijnberg.

demonstrated that related phenomena occur in catalytic reactions and formed the basis for the now widely accepted non-linear effects.

Perhaps the most decisive moment in regard to my later career was the design of chiral overcrowded alkenes that did not bear a stereogenic center but for which both the cis and trans stereoisomers consisted of enantiomeric pairs. The idea was rather simple; if a biaryl can be chiral due to hindered rotation around a single bond, the question arose "can an olefin form a stable homochiral compound exclusively due to torsion around the double bond"? Taking advantage of then newly discovered McMurry coupling of ketones, the chiral overcrowded alkenes were indeed prepared and reported in JACS 1976. How could I have realized at that moment that this discovery would later form the basis for our chiroptical molecular switches and our unidirectional rotary motors. In retrospect, the PhD period provided me with the essential atmosphere for discovery in which we were encouraged to question conventions and break paradigms. My fellow students, in particular Bert (EW) Meijer, Kees Hummelen and Henk Hiemstra, who have each made prominent academic careers over the past decades, greatly added to the stimulating and challenging atmosphere in the group. The summer of 1977 was another highly important period in my career, when I was dispatched to the US to attend the Organic Symposium in Morgantown, WA. Hans Wijnberg introduced me to many distinguished chemists but I was most impressed by the superb 2 h 20 min (a rather short lecture I was informed) evening lecture by the great Prof. R. B. Woodward. As my mentor had also arranged for me to make a short lecture tour, I had the privilege to give presentations about my PhD work at Penn State and Cornell among others and Princeton where I also had the opportunity to discuss stereochemistry with my hero Kurt Mislow. After my American journey, I was convinced that my next step was postdoctoral research in the US. But as is so often the case in life our journeys can take unexpected detours.

THE SHELL PERIOD

In the months writing up my thesis work I realized that national service, then compulsory in the Netherlands, would inevitably quench any dreams of a postdoctoral adventure. By good fortune, I was offered a position at the Royal Dutch Shell Research Laboratories (KSLA) in Amsterdam that, because of my expertise in stereochemistry, exempted me from active military service and provided the next best thing to a Postdoc period in the US; as a young academic, I was entering a highly prestigious corporate research institute, comparable to Bell Labs or DuPont central research, with a worldwide reputation in catalysis. Indeed, I experienced an amazing exposure to both fundamental and applied catalysis research during my 6.5 years at Shell. Most of my own research focused then on catalytic oxidations and novel ligand and catalyst design. In my first months, I shared an office with David Reinhoudt, who introduced me to the then rapidly emerging field of supramolecular chemistry. Although I was working on fundamental problems in catalysis, for instance photo-redox catalysis, I strongly benefitted from the interaction with process chemists also. The exposure to numerous industrially relevant projects provided me with important insights that have helped to shape my future collaborative research projects, as well as in teaching our students, the majority of whom would enter industrial careers. Definitively, my later projects on asymmetric catalysis and phosphoramidites with DSM, catalytic oxidations with Unilever and liquid crystals with Philips over the past decades, were partly rooted in my industrial research period at Shell.

Apart from the KSLA period, I spent nearly 1.5 years at Shell Biosciences center in Sittingbourne, Kent, UK, working on herbicides. This period was equally fascinating, discussing with biochemists and plant physiologists among others. Immersion in total synthesis and chemical biology further stimulated my admiration for the power of synthetic chemistry to create and the unlimited opportunities presented by molecular design. Equally stimulating were regular meetings with Sir John Cornforth and members the British chemical community. Following my return to Shell Amsterdam and the catalysis group of Piet van Leeuwen, I realized that reading the latest discoveries in the prime chemistry journals still inspired me more than delving into industrial problems. When I was approached in 1984 by my Alma Mater to consider a junior faculty position in the chemistry department, theere was no hesitation. The fact that in that year I had married my wife Betty, who then lived in Groningen and was employed by the University Medical Center there, made the decision even easier.

UNIVERSITY OF GRONINGEN

My research program over subsequent years was based firmly in synthetic organic and physical organic chemistry. Although it developed along two main lines, catalysis and molecular switches, stereochemistry remained the overarching theme. Exploring chiral space regularly provided fascinating surprises, be it a novel method to determine enantiomer excess without an external source of chirality, chiral amplification through sublimation, or DNA-based asymmetric catalysis (together with Gerard Roelfes).

Catalytic oxidation is key to many of the world's most important industrial processes, and confronted with the challenge to design selective oxidation processes we focused on anti-Markovnikov Wacker oxidation and non-heme iron and manganese based catalytic systems. As part of these programs I enjoyed superb cooperation with Larry Que (Univ. Minnesota), Ronald Hage (Unilever/Catexel) and Wesley Browne (Univ. Groningen) over many years. Building my research team in the late 80s, I became intrigued by the lack of a highly enantioselective method for conjugate addition of organometallic (alkyl-zinc and copper) reagents. The introduction of chiral phosphoramidites as a novel privileged class of chiral ligands in asymmetric catalysis resulted ultimately (in 1996) in the 1,4-addition of organozinc reagents with synthetically useful enantioselectivities. From this period on, I had the privilege to work together on highly successful projects with my close colleagues Adri Minnaard and Suzy Harutyunyan, focusing on challenging total syntheses and equally challenging problems in asymmetric catalysis. It took another 8 years before we succeeded in taming Grignard reagents for similar conjugate additions and allylic substitutions; the key was to go deep and understand at a mechanistic level both the catalyst and the reaction as a whole. Spurred on by this success, finally, after 20 years of effort, we were able to achieve catalytic asymmetric C-C bond formation with the notoriously reactive organolithum reagents. Controlling aggregation behavior and applying well defined copper complexes provided the long-awaited solution. This was the stepping stone for our current program on ultrafast organolithium cross coupling.

I was appointed as full professor in 1987, succeeding my scientific father Hans Wijnberg in 1988, and gave my inaugural public lecture at the University

Figure 5. Betty and our three daughters at younger age.

of Groningen (the academic oratie is a fine Dutch tradition) in 1989 entitled "Order and Dynamics in Synthesis." The discussion on that occasion among others centered on "intelligent molecules"; I pondered on how far we could go in building functional molecules that were designed to perform specific tasks, ultimately creating tiny molecular robots.

This event was the starting point for over 25 years of work on molecular switches and motors. The basic idea was to design molecular information storage materials taking advantage of the dormant overcrowded alkene switches from my PhD period. The excellent switching properties (photo-bistability) and inherent chirality (for non-destructive read-out) were decisive factors that enabled the birth of an entire class of chiroptical molecular switches. The merging of synthesis with mechanistic studies, photochemistry, materials chemistry and spectroscopy, in close cooperation with Wesley Browne, attracted students with distinct training and expertise who beyond doubt were highly influential in our discussions and approaches taken during the next two decades. An important collaboration on the absolute configuration of chiral overcrowded alkenes was started with Noboyuki Harada in Sendai. We extended our program on photoswitches to control biosystems such as MsCl protein channels and SecY protein transporters (with biochemist Armagan Kocer and molecular microbiologist Arnold Driessen, respectively). As our research slowly evolved from molecules into dynamic molecular systems we worked on control of organization along different length scales, i.e. gels, polymers and liquid crystals. The studies on chiroptical switches culminated in the discovery of our light-driven unidirectional rotary motor, reported in 1999. This was also the starting point for the design of several generations of motors, surface anchored rotary motors and motor-based liquid crystals (in cooperation with Dick Broer, then at Philips Research). Being a member of both the Stratingh Institute for Chemistry and the Zernike Institute for Advanced Materials at the University of Groningen was a major advantage, providing access to a wide range of facilities (in particular for surface characterization) and highly beneficial to my students working on these multi-faceted problems. The Spinoza grant was the immediate reason for the design of a four-wheel drive molecular car tackling the fundamental challenge of how to convert rotary molecular motion into translational motion across a surface. After 7 years we succeeded, in close cooperation with Kalle Ernst at EMPA, Zurich. These were fascinating years for my "motor team" as we designed single motors that could move in both directions, motors powered by visible light, multitasking chiral catalyst and self-assembled nanostructures based on rotary motors among others. In hindsight, probably the most memorable event in all these years was the direct observation by the naked eye of a micro-object rotating, while floating on a soft liquid crystal surface, by a light-driven motor.

I had the pleasure to spend a major part of my life at the University of Groningen's Chemistry Department with fine colleagues and an open border-free

Figure 6. The Feringa group at a sports event during the yearly Workweek in the 90s.

atmosphere encouraging students to cooperate and staff to discuss and work together. I enjoyed working with my group on diverse chemical problems stimulating creativity and cooperation with ample opportunities to learn and explore beyond our comfort zone. It was indeed a privilege to join my highly talented students on a fascinating journey into the largely uncharted territory of molecular motors and machines.

The long tradition of spending a week each year with my whole group abroad, visiting industry and another university or research institute, is highly valued. This "workweek" with student-organized lectures ranging from industrial innovation, ethics, chemical warfare to molecular cooking, joint symposia and sports and pub events greatly stimulated a fine team spirit.

Shortly after my appointment in Groningen, Betty and I decided to move to the village of Paterswolde just south of Groningen, giving us both the chance to enjoy a decent daily cycle to and from our respective workplaces. "Moving in Flatland" in the northern Netherlands of course gives plenty of time each day to think about the three-dimensional puzzles that we were facing in the lab. Just as memorable have been the annual BBQ's in our garden when the whole group gathers together (often during European and World Cup Soccer events) and the

Figure 7. Betty, our daughters Femke, Hannah and Emma and son-in-law Jorrit at a recent ceremony when I received a Royal decoration.

Figure 8. My current research group.

many PhD graduations, for which we have the tradition of making a movie about the candidates' time to ensure that their many unexpected talents in and out of the lab are remembered.

I enjoy long-distance skating, and as a farmer's son it is a delight to have our own piece of land with a meadow, horse and vegetable garden which allows me not only to exercise in the weekends but never lose contact with nature. Our three daughters shared the enthusiasm for learning and sports. Femke, a cell biologist, is in the final year of her PhD studies at the Netherlands Cancer Institute (NKI), Hannah just started a PhD in the area of food allergies at the Utrecht University Medical Center and Emma is a Masters student in movement sciences at the Free University of Amsterdam. The week of skiing in the Swiss Alps every winter and the sailing events on the Frisian lakes each summer provide ample opportunity for challenges beyond chemistry and are very precious moments with Betty and the children. I am extremely happy to have experienced great support during my entire career from my family and that they tolerate me being distracted by "crazy molecules" at unexpected moments. Betty always reminds us of my passion: "Being a scientist is a way of life." I could not agree more, and I am grateful that she was and is always alongside me on our journey.

The Art of Building Small: From Molecular Switches to Motors

Nobel Lecture, December 8, 2016 by
Bernard L. Feringa
Stratingh Institute for Chemistry, University of Groningen,
Groningen, The Netherlands.

AT THE START OF MY JOURNEY INTO THE uncharted territory of synthetic molecular motors I consider it apt to emphasize the joy of discovery that I have experienced through synthetic chemistry. The molecular beauty, structural diversity and ingenious functions of the machinery of life [1, 2], which evolved from a remarkably limited repertoire of building blocks, offers a tremendous source of inspiration to the synthetic chemist entering the field of dynamic molecular systems. However, far beyond Nature's designs, the creative power of synthetic chemistry provides unlimited opportunities to realize our own molecular world as we experience every day with products ranging from the drugs to the displays that sustain modern society. In their practice of the art of building small, synthetic chemists have shown amazing successes in the total synthesis of natural products [3], the design of enantioselective catalysts [4] and the assembly of functional materials [5], to mention but a few of the developments seen over the past decades. Beyond chemistry's contemporary frontiers, moving from molecules to dynamic molecular systems, the molecular explorer faces the fundamental challenge of how to control and use motion at the nanoscale [6]. In considering our first successful, albeit primitive, steps in this endeavor, my thoughts often turn to the Wright brothers and their demonstration of a flying airplane at Kitty Hawk on the 17th of December 1903 [7]. Why does mankind need to fly? Why do we need molecular motors or machines? Nobody would have predicted that in the future one would build passenger planes each carrying several hundred people at close to the speed of sound between continents. While admiring the elegance of a flying bird, the materials and flying principle of the entirely artificial airplane are quintessentially distinct from Nature's designs. Despite the fabulous advances in science and engineering over the past century, manifested most clearly by modern aircraft, we are nevertheless humbled by the realization that we still cannot synthesize a bird, a single cell of the bird or even one of its complex biological machines.

It is fascinating to realize that molecular motors are omnipresent in living systems and key to almost every essential process ranging from transport to cell division, muscle motion and the generation of the ATP that fuels life processes [8]. In the macroscopic world, it is hard to imagine daily life without our engines and machines, although drawing analogies between these mechanical machines and biological motors is largely inappropriate. In particular the effect of length scales should be emphasized when comparing, for instance, a robot in a car manufacturing plant and the biological robot ATPase. While in the first case size, momentum, inertia and force are important parameters, in the world of molecular machines non-covalent interactions, conformational flexibility, viscosity and chemical reactivity dominate dynamic function [9]. In addition, when operating at low Reynolds numbers, we go beyond the question "How to achieve motion?" and face the question "How to control motion"? In the molecular world where Brownian motion rules, and noting that biological motors commonly operate as Brownian ratchets [10], the design of molecular systems with precisely defined translational and rotary motion is the main challenge [11].

Making the leap from molecules to dynamic molecular systems while drawing lessons from life itself, an important challenge ultimately is to achieve out-of-equilibrium phenomena. Molecular switches and motors are perfectly suited to introduce dynamic behavior, reach metastable states and drive molecular systems away from thermal equilibrium. We focused on three key aspects—triggering and switching, dynamic self-assembly and organization, and molecular motion—with a future perspective directed towards responsive materials, smart drugs and molecular machines among others.

MOLECULAR SWITCHES

Chiropractical molecular switches and information storage

In our initial attempts to design molecules with the intrinsic dynamic functions that ultimately evolved into molecular rotary motors, we took inspiration from the process of vision [12]. This amazing natural responsive process is based on an elementary chemical step, the photochemical cis-trans isomerization around a carbon–carbon double bond in the retinal chromophore (Figure 1a). We envisioned the exploration of this simple switching process in the design of molecular information storage units and responsive elements in dynamic molecular systems and materials. Although molecular bi-stability can be induced by various input signals including light, redox reactions, pH changes, metal ion binding, temperature, and chemical stimuli, the use of photochemical switching has distinct advantages as it is a non-invasive process with high spatial-temporal precision [13]. Building on seminal work by Hirshberg on azobenzenes [14], Heller on fulgides [15], Irie on diarylethylenes [16] and others [17], numerous photochromic molecules have been explored in recent years in our group to achieve responsive function, including control of optical and electronic properties of materials [18], supramolecular assembly processes [19, 25] and biological function [20].

In our journey towards bistable molecules with excellent photoreversibility and high fatigue resistance, we focused on the synthesis of chiral overcrowded alkenes (Figure 1b) [21]. Non-destructive read-out of state is a central aspect of any potential molecular information storage system, and was addressed by taking advantage of the distinct right (P)- and left (M)-handed helicities in this system, enabling read-out by chiroptical techniques far outside the switching

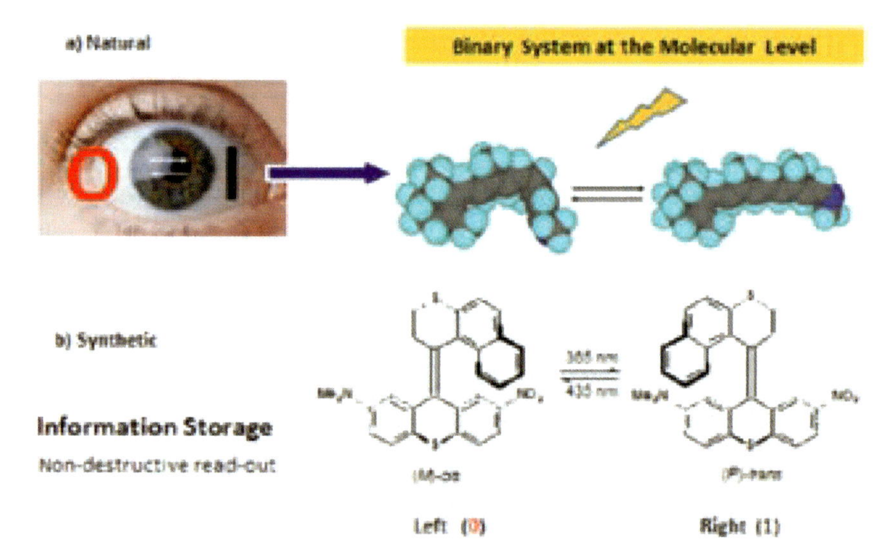

Figure 1. Optical switching systems based on bistable molecules. a) Retinal photoisomerization in the process of vision. b) Chiroptical molecular switch based on overcrowded alkenes as a molecular information storage system.

regime. The interconversion between two isomers with distinct chirality, i.e. a chiroptical molecular switch, defines a zero-one digital optical information storage system at the molecular level. Although high-density optical information storage materials based on this approach are promising, the fundamental challenge of addressing individual molecules at the nanoscale in a closely packed assembly in an all-optical device remains to be solved, despite the spectacular advances in single molecule detection techniques seen over the last decades [22].

At this point it is appropriate to emphasize two aspects of these studies. Firstly, the chiral overcrowded alkenes that formed the basis for the chiroptical molecular switches have their genesis in my PhD studies under the guidance of Hans Wijnberg on biaryl atropisomers. The idea that twisted olefins might show atropisomerism was explored, using the then recently invented McMurry coupling reaction, in the synthesis of cis- and trans-isomers of inherently dissymmetric overcrowded alkenes (see Figure 2 for a time line) [23]. The realization that these novel structures had an intrinsic chiral stilbene type chromophore that was immune from the notorious photocyclization seen in stilbenes, provided a stepping stone more than a decade later to chiroptical switches and two decades

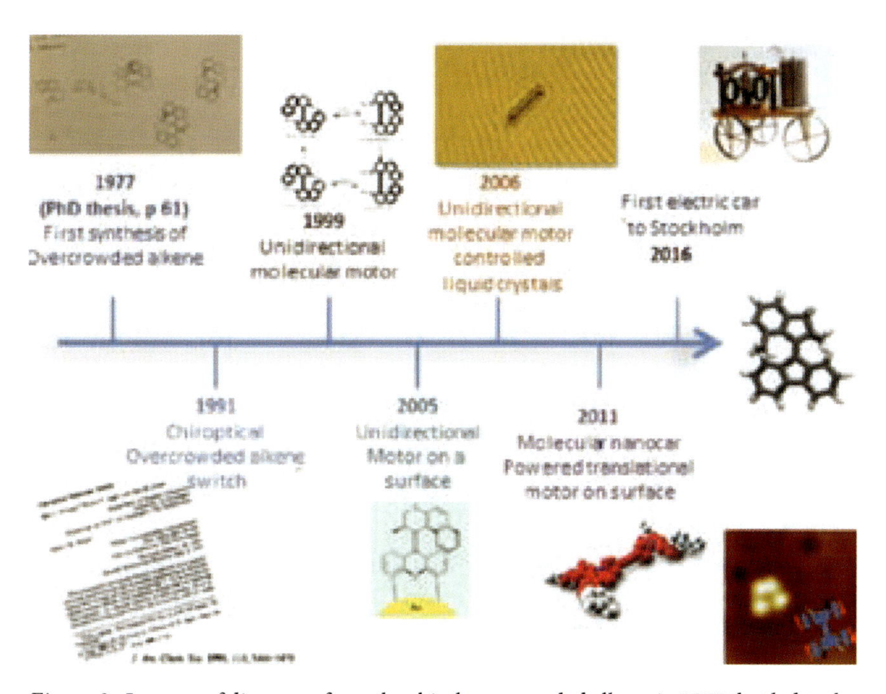

Figure 2. Journey of discovery from the chiral overcrowded alkene in 1977 that led to the light-driven molecular rotary motor in 1999 and the presentation of the first electric car (designed at the University of Groningen in 1835) and the molecular nanocar (developed in 2011) to the Nobel Museum in Stockholm.

later to light-driven rotary molecular motors. Secondly, with the photoisomerization of these chiral overcrowded alkenes, reported in 1991 [24], we demonstrated that controlled clockwise or counterclockwise motion in either direction of one half of the molecule with respect to the other half was achieved simply by changing the wavelength of irradiation. Control of directionality of rotary motion was key to the latter development of molecular rotary motors. The photoresponsive overcrowded alkenes were used as chiral dopants in mesoscopic materials to achieve chiroptical switching between cholesteric liquid crystal phases [25], as well as control elements for molecular rotors [26]and for photoswitching the handedness in circular polarized luminescence [27].

The wavelengths of switching and the stereoselectivity of the isomerization process were tuned, for instance, via donor-acceptor substituents. In a series of studies together with the Harada group at Tohoku University, we established the chiroptical properties, absolute configuration and racemization pathways of biphenanthrylidenes [28]. An important milestone was our discovery of dynamic control and amplification of molecular chirality by circular polarized light (CPL) [25]. Here CPL irradiation shifted the equilibrium between P or M helices of chiroptical switches to achieve a tiny chiral imbalance that was amplified through formation of a twisted nematic liquid crystalline phase. This discovery strengthened the idea that unidirectional rotary motion was in principle possible using CPL irradiation although, on the basis of the Kuhn anisotropy factor for such systems, the efficiency and directionality parameter will be very low [29].

Responsive materials and self-assembly

Molecular switches offer tremendous opportunities to introduce dynamic behavior into materials and as part of our program on responsive functions, over the past 30 years, we have explored a wide variety of both photochemical and redox switches far beyond the initial chiral overcrowded alkenes. The few examples discussed here illustrate the potential in areas ranging from soft materials to biomedical applications. Modulation of electronic properties through photoswitching has potential in integrating optics and electronics in molecular based devices provided that the molecular components operate properly when incorporated in semiconductor based systems. For instance, self-assembly of diarylethene photoswitches in mechanically controlled break junctions enabled single molecule optoelectronic switching, although bistability was initially compromised [19]. In later designs, large array devices were fabricated using an inorganic semiconductor and conducting polymer hybrid system in combination with monolayers of photoswitches [30]. In the bottom-up approach to molecular electronics [31] numerous other approaches have been explored [32]. The pioneering work by the Heath and Stoddard team on rotaxane based devices [33] and the use of alternative switches such as azobenzenes and spiropyrans spring immediately to mind [34]. It is now apparent that photo- and redox-switchable molecules are a fertile test ground for potential information storage, sensing, molecular electronics, imaging and responsive optical systems and smart materials.

The introduction of optical switches in components that are designed to undergo self-assembly allows the construction of supramolecular systems that can adapt and reconfigure in response to an external light signal. For instance, a photo- and redox active bisthioxanthylidene unit formed the core of amphiphiles specifically designed to form highly stable nanotubes (Figure 3a). Following this approach self-assembled multicomponent nano-objects, i.e. vesicle capped nanotubes and vesicles embedded in nanotubes, were obtained and the disassembly of these responsive supramolecular systems can be controlled by with

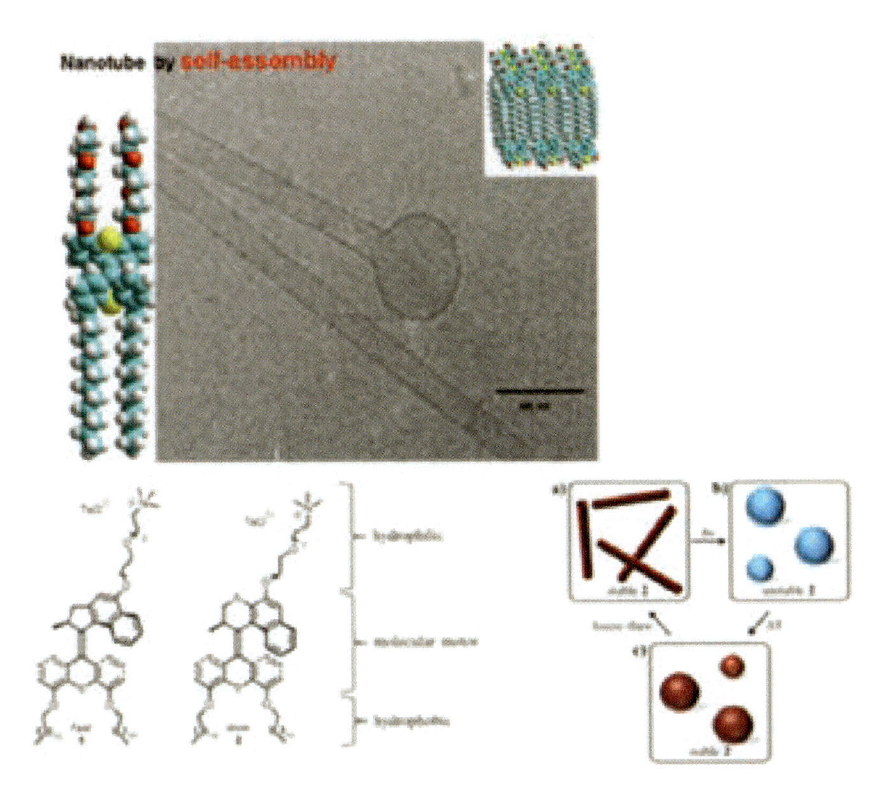

Figure 3. Light-responsive self-assembled nanoobjects. a) Bisthioxanthylidene-based amphiphiles that self-assemble into nanotubes and vesicle capped nanotubes. b) Overcrowded alkene based amphiphiles that can undergo nanotube to vesicle to vesicle to nanotube transitions.

light [35]. Slight structural modification of these photo-responsive amphiphiles resulted in bidirectional optical control of surface tension in Langmuir layers. [36] Taking this design a step further we have recently used overcrowded alkenes to achieve nanotube to vesicle to vesicle to nanotube transitions illustrating a more complex adaptive behavior as the system is responding to light and heath in a fully reversible behavior (Figure 3b) [37]. Small molecule gelators are another class of fascinating structures which we studied in the context of responsive self-assembly. For instance, bisamide based gelators with diarylethene photo-switchable core units allowed modulation between several distinct gel-states [19, 38]. An intriguing aspect of these light-responsive gels is the observation of metastable aggregates that are formed in a non-invasive manner (in response to irradiation with light) setting a stage for out-of-equilibrium assembly of soft materials. Embedding intrinsic switching functions in supramolecular systems and macromolecules will likely provide fascinating opportunities for responsive materials and smart surfaces for future applications such as drug delivery, cell growth or responsive coatings.

The construction of a nanovalve by which we might be able to control transport through artificial membranes or deliver on demand molecules from vesicles or other capsules was another appealing target in our program on molecular switches. Towards this goal we focused the mechanosensitive channel MsCl protein complex of large conductance from the cell membrane of E-Coli (Figure 4) [39]. This pentamer peptide system is sensitive to osmotic pressure opening a 3–4 nm pore allowing material to flow out of the cell, preventing cell damage. Using genetic modification five cysteine moieties were introduced at specific sites in the constriction zone of the protein complex and the thiol moieties enabled the attachment of photoswitches. After initial failures to achieve a proper response in the biohybrid system we focused on spiropyran photoswitches. The reasoning was that light-induced switching resulted in opening of the rigid spiropyran units to the zwitterionic and more flexible merocyanine form, simultaneously enhancing hydrophilicity. Electrostatic repulsion of the five zwitterionic units and the enhanced propensity to recruit water molecules near the constriction zone of the protein complex was anticipated to result in sufficient conformational change to open the pore of the MsCl protein complex. The successful incorporation of the spiropyran photochromic units and the proper functioning of the photo-switches in the modified MsCl protein were readily demonstrated, but it required extensive electrophysiology studies using patch-clamp techniques to establish photochemically induced opening and closing (using distinct wavelengths of light) of the MsCl nanopore. The critical test came with a system in which the photoresponsive MsCl hybrid was embedded in the membrane of a giant vesicle.

Calcein efflux measurements showed transport out of vesicles upon triggering with light and proper functioning of the modified MsCl as a photoresponsive nanovalve was demonstrated. Follow up studies focused on pH sensitive MsCl channels [40] and the incorporation of photoswitches in Sec-Y channels to control protein transport through membranes with light [41]. The ability to control molecular transport from capsules, such as the vesicles discussed here, through photoresponsive nanopores provides ample opportunities to design responsive systems for control drug delivery or self-healing materials.

Photopharmacology

Light offers superb opportunities as a noninvasive regulatory element in biological and biomedical applications. With a variety of molecular photoswitches available, a novel approach to control drug activity dynamically is within reach with the potential to bypass key issues associated with drug selectivity [20, 42, 43]. Light can be delivered with high spatial temporal precision, a key feature for tuning the action of bioactive molecules. It shows a high degree of orthogonality and usually low toxicity, which are attractive aspects in order to regulate

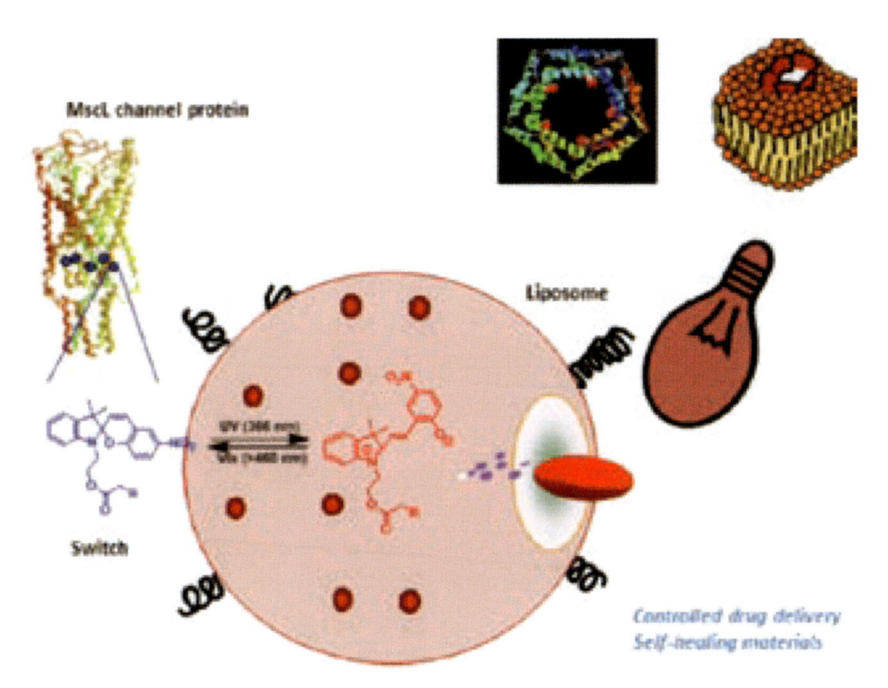

Figure 4. Giant vesicle with photoresponsive nanopore based on engineered MsCl protein complex with intrinsic spiropyran photochromic units as delivery system.

biological processes. By adjusting wavelength and intensity, switching processes can be readily controlled in a quantitative manner. The term photopharmacology was coined for this approach [42], as it is based on small molecule bioactive compounds with intrinsic photoswitchable functions; indeed, a drug that can be activated/deactivated with light (Figure 5). Of course a clear perspective on photopharmacology necessitates the realization that light-responsive molecules have seen extensive application in biomedicine Photodynamic therapy and the use of sophisticated fluorescence imaging techniques are now routine in the clinic while optogenetics, in particular for the control of neural functions, and photocleavable groups to activate prodrugs for precision therapy offer exciting opportunities.

Antibiotic resistance is an increasingly urgent global societal problem, with many strategies now being pursued to overcome and avoid it. The conceptual approach we took was to switch antibiotic activity on (cis-isomer) and off (trans-isomer) using light by incorporated azobenzene switching motifs in quinolone based broad spectrum antibiotics [44]. This design enabled the photoactivation of the responsive antibiotic and demonstrated in patterning of bacterial growth on plates using photomask techniques. The wealth of experience the organic photochromism community has built up over the last century is essential in such efforts, with rational tuning of the thermal stability of the cis-isomer through structural modifications to allow time taken to switch back to the off state to be controlled precisely. The proof of principle of light-activated antibiotics offers the prospect of enhanced efficacy by high-precision treatment at the point of infection and avoiding the harmful effects of antibiotics to beneficial bacteria in the organism. Arguably a more important possibility is that antibiotic activity is automatically switched off within a given time after treatment, providing an unconventional way to fight build-up of bacterial resistance towards antibiotics.

Having established the principle of photoswitchable antibiotics and applying this to patterning of bacterial growth using photomask techniques, we were excited by the prospect of non-invasive interference with bacterial communication [45]. Bacteria rely on communication through quorum sensing (QS) to synchronize the gene expression processes that are essential for, e.g., biofilm formation. We incorporated azobenzene photochromic units in N-acyl homoserine lactones, which are an important class of small molecule QS auto-inducers that play a role in the communication system of gram-negative bacteria. Two switchable QS molecules were identified that show opposite effects under UV irradiation in bioluminescence assays with E-coli; either gaining or losing QS activity upon trans-cis isomerization of the azobenzene unit. These compounds were also used to control the expression of virulence genes in Pseudomonas

Photopharmacology

0 ⇌ 1
(off) (on)

Triggering
activation with non-invasive light signal

Response
Spatio-temporal precision in action

Smart Pharmaceuticals

a)

Photocontrolled Antibiotic

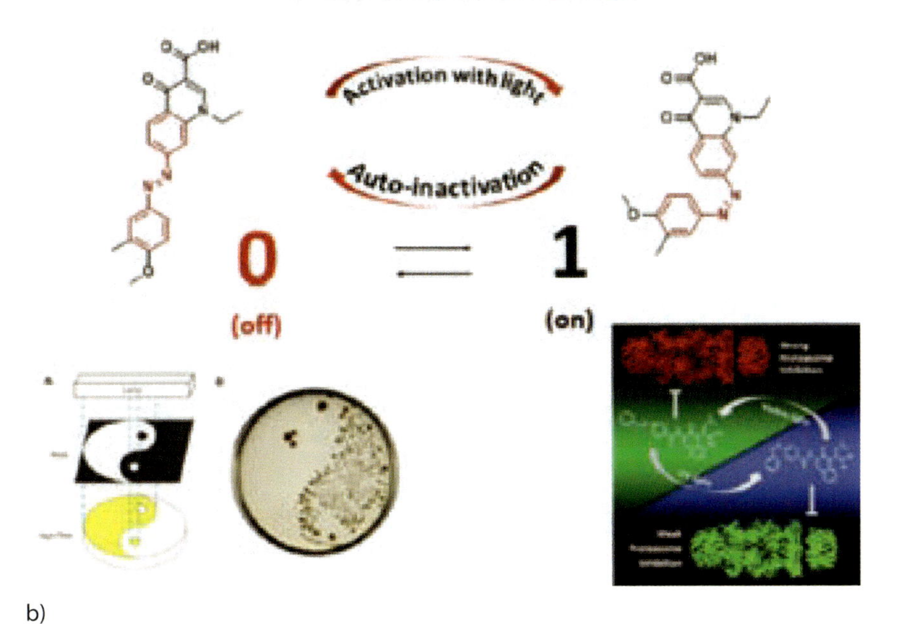

b)

Figure 5. a) Photopharmacology, on-off switching of the biological activity of a small molecule drug. b) A ciprofloxacin-based photoresponsive antibiotic. c) patterning of bacterial growth and photoresponsive analogue of Bortezomib® proteasome inhibitor.

aeruginosa by light. These findings offer a new approach to control bacterial growth and biofilm formation.

Photodynamic therapy has a long history in oncology, primarily through singlet oxygen generation strategies. We imagined that photoresponsive antitumor agents where the use of light is combined with molecular switching of drug activity could offer tremendous opportunities for precision therapy through control of drug function. As a proof of principle study, we focused on Bortezomib, a chemotherapeutic agent in clinical use, which was modified with an azobenzene motif [46]. The biological activity could be switched between strong (trans-isomer) and weak (cis-isomer) proteasome inhibition using UV and visible light, respectively. Instead of switching antitumor activity off with light, a much more desired function is on-switching of biological activity. This was realized with an azobenzene modified version of SAHA, a histone deacetylase (HDAC) inhibitor used in anti-cancer chemotherapy [47]. Here the photochemically accessible less-stable cis isomer is nearly as active (in vitro) as the clinically applied drug and it reverts to the inactive form, either by visible light irradiation or a thermal isomerization process, the rate of which can be controlled by molecular design. These approaches could provide unconventional solutions to mitigate the often severe side effects of commonly used chemotherapeutic agents. A particular attractive scenario is to directly use the information acquired by modern imaging techniques to guide the light activation of the switchable chemotherapeutic agent for high precision treatment of, e.g., inaccessible and small tumors. Of course, it should be emphasized that, prior to clinical use of such drug switching strategies, many hurdles need to be overcome.

We identified several of the challenges including high drug efficacy of photoresponsive analogs, drug delivery and most importantly the wavelengths of light that need to be applied i.e. irradiation with visible/near-infrared light is needed to avoid side effects and enable deep tissue penetration Recently several groups focused on the design of photoswitches that operate in the therapeutic window of interest in biomedical applications [48]. Using such principles we have designed potent photoswitchable mast cell inhibitors [49] while other groups have reported photoswitchable nociception, human carbonic anhydrase inhibition, cell division and control of neural processes among others, demonstrating the broad scope and potential of photopharmacology [42, 43, 50]. A next step in addressing future challenges and arriving at more effective medical therapies might be the design of more complex responsive systems in which sensing, transport and delivery and therapeutic action are combined and with multiple functions that can be addressed orthogonally with external stimuli. Recently, we have taken the first steps towards highly selective orthogonal control in multifunctional systems using photocleavable or photoswitchable groups [51]. It should be emphasized that there are ample opportunities to combine photochemical switches with various other switching functions. The noninvasive up- and down-regulation of competitive chemical and biological pathways in complex (bio-) molecular networks will open fascinating opportunities in chemical biology and the study of dynamic molecular systems [50].

MOLECULAR MOTORS

Our work on chiral overcrowded alkenes [23] and chiroptical molecular switches [24] paved the way for the discovery of the first light-driven unidirectional rotary motor [52]. See the time line in Figure 2. Chirality is central to function, and it is pertinent that a few lines are devoted to the magnificent phenomenon that is stereochemistry, which has fascinated me over my entire scientific career. Standing on the shoulders of the first Nobel Laureate in Chemistry, Jacobus van 't Hoff, who together with LeBel was a founding father of stereochemistry, and taking inspiration from scholars such as Cram, Mislow, Prelog, Wijnberg and Eliel, I was driven to explore chirality as a handle to control structure and function ranging from asymmetric catalysis to molecular machines. Here again Mother Nature sets the stage, with homochirality playing a central role in its essential molecules as emphasized by Albert Eschenmoser: "Chirality is a signature of life". To build a molecular rotary motor, the fundamental questions we were facing was how to induce rotary motion and how to control right- (clockwise) or left- (counter-clockwise) handed rotation at the nanoscale. The unique stereochemistry of the motor molecules allowed us to continue our exploration in the right direction.

First Generation Light-driven Rotary Motors

The first light-driven unidirectional rotary motor reported in 1999, shown in Figure 6 [6], has two distinct stereochemical elements: a helical structure (P or M helicity as in the chiroptical switches) and stereocenters (R or S) both in upper and lower halves [52]. The methyl substituents, originally introduced for the purpose of absolute stereochemical determination, can adopt a pseudo-axial or pseudo-equatorial orientation. Photochemical switching experiments revealed a surprising result; helix inversion as detected by CD spectroscopy was commonly associated with trans-cis isomerization in our chiroptical switches but in this case CD measurements indicated the same helicity for starting material and product. NMR, chiroptical and kinetic studies, supported by calculations, revealed "the missing isomer" and a sequential process of photoisomerization from stable trans to unstable cis followed by a thermal helix inversion to stable cis. We could show that the photochemically generated unstable cis isomer has the methyl groups in a sterically crowded pseudo-equatorial orientation and by helix inversion restoring the pseudo-axial orientation, strain is relieved. With this serendipitous discovery of a 180-degree unidirectional rotary process, based on energetically uphill photochemical alkene isomerization followed by an energetically downhill thermal helix inversion, we quickly realized that a full unidirectional rotary cycle was within reach by simply repeating the two-step process. The combination of four steps, two ultrafast photochemical steps [6, 53] each followed by a rate determining thermal step, add up to a 360-degree unidirectional rotary cycle that can be repeated many times. This system has all characteristics of a power-stroke rotary motor [6, 52]; rotary motion is achieved, fueled by light energy, shows control over directionality, and is a repetitive rotary process.

It is interesting to note here that the mechanism of the Anabaena sensory rhodopsin photoresponsive systems is closely related to that of our synthetic motor, as revealed recently by the team led by Olivucci [54]. Again, two olefin photoisomerizations and two thermal interconversions of helical conformations are involved in a four-step rotary cycle in this biological realization of a rotary molecular motor, emphasizing Nature's seemingly limitless number of elegant designs towards achieving complex functions. After our initial discovery, a large number of first generation rotary motors were synthesized in our group [55] in order to enhance rotary speed, shift absorption wavelengths into the visible

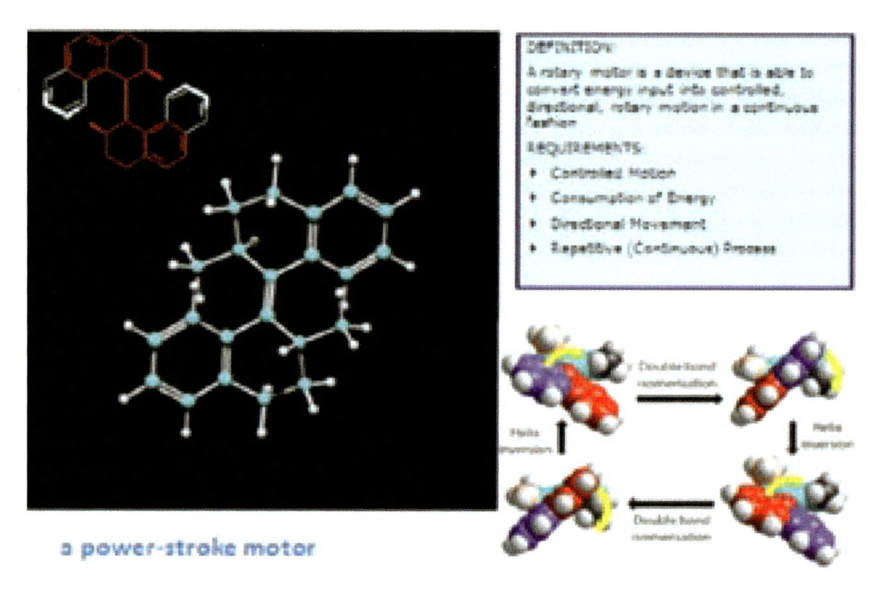

Figure 6. First generation light-driven rotary molecular motor and four stage rotary cycle.

region and attach functional groups [6, 56]. Through systematic change in steric parameters, especially by widening the "fjord region" to facilitate the rate determining thermal helix inversion and by changing the size of the substituents at the stereogenic centers, the rotary speed was enhanced from one cycle per hour to seconds. However, it should be noted at this stage that overall rotary speeds and efficiency of light-driven molecular motors are strongly dependent on parameters such as energy input, quantum yield, medium effects and surface confinement.

An important issue we were facing in view of potential application of these rotary motors controlling function is to what extent the medium and size will affect rotary behavior. A series of first generation motors with pendant rods of different lengths and flexibility were prepared and kinetic and thermodynamic parameters of the thermal isomerization processes determined [57]. These studies revealed that solvent viscosity is the dominant factor showing strong retardation for longer rigid arms. Analysis of the fraction of the molecule involved in the rotary process in terms of free volume model and solvent displacement shows a rather exceptionally high alpha factor for these motors. Extending these studies to excited state dynamics of the photochemical isomerization process, in cooperation with the teams led by Meech and Browne, confirmed that isomerization and relaxation to the ground state is largely polarity independent but governed by solvent viscosity [53].

Molecular motors are perfectly suited to drive far-from-equilibrium systems. Recently, we developed motor-driven responsive self-assembled helicates that can reconfigure between distinct supramolecular states. Taking inspiration from the self-assembled double-stranded copper helicates pioneered by Lehn, we have introduced functional rod like (oligo-)bipyridine ligands to the first-generation motors [58]. Upon copper(I) binding both monomer and oligomer copper helicates are obtained, and photochemical and thermal isomerization processes enable interconversion between different aggregation states and helicities in these complex dynamic assemblies.

Second Generation Light-driven Rotary Motors

As the two thermal isomerization steps in the first-generation motors typically have very distinct barriers, we designed a large series of second generation motors to achieve more uniform rotary behavior and to facilitate chemical modification [59]. A single stereocenter is present in the upper rotor half of these systems and, as in the first-generation motors, photochemical isomerization around the double bond axle generates an unstable isomer with the methyl-substituent in a higher energy-pseudo-equatorial conformation. Strain is released in the subsequent thermal isomerization, with the methyl group again adopting a favorable pseudo-axial orientation. It was highly rewarding and an essential point in our motor program, to establish that a single stereogenic center bearing a small methyl substituent is sufficient to govern a unidirectional rotary cycle feature, four helix inversion steps and four pseudo-enantiomeric states as revealed by NMR and CD spectroscopy. In the second-generation motor design, the lower stator half is derived from a symmetric (except for substituents) tricyclic unit, which offers distinct advantages. First, the barrier for helix inversion is nearly the same in both thermal steps of the rotary cycle, drastically reducing complexity in our efforts to accelerate overall rotation rates. Second, the inherent difference between stator and rotor facilitates selective functionalization, for instance, for surface assembly (see below). A third important aspect is that both rotor and stator parts can be synthesized independently, which proved especially important for the synthesis of complex (functional) motors. This also allowed the use of the Barton-Kellogg modification of a Staudinger diazo-thioketone olefination as the method of choice for the late stage introduction of the sterically demanding central double bond (rotary axle) in the total synthesis. Using various classes of second generation motors, a systematic structural variation was performed to elucidate parameters that govern rotary speed [60]. The example of fluorene-based second-generation motors is illustrative for the accelerations that can be achieved by modification of ring size and substituents resulting in, for instance, motor 9 with a half-life of 5.7 ms at room temperature (Figure 7).

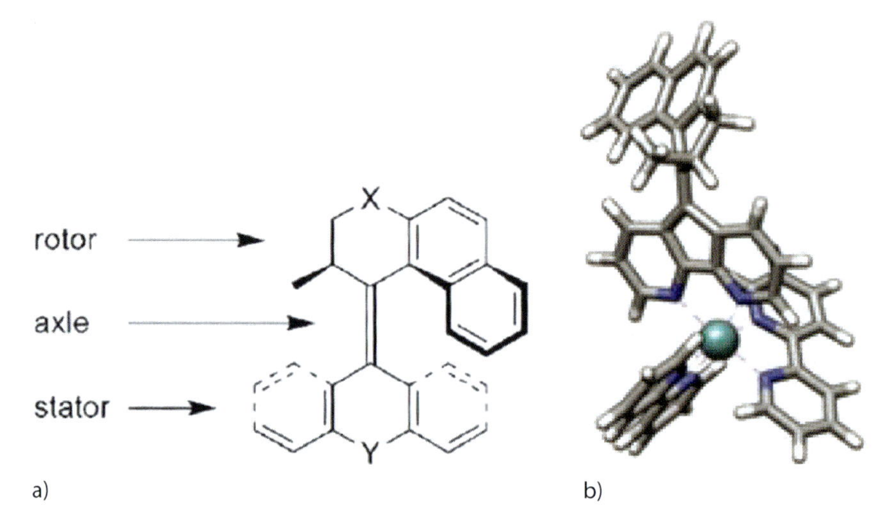

Figure 7. a) Second generation rotary molecular motor. b) Visible light driven Ru(II)-bipyridine based second generation motor.

Recently we introduced an alternative way to control the rotary speed of molecular motors by replacing the fluorene stator part by introducing a 4,5-diazafluorenyl-ligand moiety [61]. This allowed binding of metal ions of different sizes and as a consequence of metal-coordination the bond angles change as well as the barrier for thermal helix inversion. Fine tuning of rotary speed upon binding of metals of different sizes had the additional benefit that we can induce photoisomerisation with visible light.

A different approach to achieve visible light driven molecular motors was to use metallo-porphyrin sensitizers, including a Pd-porphyrin covalently attached as an antenna to the motor, taking advantage of inter- or intra-molecular energy transfer to drive rotary motion [62].

Dynamic Control of Function

We considered that a key next challenge in our motor program, on the way to molecular machine-like behavior, was how to dynamically control function and allow specific tasks to be performed. The structure of first- and second-generation motors is particular suited to the introduction of functional groups that allow, e.g., physical properties, distance, cooperativity and stereochemistry to be modulated in a directional and sequence controlled manner. An illustrative example of a responsive chiral catalyst based on a rotary motor is shown in Figure 8 [63], which was inspired by Jacobsen's chiral organocatalysts with

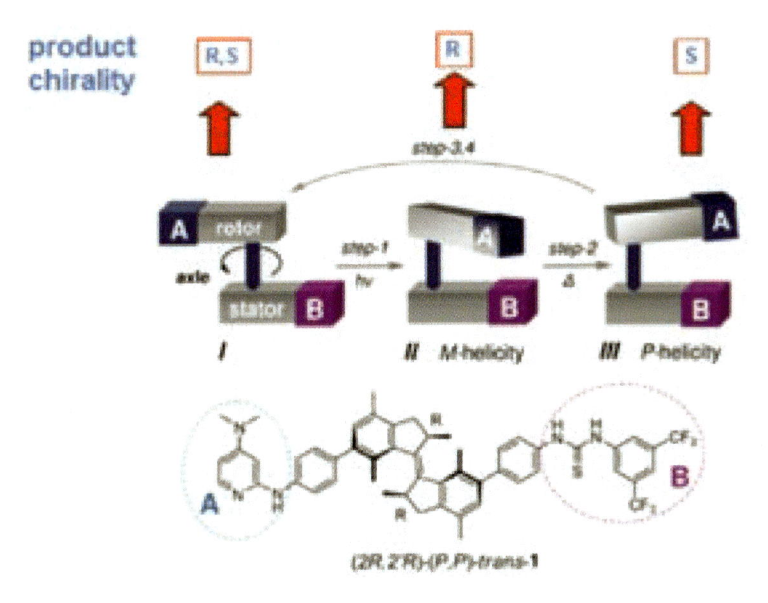

Figure 8. Dynamic control of chiral space in a molecular motor-based organocatalyst.

DMAP and thiourea moieties introduced in the trans isomer of a specific first-generation motor.

Here, the hydrogen donor and acceptor moieties do not cooperate effectively, resulting in low catalyst activity and a racemic product of a thiol 1,4 addition. Irradiation results in the formation of the cis-isomer with M-helicity and the catalytic moieties can cooperate. As a consequence, catalytic activity is dramatically enhanced as well as preferential formation of the R-product enantiomer. The next thermal step in the rotary cycle leads to cis-isomer with P-helicity and the S-enantiomer of product of the catalytic reaction. In this case the motor-based chiral organocatalyst functions as a multi-state switch, allowing not only the modulation of catalytic activity but also formation of racemic (R,S) or either enantiomer (R and S) in a sequence dependent manner. The sequence of events is strictly controlled by the clockwise or counterclockwise rotation of the motor unit. These concepts were subsequently extended in the design of responsive organocatalysts for asymmetric Michael and Henry reactions [64]. An important next step was the proof of concept of switchable chiral phosphines based on rotary motors as shown in highly enantioselective Pd-catalyzed desymmetrization reactions [65]. Again, depending on an external input signal (light or heat), distinct product stereoisomers are accessible with a single (responsive) catalyst. Bringing the principle of switchable chiral catalysts into the realm of transition metal catalysis opens many new avenues including multitasking and cascade transformations, adaptive and responsive behavior and ultimately up-down regulation of catalytic activity in complex catalytic networks. It should be noted that dynamic control of function is not limited to catalysis as we demonstrated for instance in modulation spin-spin interactions [66], and fluorescence [67], gel [68] and amyloid fiber formation [69], and chiral recognition and phosphate binding [70]. The recent demonstration of intramolecular cargo transport [70] and a variety of other mechanical tasks—elegantly shown by the Sauvage, Stoddart, Leigh, Guiseppone, Harada and Aida groups and others—illustrate the potential of molecular machine-like functions.

Motion at Different Length Scales

A major part of our research program on molecular motors has been devoted to the control, use and visualization of motion at different length scales (Figure 9). As is evident from the ATPase rotary motor embedded in the cell membrane and myosins moving along actin filaments, most biological motors operate at interfaces. We considered as a crucial step in the design of molecular devices based on rotary motors their assembly on surfaces and interfacing to macroscopic systems.

Second generation motors are particularly suited, as the stator part allows the introduction of various "legs" for surface anchoring, leaving the rotor part free to undergo light-driven rotary motion (Figure 9) [71]. Our initial attempts with short legs and thiol groups for self-assembly on Au failed due to quenching of the excited state isomerization pathways of the motor by the surface, but extending the legs with hydrocarbon moieties (lifting the motor from the surface) solved the problem. The presence of two legs prevented uncontrolled motion of the entire motor molecule, while sufficient conformational flexibility allowed uncompromised rotor movement. This design enabled self-assembly of rotary motors on Au nanoparticles and flat Au surfaces, resulting in our first "nanoscale windmill park" powered by light [72]. It was also the basis for several years of synthesis and surface science studies in order to design a variety of responsive interfaces. This included the assembly of motors in azimuthal and altitudinal orientations on quartz, Au, etc. and the anchoring with bis-, tris-, or tetrapodal-units to the surface to control rigidity, orientation with respect to the surface and spacing between individual motors on the surface. The surface bound motor shown in Figure 9 illustrates the concept elegantly; the tripodal anchoring, its size and the altitudinal orientation enables not only proper functioning of individual motors but also dynamic orientation of the hydrophobic perfluoroalkyl moiety towards or away from the surface. In this way photoresponsive behavior of the surface is readily achieved and precisely controlled, allowing both thickness and surface wettability to be modulated by light [73]. Currently we are investigating the rotary function of individual motors assembled on surfaces, using single molecule fluorescent techniques, to mimic the elegant experiments on visualization of rotation motion of the single ATPase protein motors [74].

Our next goal was the amplification of motion from the molecular to the mesoscopic and microscopic level. Overcrowded alkene-based rotary motors, due to their inherent dissymmetric structure and helical chirality, turned out to be excellent chiral dopants for nematic liquid crystal (LC) materials. Twisted nematic (cholesteric) LC films were obtained using small amounts (1 wt. %) of rotary motors and upon irradiation the change in helical chirality of the motors was amplified to induce dynamic changes in the supramolecular organization in the mesoscopic film as well as the surface structure at the LC-air interface. These responsive LC films allowed color change through the entire visible spectrum (color pixel formation) and rotation of micro-objects floating on its soft surface in a unidirectional sense when illuminated, resulting in an amplification over four orders of magnitude [75]. These discoveries marked a milestone in our motor research; for the first time, we observed the manifestation of autonomous rotary motion with the naked eye, induced by the dynamic function of a molecular rotary motor. It also laid the foundation for dynamic reorganization inside and at the surface of LC microdroplets triggered by light [76].

A second approach to amplify motion is via dynamic macromolecules with the perspective to design responsive and mechanical polymer materials i.e. fibers, networks, gels and films. For instance, amide-functionalized second-generation motors were applied as initiators in the polymerization of hexylisocyanate to provide a photoreponsive helical polymer [77]. Upon irradiation, the unidirectional rotary cycle of the single motor unit at the terminus of the polymer induces helix reversals in the polymer chain. This amplification of motion mimics a kind of flagellar function, while continuous irradiation drives the system to a steady state out-of-equilibrium. Large array surface patterning by self-assembly and responsive polymer LC films were obtained depending on the anchoring position of rotor and stator to the helical polymer chain. This design allows the transmission of motion and helical chirality over different length scales e.g. from the molecular, to macroscopic and finally mesoscopic hierarchical level. The use of rotary motors in polymer gel networks by Giuseppone is another elegant example showing the potential of molecular motors controlling mechanical functions in soft materials [78].

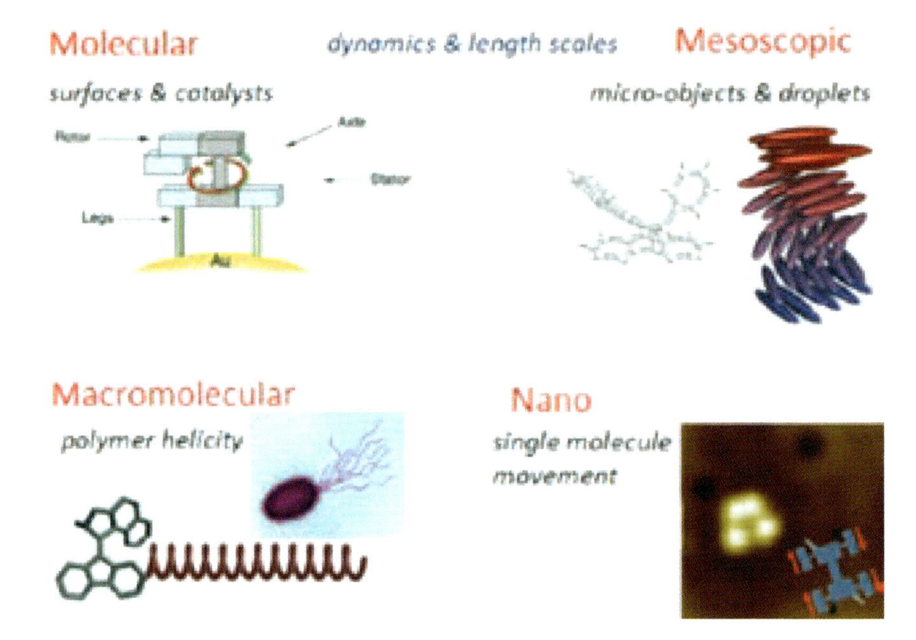

Figure 9. Control of motion across different length scales.

From rotary motion to translational motion

The idea of building a "four-wheel-drive molecular nanocar" started at the point where we were confronted with two fundamental questions; i) how to demonstrate single molecule motion? ii) How to convert rotary motion into translational motion? At the start of our lengthy journey, which ultimately resulted in the realization of a nanocar moving autonomous over a Cu surface, critical design features that we explored were a rather rigid frame with four second generation rotary motor units functioning as "wheels" [79]. We envisioned cooperativity of the motors which, due to their helical structure, also could

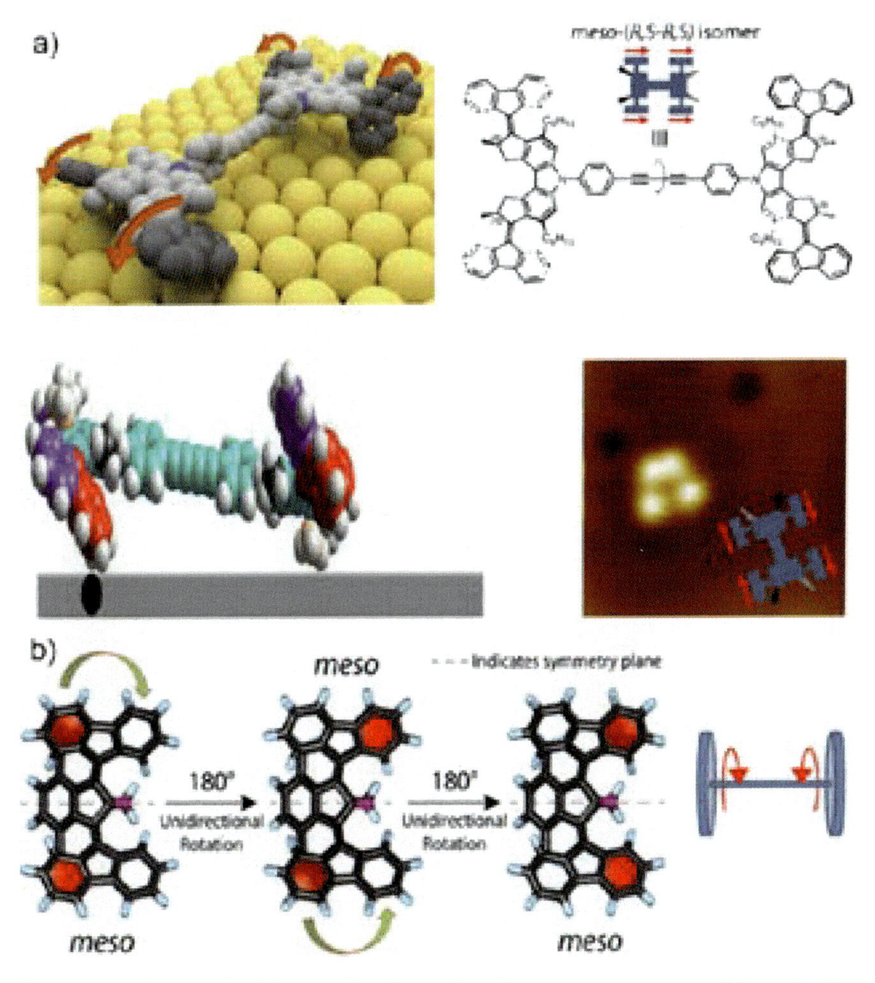

Figure 10. a) Four-wheel drive molecular car based on rotary motors; models, molecular structure and STM image. b) Third generation symmetric molecular motor.

lift the entire molecule a little from the surface, but sufficiently to overcome the strong adhesive interactions. In a combined effort with the Ernst group at EMPA Zurich, it was found that electrical excitation with an STM tip (at low temperature) of the meso-(R,S-R,S) isomer of the nanocar deposited on a Cu(1,1,1) surface induced propulsion over the surface along a more or less linear trajectory. Changing the stereochemistry of the "wheels", a single enantiomer of the nanocar was prepared with all the motor units having the same (R,R-R,R) chirality. Now the motion on the surface changed from more linear to random or rotary motion without significant translation in accordance with expectation on the basis of symmetry considerations (*see below*). It should be noted that molecular modeling indicates a "walking type" of motion for the nanocar reminiscent of the movement of kinesin proteins motors on actin filaments. Exploring these molecular propulsion systems, we demonstrated intrinsic motor function, cooperative action, autonomous movement on electrical excitation and control to some extent of directionality of movement at the single molecule level. With these findings, the stage is set for autonomous directional movement along tracks and cargo transport.

These results brought us also to another fundamental question: Is intrinsic molecular chirality needed to achieve unidirectional motion in a molecular rotary motor? To avoid an equal probability of clockwise and counterclockwise rotation around a single rotary axle connecting stator and rotor, our rotary motors rely on the chirality of the system [52, 59, 60]. It should be remembered that in a mechanically interlocked system, directionality in rotary motion has been achieved due to a specific sequence of chemical steps [80] while a non-symmetric environment can govern directionality in surface assembled rotors [81]. In the overcrowded alkene motors the directionality of rotary motion is controlled by point chirality as it dictates the thermodynamically preferred helical chirality. To guide our design of third generation motors we started with symmetry considerations of rotary motion at macroscopic length scales, e.g., the disrotary motion of two (car) wheels on an axle. [82]. The directionality of rotary motion from an observer at the symmetry plane is opposite (Figure 1A) while, despite the entire system being symmetric (C_s, with a mirror plane of symmetry), the rotary motion of the two wheels on an axle with respect to the surrounding is identical (e.g., both forward rotation for an external observer) enabling concerted rotary motion to induce directional linear motion. Translating these symmetry considerations to a stereochemical design featuring two integrated rotor moieties in a meso compound, we demonstrated that a symmetric (achiral) light-driven molecular motor is indeed feasible. The presence of a pseudo-asymmetric carbon atom bearing a methyl and fluor substituent, which proved to be of sufficiently different size to govern directionality, exclusive disrotary motion of two appending rotor moieties was achieved. Besides providing important insight in how to control nanoscale movement, these third-generation motors are particularly suited to build molecular dragsters and responsive materials.

Catalytic Motors and Propulsion Systems

Although our research started with light-induced switching and motion, inspired by the process of vision, part of our program has been devoted to catalytic motors and propulsion systems. Typically, biological motors such as ATPase, kinesin or bacterial flagella motors rely on catalysis, converting the chemical fuel ATP into kinetic energy. Proof of principle of a chemical driven rotary motor was demonstrated with the biaryl rotor system. The underlying dynamic stereochemical features are: First, hindered rotation in a tetrasubstituted biaryl prevents interconversion of enantiomers, although there is sufficient conformational freedom in the molecule to position ortho substituents at the two aryl units in proximity or remote from each other. Second there is a sufficiently low barrier for helical interconversion via a planar transition state of the lactone bridged biaryl. Using asymmetric CBS oxaborolidine catalyzed ring opening of the lactone as the key step governing > 90 % unidirectionality and a sequence of orthogonal (de-) protection steps a four stage unidirectional rotary cycle was accomplished [83].

Although not yet fully catalytic, an additional benefit of this system is that the direction of rotary motion can be reversed by simply switching the chirality of the catalyst. Recently we have extended these basic principles of control of dynamic stereochemistry in combination with chemical driven directional motion in a biaryl motor to a metal-mediated system [84]. The presence of both axial chirality and a stereogenic centre in combination with Pd(0), Pd(II) redox cycles enabled for the first time unidirectional rotary motion induced by sequential transition metal catalyzed conversions of chemical fuels. Autonomous translational motion based on the catalytic conversion of chemical fuels was also achieved. In contrast to the use of metal-based micro/nano-rods for hydrogen peroxide decomposition, as shown by Whitesides and others [85] to achieve autonomous propulsion, we followed a molecular approach. For instance, bimetallic Mn-catalysts were designed as functional mimics of the active site of catalase enzymes followed by covalent attachment of these catalysts to various microparticles, including polymers [86]. These supported catalysts enabled autonomous swimming motion of particles by converting hydrogen peroxide as a fuel. In a more elaborate design carbon nanotubes were covalently modified with two enzymes, catalase and glucose oxidase [87]. The concerted action of

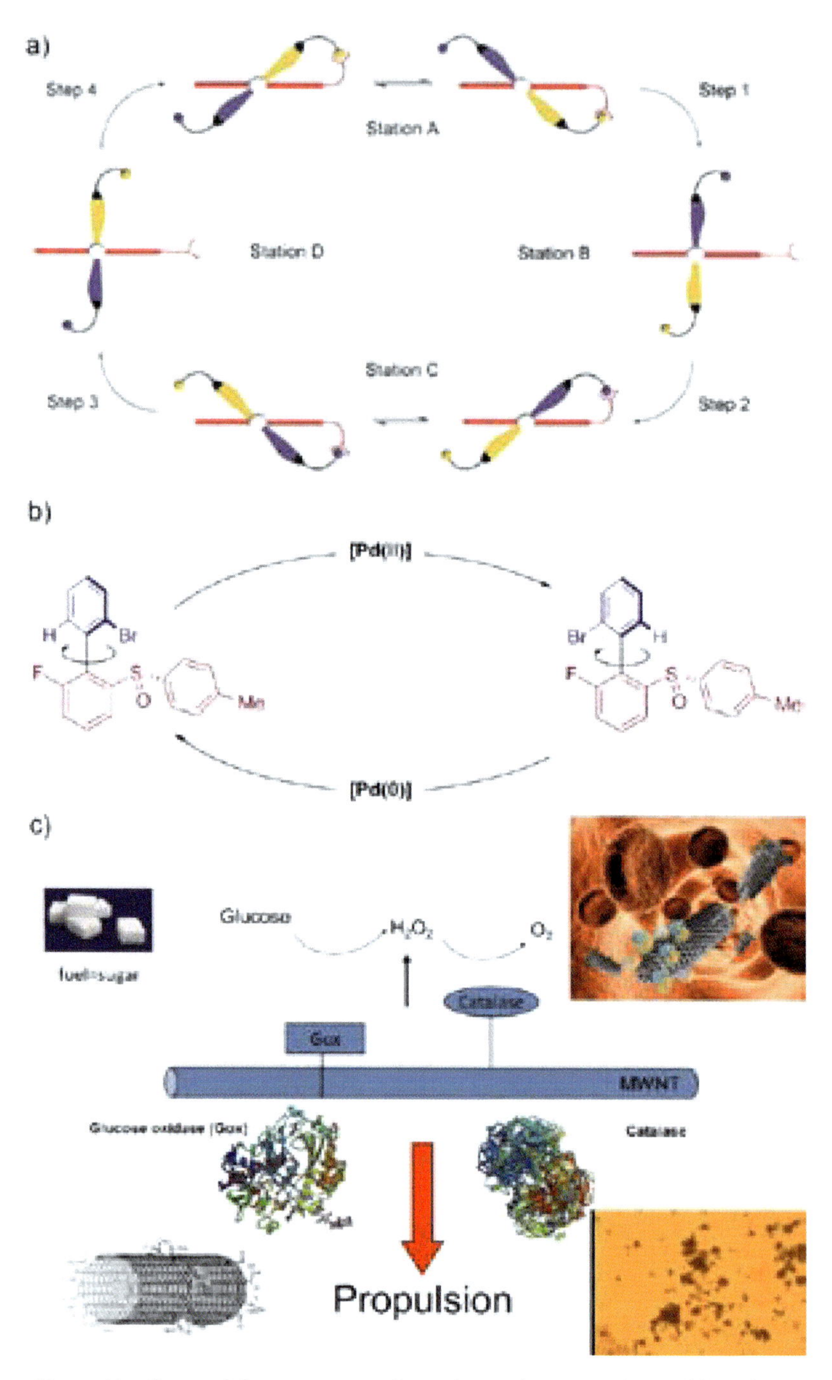

Figure 11. Chemical driven rotary and translational motion. a) Biaryl-based 4-step unidirectional rotary motor. b) Pd-mediated rotation in biaryl. c) Catalytic nanotube propulsion system powered by glucose.

these two enzymes, converting glucose and generating oxygen, induced autonomous movement of carbon nanotube aggregates in water, albeit with no control over directionality.

Although still rather remote from nanopropulsion systems carrying loads under physiological conditions in a highly controlled manner, our catalytic propulsion systems and related designs will likely guide the molecular motorist on a "fantastic voyage" in the world of autonomous operating molecular machines.

CONCLUDING REMARKS

The development of molecular motors arguably offers a fine starting point for the construction of soft robotics, smart materials and molecular machines. Our ability to design, use and control motor-like functions at the molecular level sets the stage for numerous dynamic molecular systems. Starting with the "synthesis of function", our focus was to program molecules by incorporating responsive and adaptive properties and being able to control motion. Molecular information systems, responsive materials, smart surfaces and coatings, self-healing materials, delivery systems, precision therapeutics, adaptive catalysts, roving sensors, soft robotics, nanoscale energy converters and molecular machines are just a small fraction of the systems where fascinating discoveries can be expected and where the ability to control dynamic functions will be essential. The practitioner of the art of building small will have to reach out to new levels of sophistication when dealing with complex dynamic molecular systems. In this endeavor, while trying to imagine the unimaginable, Nature's motors and machines can to some extent guide the molecular explorer. However, at the start of our next journey we should not forget the words of Leonardo da Vinci [89]: *"Where Nature finishes producing its own species man begins, with the help of Nature, to create an infinity of species."*

ACKNOWLEDGMENTS

I am extremely grateful to all the past and present group members that made our research possible. I had the good fortune to work with exceptionally talented undergraduates, PhD students and postdocs over the course of my career. The staff and colleagues in the Stratingh Institute for Chemistry at the University of Groningen, and all our collaborators in the Physics and Biology Departments, the University Medical Centre Groningen and Zernike Institute for Advanced Materials are gratefully acknowledged. I had the good fortune also to collaborate with great experts around the world in the supramolecular chemistry, molecular machines, catalysis and materials communities and I am grateful for their contributions. Finally, I would like to acknowledge the financial support from many granting organizations that kept our experimental programs moving forward over the past decades.

REFERENCES

1. (a) D. S. Goodsell, *Our Molecular Nature: The Body's Motors, Machines and Messages*, Copernicus, John Wiley & Sons: New York, 1996; (b) D.S. Goodsell, *The Machinery of Life, 2nd ed.*, Springer, New York, 2009.
2. J. M. Berg, J. L. Tymoczko, L. Stryer, *Biochemistry, 5th ed.*, New York, W. H. Freeman, 2002.
3. K. C. Nicolaou, E. J. Sorensen, *Classics in Total Synthesis: Targets, Strategies, Methods*, Wiley-VCH, Weinheim, 1996; E. J. Corey, X.-M. Cheng, *The Logic of Chemical Synthesis*, Wiley-VCH, Weinheim, 1995.
4. E. N. Jacobsen, A. Pfaltz, H. Yamamoto, *Comprehensive Asymmetric Catalysis*, Springer, Berlin, 1999.
5. (a) J.-M. Lehn, *Supramolecular Chemistry, Concepts and Perspectives*, Wiley-VCH, Weinheim, 1995; (b) K. Kinbara, T. Aida, *Chem. Rev.* 2005, **105**, 1377; (b) G. M. Whitesides, M. Boncheva, *Proc. Natl. Acad. Sci. USA* 2002, **99**, 4769; (c) T. Aida, E. W. Meijer, S. I. Stupp, *Science* 2012, **335**, 813. (d) J. W. Steed, P. A. Gale, *Supramolecular Chemistry: From Molecules to Nanomaterials, Vol.6: Supramolecular Materials Chemistry*, Wiley-VCH, Weinheim, 2012.
6. W. R. Browne, B. L. Feringa, *Nature Nanotechnol.* 2006, **1**, 25.
7. D. Mc Cullough, *The Wright Brothers*, Simon and Schuster, 2016.
8. M. Schliwa, *Molecular motors*, VCH-Wiley, Weinheim, 2006.
9. E. R. Kay, D. A. Leigh, F. Zerbetto, *Angew. Chem., Int. Ed.* 2007, **46**, 72.
10. R. D. Astumian, *Proc. Natl. Acad. Sci. USA* 2005, **102**, 1843.
11. (a) H. Iwamura, K. Mislow, K. *Acc. Chem. Res.* 1988, **21**, 175 (b) V. Balzani, M. Gomez-Lopez, J. F. Stoddart, *Acc. Chem. Res.* 1998, **31**, 405; (b) J. F. Stoddart, Ed. *Molecular Machines*, Special Issue. *Acc. Chem. Res.* 2000, **100**, 409 (c) V. Balzani, M. Venturi, A. Credi, *Molecular Devices and Machines—A Journey into the Nanoworld*; Wiley-VCH, Weinheim, 2003; (d) K. Kinbara, T. Aida, *Chem. Rev.* 2005, **105**, 1377; (e) V. Balzani, A. Credi, M. Venturi, *Chem. Soc. Rev.* 2009, **38**, 1542; (f) A. Coskun, M. Banaszak, R. D. Astumian, J. F. Stoddart, B. A. Grzybowski, *Chem. Soc. Rev.* 2012, **41**, 19; (g) J.-P. Sauvage, *Molecular Machines and Motors. Structure and Bonding*, Springer-Verlag Berlin Heidelberg, 2001; (h) G. S. Kottas, L. I. Clarke, D. Horinek, J. Michl, *Chem. Rev.* 2005, **105**, 1281; (h) B. L. Feringa, *J. Org. Chem.* 2007, **72**, 6635.
12. J. T. Mc Ilwain, *An Introduction to the Biology of Vision*, Cambridge, 1997.
13. B. L. Feringa, W. F. Jager, B. de Lange, Tetrahedron 1993, **49**, 8267.
14. Y. Hirshberg, *C. R. Acad. Sci.* 1950, **231**, 903.

15. H. G. Heller, In *Fulgides and Related Systems, CRC Handbook of Organic Photochemistry and Photobiology*, W. M. Horspool, P.-S. Song, Eds. Ch.13, Boca Raton, 1995.
16. (a) M. Irie, Photochromism: Memories and Switches, Special Issue. *Chem. Rev.* 2000, **100**, 1683; (b) M. Irie, T. Fukaminato, K. Matsuda, S. Kobatake, *Chem. Rev.* 2014, **114**, 12174.
17. (a) B. L. Feringa, Ed. *Molecular Switches*, Wiley-VCH, Weinheim, 2001; (b) *Molecular Switches*, W. R. Browne; B. L. Feringa, Eds, Wiley-VCH, Weinheim, 2011, vol. I, II.
18. (a) D. Dulic, S. J. van der Molen, T. Kudernac, H. T. Jonkman, J. J. D. de Jong, T. N. Bowden, J. van Esch, B. L. Feringa, B. J. van Wees, *Phys. Rev. Lett.* 2003, **91**, 207402; (b) T. Kudernac, N. Katsonis, W. R. Browne, B. L. Feringa, *J. Mater. Chem.* 2009, **19**, 7168.
19. J. J. D. de Jong, L. N. Lucas, R. M. Kellogg, J. H.van Esch, B. L. Feringa, *Science.* 2004, **304**, 278.
20. (a) W. Szymanski, J. M. Beierle, H. A. V. Kistemaker, W. A. Velema, B. L. Feringa, *Chem. Rev.* 2013, **13**, 6114; (b) W. Szymanski, D. Yilmaz, A. Kocer, B. L. Feringa, *Acc. Chem. Res.* 2013, **46**, 2910.
21. B. L. Feringa, *Acc. Chem. Res*, 2001, **34**, 504.
22. M. Sauer, J. Hofkens, J. Enderlein, *Handbook of Fluorescence Spectroscopy and Imaging*, VCH-Wiley, Weinheim, 2010.
23. B. L. Feringa, H. Wijnberg, *J. Am. Chem. Soc.* 1977, **99**, 602.
24. B. L. Feringa, W. F. Jager, B.de Lange, E. W. Meijer, *J. Am. Chem. Soc.* 1991, **113**, 5468.
25. N. P. M. Huck, W. F. Jager, B. de Lange, B. L. Feringa, *Science* 1996, **273**, 1686–1688.
26. A. M. Schoevaars, W. Kruizinga, R. W. J. Zijlstra, N. Veldman, A. L. Spek, B. L. Feringa, *J. Org. Chem.* 1997, **62**, 4943.
27. R. A. van Delden, N. P. M. Huck, J. J. Piet, J. M. Warman, S. C. J. Meskers, H. P. J. M. Dekkers, B. L. Feringa, *J. Am. Chem. Soc.* 2003, **125**, 15659; (b) N. P. M. Huck, B. L. Feringa, *Chem. Comm.* 1995, 1095.
28. (a) N. Harada, A. Saito, N. Koumura, H. Uda, B. de Lange, B; W.F. Jager, H. Wynberg, B. L. Feringa, *J. Am. Chem. Soc.* 1997, **119**, 7241; (b) N. Harada, A. Saito, N. Koumura, D. C. Roe, W. F. Jager, R. W. J. Zijlstra, B. de Lange, B. L. Feringa, *J. Am. Soc. Chem.* 1997, **119**, 7249; (c) N. Harada, N. Koumura, B. L. Feringa, *J. Am. Chem. Soc.* 1997, **119**, 7256; (d) E. M. Geertsema, A. Meetsma, B. L. Feringa, *Angew. Chem. Int. Ed.* 1999, **38**, 2738.
29. (a) W. Kuhn, E. Braun, *Naturwissenschaften* 1929, **17**, 227; (b) W. Kuhn, E. Knopf, *Naturwissenschaften* 1930, **18**, 183.
30. A. J. Kronemeijer, H. B. Akkerman, T. Kudernac, B. J. van Wees, B. L. Feringa, P. W. M. Blom, B. de Boer, *Adv. Mat.* 2008, **20**, 1467.
31. A. Nitzan, M.A. Ratner, *Science* 2003, **300**, 1384.
32. (a) S. J. van der Molen, J. Liao, T. Kudernac, J. S. Agustsson, L. Bernard, M. Calame, B. J. van Wees, B. L. Feringa, C. Schoenenberger, *Nano Lett.* 2009, **9**, 76; (b) Arramel, T. Pijper, T. Kudernac, N. Katsonis, M. van der Maas, B. L. Feringa, B. J. van Wees, *Nanoscale* 2013, **5**, 9277.
33. J. E. Green, J. W. Choi, A. Boukai, Y. Bunimovich, E. Johnston-Halperin, E. DeIonno, Y. Luo, B. A. Sheriff, K. Xu, Y. S. Shin, H.-R. Tseng, J. F. Stoddart, J. R. Heath, *Nature* 2007, **445**, 414.

34. (a) S. Kumar, J. T. van Herpt, R. Y. N. Gengler, B. L. Feringa, P. Rudolf, R. C. Chiechi, *J. Am. Chem. Soc.* 2016, **138**, 12519; (b) W. R. Browne, B. L. Feringa, *Ann. Rev. Phys. Chem.* 2009, **60**, 407; (c) J. Areephong, T. Kudernac, J. J. D. de Jong, G. T. Carroll, D. Pantarotti, J. Hjelm, W.R. Browne, B. L. Feringa, *J. Am. Chem. Soc.* 2008, **130**, 12850.
35. A. C. Coleman, J. M. Beierle, M. C. A. Stuart, B. Maciá, G. Caroli, J. T. Mika, D. J. van Dijken, J. Chen, W. R. Browne, B. L. Feringa, *Nat. Nanotech.* 2011, **6**, 547.
36. J. Cheng, P. Štacko, P. Rudolf, R. Y. N. Gengler, B. L. Feringa, *Angew. Chem. Int. Ed.* 2017, **56**, 291.
37. D. J. van Dijken, J. Chen, M. C. A. Stuart, L. Hou, B.L. Feringa, *J. Am. Chem. Soc.* 2016, **138**, 660.
38. (a) J. H. van Esch, B. L. Feringa, *Angew. Chem Int. Ed.* 2000, **39**, 2263; (b) J. T. van Herpt, M.C.A. Stuart, W. R. Browne, B. L. Feringa, *Chem. Eur. J.* 2014, **20** (11), 3077; (c) J. J. D. de Jong, T. D. Tiemersma-Wegman, J. H. van Esch, B. L. Feringa, *J. Am. Chem. Soc.* 2005, **127**, 13804; (d) M. de Loos, B. L. Feringa, J. H. van Esch, *Eur. J. Org. Chem.* 2005, 3615; (e) D. J. van Dijken, J. M. Beierle, M. C. A. Stuart, W. Szymanski, W. R. Browne, B. L. Feringa, *Angew. Chem. Int. Ed.* 2014, **53**, 5073.
39. A. Kocer, M. Walko, W. Meijberg, B. L. Feringa, *Science* 2005, **309**, 755.
40. A. Kocer, M. Walko, E. Bulten, E. Halza, B. L. Feringa, W. Meijberg, *Angew. Chem. Int. Ed.* 2006, **45**, 3126.
41. F. Bonardi, E. Halza, M. Walko, F. Du Plessis,N. Nouwen, B. L. Feringa, A. J. M. Driessen, *Proc. Natl. Acad. Sci. USA* 2011, **108**, 7775.
42. W.A. Velema, W. Szymanski, B.L. Feringa, *J. Am. Chem. Soc.* 2014, **136** (6), 2178–2191.
43. J. Broichhagen, J. A. Frank, D. Trauner, *Acc. Chem. Res.* 2015, **48**, 1947.
44. (a) W. A. Velema, J. P. van der Berg, M. J. Hansen, W. Szymanski, A. J. M. Driessen, B. L. Feringa, *Nature Chem.* 2013, **5**, 924; (b) W. A. Velema, J. P. van der Berg, W. Szymanski, A. J. M. Driessen, B. L. Feringa, *Acs Chem. Biol.* 2014, **9**, (9), 1969; (c) W. A. Velema, M. J. Hansen, M. M. Lerch, A. J. M. Driessen, W. Szymanski, B. L. Feringa, *Bioconj. Chem.* 2015, **26**, 2592.
45. J. P. Van der Berg, W. A. Velema, W. Szymanski, A. J. M. Driessen, B. L. Feringa, *Chem. Sci.* 2015, **6**, 3593.
46. M. J. Hansen, W. A. Velema, G.-J. de Bruin, H. S. Overkleeft, W. Szymanski, B. L. Feringa, *ChemBioChem.* 2014, **15**, 2053.
47. W. Szymanski, M. E. Ourailidou, W. A. Velema, F. J. Dekker, B. L. Feringa, *Chem. Eur. J.* 2015, **21**, 16517.
48. (a) A. A. Beharry, O. Sadovski, G. A. J. Woolley, *J. Am. Chem. Soc.* 2011, **133**, 19684; (b) D. J. Bléger Schwarz, A. M. Brouwer, S. Hecht. *J. Am. Chem. Soc.* 2012, **134**, 20597; (c) D. B. Konrad, J. A. Frank, D. Trauner, *Chem. Eur. J.* 2016, **22**, 4364; (d) M. M. Lerch, S. J. Wezenberg, W. Szymanski, B. L. Feringa, *J. Am. Chem. Soc.* 2016, **138**, 6433; (e) M. J. Hansen, M. M. Lerch, W. Szymanski, B. L. Feringa, *Angew. Chem.Int. Ed.* 2016, 13514.
49. W. A. Velema, M. van der Toorn, W. Szymanski, B. L. Feringa, *J. Med. Chem.* 2013, **56**, 4456.
50. M. M. Lerch, M. J. Hansen, G. M. van Dam, W. Szymanski, B. L. Feringa, *Angew. Chem. Int. Ed.* 2016, **55**, 10978.
51. M. J. Hansen, W.A. Velema, M. M. Lerch, W. Szymanski, B. L. Feringa, *Chem. Soc. Rev.* 2015, **44**, 3358.

52. N. Koumura, R. W. J. Zijlstra, R.A. van Delden, N. Harada, B. L. Feringa, *Nature* 1999, **401**, 152.
53. (a) M. Klok, N. Boyle, M. T. Pryce, A. Meetsma, W. R. Browne, B. L. Feringa, *J. Am. Chem. Soc.* 2008, **130**, 10484; (b) J. Conyard, K. Addison, I. A. Heisler, A. Cnossen, W. R. Browne, B. L. Feringa, S. R. Meech, *Nature Chem.* 2012, **4**, 547.
54. A. Strambi, B. Durbeej, N. Ferré, M. Olivucci, *Proc Natl Acad Sci USA* 2010, **107**, 21322.
55. (a) M. K.J. ter Wiel, R. A. van Delden, A. Meetsma, B. L. Feringa, *J. Am. Chem. Soc.* 2005, **127**, 14208; (b) D. Pijper, R. A. van Delden, A. Meetsma, B. L. Feringa, *J. Am. Chem. Soc.* 2005, **127**, 17612.
56. M. K. J. ter Wiel, B. L. Feringa, *Synthesis-Stuttgart* 2005, 1789.
57. J. Chen, J. C. M. Kistemaker, J. Robertus, B. L. Feringa, *J. Am. Chem. Soc.* 2014, **136**, 14924.
58. D. Zhao, T. van Leeuwen, J. Cheng, B. L. Feringa, *Nature Chem.* 2017, **9**, 250.
59. N. Koumura, E. M. Geertsema, A. Meetsma, B. L. Feringa, *J. Am. Chem. Soc.* 2000, **122**, 12005.
60. (a) J. Bauer, L. Hou, J. C. M.Kistemaker, B. L. Feringa, *J. Org. Chem.* 2014, **79**, 4446; (b) N. Koumura, E. M. Geertsema, M. B. van Gelder, A. Meetsma, B. L. Feringa, *J. Am. Chem. Soc.* 2002, **124**, 5037; (c) M. K. J. ter Wiel, R. A. van Delden, A. Meetsma; B. L. Feringa, *J. Am. Chem. Soc.* 2003, **125**, 15076; (d) J.Vicario, M. Walko, A. Meetsma, B. L. Feringa, *J. Am. Chem. Soc.* 2006, **128**, 5127; (e) N. Ruangsupapichat, M. M. Pollard, S. R. Harutyunyan, B. L. Feringa, *Nature Chem.* 2011, **3**, 53.
61. (a) A. Faulkner, T. van Leeuwen, B. L. Feringa, S. J. Wezenberg, *J. Am. Chem. Soc.* 2016, **138**, 13597; (b) S. J. Wezenberg, K.-Y. Chen, B. L. Feringa, *Angew. Chem. Int. Ed.* 2015, **54**, 11457.
62. A. Cnossen, L. Hou, M. M. Pollard, P. V. Wesenhagen, W. R. Browne, B. L. Feringa, *J. Am. Chem. Soc.* 2012, **134**, 17613.
63. J. Wang, B. L. Feringa, *Science* 2011, **331**, 1429.
64. D. Zhao, T. M. Neubauer, B. L. Feringa, *Nature Comm.* 2015, **6**, 6652.
65. M. Vlatković, B. S. L. Collins, B. L. Feringa, *Chem. Eur. J.* 2016, 17080.
66. J. Wang, L. Hou, W. R. Browne, B. L. Feringa, *J. Am. Chem. Soc.* 2011, **133**, 8162.
67. W. R. Browne, M. M. Pollard, B. de Lange, A. Meetsma, B. L. Feringa, *J. Am. Chem. Soc.* 2006, **128**, 12412.
68. G. T. Carroll, M. G. M. Jongejan, D. Pijper, B. L. Feringa, *Chem. Sci.* 2010, **1**, 469.
69. C. Poloni, M. C. A. Stuart, P. van der Meulen, W. Szymanski, B. L. Feringa, *Chem. Sci.* 2015, **12**, 7311.
70. M. Vlatković, B. L. Feringa, S. J. Wezenberg, *Angew. Chem. Int. Ed.* 2016, **55**, 1001.
71. J. Chen, S. J. Wezenberg, B. L. Feringa, *Chem. Comm.* 2016, **52**, 6765.
72. (a) J. Vachon, G. T. Carroll, M. M. Pollard, E. M. Mes, A. M. Brouwer, B. L. Feringa, *Photochem. Photobiol. Sci.* 2014, **13**, 241; (b) G. T. Carroll, M. M. Pollard, R. A. van Delden, B. L. Feringa, *Chem. Sci.* 2010, **1**, 97; (c) G. London, G. T. Carroll, T. Landaluce, T. Fernandez; M. M. Pollard, P. Rudolf, B. L. Feringa, *Chem. Comm.* 2009, 1712; (d) M. M. Pollard, M. K. J. ter Wiel, R. A. van Delden, J. Vicario, N. Koumura, C. R. van den Brom, A. Meetsma, B. L. Feringa, *Chem. Eur. J.* 2008, **14**, 11610.
73. R. A. van Delden, M. K. J. ter Wiel, M. M. Pollard, J. Vicario, N. Koumura, B. L. Feringa, *Nature* 2005, **437**, 1337.

74. (a) K.-Y. Chen, O. Ivashenko, G.T. Carroll, J. Robertus, J. C. M. Kistemaker, G. London, W.R. Browne, P. Rudolf, B. L. Feringa, *J. Am. Chem. Soc.* 2014, **136**, 3219; (b) G. London, K.-Y. Chen, G. T. Carroll, B. L. Feringa, *Chem. Eur. J.* 2013, **19**, 10690.
75. H. Noji, R. Yasuda, M. Yoshida, K. Kinosita Jr., *Nature* 1997, **386**, 299.
76. R. Eelkema, M. M. Pollard, N. Katsonis, J. Vicario, D. J. Broer, B. L. Feringa, *J. Am. Chem. Soc.* 2006, **128**, 14397.
77. A. Bosco, M. G. Jongejan, R. Eelkema, N. Katsonis, E. Ernmanuelle, A. Ferrarini, B. L. Feringa, *J. Am. Chem. Soc.* 2008, **130**, 14615.
78. D. Pijper, B. L. Feringa, *Angew. Chem. Int. Ed.* 2007, **46**, 3693.
79. Q. Li, G. Fuks, E. Moulin, M. Maaloum, M. Rawiso, I. Kulic, J. T. Foy, N. Giuseppone, *Nature. Nanotechnol.* 2015, **10**, 161.
80. T. Kudernac, N. Ruangsupapichat, M. Parschau, B. Macia, N. Katsonis, S. R. Harutyunyan, K.-H., Ernst, B. L. Feringa, *Nature* 2011, **479**, 208.
81. D. A. Leigh, J. K. Y. Wong, F. Dehez, F. Zerbetto, *Nature* 2003, **424**, 174.
82. H. L. Tierney, C. J. Murphy, A. D. Jewell, A. E. Baber, E. V. Iski, H. Y. Khodaverdian, A. F. McGuire, N. Klebanov, E. C. Sykes, *Nat. Nanotechnol.* 2011, **6**, 625.
83. J. C. M. Kistemaker, P. Štacko, J. Visser, B.L. Feringa, *Nature Chem.* 2015, **7**, 890.
84. S. P. Fletcher, F. Dumur, M. M. Pollard, B. L. Feringa, *Science* 2005, **310**, 80.
85. B. S. L. Collins, J. C. M. Kistemaker, E. Otten, B. L. Feringa, *Nature Chem.* 2016, **8**, 860.
86. (a) R. F. Ismagilov, A. Schwartz, N. Bowden, G. M. Whitesides, *Angew. Chem. Int. Ed.* 2002, **41**, 653; (b) T. E. Mallouk, A. Sen, "Powering nanorobots," *Scientific American* 2009, 72.
87. (a) N. Heureux, F. Lusitani, W. R.Browne, M. S. Pshenichnikov, P. H. M. van Loosdrecht, B. L. Feringa, *Small* 2008, **4**, 476; (b) C. Stock, N. Heureux, W. R. Browne, B. L. Feringa, *Chem. Eur. J.* 2008, **14**, 3146.
88. D. Pantarotto, W. R. Browne, B. L. Feringa, *Chem. Comm.* 2008, 1533.
89. M. White, *Leonardo: The First Scientist*, St. Martin's Griffin, 2001.

Chemistry 2017

Jacques Dubochet, Joachim Frank and Richard Henderson

"for developing cryo-electron microscopy for the high-resolution structure determination of biomolecules in solution"

The Nobel Prize in Chemistry, 2017

Presentation speech by Professor Peter Brzezinski, Member of the Royal Swedish Academy of Sciences; Member of the Nobel Committee for Chemistry, 10 December 2017.

Your Majesties, Your Royal Highnesses, Esteemed Nobel Laureates, Ladies and Gentlemen,

What lovely music we have been hearing flowing from the choir loft. In our ears, electrical signals are being generated, nerve cells are activated, and molecules causing pleasure are released and bound to their receptors. These molecules are conducted inside the cells, out of reach of our senses. Even so, these processes allow us to hear the music.

A long-held dream would be fulfilled if we were able to dive into a cell and see the molecules. The world of molecules is the core of what we call life. But molecules are small. If we were able to enlarge your ear sensory organ to the size of this hall, hundreds of protein molecules would fit into each comma in the programme you are holding in your hand.

The electron microscope allows us to "see" these molecules – and even their components: the atoms. This instrument was first invented nearly 100 years ago. But it would take many decades before it could be used to study biological molecules, since they tended to dry up in the microscope's vacuum and be burned by the electrons used to visualize the samples.

To examine biological samples in an electron microscope, they need to be frozen. Molecules in our cells, however, are surrounded by water that forms ice crystals when it freezes. These crystals are beautiful to look at, but the electrons lose their way in the crystal labyrinth and the images come out black.

Freezing water without forming ice crystals was thought to be impossible. Jacques Dubochet moved to Heidelberg to try to do just this. Out of pure curiosity, he wanted to do something that was considered impossible, something that even lacked practical applications. How would one write a grant proposal for a project of this type? Who would be willing to finance such a project?

In the early 1980s, when Dubochet flash froze water at minus 200 degrees Celsius in cold ethane, he gazed at the impossible. The water was both liquid and solid at the same time. We call this state a glass. Dubochet had thus created a window into the world of molecules.

At around the same time, Richard Henderson was studying a microorganism living in salt-saturated lakes in the desert. This organism has small, well-organised solar collectors made from protein molecules on its surface. In 1990, after having sought out the best electron microscopes in the world, Henderson was able to show the structure of these solar collectors at atomic resolution for the first time. In doing so, Henderson showed that the microscope could in principle provide detailed structures of the thousands of molecules found in our cells.

But finding these molecules in microscope images can be compared to finding a shadow of a bird in the weak glow of the moon. Joachim Frank succeeded in developing methods that made it possible to find these faint molecular shadows. The challenge was then to determine how the molecules are positioned in relation to each other. After more than 10 years work, Frank finally was able to present sharp, three-dimensional models created from thousands of these faint molecular shadows.

In recent years, advances have come quickly. The scientific literature describes the progress as an ongoing revolution. Those of you in the front rows need not be concerned, however. Within science, revolutions are international in nature, they come through collaboration between individuals, and the results confer benefit to mankind.

Some say that it is a miracle that we now, thanks to the contributions of the Laureates, can look into ourselves, thereby tricking our senses. But to quote a famous chemist, "miracles sometimes occur, but one has to work terribly hard for them."

Jacques Dubochet, Joachim Frank and Richard Henderson:

Your work has led to the development of cryo-electron microscopy for the high-resolution structure determination of biomolecules in solution. That is a truly great achievement. On behalf of the Royal Swedish Academy of Sciences I wish to convey to you our warmest congratulations. May I now ask you to step forward and receive your Nobel Prizes from the hands of His Majesty the King.

Jacques Dubochet. © Nobel Prize Outreach AB. Photo: A. Mahmoud

Jacques Dubochet

Biography

MY CURRICULUM VITAE has been on my personal page of the University of Lausanne's website for a long time. Few people had looked at it so far. Suddenly, with the news of the Nobel Prize, it became a worldwide buzz almost overnight. All of this because people found it, let's say, "unusual". But why is it so unusual? Of course, there is little place for creativity in a resume sent to apply for a position in some political institution or international firms – but why not being a little bit imaginative when presenting yourself on your own, personal web page?

This old CV has been rejuvenated to fit recent developments and has been enriched with commentaries. It is presented below.

OCTOBER 1941
Conceived by optimistic parents.
This was a bad time. The Germans were approaching Moscow. Switzerland was encircled by countries under the Nazi or Fascist regime. My father, a civil engineer, was building fortifications for the army. My mother Liliane was taking care of my sister Michèle, 3, and my brother Emmanuel, 2.

BORN JUNE 8, 1942

1946
No longer scared of the dark, because the sun comes back; it was Copernicus who explained this.

To make it simple – too simple almost – two solutions were offered to me: prayers with my Protestant mother or logical explanation from my atheist father. As time passed, the second option seemed more and more alluring.

1948–1955
1st part of an experimental scientific career in Wallis and Lausanne (instruments: knives, needles, strings, matches).

My father was building a dam, high in the mountain. We were living in a small village where electricity was recently brought in. At school, there were two classes for the boys, each with a wooden stove in the middle. The good boys were allowed to sit close to the stove, while the bad boys had to sit by the window. Since we were the engineer's children, our place was by the stove of course! We spent the six-month-long summer holidays in a chalet further up, closer to Dad's work. We had no electricity, and no shops close by. Rye bread was getting hard after a few weeks. There was a big rock, too big for me – but not so big as I realized when I came back as an adult – on which my brother and sister were spending hours climbing and playing. There were thousands of other adventure grounds and experimenting places all around and down by the river.
Then we went to the big city of Sion and to the even larger capital of Lausanne where I had to find my way – with difficulty – through a more standard education system. I succeeded somehow in passing the college examination (normally passed at 11, but I was already one year late).

1955
First official dyslexic in the canton of Vaud – this licensed me to be bad at everything ... and allowed me to understand those with difficulties.

It didn't take long for my parents to find out that my grades were not promising, but they noticed that my spelling mistakes – as those of my brother - were unusual. They drew the attention of the college's director to this. He decided to take the case further and this is how I became the first recognized dyslexic child of the Canton. This meant that I was allowed to pass from one class to the next in spite of more and more catastrophic grades. This was a bad time. From being bad in spelling I soon became very bad in everything, because dyslexia was my "laziness pillow". Not completely though; following the instructions of the book by Jean Texereau, I was building a 15 cm aperture telescope. My handwork teacher spent more time helping me than he spent with all my classmates put together. The college director retired shortly before I reached the end of the compulsory school program. It didn't take long until I was dismissed. Still optimistic, and creative, my parents sent me to the boarding school of Kantonschule Trogen, deep in Swiss-German speaking central Switzerland. The message was clear: either I move on, or I get stuck. One year later, the German teacher asked me to give a talk to the class. I spoke about rockets, and it was good. I knew I was on my way to becoming a scientist. And that was the end of the central-Switzerland episode.

1962
Federal maturity exam.

After the salutary shake-up in Trogen, my parents sent me to a private school in Lausanne where I could prepare the examination for entering Uni-

versity. It was a time of intense catching up. I am still surprised by how much a teenager or young adult can learn when he is motivated. My cultural background of poetry, music, history, and geography is still strong – but it's not as much about language and spelling. The maturity examination went well.

Shy and polite, but socially unskilled, I gained preliminary social experience in homes for disabled children where my sister – a work therapist – brought me during the holidays. Then it was the military service. I still have nightmares from this time, but I benefitted there from meeting regular human beings. I became an officer, even though I wasn't exactly fit for the job.

1967

Physicist-engineer at EPUL, with the intention to become a biologist.

I wanted to understand more about the world, the living world in particular, and to become a scientist. It was a time during which Physics were shaping Biology. Watson, Crick, Kendrew and Perutz had won their Nobel Prizes. Quite obviously, I chose to study Physics at EPUL, École Polytechnique de l'Université de Lausanne (now federalized as EPFL), where my father had studied Civil Engineering. I found calculus difficult during the first year and, contrary to some of my admired classmates, I never became a skilled mathematician. Nevertheless, I tremendously enjoyed everything I learned. I felt more and more at home in Physics, mostly thanks to my professor Jean-Pierre Borel and to the three volumes of Feynman's "Lectures on Physics".

During my second year I went up to my preferred professor and I asked him for advice. "Where shall I go for a PhD in Biology when I am finished with my diploma?" He had the answer: "Prof. Édouard Kellenberger, at the laboratory for Biophysics in Geneva." So there I went. Édouard was very friendly and he offered me a position as a doctoral assistant. "Oh, not so fast," I replied, "I have 3 more years to go with my studies in Lausanne." "OK, come back in 3 years." Three years later, I was there again. In the meantime, Édouard had been to the States and married Cornelia, and he had forgotten me. I got the doctoral assistant position anyway.

The Laboratory for Biophysics at the University of Geneva was a remarkable place (Strasser, 2006) – one of those in which Molecular Biology was introduced in Europe. Science was practiced there in a most enthusiastic, creative, and open way. Mountain touring and climbing the Salève were the only limitations on the long working hours in Biology courses and in the lab with my electron microscope – an old RCA EMU2.

1968

Very important.

Then came the student revolution. We couldn't escape. We didn't. Unprepared, I played along the game of being politically active in the midst of big turmoil. We were left-oriented of course, but our group 2002 (that was its name) was not along the general line. We had a strong involvement in environmental protection. I cherish the memory of the moment when, having climbed high on a pole to plaster a poster against a car exhibition, I saw, down below on the street, two smiling policemen waiting for me to come down. That stunt cost me a major part of my meager salary.

A friend, more committed to the revolution than me, gave up his studies and rejected his family. His father, a banker driving a big black car, told me – perhaps because I still looked a bit reasonable – "Don't worry, he will soon become normal again." I told myself, "For sure, I'll never be 'normal again' as he means it".

1969

Certificate of Molecular Biology in Geneva to become a biophysicist. Began to study electron microscopy of DNA, which remains my main topic.

My diploma in Physics didn't bring me much in Biology. The certificate was designed to bridge the gap in order to form this new kind of scientist: the biophysicist. Namely: those who are biologists but with the spirit of a physicist. I took courses with Biology students and, more importantly, I discovered the strange way of living of those dedicated to the observation of natural life. With them, I woke up at dawn for bird watching and digging the soil to count earthworms.

1973

Thesis in biophysics at Geneva and Basel with Édouard Kellenberger who taught me Biophysics, ethical responsibility and durable friendship.

Édouard Kellenberger was called from Geneva to lead the final construction and early operation of the new Biocenter at the University of Basel. He took with him a group of colleagues and students. Most of us were still politically active. My bias was still towards environmental protection and durability, but the work in the laboratory was my major activity. I became the first Philosophy II graduate from the Biocenter with a PhD entitled "Contribution to dark-field electron microscopy". In fact, dark field was a minor part of the PhD and the conclusion was that it is not very useful for biological observation. However, I learned how to operate an electron microscope and a lot about the strange behavior of matter at small dimension.

1970–1976

Very classic psychoanalysis.

As it should be, my affective life was quite intense during my psychoanalysis. Toward the end of this period, I met Christine. Our second encounter was during a manifestation against a planned nuclear power plant near Basel (the plant was never built). Christine is an art historian from Basel and Paris.

She was teaching art at school. We settled in together and got married when she decided to move with me to Heidelberg.

What did I get from the unreasonable effort of a Freudian psychoanalysis? I asked myself this question, walking along the Rhine after my last session. The answer I gave to myself was "I don't know yet, but in ten years' time I will come back to this". Ten years later, I thought the decision was pretty good. Ten more years later, I thought it was very good. At present I do believe that it was the best decision of my life, together with the other one – living with Christine.

1978

Group leader at EMBL (Heidelberg); how to deal with water in electron microscopy. Discovery of water vitrification and development of electron cryo-microscopy.

The newly formed European Molecular Biology Laboratory, hidden in a beautiful forest above the old city of Heidelberg, was a kind of paradise for research. John Kendrew, the initiator of the laboratory and first General Director, appointed a host of young scientists with ambitious projects. Everything was arranged for us to work freely under the best conditions, with the sole expectation of producing knowledge of significance. My project consisted in learning how to deal with water in electron cryo-microscopy. It didn't start well but we have been lucky for the rest. The story has been told elsewhere (Dubochet, 2011).

At this time, we were living in a small village in a vineyard south of Heidelberg. Christine gave birth to a boy, Gilles, and 18 months later to a girl, Lucy. I was used to working early in the morning and coming back in the middle of the afternoon. I had the opportunity of participating closely in family life. We also had a good group of parents sharing the care of the children as well as their education. It was a great time!

1987

Professor at the University of Lausanne (UNIL), Department of Ultrastructural Analysis.

I was among the lucky few who had a permanent contract at EMBL. Nevertheless, I was attracted by teaching and I doubted that I could be creative all my remaining professional life in pure research only. I didn't hesitate to accept the offer for a professorship in Lausanne, which involved the management of the well-established Electron Microscopy Center with its service duty, and the chance to install a brand new Laboratory for Ultrastructural Analysis where I could pursue my own research under favorable conditions.

During the 20 years as professor in Lausanne, I also had the chance to extend my research work in the field of science and society. We developed a compulsory curriculum whose aim was to make sure that our students are as good citizens as they are good biologists.

1998

President of the Biology section with the chance to perform this assignment with Nicole Galland and Pierre Hainard, and to live at a moment when interesting things were happening in Biology in Lausanne.

Yes, interesting things indeed. This was the time when a major rearrangement took place between UNIL and EPFL. The principle was simple. At that time, Biology was the exclusivity of UNIL but departments of Mathematics, Physics and Chemistry existed both at UNIL and EPFL. This seemed unreasonable. It was decided to concentrate these three activities exclusively at EPFL and to reinforce Biology accordingly at UNIL. The continuation was more complicated. I discovered what real politics were. The result was, indeed, the move of Mathematics, Physics and Chemistry to EPFL but, in an unexpected twist, EPFL also developed a strong department of Life Sciences and, at UNIL, what was left of the Faculty of Sciences merged with Medicine into the new Faculty of Biology and Medicine. The result is probably better than the original plan, but what a stir it all was!

2002

End of the assignment. Sabbatical in Australia, Germany and Paris.

2004–2007

Maturation of CEMOVIS (cryo-electron microscopy of vitreous sections).

The success of electron cryo-microscopy relies on the observation of very thin specimens, in the sub-μm range. This is even too thin for the observation of a single normal cell, without speaking of a tissue or of a complex organism. From the start, our electron cryo-microscopy project included the observation of bulky specimens. For that aim, the strategy consists in vitrifying a volume as large as possible and then cutting it into vitreous sections that can be directly observed in the electron cryo-microscope. The method faces a number of difficulties that we summarized with the acronym SIVEMCATOR (Al-Amoudi, Studer and Dubochet, 2004) which, for some, is the symbolic expression of the hopeless task that I imposed on a number of my collaborators. I think they are wrong. The need for electron cryo-microscopy of bulky specimens is obvious and CEMOVIS is the most direct avenue to solve it. My guess is that the success of the thin film vitrification method applied to macromolecular complexes or small organels has depleted the group of those ready to accept the most challenging task of studying large objects. This will change. The future of CEMOVIS is bright.

2007

June Retirement Colloquium.

2007 =>

Host of the Department of Ecology and Evolution. Science and Society for the elderly.

Retirement in Swiss universities is compulsory at age 65. Some try to find a solution to continue the work they are trained for and good at. I thought that, with a bit of luck, 65 years would not prove to be so old. Statistically, it leaves you with about 20 years of creative life. I decided to cultivate my 4 "S". The first S stands for Self, taking good care of oneself. The second S is for Social, living together. I started teaching mathematics – that is 2+3 – to young migrants; the effort broadened and I went into politics in my small city and, back like in the old days, to the movement for environmental protection. The 3d S stands for Science, because I love it. I have the chance to keep my mind on it, through direct contact with my colleagues at the university, where they generously left me an office desk. The last S means Service, because the fruits of the quince tree are better as marmalade than rotting on the ground, and because dishes must be placed in the dishwasher. My sister gave me the advice to devote the first year of retirement to learning this new job. At the end of the year, I found that a second year of training was necessary. Ten years later, the work is still in progress.

The children are grown up. We have a son-in-law from India. They are all working for the common good or development help. They have not yet made us grandparents, even if we are active members of the association "Grandparents for the climate" (https://www.gpclimat.ch/fr/).

OCTOBER 4, 2017

Ouch! A Nobel Prize

Christine says: "It's a good thing for us that you got it late and that you had 10 years of retirement to broaden your scope."

BIBLIOGRAPHY

Al-Amoudi, A., Studer, D., & Dubochet, J. (2005). "Cutting artefacts and cutting process in vitreous sections for cryo-electron microscopy" *J Struct Biol*, 150(1), 109–121.

Dubochet, J. (2011). "Cryo-EM – The first thirty years" *J Microscopy*, 245(3), 221–224.

Strasser, B. J. (2006). *La fabrique d'une nouvelle science: La biologie moléculaire à l'âge atomique (1945–1964)*. Leo S. Olschki, Geneva.

Early Cryo-Electron Microscopy

Nobel Lecture, December 8, 2017 by Jacques Dubochet
University of Lausanne, Lausanne, Switzerland.

IN THIS VERY SPECIAL moment my first feeling is thankfulness. It goes out to my late parents, to my family that I have the pleasure of seeing in the audience, to my scientific colleagues, to whom I owe being here, to my friends, and to all those who contributed to making me who I am, here and now. I want to pay a special tribute to Prof. Édouard Kellenberger, my "doctor father" and lasting friend, who taught me how to be a scientist and passed onto me the sense of responsibility that should be associated with this profession. My special thanks also go out to Sir John Kendrew, first General Director of the European Molecular Biology Laboratory (EMBL) in Heidelberg, who gave me the chance to conduct, under ideal conditions, our ambitious project on water in electron cryo-microscopy (ecm).

WHY ELECTRON CRYO-MICROSCOPY?

Like any living organisms, we are a bag of water, formed from billions of cells which all are small bags of water. Since air is not transparent to electrons, an electron microscope must operate under vacuum – which means that any observed biological specimen must be dry. This is not good. The original structure can't be preserved in these conditions. When water is removed, floating molecules stick to each other. Skilled microscopists know how to minimize the damage, but they will never prevent some forms of aggregation since "fishes never fly". Even objects supposed

Figure 1. Prof. Édouard Kellenberger, Sir John Kendrew.

to be resistant, like bacteriophage T4, look terrible when they are dried on a solid surface without particular precautions (figure 2a). For decades, electron microscopists have invented methods improving the structural preservation of every possible dry specimen. Negative staining has proved especially effective, as the subtle details visible on the micrograph demonstrate (figure 2b). Freeze-drying is a bit more complicated, but also has its advantages (Figure 2c). Nevertheless, it is obvious that the head of the virus does not look healthy.

Since the end of the 60s, I was among those working hard to find better methods for preparing and observing delicate biological specimens. At

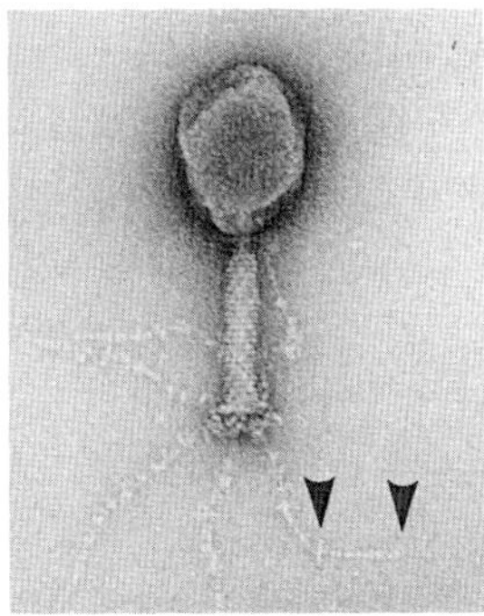

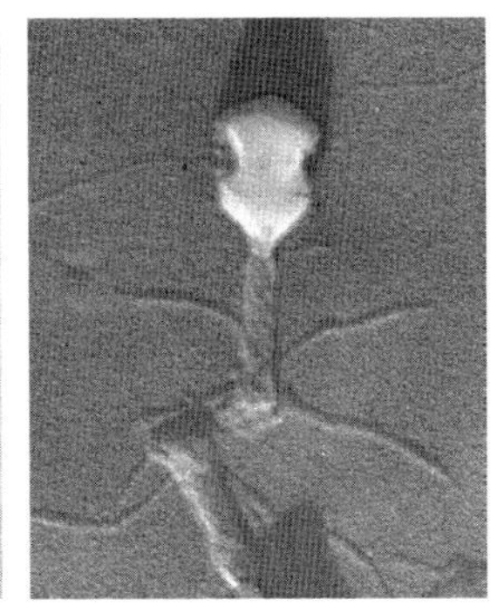

Figure 2. T4 bacteriophages prepared by different methods. a) Direct drying. b) Negative staining; the specimen is in a solution of heavy metal salt that forms a protective coat around the particle when water evaporates. c) Freeze-drying; a thin layer of suspension is frozen on a supporting film. Ice is removed by sublimation under vacuum.

the time, my hero was Nigel Unwin. I was impressed by his creativity and skill. Beside the similarity of our research's direction, I discovered – having been invited in his home the first time I was in Cambridge – under the bed, a self-made telescope of the same type that the one I had built myself in my late adolescence. Would the Nobel Prize by-law allow four persons to share the prize, I do believe that we would be standing here together. In the 70s, Nigel had a brilliant idea. He realized that drying a biological specimen in a heavy metal salt was not the best environment for preserving a delicate structure; friendlier surroundings would be better. He tried to do it in sugar. It worked. Of course, the contrast in sugar is much lower than in metal, but Nigel realized that contrast is not the limitation – that is the signal-to-noise ratio. This can be improved by means of methods used in X-ray crystallography, taking advantage of the redundancy of the information in a crystal. He joined forces with his friend and colleague Richard Henderson, an experienced crystallographer, and together in 1975 they solved the first 3-dimensional structure of a membrane protein (Henderson and Unwin, 1975).

Bob Gleaser is another person of great importance for me. I worked in his steps for a good part of my PhD, and it was a micrograph he published in 1976 (Taylor and Glaeser, 1976) that redirected my working plans. It was a sample of broken bacteria containing a rich collection of their various substructures. Some of them were also subjects of our own research. This micrograph was special because it was a thin frozen layer of the aqueous sample observed at −170°C in a specially cooled specimen holder. The specimen was in ice and the biological material was more beautiful than anything I had ever seen before. I was immediately convinced that cold water was the future. Two years later, Sir John Kendrew offered me a position as group leader at the newly formed EMBL for a project entitled "How to Deal with Water in Electron Cryo-Microscopy". As it has been explained elsewhere (Dubochet, 2011), it didn't start well at all. But the continuation proved to be better.

The problem with water is that it crystallizes into ice when it is cooled at a temperature in which it does not evaporate in the vacuum of the electron microscope. So, we had to learn more about water, cooling, freezing, and observing. We tried everything we could think of, and learned from all our predecessors in the field. As it turns out, we started experimenting with the sophisticated machine presented in Figure 3 (a copy of it is presently exposed in the Nobel Museum). On the right, not visible, is a nebulizer throwing a stream of microdroplets of water through a small slit in the cupboard. The mobile tweezer, above the dewar filled with liquid nitrogen at −188°C, is holding a grid covered with a thin specimen supporting film. We let the tweezer fall and the grid, having harvested some droplets while crossing the stream, is immediately frozen. The frozen ice

Figure 3. The apparatus for freezing water microdroplets.

droplet has the characteristic aspect shown in Figure 4a. One day, my colleague Alasdair McDowall (Figure 4b, inset) decided to place a little beaker in the liquid nitrogen dewar and condense in it liquid ethane, because it was known that it is a better coolant than liquid nitrogen. He called me to the microscope, something unexpected was there (Figure 4b). It was a "frozen" microdroplet; it was not ice, it was amorphous. We didn't know what it was. We let the specimen warm up slowly – at that moment it was at about –160°C – hoping that the evaporation of the droplet would tell us more about its nature. Suddenly, at –135°C, in a few seconds, the droplet turned into a multi-crystal of a substance we knew well from previous experiments. It was cubic ice, a form of ice which is typically formed by condensation of water vapor at low temperature. The conclusion was obvious. We were seeing ice originating from an amorphous substance: it was vitreous water. I told Alasdair: "Aha! We have something great!"

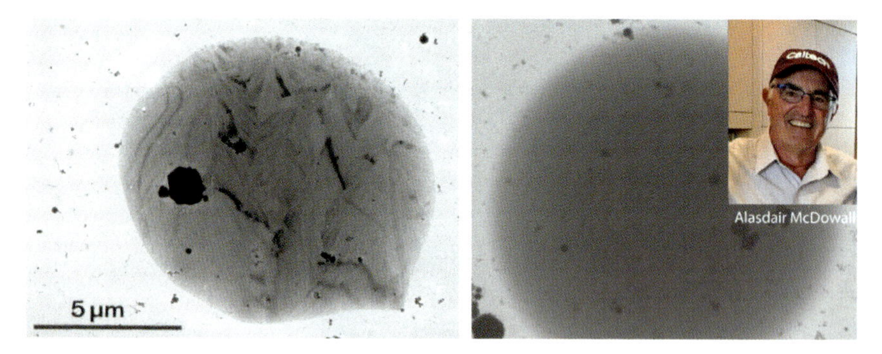

Figure 4. a. Frozen ice microdroplet. b. Vitrified water microdroplet.

The trouble was that vitrification of liquid water should have been impossible. This was demonstrated on solid thermodynamics grounds over decades of previous work. Basile Luyet, acknowledged father of cryobiology and Catholic priest in the congregation of St-François de Sales was among the major contributors to this body of work. I like Basile Luyet because of his strange combination of strict Catholic faith with uncompromised scientific mindset. I also like him because he was born in Savièse, a village in the Wallis Alps of Switzerland to which he remained attached all his life. It is only a few kilometers away from the village of Nendaz where I spent my first school years, at a time when understanding things of nature became important to me.

As a consequence of the accepted impossibility of vitrifying water, the report of our observation was rejected from publication. The editor was doubly wrong.

Firstly, because at the very moment our article was rejected, the same journal had in press the article of an Austrian group demonstrating that vitreous water can be obtained by rapid cooling of the liquid. Their experiment was similar to ours but they used X-ray diffraction to demonstrate the nature of the observed substance and its transformation into ice upon warming (Mayer and Bruggeler, 1980).

Secondly, they were also wrong because – as we reported – vitrification is rapid, reproducible and easily repeated.

So why is it possible to obtain vitreous water when it should be impossible? Is this one more illustration that science sometimes fails? Not so fast! The work of Luyet and of those of the field is solid, and thermodynamics should not be taken lightly. At present, we still don't really understand the nature of the vitreous water we observe. We know that it is not simply immobilized liquid water, but some other form of amorphous solid. The science of water still has shadowed regions. I can imagine that, when the light comes, it will have consequences on a larger scale – for biology also. For now, we are pleased to observe that biological objects vitrified by rapid cooling seem to be indiscernible from those floating in good bona fide liquid water. Electron cryo-microscopists feel safe in their knowledge for now, but they prudently keep an eye on the real nature of vitreous water.

Knowing how to vitrify a droplet of water is one thing; preparing a biological sample for biological observation is another. The major problem comes from the high surface tension of liquid water, which makes water droplets spherical. Spreading a thin layer of liquid on a supporting film requires that the interaction's energy between the drop and the surface must exactly compensate the surface tension of water. I was an expert on how to treat supporting film for optimal wetting. We were combining this knowledge with our newly acquired competence in vitrification. At that time, late Dr. Marc Adrian (Figure 5 inset), a French microscopist of great culture and strong mind, had joined the group. He didn't like our subtle and poorly reproducible spreading procedures. He wanted to get rid of the supporting film altogether. I tried to discourage him, but Marc was not one to easily give up. A while later, he came up with the kind of image shown on Figure 5. It was a vitrified layer of a Semliki Forest Virus (SFV) suspension stretched over the 18µm holes of a grid. There is no supporting film, just a thin layer of suspension with perfectly preserved virus floating immobilized in their vitrified aqueous medium. Indeed, the ideal specimen for electron microscopy observation. Adrian's method is simple. The grid is held on a tweezer, itself mounted on a plunger. A drop of suspension is put on the grid and most of it is then sucked away with a blotting paper. This takes about one second. The surprising thing is that the last fraction of a micrometer takes another full second before it breaks and vanishes. This leaves ample time for the operator to liberate the plunger and let the grid fall freely into an ethane beaker some 10 cm

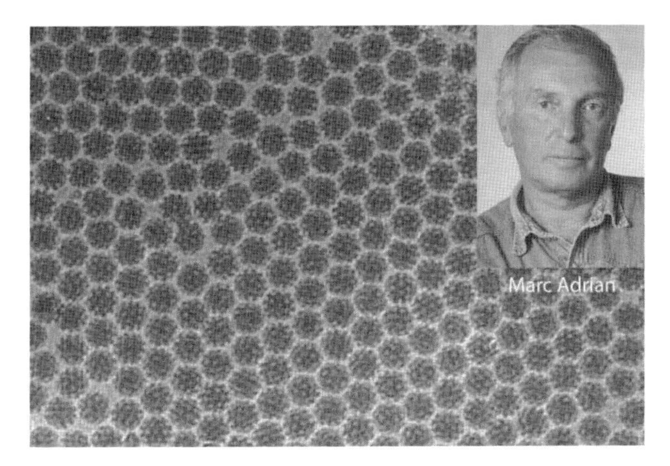

Figure 5. Unsupported thin film of vitrified suspension of SFV. Inset: Marc Adrian

below. The preparation is simple, it only takes a few seconds and it is easily reproducible. It came as no surprise that the rumor of this elegant preparation method spread rapidly. It was a great time in our laboratory. Water specialists, some of them quite incredulous, came to observe the strange phenomena of vitrification. We learned a lot from them. Electron microscopist colleagues wanted to adopt the method. The first electron cryo-microscopy course was organized. The result was broadly published (Adrian et al., 1984). One day, I got a phone call from Pierre-Gilles de Gennes, the world leader on entangled polymers and spreading viscous fluids. One of his books was a difficult read I kept on coming back to, time and time again. That early morning, he was teasing me. "I am sure that you do not know why your thin layer can survive the final step of the preparation!" He was right. He explained that, in order to break, the two surfaces of the thin layer must fuse together and entropy prevents that – for a moment. He could even articulate a number: one second.

More than thirty years later, the basic principle of Adrian's method is still being used, unchanged. The bare grid was soon abandoned for a grid covered with a film with μm-sized holes. The biological suspension is then stretched through the holes. Nowadays, only the older generation is still using a manual plunger, as full automatic devices are making cryo-specimen preparation simpler and more reproducible. But democratization has a price.

Our results were soon published. When he saw the micrographs of his pet adenovirus in unprecedented details, Lennart Philipson, who succeeded Sir John Kendrew as General Director of the European Molecular Biology Laboratory (EMBL), became convinced that this project by physicists about water was valuable and so it gained his full support. We joined forces with R.H. Vogel and S.W. Provencher, specialists in 3-d reconstruction from 2-d images, and in 1986 we published a 3-dimensional model of the SFV at 35 Å resolution (Vogel et al., 1986) (Figure 6).

Building on previous work and continuously improved by the creative efforts of many scientists – the long-lasting efforts of Joachim Frank for 3-dimensional reconstruction of single particles were of seminal importance – electron cryo-microscopy progressed smoothly throughout the years. The thirty-five Ångstrom of 1986 was well and good but some specialists in X-ray diffraction – for long the dominant method for molecular structure determination – jokingly invented the word "blobology" to describe our work. Thirty years later, 3.5 Å is achieved on a nearly routine basis. Who could imagine this in the 80s? Richard Henderson was perhaps the only one who had this clear vision. He worked continuously to make it become real.

Three and a half Å is certainly not an impressive number for non-specialists. Nevertheless, everyone can understand that the resolution was

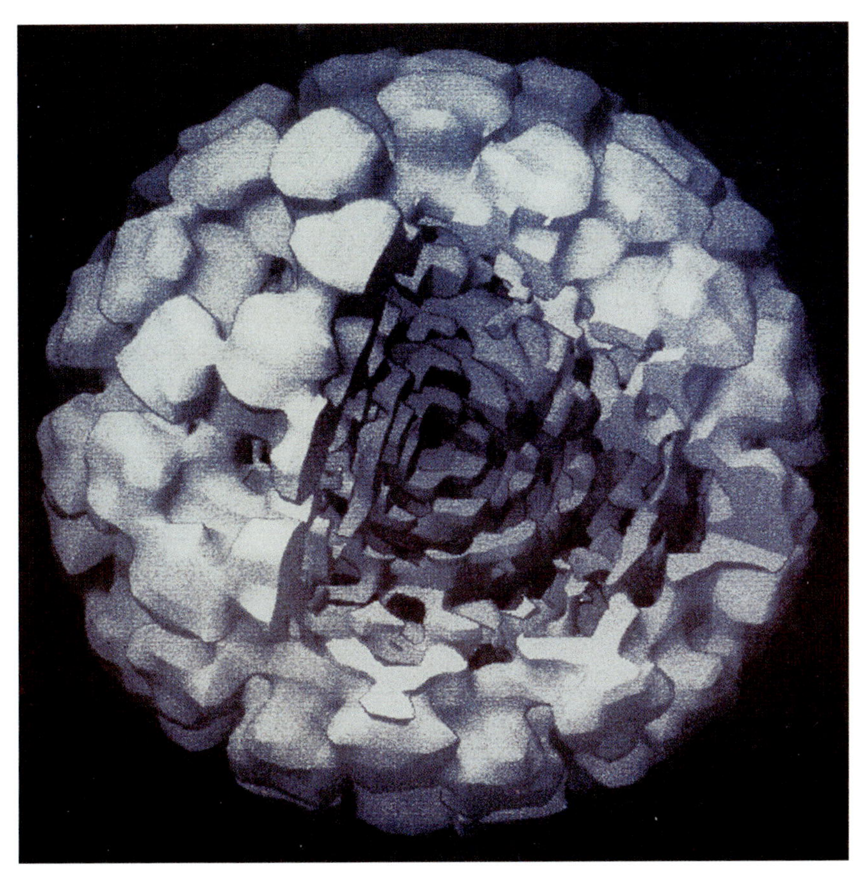

Figure 6. 3-dimensional model of a SFV at 35 Å resolution. Nature 320 (6062), 10 April 1986.

improved by a factor of 10 since the 80s. This means that the volume element resolved at present is one thousand times smaller than before; the density of information that can be harvested from the specimen is now multiplied by one thousand. This is truly a remarkable achievement. Bravo!

But the real breakthrough came from the fact that, around 3.5 Å, atoms become visible. Or, in other words: blobology becomes chemistry. This is the reason why three biophysicists who never thought of themselves as good chemists are gratified with a Nobel Prize in Chemistry. This is not because we have reached our level of incompetence, as promised by Peter's principle; it is a testimony to the unity of science. Physics, biology, chemistry; all is just science.

At present, electron cryo-microscopy has not yet brought an important result that could be translated into practical applications whether in medicine or in technology, but this will come, soon! Chemistry is a powerful science. When the arrangement of atoms can be visualized, the possibility to act on them is not very far from our reach. For example, will it soon be possible to prevent the pathological entangled binding of proteins that seems to be associated with Alzheimer's disease and numerous other neurological disorders? Many of us would be interested in such progress "for the greatest benefit to mankind", along the line of Alfred Nobel's expectations.

And science will continue. It will take time to explore the brain. We may understand how we think. Perhaps conscience will emerge. This is knowledge without limits.

But knowledge also has its practical consequences. It shapes our lives.

I got my first personal computer in 1984 as Adrian's method was being implemented. Nowadays, billions of people are sitting in front of a computer screen for the major part of their days, and communication between individuals is fundamentally changed.

My grandfather's father was living in scarcity. He was never sure he could bring home the minimum required for a decent life for his family.

=> Now we are submerged with excess
=> and the world's climate is collapsing
=> as is the glacier just above our mountain hut.

Figure 7. Extraordinary ice collapse in the glacier of Ferpècle (Wallis, Switzerland). Photo: Gerard Stampfli.

Five hundred years ago, François Rabelais wrote,

> *Science sans conscience n'est que ruine de l'âme.*
> *Science without conscience is but the ruin of the soul.*

The problem is not new but now, it is urgent.

What can we do?

One thing is for certain: we scientists must come down from our ivory tower and be involved in the society for which we produce knowledge. That knowledge can have equally good or bad consequences, and we must become more aware and responsible.

This is the reason why, more than 20 years ago, we introduced a compulsory curriculum in our university: "Biology and Society". We want our students to be as good citizens as they are biologists.

This is good, but it is not enough.

How can we be as good in using our knowledge for the well-being of all as we are in producing it?

I don't know the solution, but I know the value of knowledge. It is our most precious common good. We must protect it, develop it and make the best of it for the well-being of mankind, now and for future generations.

Imagine,
It's easy if you try.

Imagine, for example, that we think about health.

Imagine that we empower the World Health Organization, United Nations WHO, with all we know about medicine and medical treatments, and trust this institution with the duty and the competence to use this knowledge for the well-being of all. Of course, we will give those who produce the knowledge the rightful reward for their efforts.

It isn't hard to do.
Imagine...
You may say I'm a dreamer
But I'm not the only one
I hope someday you'll join us
And the world will live as one
John Lennon

REFERENCES

Adrian, M., Dubochet, J., Lepault, J. & McDowall, A. W. (1984), "Cryo-electron microscopy of viruses," *Nature* **308**, 32–36.

Dubochet, J. (2011), "Cryo-EM – The first thirty years," *J. Microscopy* **245**, 221–224.

Henderson, R. & Unwin, P. N. T. (1975), "Three-dimensional model of purple membrane obtained by electron microscopy," *Nature* **257**, 28–32.

Mayer, E. & Brüggeller, P. (1980), "Complete vitrification in pure liquid water and dilute aqueous solutions," *Nature* **288**, 569–571.

Taylor, K. A. & Glaeser, R. M. (1976), "Electron microscopy of frozen hydrated biological specimens," *J. Ultrastruct. Res.* **55**(3), 448–456.

Vogel, R. H., Provencher, S. W., Von Bonsdorff, C.-H., Adrian, M. & Dubochet, J. (1986), "Envelope structure of semliki forest virus reconstructed from cryo-electron micrographs," *Nature* **320**, 533–535.

Joachim Frank. © Nobel Prize Outreach AB. Photo: A. Mahmoud

Joachim Frank

Biography

I WAS BORN ON SEPTEMBER 12, 1940 in Weidenau/Sieg, Germany. Since 1972 the town has been part of Siegen, a city with currently some 100,000 inhabitants, situated at the southern tip of North Rhine Westphalia. The mountainous area around it is called *Siegerland*, for centuries home of the iron mining, processing and manufacturing industry. Mining of iron went all the way back to the Celts, two thousand years ago. After mining and processing moved to the Ruhr area, only the iron manufacturing part remained – boilers, metal pipes, railroad tracks, buckets and many other parts made of iron and steel. Weidenau's most prominent landmark is called the Fujiyama, a giant heap of slag from iron mining, matching the shape of the famous mountain in Japan. Siegen was also the seat of the House of Orange Nassau, related to the Dutch royal family.

The city of Siegen prides itself of being the birthplace of Peter Paul Rubens. However, the only reason he was born there and not in Cologne, his parents' place of residence, was that his father got arrested as he was passing Siegen in a carriage with his pregnant wife, in a case of mistaken identity. The dispute between three cities claiming Ruben as their son – Siegen, Cologne and Antwerp – is immortalized in a fountain sculpture at Siegen's Upper Castle showing three mothers holding, and fighting over, baby Peter Paul.

My father Wilhelm Frank was a judge (*Amtsgerichtsrat*) at the District Court in Siegen. He was born in 1896 in Weidenau. His study of law was interrupted by the draft to fight in Verdun in WWI where he was wounded, losing most of his left hand. His mother came from a wealthy local family, the Schleifenbaums, that owned a flourishing iron-manufacturing business, and his father was a high school teacher who came from a rural family in Banfe, in nearby Wittgenstein. My mother Charlotte came from the distinguished Manskopf family that traced its origins in Siegen back to the 15th century. A branch of her family settled in Frankfurt in the 18th century and gained wealth and notoriety from international wine trad-

Figure 1. Left: with my mother Charlotte, father Wilhelm, his sister Elisabeth, my grandmother Amalie Schleifenbaum, brother Helmut, and sister Ingeborg. (1940). My sister Renate would be born four years later. Right: my parents' home in Weidenau, Engsbachstrasse 3. Architectural drawing from 1905 shows original two-story veranda.

ing. They were friends with Johann Wolfgang von Goethe's family around the beginning of the 19th century.

My mother, educated in *Stift Keppel*, a high school for girls with a history going back to the 13th century, stayed at home, taking care of her four children – myself, my four year younger sister Renate and two older siblings: Ingeborg and Helmut. Our house was large and stately, solidly built from red double-glazed bricks by my grandparents in 1905. It was set on a good-sized plot bordered from the street by a wrought-iron fence. The house had verandas on the first and second floor overlooking the backyard. Walking paths were seamed with boxwood and covered with ornamental gravel.

THE WAR

As I was born during WWII, my whole childhood was marked by the war. Siegen's iron manufacturing industry made the city a target of the Allied raids, and eventually, by the end of the war, led to the destruction of 80% of all buildings. My first memories, at age four, were of houses in the neighborhood going up in flames. In one of the early raids in February of 1944, my parents' house was fire-bombed. Since the roof and upper floor of our house were destroyed, and the rest rendered uninhabitable by extensive water damage, we had to move for over a year to Hilchenbach, a town 20 kilometers to the north, where a colleague of my father offered us room in his large apartment. This apartment was located in the *Williamsburg*, an 18th century water castle that served as the court building at the time. The memory of sitting in the bomb shelter, in the basement of the large building, surrounded by crying babies, and listening to the sounds of planes, air raids, and radio announcements were the stuff of nightmares well into my adolescence.

The time immediately after the war was one of great hardship. My mother went on "hamstering" trips into the country by train, traveling with zinc-plated buckets, manufactured by the company of her in-laws and much treasured by the farmers. She bartered them for butter, ham, big loafs of bread, flour, and eggs. Back home, she would mix "real butter" with margarine in a large bowl to stretch out the experience of tasting traces of real butter into weeks. We had a good-sized garden with apple, pear and cherry trees. For a while we grew sugar beets to make syrup and tobacco plants to support my father's smoking habits. We kept chickens in the backyard, and even kept a pig at one time in the space under the veranda. Helping my 10-year older brother with the garden work and spending time with the chickens made me appreciate nature from a close range.

The sight of rubble of houses burned and collapsed in our neighborhood had a peculiar effect on me, a mixture of fright and fascination. The fright was the natural reaction to the sights of chaos and destruction, which implied to a child that nothing still standing is safe. The fascination part came from the experience of playing, together with other boys my age, on desolate lots filled with bricks, pots, twisted wires and Bakelite insulators. Here and there we uncovered a family of mice with pink still-blind babies.

SCHOOL YEARS

My elementary school, where I spent my first four years, was right across the street from our house. I was eight years old when I started my first experiments in the dark place underneath the veranda where our little pig once roamed. It was natural curiosity that made me do it, before I had any concept of science. I built a shelf, collected little *Magenbitter* liqueur bottles and filled them with every liquid I could get hold of: oil, water, gasoline and, when I was a little older, hydrochloric acid. In bouts of intuition I mixed the fluids, exposed metals to them and recorded the results. I watched calcium carbide dissolve in water and enjoyed watching the violent reaction and the smell of the escaping gas. I watched zinc dissolve and bubble up in hydrochloric acid. I heated up coal in a metal container connected to a tube since I'd heard that a flammable gas would escape.

My parents' house contained one amazing treasure that would accompany me through all of my boyhood and adolescence, as soon as I was able to read: Meyers *Konversations-Lexikon*, an encyclopedia in 20 volumes published in 1905. Each volume measured about 1000 pages, filled with scholarly articles, technical drawings, colorful photogravures, and maps from all over the world. Quite possibly I read them all over the course of the years. As a whole, this large encyclopedia reflected the belief that all that ever needed to be studied was known already, and that progress from then on would be at most incremental. Ironically, 1905 hap-

Figure 2. In science class in high school. Behind me, further to the right are Horst Schmidt-Böcking and Ulrich Mebold, who would also study to become physicists. (Photograph thanks to Friedhelm Schick.)

pened to be the exact year when Albert Einstein published a paper on the photoelectric effect, with its evidence for the quantization of energy, the precursor of Quantum Mechanics. It would leave little of the old wisdom untouched. I would later claim the collection as my only bid for a tangible heirloom in my parents' house.

Starting with fifth grade I went to the *Fürst Johann Moritz Gymnasium*, named after one of the prominent Orange Nassau dukes. I was one of only four to step up from my elementary class of twenty students. In the Gymnasium, which in the German system combines middle and high school, I immediately took an interest in science classes, particular physics. Meanwhile my tinkering at home had migrated from the place under the veranda to the attic of our house and expanded to include radios, which I rebuilt from used and mail-ordered parts. This obsession with radios started after my brother showed me how to build a crystal radio. I constructed several fancy miniature radios fitting in soap boxes. Most of my savings went into the purchase of valves, transistors, resistors, and capacitors. The attic was filled with the exciting smell of vapor from the soldering rosin. I made a friend in school who shared my hobby and lived across the street.

I should add at this point that all three of my siblings went to the same Gymnasium.

After receiving his Abitur (high school diploma) my brother finished his Ph.D. in Engineering and became a civil servant for Occupational Safety. Both my sisters left at the "Einjaehrige," an early departure point from high school for a switch to a trade school, in their case a school for physical therapists. My elder sister finished her Abitur many years later after she married and her kids had grown up, proceeding to college and obtaining a Ph.D. in Biochemistry. My younger sister, after working as a physical therapist, became an artist and made many beautiful quilts until her early death in 1998, from cancer.

COLLEGE

For me the choice of physics for study in college was always a foregone conclusion, though my father needed extra reassurance, as he doubted this would lead to a career that would ever earn me a living. In 1960, after finishing my Abitur I went to the University in Freiburg. The move into the little quiet university town with its large Gothic cathedral and charming medieval buildings was nevertheless a huge step for me coming from the provincial town I grew up in. I took Calculus and Linear Algebra, and learned how to write rigorous mathematical proofs. I also took courses in Special Functions of Mathematical Physics and Statistical Mechanics.

Following the example set by my brother, who had studied Engineering in Aachen, I joined a fraternity, Corps Suevia, and made several friends there. Later, though, influenced and enlightened by the political upheavals in the sixties, I decided to quit the Corps since I recognized the nationalistic, right-wing roots of German student organizations. Freiburg was also the place where *Martin Heidegger*, as rector of the university, had infamously aligned himself with the Führer. During my time there I saw the aged Heidegger, a little man, give one of his rare public speeches in front of the University, barely visible as a throng of students surrounded him.

Based on my performance in the Vordiplom (B.S. equivalent) exam, I was nominated for the *Studienstiftung des Deutschen Volkes*, a special fellowship that would prove instrumental in widening my horizon to include other fields of science and humanity. The Studienstiftung fostered interdisciplinary discourse by organizing meetings at the forefront of science. In one of these meetings, in 1964, I first learned about the tenets of the Central Dogma and the structure of DNA. I was also here that I met Wolf Singer, a neurophysiologist, starting a close friendship that would last until today. With him and like-minded students, I founded a discussion group focused on Cybernetics, the hot subject of the time.

GRADUATE STUDIES

I went to the Physics Department at the University of Munich to do work toward my Diploma thesis, the equivalent of a Masters degree. The thesis project had to do with the back-scattering of electrons on gold in the liquid phase, an esoteric subject vaguely related to the then-emerging technology of machining with high-intensity electron beams. My mentor, Ernst Kinder, had done early work with the electron microscope, tracing the colorful patterns of butterfly wings to light interference created by submicroscopic arrangements of tiny scales. He still kept an ancient electron microscope in his office.

After finishing my diploma, when it came to choosing a Ph.D. thesis mentor, I was therefore prepared and open to the idea of working on a project that involved electron microscopy. The mentor I chose was Walter Hoppe, an X-ray crystallographer-turned electron microscopist at the *Max-Planck Institut für Eiweiss- und Lederforschung* on the Schillerstrasse in the center of Munich, which later relocated to Martinsried and became the Max-Planck Institute for Biochemistry. Hoppe looked for ways to use the electron microscope for imaging biological molecules in three dimensions. My thesis focused on an exploration of the properties of electron micrographs using methods gleaned from other fields, such as Statistical Optics. My first paper, in the journal Optik, examined the optical diffraction patterns of micrographs affected by specimen drift, and interpreted the stripes observed in terms of Fourier theory (Frank, 1969). I was proud when Hoppe refused to put his name on it, recognizing it as a totally independent piece of work.

My first experience in computer programming was with the programming language ALGOL, and involved a 20-minute walk to the Technical University for every compilation and every run of a newly written program. I later learned to program in FORTRAN on an IBM 1130 machine tucked in a little basement room of our Institute, where I sometimes worked late into the night. The Institute, located just minutes' walk away from the *Wiesn*, the site of the *Oktoberfest*, had its own social life with distinct Bavarian color. Early-morning mushroom picking raids were organized when they were in season. The porcinos and pfefferlings brought back by the teams of three or four students – always including at least one expert – wound up boiling in Erlenmeyer flasks in the exhaust hood. They were sprinkled with salt and served with pieces from a big loaf of Bavarian bread. We celebrated the acceptance of papers with a keg of beer and big hunks of meatloaf in the library room.

Munich at the time, as now, was a city rich with cultural events. There were so many venues; it was possible to go to a classical concert every day. One of my friends, who also made the move from Freiburg to Munich, was a classical music aficionado and lured me to many outstanding performances. It was then that I learned to recognize many classical symphonies from a few opening notes. Little experimental theaters were abounding. The Munich Opera House offered a grand experience for affordable ticket prices. I spent my time with two circles of friends, one around Wolf Singer, whom I'd met through the Studienstiftung, the other around Jan Groneberg, a firebrand college dropout with utopian ideas who lived in a little cottage outside of Munich. It was in Wolf Singer's circle of friends where I met my first wife, Cathy Engelberger. We married in 1969, but the marriage would last less than 10 years.

During this time, a meeting in Hirschegg, in 1968, gave me the opportunity to meet several people who later became important in the field. The workshop (as well as later ones) was co-organized by Walter Hoppe and

Max Perutz from the Laboratory of Molecular Biology of the MRC in Cambridge, known for his pioneering work in protein X-ray crystallography. Among the people I met there were Harold Erickson, Richard Henderson, Ken Holmes, Hugh Huxley, and Nigel Unwin. With afternoons free for skiing, and both mornings and evenings reserved for the lectures and discussions, the format resembled that of the Gordon Conferences. Two papers (in German) related to my thesis were later published in the proceedings of the meeting, in a special issue of *Berichte der Bunsengesellschaft für Physikalische Chemie* (Frank et al., 1970; Langer et al., 1970).

POSTDOCTORAL STUDIES

After my thesis defense at the Technical University Munich in early summer of 1970 I was awarded a *Harkness Fellowship*, which allowed me to spend two years in the USA at labs of my choice. I chose the Jet Propulsion Lab (JPL) at Caltech in Pasadena, the Donner lab in Berkeley, and Cornell University. Coming from Europe, the culture shock of being placed into the Hollywood-like landscape of Pasadena with its restless freeways and little houses with palm trees and little old ladies with tennis shoes could not be greater.

In hindsight, all three labs gave me important impulses toward my future direction. The JPL at the time had the world's best image processing equipment, and had developed a modular image processing system, VICAR, that I could hook my own programs to. This package would later serve as a model for developing my own system, SPIDER. At Donner lab, which was part of the Lawrence Berkeley labs on the hill, I spent time with the group of Bob Glaeser, who focused on two quintessential problems faced by structure research with the EM: radiation damage, and the need for a hydrated environment. He and his student Ken Taylor were already experimenting with the preparation of frozen-hydrated samples, but the decisive invention of the vitrification technique in Jacques Dubochet's hands had yet to come. At Cornell University, in the group of Benjamin Siegel in Clark Hall, I made the acquaintance of Ken Downing and William Goldfarb. I later asked William to join me in Albany as part of my team.

While in Ithaca, in 1972, my son Hosea Jan Frank was born.

Returning from the USA, I spent a brief time back at the Max-Planck Institute, in the winter of 1972/73, working on the theory of partial coherence in electron microscopy. This work brought me in contact with Peter Hawkes, a world expert in Electron Optics. In 1973 I joined the group of Vernon Ellis Cosslett at the Cavendish Laboratory in Cambridge, still at its old location in Free School Lane, as a Senior Research Assistant. Among the people I interacted with were Owen Saxton and Peter Hawkes.

During my years at the Cavendish I worked further on partial coherence and found a way to obtain the signal-to-noise ratio of electron micrographs by computing the cross-correlation of two successive images of the same field.

This was the time when the vision of single-particle averaging and reconstruction took hold in my mind – the idea of spreading out electron dose among multiple "copies" of a molecule randomly arranged on the grid. In 1975 I published a concept paper presenting the idea that the structure of a molecule could be retrieved by taking advantage of multiple occurrences of this molecule in solution (Frank, 1975). Together with Owen Saxton I analyzed the conditions under which bright field images of biological molecules can be aligned with sufficient accuracy for the image average to reach a given resolution. The result of this study, which we jointly published in 1977 (Saxton and Frank, 1977), gave me confidence that the single-particle approach would work even under weak native contrast (i.e., protein vs. water) conditions.

ALBANY, AND THE WADSWORTH CENTER

In 1975 I received a job offer from Don Parsons at the Division of Labs and Research (later renamed Wadsworth Center) of the New York State Department of Health in Albany, New York. While the original mission was tomographic reconstruction of cell sections, I focused on the implementation of the single-particle approach. In both areas I recognized the need for a workbench of programs to gain flexibility in the design of programs, and started on the development of SPIDER, an image processing system of modular design (Frank et al., 1981). As the single-particle techniques developed, SPIDER became the vehicle for disseminating the technique to the community. It was initially distributed under a license agreement, for a one-time fee, and later became available for free under a creative commons license.

It would still take a few years until proof of concept would be available with actual images of biological molecules. These were glutamine synthetase, provided by David Eisenberg at UCLA, acetylcholine receptor, by Peter Zingsheim in Goettingen, and ribosomes, by Miloslav Boublik, at Roche in New Jersey. Martin Kessel, an early convert and close friend, helped me in some of these studies as he took a Sabbatical leave from Hadassah University Medical Center. In each case, reproducibility of two-dimensional averages demonstrated that the approach was sound. Still, there was a lot of skepticism among practitioners of electron microscopy. A turning point came with the addition of a method addressing the problem of heterogeneity, which I developed jointly with Marin van Heel, a student visiting from the Netherlands in 1980 (van Heel and Frank,

1981). Looking for other suitable challenging molecules to try the technique on, I started a collaboration with Jean Lamy and his student Nicolas Boisset in Tours, France, to image a variety of arthropod hemocyanins. (I stayed in touch with Nicolas over the years until his untimely death in 2008. He had a meticulous way of record keeping and developed beautiful slides for teaching the principles of single-particle reconstruction).

Albany is the capital of New York State but has a distinctly provincial character, especially lying as it does in the shadow of New York City. The town is surrounded by beautiful countryside, and hikes into the Adirondack mountains are not far away.

The move to Albany not only gave me my first independent position, it also unleashed an urge in me for creative expression in areas not associated with science. I joined an artists' collective, called WORKSPACE, founded by Jacy Garrett. At the time, performance art was being redefined across the country, and artists' collectives were springing up everywhere. The FLUXUS movement directed attention to the peripheral, the accidental. I enjoyed being accepted by the collective without formal credentials, just by virtue of my creative contributions. I participated in mail art correspondence and, for several years, either edited or co-edited a small literary magazine called PROP.

At the end of the 70s my first marriage ended. The divorce agreement gave us joint custody over our son, an arrangement that would keep me in town for quite some time, as I would see Hosea grow and become a multitalented artist who renamed himself Ze. I was to meet my present wife, Carol Saginaw, in Albany in 1982. Carol worked initially at the New York State Office of Mental Health, then over the years was executive director of several statewide non-profit organizations in mental health and later in early care and education. Carol was from Michigan, from a Jewish family that had lost many of their members in death chambers constructed by the country I was from. Although our diverse backgrounds presented a challenge, we were married in 1983, and we have been happily together ever since. In good part it was Carol's continuous support, and her faith in me, that made me prevail and come to the present point in my career.

At that time, I also started writing fiction in English and was quite flattered when William Kennedy, and later Steven Millhauser and Eugene Garber, gave me very positive feedback on my manuscripts. To me the idea that I might be able to express myself creatively in my second language was thrilling as I was unsure at this stage if I would ever return to live in Germany. After a course in fiction writing with Eugene Garber at SUNY Albany, the participants of his class, myself included, decided to continue meeting as a writers' group. Constructive criticism by this group, and other groups that I joined later on, honed my writing and helped me recognize "my voice," and writing became part of my life (*www.franxfiction.com*).

Looking back now, I see that my early contributions to single-particle EM were made possible mainly by three factors: the peace and quiet of the place where I worked, the absence of any teaching requirements, and the steady support by the National Institute of General Medical Sciences, of NIH, which lasts until the present day. I was quite fortunate to have Michael Radermacher join my team in 1982, a German student who had also trained under Walter Hoppe and had a special background in three-dimensional reconstruction with arbitrary geometries. Michael was the one who single-handedly designed the random-conical reconstruction programs in my lab that, in 1986, yielded the first three-dimensional reconstruction of a totally asymmetric molecule, the large subunit of the *E. coli* ribosome. By adopting the novel plunge-freezing and vitrification technique of Jacques Dubochet, we were soon able to reconstruct biological molecules in their hydrated, native state as well. From that point on, in the late 1980s, the technique we had been working on so hard was evidently headed for success, though it was still uncertain if it would ever be able to compete with X-ray crystallography in resolution and propensity to yield atomic structures.

Figure 3. Albany reunion of former and then-present lab members at the first Gordon Conference on 3DEM, in 1985. From left to right: Marin van Heel, Jean-Pierre Bretaudiere, Adriana Verschoor, Bruce McEwen, Joachim Frank, Terry Wagenknecht, Michael Radermacher, and Martin Kessel.

Meanwhile, in 1985, our daughter Mariel Beth was born and became the center of our life. When she was two years old, I received the invitation for a Sabbatical stay at the Medical Research Council (MRC)'s Laboratory for Molecular Biology (LMB) in Cambridge, England, with Richard Henderson as host. We rented a charming little home, King's Cottage in Little Shelford, with a flower garden where Mariel played with other children. We made punting trips on the river Cam and walked in the beautiful parks in the surroundings of Cambridge. Most of my interactions in the lab were with *Wah Chiu*, whom I first met as a student during my visit at Bob Glaeser's lab, and who visited the LMB at that same time. With his help, and using his data on two-dimensional crystals of crotoxin, I developed and demonstrated *patch averaging*, a method of structure recovery that made use of "local" averages of a crystal divided into small areas – essentially the single-particle approach applied to pieces of a crystal.

Back in Albany, among the first molecules we reconstructed in 3D were hemocyanins, in continuation of our collaboration with Jean Lamy's group in France. Another collaboration, with Sydney Fleischer of Vanderbilt University, gave me the first opportunity to work on the structure of the ryanodine receptor. Still, the work on the structure of the ribosome continued to fascinate me most. As early as 1990 I became convinced that my lab would be able to contribute significantly to the structure and function of the ribosome, and I started hiring biochemists with ribosome background. Rajendra Agrawal, trained in the Burma lab at the Baranas Hindu University, was the first to bring real "ribosomologist" expertise into the lab. Others would follow later, among these Christian Spahn, trained in the lab of Knud Nierhaus in Berlin.

A major factor promoting discussions in the growing cryo-EM community and the dissemination of the new technologies of sample preparation, instrumentation and data processing has been the Gordon Conference on three-dimensional electron microscopy (3DEM). Established in 1985, it met every two years initially and later switched to the present annual cycle. My election in 1987 to be Vice-chair in 1989 with David DeRosier and to be Chair in 1991 was a big step marking recognition of the single-particle techniques by the whole community.

A Humboldt-funded Sabbatical stay in 1994 at the Max-Planck Institute for Medical Research in Heidelberg, hosted by Ken Holmes and Rasmus Schröder offered me the first opportunity to work in Germany again. Through the efforts of my graduate student Jun Zhu and my postdoc Pawel Penczek, the first detailed map of the *E. coli* ribosome emerged, well before the X-ray structures came out. The putative placements, by Raj Agrawal, of tRNAs and mRNA into this map of the ribosome have stood the test of time. It was also in Heidelberg that I completed writing

Figure 4. Albany reunion picture at the 2005 Gordon Conference on 3DEM. From left to right: Bill Baxter, Martin Kessel, Nicolas Boisset, Joachim Frank, Christian Spahn, Tanvir ("Tapu") Shaikh, Pawel Penezek, Rajendra Agrawal, Zheng Liu, and Jose-Maria Carazo.

my book on 3D electron microscopy, which would be published in 1996 and, in a second edition, in 2006 (Frank, 2006).

In 1998 I was appointed a Howard Hughes Medical Institute (HHMI) investigator, a position that would last for 19 years and was only recently terminated. The funding by HHMI during these years was absolutely crucial for my lab to continue development of cryo-EM and realize very challenging biological projects with several collaborators. At about that time the Wadsworth Center joined eight institutions in New York City to form a consortium for structural research, called the *New York Structural Biology Center,* which supports NMR, X-ray crystallography, and cryo-electron microscopy. This connection provided entrées for me at Columbia University and other leading institutions in New York.

In 2000, on my 60th birthday, I organized a meeting in Rensselaerville, in continuation of a series of conferences started by Anders Liljas in Sweden on the Structural Basis of Translation. The conference site in Rensselaerville is set in a beautiful park, an hour driving distance from Albany. As I spent time during this meeting with Måns Ehrenberg we made concrete plans for collaboration on ribosome structure and function. This turned out to be the beginning of an exhilarating journey that has lasted until now, as we investigated the structural basis for initiation, decoding, mRNA-tRNA translocation, termination, and the recycling process, thereby contributing to the rich knowledge base on the mechanism of translation available today.

My children at this point were grown and on their own. My son Ze Frank had majored in neuroscience at Brown University and started a band, playing the guitar. His special talents for music and the arts had been in evidence early on. He subsequently moved to New York and began doing web design. Through a fortuitous route, which he recounted in his first TED Talk, he became an internet personality virtually overnight. Most recently he was a media executive at Buzzfeed. He now lives with his wife and two children in Los Angeles. My daughter Mariel Frank majored in linguistics at Barnard College. Speaking multiple languages, she taught English in Japan, worked for a Latinx non-profit organization and is now a programmer and curriculum developer at Code academy. She is married and lives in Brooklyn.

COLUMBIA UNIVERSITY

In 2008 I joined Columbia University as a faculty member of both the Department of Biochemistry and Molecular Biophysics, and the Department of Biological Sciences. After more than 30 years, this move from pastoral Albany to New York City was quite exciting as it offered many opportunities for collaborations. I brought the HHMI-owned FEI Polara

Figure 5. My family at our house in Alford, in the Berkshires, in summer of 2017. Top, left to right: Hosea (Ze), son Jonah, Tom Murphy (Mariel's husband), J.F. Bottom: Ze's wife Jody Brandt, daughter Rose, Mariel, my wife Carol Saginaw.

microscope with me and, together with the FEI F20 microscope purchased as part of the startup, established cryo-EM at Columbia University. One area of collaboration I was immediately attracted to was single-molecule FRET, which had just been set up by Ruben Gonzalez coming from the Puglisi lab at Stanford.

For the first four years at Columbia, progress with our cryo-EM projects was slow as it was still limited by the poor quality of recording media. The situation changed radically when direct electron detection cameras were introduced commercially, transforming the field profoundly and opening up many new avenues in my lab for exciting collaborations, particularly on channel structures. A Columbia-wide cryo-EM resource facility has been recently created and, thanks to generous gifts by donors and the cooperation by the deans of all three campuses, Columbia is now headed toward becoming one of the world's leading centers for cryo-EM.

Beyond the benefits to Columbia and the USA, looking at the way the new technology has recently spread across the entire industrialized world, I'm gratified to see that single-particle cryo-EM is now able to fill a huge gap in molecular structure research as membrane-bound channels and receptors, and also many large molecules with flexible regions can be tackled, promising to add significantly to the war chest of human medicine in years to come.

ACKNOWLEDGMENTS AND FINAL NOTE

The emergence of many near-atomic structures in the last five years in several labs has drawn the world's attention to the many preceding years of work not just by the Nobel Laureates now honored and their groups, but by the whole cryo-EM community. Because, for perspective, it is necessary to note that since about 1990, when the technique started to receive recognition within the cryo-EM community, there have been major contributions by many groups to every aspect: sample preparation, automation of data collection, computation, validation and atomic model building. These contributions are too numerous to list, but instead I refer to a recent review of the whole field (Frank, 2015).

I have been very fortunate throughout the journey that has brought me to this point. I would like to express my gratitude to my family, particularly my wife Carol, for their steady support over this long time. My sister Ingeborg Berg, with her training in biochemistry, was the one person in my family who could appreciate the enthusiasm I expressed in letters to her early on. Among my friends I need to single out Martin Kessel for his early encouragement and support and Jose-Maria Carazo for so many brainstorms along the way.

As a final note, the recognition by the Nobel Prize is an experience both extremely thrilling and humbling. A lifetime achievement benefiting the whole of humankind – in the words of Alfred Nobel's will – seems an idea so grandiose that only few can live up to it. Ernest Rutherford? Linus Pauling? Marie Curie? Their shoes are difficult to fill. The event is also transformative since all of a sudden my life is defined, or confined, by the perceptions of many people I have never met. It dawns on me that there is no way back – for better or worse – to the life I was living before. This is why I'm grateful for this opportunity to tell my own story, in my own words.

REFERENCES

Frank, J. (1975). "Averaging of low exposure electron micrographs of non-periodic objects." *Ultramicroscopy* **1**, 159–162.

Frank, J., Shimkin, B., and Dowse, H. (1981). "SPIDER — A modular software system for image processing." *Ultramicroscopy* **6**, 343–358.

Frank, J. (2006). *Three-Dimensional Electron Microscopy of Macromolecular Assemblies* (New York, Oxford U. Press).

Frank, J. (2015). "Generalized single-particle cryo-EM – a historical perspective." *Microscopy* **65**, 3–8.

Single-Particle Reconstruction of Biological Molecules — Story in a Sample

Nobel Lecture, December 8, 2017 by Joachim Frank[1,2]
[1]Department of Biochemistry and Molecular Biophysics, Columbia University Medical Center; [2]Department of Biological Sciences, Columbia University, New York, NY, USA.

THE BACKGROUND

I developed an interest in electron optics when I worked with *Ernst Kinder* on my masters thesis project in Physics at the University of Munich. The subject of my thesis was backscattering of electrons on the surface of liquid gold, an ambitious undertaking that forced me to construct a vacuum chamber, a crucible to heat up the gold, a detector, and an electron gun. In 1943, working with the electron microscope, Kinder had studied butterfly wings, which as he realized gained their brilliant colors from interference of light on gratings formed by tiny scales, arranged in regular order.

I signed on to a graduate project with *Walter Hoppe* at the Max-Planck Institute in Munich (Fig. 1), an X-ray crystallographer whose interest had turned to electron microscopy (EM) as a means to study biomolecules. He viewed the electron microscope as a diffractometer that, unlike the one employed in X-ray crystallography, could record not just amplitudes of diffracted electrons, but their phases as well. This was a fancy way of saying electron microscopes were able to form images.

Figure 1. Walter Hoppe (1917–1986) with the Siemens Elmiskope 102.

It is necessary at this point to look back at the state of the art of molecular EM in the late 60s and early 70s. During the initial years, from the 1930s to the 1950s, the contributions of EM to biology had been confined mainly to the investigation of tissue at relatively low magnification. Serious forays into the quantitative visualization of molecular structure did not commence until the 1960s and were concentrated in three groups: *Aaron Klug's* at the Laboratory of Molecular Biology of the Medical Research Council (MRC) in Cambridge, England, my mentor *Walter Hoppe's* at the Max-Planck Institute in Munich and *Edward Kellenberger's* at the Biozentrum in Basel. (Photograph from archives of the Max-Planck Society, Berlin).

Unless symmetries are present, three-dimensional reconstruction of an object requires the combination of its projections from a wide angular range. The first pioneering achievements in molecular structure research with the electron microscope were the three-dimensional (3D) reconstruction of the bacteriophage tail with helical symmetry in 1968 by DeRosier and Klug (1968) and the first reconstruction of an icosahedral virus in 1970 by *Tony Crowther* (Crowther et al., 1970; Crowther, 1971).

At that time biological molecules could not be imaged in a close to native state. Negative staining – which amounts to embedding the molecule in a puddle of heavy metal salt as it is air-dried from solution – was the only means available to produce contrast. On the other hand, biological molecules were known to be quite fragile, and maintenance of their integrity would require a fully hydrated environment.

Graduate Studies and Harkness Fellowship

As I started my work as graduate student under Walter Hoppe, in 1967, I was exposed to discussions in a Workshop in *Hirschegg* in the Tyrolean Alps, co-organized by Walter Hoppe and *Max Perutz* in 1968, later to be continued in meetings in Hirschegg in 1970 and *Alpbach* in 1976. These were the first meetings that brought together protein crystallographers and people working in EM (see Holmes, 2017).

In my thesis project I analyzed electron micrographs with the optical diffractometer and explained patterns observed in case of drift in terms of Thon rings (Thon, 1966) modulated by a sinc function or Young's fringes (Frank, 1969). I also examined the statistical properties of digitized micrographs. For digitization I used a densitometer built in-house, which rendered the images on punched tape to be fed into the computer. In my first applications of digital image processing in EM, I explored the use of correlation functions for alignment of images (Langer et al., 1970). Another topic of my dissertation was the distortion of information by the contrast transfer function (CTF), caused by the lens aberrations of the electron microscope (Scherzer, 1948), and its recovery by CTF correction (Frank et al., 1970).

After finishing my Ph.D., in 1970, I went to the United States for two years under a Harkness Fellowship. The visit to three labs I chose was an eye-opener in several regards. The Jet Propulsion Lab (JPL) in Pasadena, at the time, was arguably the most advanced place in image processing hardware and software. In a project aimed to correct the contrast transfer function from a defocus series, I used their scanner to digitize micrographs of negatively stained DNA that were given to me by Walter Stoeckenius at UCSF, and I adapted my programs to interface with JPL's VICAR system. VICAR, used to process images from the Jupiter fly-by mission, was a modular image processing system that would later serve as a model for the development of SPIDER. The second lab I visited was the Donner Lab in Berkeley, where R*obert M. Glaeser* studied the effects of radiation damage on biological molecules under the electron beam (Glaeser, 1971). He also started developing techniques to render molecules frozen-hydrated in the EM (Taylor and Glaeser, 1974). The third lab was Benjamin Siegel's at Clark Hall, Cornell University, where an experimental microscope in the mid-voltage (600 kV) range was being built. It was here that I first met *Ken Downing*, who worked on optical methods of information retrieval such as single-side band holography, and *William Goldfarb*, who would later join me in Albany.

The numerous problems faced by people attempting to image biological molecules in the EM were discussed at a workshop organized by *Edward Kellenberger* in *Gais*, in the Swiss Alps in 1973. The state of the art at the time was reflected in the title of a proceedings paper (Beer et al., 1974) as "high resolution" was equated with any results at better than

30Å. Paramount at the Workshop was the search for a method that would keep the molecule fully hydrated while exposed to the electron beam. In addition, following the pioneering studies by *Glaeser* (Glaeser, 1971) – just at the time I visited his lab as a Harkness Fellow – radiation damage was recognized as a major obstacle in the efforts to achieve high resolution. Averaging over a large number of repeats of a structure exposed to very low dose was seen as a general solution to this problem. Thus this meeting set the stage for a ground-breaking study by *Richard Henderson* and *Nigel Unwin* (1975): the reconstruction of bacteriodhopsin from the purple membrane of *Halobacter* embedded in glucose under near-native conditions. The confluence of novel approaches to three areas, namely sample preparation, data collection at extremely low electron dose ($< 1\ e^-/A^2$), and merging a tilt series of these noisy images into a 3D image, made this work a towering achievement. The time after this was marked by general excitement in the community, and many attempts to use the same or similar methods in the study of other proteins amenable to 2D crystallization. However, the intrinsically low contrast between proteins and glucose and residual disorder in the crystals formed made these attempts difficult, and embedment in ice remained the agreed general goal.

In the 70s, and well into the 80s, most researchers in the field that was to be called Structural Biology were united in the belief that serious structure determination required highly ordered samples, such as 2D crystals, helical arrangements, or viruses with high symmetry. Attempts to extract structural information from free-standing, single, asymmetric molecules were not taken seriously. One such alternative approach, pursued by *Walter Hoppe*, was to tilt the EM grid to which isolated molecules were attached into multiple angles while collecting the projections, from which the molecules could then be reconstructed (Hoppe et al., 1974).

While the angles are exactly known in Hoppe's "tomographic" approach, the accumulation of radiation damage over multiple exposures, to more than 1000 electron per Angstrom square, rendered the end result essentially meaningless (Baumeister and Hahn, 1975). The other approach was the one I formulated in a concept paper in 1975 (Frank, 1975), to be elaborated in the following.

POSTDOCTORAL WORK AT THE CAVENDISH LAB: THE CONCEPT OF SINGLE-PARTICLE AVERAGING AND RECONSTRUCTION

The concept of the single-particle approach began to take shape while I was working at the Old Cavendish Laboratory, Free School Lane, Cambridge, starting in 1973 – at the very place where Max Perutz had started the Laboratory of Molecular Biology in a courtyard barrack. I had accepted a postdoc position in the group of *Vernon Ellis Cosslett*.

The approach takes advantage of the fact that biological molecules purified from cell extracts and suspended in solution typically exist in thousands or even millions of "copies" (that is, separate realizations) with virtually identical structure. Hence, if a grid to which such a sample has been applied is put into the EM, a large number of projection images of the same molecule lying in random orientations are encountered (Fig. 2). The important point is that instead of collecting multiple images from *one* molecule by tilting it in the EM, one merely has to take a snapshot of *multiple* copies of that molecule with a very small radiation dose. The advantage in the latter case is that the 3D image obtained relates to a molecule that has "seen" only one single low exposure, and has remained practically undamaged. The later-coined terms "single-particle averaging" and "single-particle reconstruction" make reference to the fact that the molecules in the sample are free-standing, not attached to one another as in a crystal.

At this point it is important to emphasize that no symmetries are assumed. Rather, the idea was that entirely asymmetric molecules could be reconstructed in this way. Symmetries obviously simplify the problem as in those cases the image contains multiple projections of the repeating unit in different, known orientations, allowing data from different molecules to be readily merged.

Simple as the concept sounds, the realization of single-particle reconstruction required several problems to be solved, all in the area of image processing. These can be stated in the form of five questions: (1) Is it at all feasible to align noisy molecule images with one another with sufficient accuracy? (2) How do we estimate the resolution of an average (or reconstruction) obtained by combining all images? (3) How do we sort molecule images by appearance related to view angle or conformation? (4)

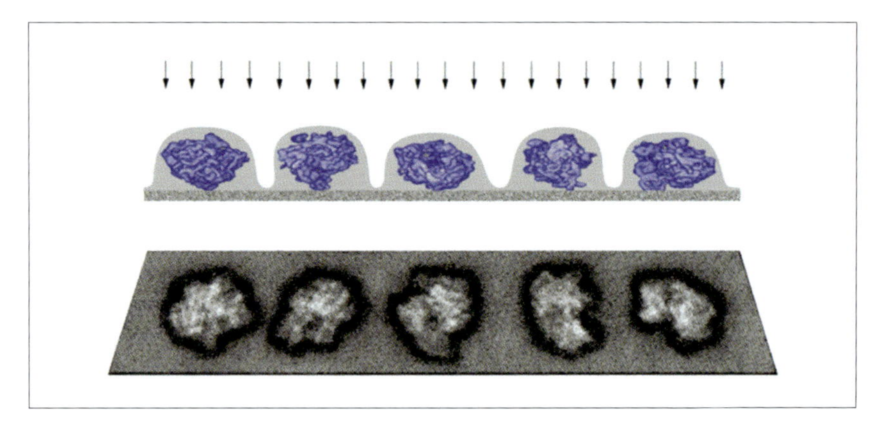

Figure 2. Schematic of single-particle data collection for air-dried, negatively stained molecules. The molecules are randomly oriented and each is embedded in a layer of heavy metal salt, which provides high contrast in the electron beam. Air-drying results in partial collapse of the molecule in z-direction.

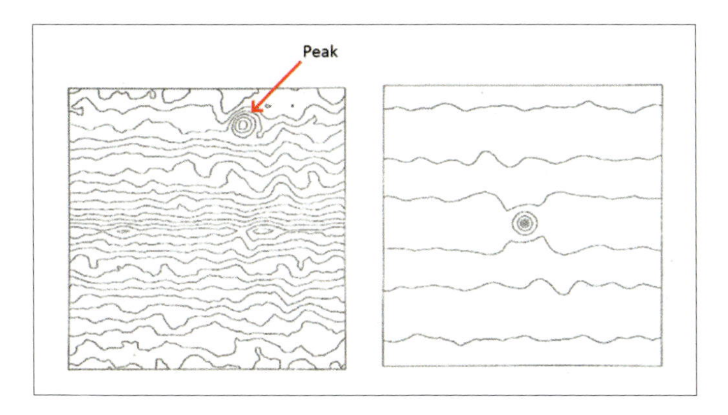

Figure 3. Left: Cross-correlation function of two micrographs of carbon foil, taken successively in the same area. Right: Autocorrelation function of one of the micrographs. In the autocorrelation function, the peak reflects superimposition of both signal and noise parts, hence its sharpness. In the cross-correlation function, the noise term is eliminated and the peak reflects only the signal part. (Reproduced from J. Frank, Ph.D. thesis 1970)

How do we find the viewing angles of projections *a posteriori*? (5) How do we reconstruct the molecule from projections at randomly spaced view angles?

It was in the nature of these problems and their novelty that progress was slow and piecemeal, one step at a time. During the work on my Ph.D. thesis (1968–1970) I had already explored the use of the cross-correlation function for aligning electron micrographs. I discovered that two successive images of the same area of carbon could be aligned with precision better than 3 Angstroms, the resolution of the electron microscope (Frank, 1970; Langer et al., 1970) (Fig. 3).

Actually, the process of 2D alignment is a bit more complicated, since molecules picked from the micrograph differ *both* in shift and in-plane orientation. The solution I found to determine both relative rotation angle and shift simultaneously makes use of the fact that the autocorrelation function of a molecule image is shift-invariant – it does not depend on its position within the image frame (Frank, 1980). Thus the determination of rotation can be entirely decoupled from that of shift (Fig. 4).

Working with *Owen Saxton*, I was able to show that alignment via cross-correlation would be accurate enough for the purpose of aligning noisy images of single molecules if the dose exceeded a certain level, which depended on the molecule's size and contrast against the background (Saxton and Frank, 1977). Using the ribosome as an example, it became clear from the formula we obtained that the single-particle approach to structure research was indeed feasible for molecules of sufficient size:

PARTICLE SIZE > 3/[CONTRAST^2 x RESOLUTION (in Å) x CRITICAL ELECTRON DOSE]

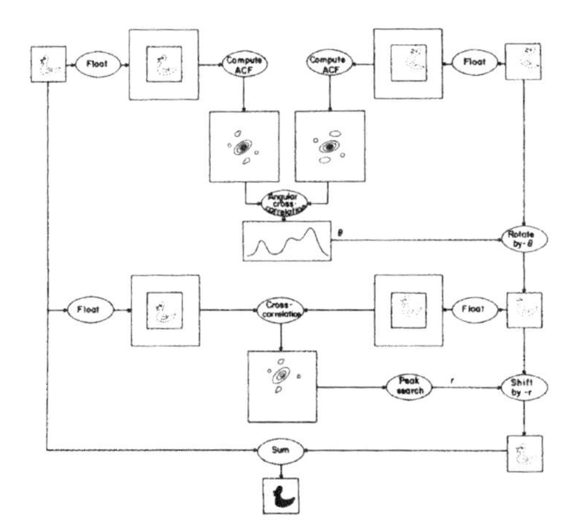

Figure 4. Schematic of image alignment making use of the translation-invariance of the autocorrelation function. (reproduced with permission from Kessel et al., 1980).

MOVE TO THE WADSWORTH CENTER: FROM CONCEPT TO PRACTICE

My appointment in 1975 as Senior Research Scientist at the Division of Laboratories and Research of the New York State Department of Health (DLR, later named *Wadsworth Center*) in Albany, New York offered me the opportunity to explore this idea with practical applications. (I had been asked to start an image processing group at DLR by Donald Parsons, a Roswell Park, Buffalo research scientist who was in the process of moving the Albany and setting up a high-voltage EM facility there).

With the help of micrographs provided by *David Eisenberg, Tim Baker, Peter Zingsheim* and *Miloslav Boublik* I was able to demonstrate the feasi-

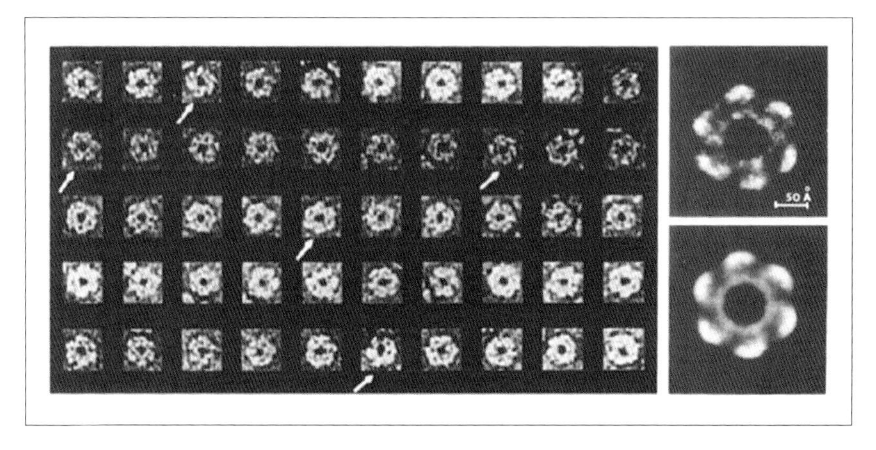

Figure 5. Single-particle averages obtained from images of negatively stained glutamine synthetase. Left: gallery of particles selected from the micrograph and aligned. Right: averages with and without six-fold symmetrization. (Reproduced from Frank et al., 1978).

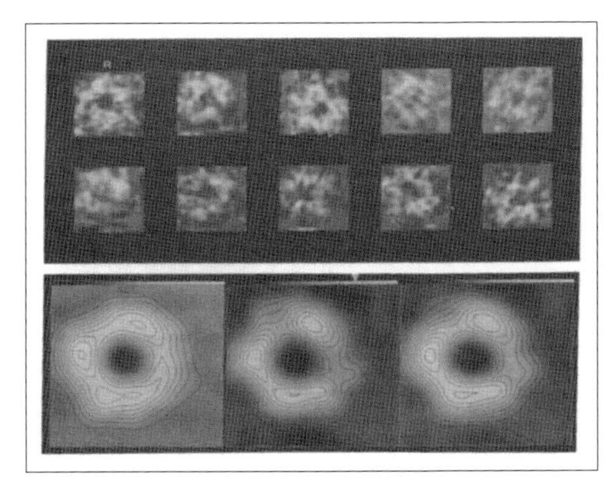

Figure 6. Single-particle averages obtained from images of negatively stained acetylcholine receptor of *Torpedo marmorata*. Top: examples for images selected from micrographs. Bottom: two half-averages and one full average (right). The average shows distinct departure from 5-fold symmetry deduced from low-resolution 2D crystals averages by other groups. (Reproduced, with permission, from Zingsheim et al., 1980).

bility of obtaining two-dimensional averages showing enhanced features of molecules with images of *glutamine synthetase* (Frank et al., 1978; Kessel et al., 1980) (Fig. 5), acetylcholine receptor (Zingsheim et al., 1980) (Fig. 6), and 40S ribosomal subunits from HeLa cells (Frank et al., 1981a) (Fig. 7).

Among these, the 40S subunit averages were arguably the most striking in showing the potential of the single-particle averaging technique, results that proved instrumental for gaining funding from the National Institutes of Health. Nonetheless, presentations of the results for glutamine synthetase, acetylcholine receptor and ribosome by myself and two of my collaborators, *Martin Kessel* and *Peter Zingsheim*, at the meeting organized by Wolfgang Baumeister in Burg Gemen, Germany (1979) were greeted with a great deal of skepticism.

One issue to be addressed, as mentioned before, was the fact that due to the absence of crystal order, the average of aligned molecule images

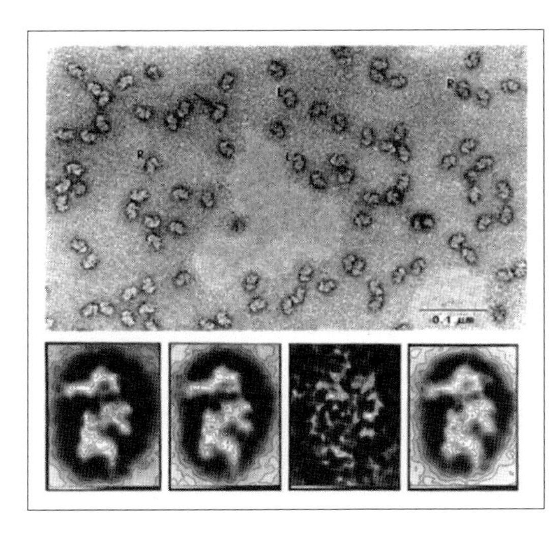

Figure 7. Single-particle averages obtained from images of 40S ribosomal subunits of HeLa cells. Top: micrograph showing 40S subunits in two views, left-facing (L) and right-facing (R). Bottom, from left to right: two half-averages, variance map, and full average of 81 L-view particles. (Reproduced, with permission, from Frank et al., 1981a).

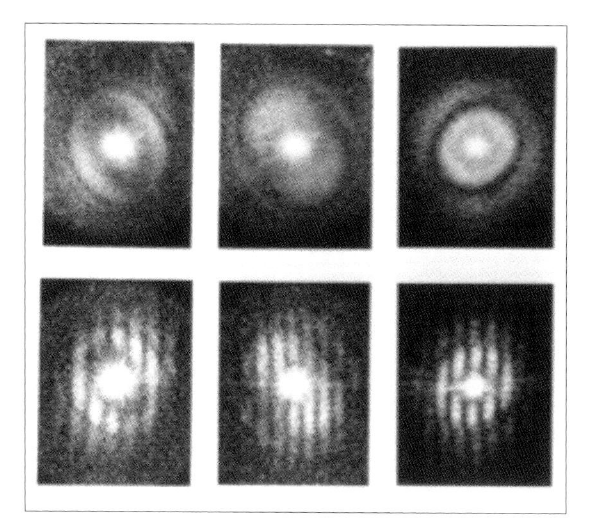

Figure 8. Reproducibility of the signal content in two successive electron micrographs of carbon film, demonstrated for three different defocus settings. Upper row: optical diffraction pattern of one of the micrographs, showing Thon patterns. Lower row: Young's fringes obtained by first aligning the micrograph pairs and then translating them relative to each other by a slight amount. (Reproduced from J. Frank, Ph.D. thesis, 1970).

shows no diffraction spots in its Fourier transform, and therefore lacks an inherent measure of resolution. Without such a measure, progress in quality could not be tracked and compared among different groups. From the earlier study, during my dissertation work, on the effects of drift on an electron micrograph (Frank, 1969), I realized that signal bandwidth is reflected by the extent of *reproducible* information in Fourier space (Frank and Al-Ali, 1975; Frank, 1976). This extent of reproducible information is apparent from the extent of Young's fringes that show up in the optical diffraction pattern when two successive micrographs of the same specimen field are superimposed with a slight shift (Fig. 8).

How could this idea be translated into a quantitative measure? The extent of reproducibility in Fourier space can be quantified computationally by dividing the data going into an average randomly in half, then comparing the Fourier transforms of half-averages over rings in Fourier space. Resolution is then defined as the Fourier ring radius where a measure of comparison, such as phase residual, or R-factor (Frank et al., 1981a), or cross-correlation ("Fourier ring correlation") (Saxton and Baumeister, 1982; van Heel et al., 1982), passes a critical threshold (Fig. 9). The same measures, computed over shells, would later prove important in estimating resolution of 3D reconstructions, as well (Harauz and van Heel, 1986).

These first studies of image averaging immediately brought up the problem of heterogeneity – only those molecule images could be reasonably combined in an average if they originated from molecules of identical structure and presented the same view. At that time, one of *Ernst van Bruggen's* students, *Marin van Heel*, visited my lab bringing with him images of *Limulus polyphemus* hemocyanin – an oligomer with distinct architecture showing multiple preferred views when negatively stained

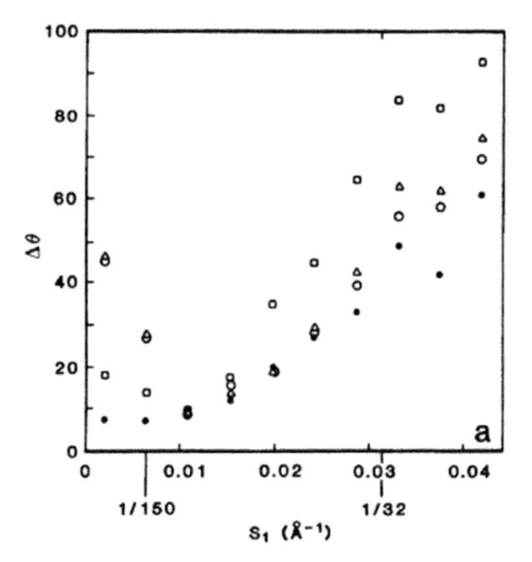

Figure 9. Resolution of single-particle averages defined by reproducibility of half-averages in Fourier space. Shown is the differential phase residual as a function of Fourier ring radius for the half-averages of 40S ribosomal subunits of HeLa cells. Resolution is then defined by the ring radius where the phase residual first exceeds 45 degrees. (Reproduced with permission from Frank et al., 1981a).

and imaged in the electron microscope (Fig. 10a). These images therefore presented a perfect example of heterogeneity. Before attempting to average those images, they had to be sorted, or classified into their subsets. The solution to this problem (van Heel and Frank, 1981; Frank 1984) came from the insight that images, once aligned with one another, may be regarded as vectors in a space of *N* dimensions, where *N* is the number of pixels. Groups of images that are similar will then show up as clusters of vectors in that space. Equivalent problems of finding clusters in high-dimensional space had been encountered in many fields of science, and gave rise to multivariate statistical analysis, a procedure which determines a compact low-dimensional subspace tailored to the problem. With the help of *Jean-Pierre Bretaudiere*, a Wadsworth Center scientist working in Laboratory Medicine, we were able to use a program meant to sort blood samples to sort images instead (see Mossman, 2007, where this episode is recounted). Application to hemocyanin proved an immediate success (Fig. 10b, c).

Early on, as I set out on the single-particle approach to recovering structure, it became clear to me that in order to make systematic progress in the development of algorithms and computer programs with ever-changing and expanding goals required a workbench with a large set of tools. To this end I developed a modular image processing system called *SPIDER* (for *S*ystem for *P*rocessing of *I*mage *D*ata in *E*lectron microscopy and *R*elated fields) (Frank et al., 1981b), which made it possible to design complex programs from pre-coded building blocks using a simple script language. For example, the command WI would invoke a routine for extracting a rectangular portion of an image, FT would invoke Fourier transformation, and AC would compute the autocorrelation of an

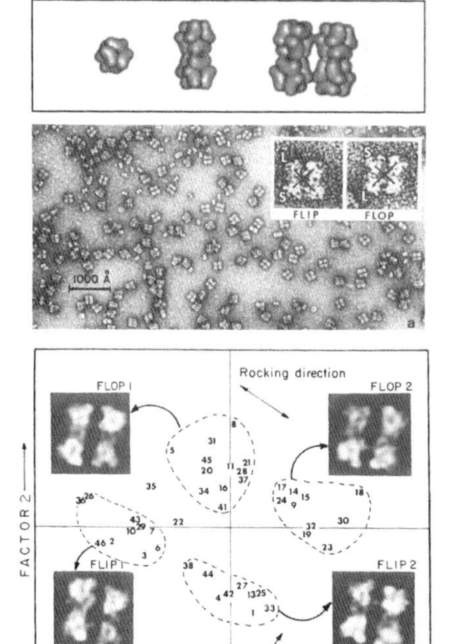

Figure 10. Sorting of hemocyanin images by Correspondence Analysis, a branch of multivariate statistical analysis. Top: Make-up of the dodecameric molecule of Limulus polyphemus hemocyanin. The slightly rhombic, twisted arrangement of the subunits creates a nonplanar architecture, reflected by the rocking of the molecule on the grid. Middle: Micrograph of negatively stained molecules showing them in different three-dimensional positions, related by flipping and rocking. Bottom: Factorial map, obtained by multivariate data analysis of the aligned molecule images, separates the images into four clusters. (Reproduced, with permission, from van Heel and Frank, 1981).

image. Hundreds of commands were implemented over the course of the next few years.

All programs were coded in FORTRAN, the most advanced language at the time. In most of the initial programming I was assisted by Helen Dowse, a SUNY Albany student of Computer Science, and Brian Shimkin, an undergraduate. As the functionality of SPIDER expanded, its script language became literally the *lingua franca* in my lab and, as the suite was disseminated to other labs, within a growing community of users. As noted earlier, I trace the idea underlying the SPIDER system and its modular design back to my stay at the *Jet Propulsion Lab* in 1970 under the Harkness Fellowship, where I became familiar with JPL's own VICAR image processing system.

DETERMINATION OF ANGLES AND THREE-DIMENSIONAL RECONSTRUCTION

For computing the 3D structure of an object from its projections, one requires a fairly even coverage of the whole view range, and the angles of each projection must be known. Thus, in the single-particle approach, determination of the angles of randomly oriented molecules recorded in a micrograph was the most important yet most difficult problem to be

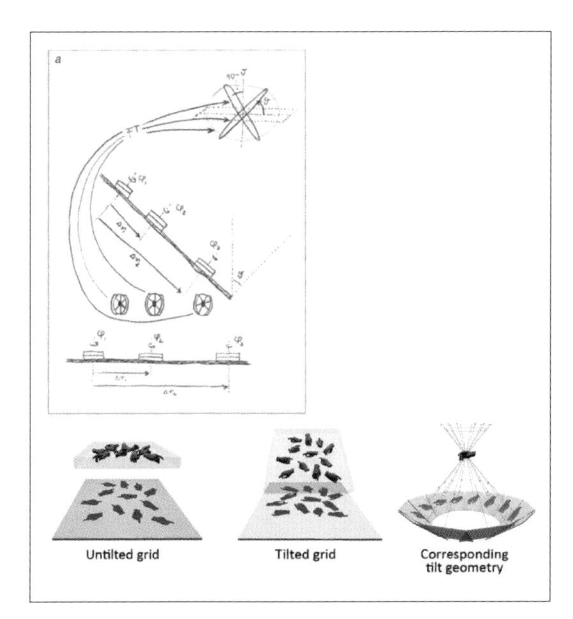

Figure 11. Random-conical data collection geometry.
Top: Concept. Untilted grid is shown with molecules attached with the same face but different azimuths. Tilting of the grid by a large angle results in a unique direction of projection for each molecule. In Fourier space these correspond to intersecting central sections. (Frank, 1979; hand-drawn sketch on an overhead transparency, unpublished).
Bottom: Illustration of data collection, and equivalent conical geometry. (Reproduced with permission from Frank, 1998).

solved. The solution came from the insight that two micrographs, one of a field of untilted particles, one of the same field tilted by a large angle, contained all the information required to assign Eulerian angles to each tilted particle (Frank et al., 1978; Radermacher et al., 1986; 1987a). In this geometry (Fig 11a), the angles of the tilted projections lie on a cone with random azimuths (Fig. 11b), a feature which would later give rise to the term "random-conical reconstruction."

In 1982 I was joined by physicist *Michael Radermacher*, also a student of Walter Hoppe, who had worked in his dissertation project on algorithms for 3D reconstruction from projections arranged in a regular conical geometry. Thus, he had the perfect background required to develop computer programs that implemented the concept of the random-conical reconstruction. One important step was still missing, though: the generalization of the 3D reconstruction algorithm, which assumed regularly spaced conical tilting, to the general case of random angles. Once this had been accomplished, as reported in a short communication in 1986 (Radermacher et al., 1986), we obtained the first single-particle reconstruction using the random-conical method: the 50S subunit of the *E. coli* ribosome (Radermacher et al., 1987b) (Fig. 12). It is now on permanent exhibit in the Nobel Museum in Stockholm in the form of a transparent contour stack mounted in a wooden frame.

This reconstruction was limited in quality by two factors: one was the missing cone of information in the 3D Fourier transform, the source of unidirectional artifacts in the 3D density map, and the other were the arti-

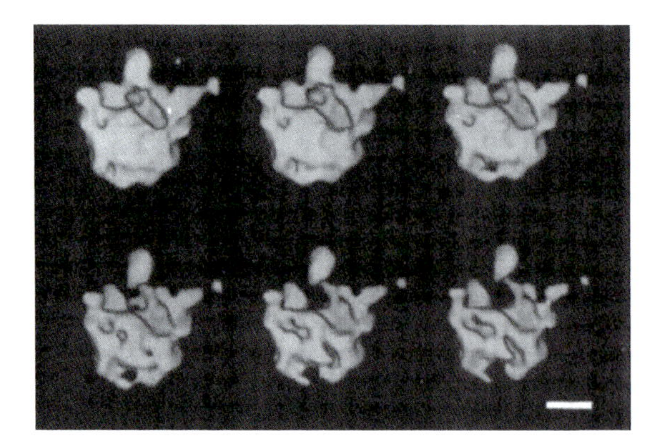

Figure 12. First single-particle reconstruction of an asymmetric molecule: the 50S subunit of the E. coli ribosome, prepared by negative staining. Scale bar is 100 Å. The panels A–F depict the molecule with increasing density threshold, using a then-novel surface representation technique (Radermacher and Frank., 1984).

facts due to the preparation of the sample by air-drying and negative staining. Both limitations were readily overcome within a short period of time: the missing cone problem was solved by merging datasets obtained with three or more different zero-degree views (Penczek et al., 1992), and the preparation of the sample with negative staining was replaced by cryo-embedding in vitreous ice, following the spectacular success of Jacques Dubochet's vitrification method by plunge-freezing into liquid ethane (Dubochet and McDowall, 1981) in the application to viruses (Adrian et al., 1984).

Yet another important problem to be addressed, which affected the quality of all reconstructions from EM data, was the modulation of the image transform by the CTF. I had first worked on this problem during my dissertation work, then through contributions to specific issues

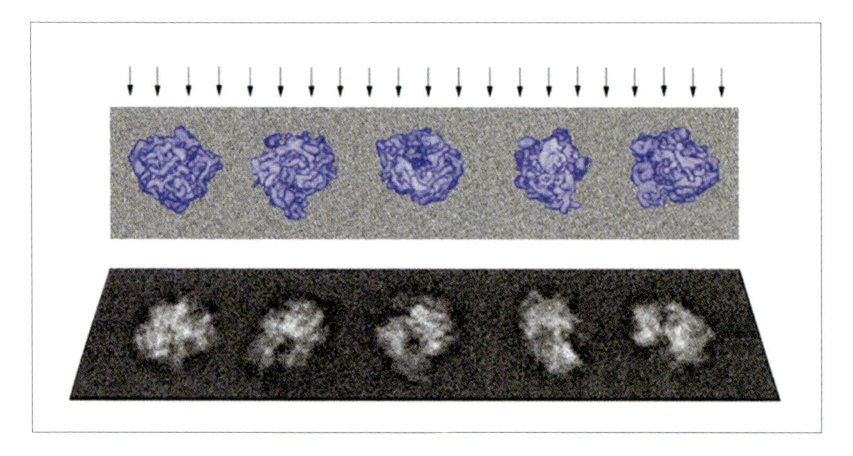

Figure 13. Schematic of single-particle data collection for molecules randomly oriented and embedded in vitreous ice. The molecules are fully hydrated in an aqueous medium and, in contrast to the preparation with negative staining and air-drying shown in Fig. 2, they exhibit no shrinkage in the direction normal to the grid plane.

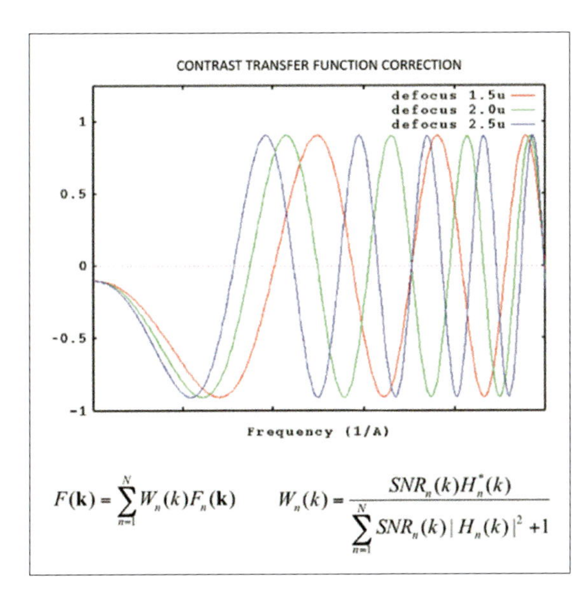

Figure 14. Contrast transfer function correction using Wiener filtering. The Fourier transform of the corrected image is obtained by a weighted sum of Fourier transforms of the defocus series. The weights Wn are given by the Wiener filter, which is proportional to the CTF Hn weighted by the signal-to-noise ratio SNRn. (Damping due to partial coherence is not shown here for simplicity). (Reproduced, with permission, from Penczek et al., 1997).

(Frank, 1972; Frank, 1973; Wade and Frank, 1977). This problem and its resolution-limiting effects on the reconstructed density maps were eventually overcome through the merging of data obtained with different defocus settings using a Wiener filtering algorithm (Zhu et al., 1997; Penczek et al., 1997) (Fig. 14).

The first molecules we visualized by cryo-EM and reconstructed in three dimensions with the single-particle methods described above were the *E. coli* ribosome (Frank et al., 1991; Penczek et al., 1992; Frank et al., 1995), hemocyanin (Lambert et al., 1994), and calcium release channel (Radermacher et al., 1992; Rademacher et al., 1994; Wagenknecht et al., 1994) (Fig. 15).

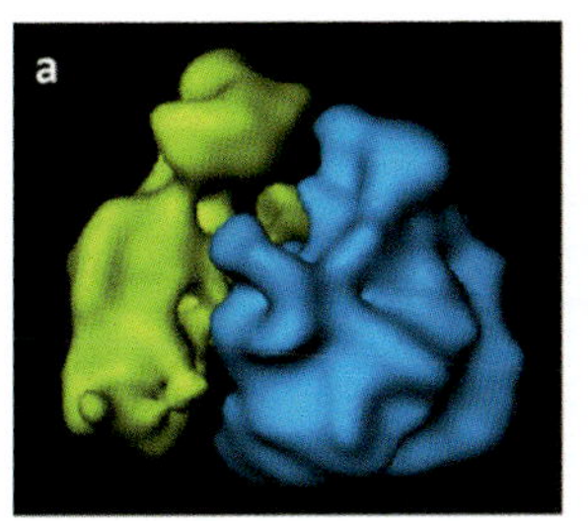

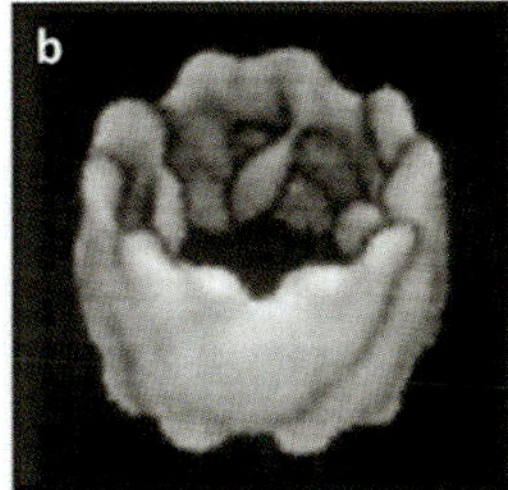

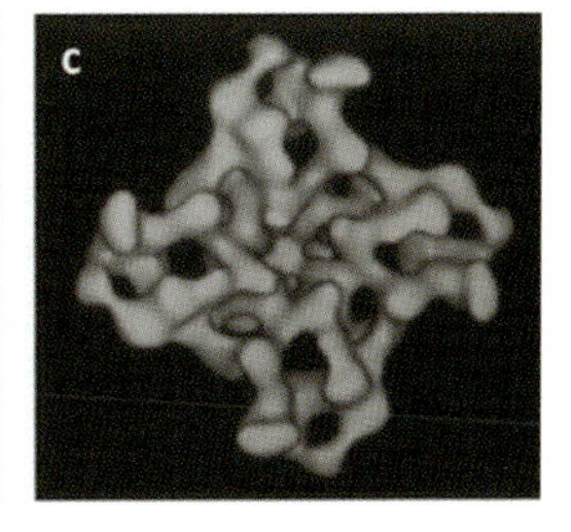

Figure 15. Cryo-EM reconstructions of three molecules obtained with the matured single-particle reconstruction method: (a) E. coli ribosome (Frank et al., 1995), (b) Octopus hemocyanin (Lambert et al., 1994), (c) calcium release channel/Ryanodine receptor (Radermacher et al., 1994).

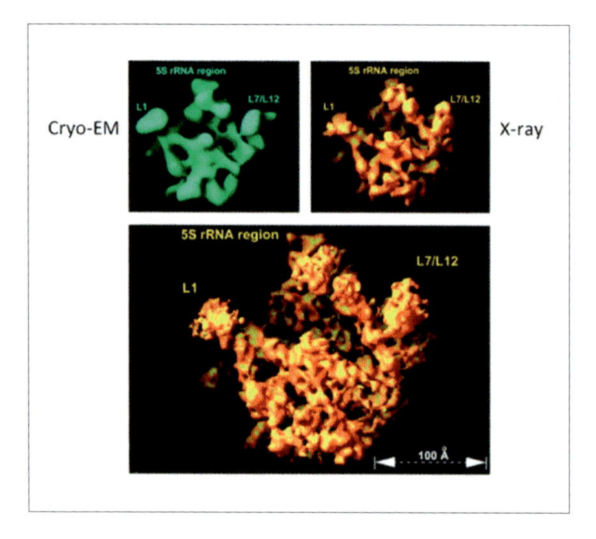

Figure 16. Cryo-EM reconstruction of the large ribosomal subunit from Haloarcula marismortui, used for solving a phasing ambiguity in solving the X-ray structure. (Reproduced, with permission, from Ban et al., 1998).

A cryo-EM reconstruction of the H*aloarcula marismortui* ribosome (Fig. 16) proved to be helpful in the phasing of the first X-ray structure of the large ribosomal subunit (Ban et al., 1998; Ban et al., 2000; Steitz, Nobel lecture 2009).

The quality of final reconstructions benefits from iterative angular refinement that starts out with the first rough reconstruction. In any project dealing with a molecule whose structure is unknown, it is practical to make a distinction between two phases, a *bootstrap phase* and a *refinement phase.* In the bootstrap phase, a first rough reconstruction is obtained – either by the random-conical method, or by an alternative method, developed by Marin van Heel and others, in which common lines in Fourier space are employed (Goncharov et al., 1987; van Heel, 1987; Penczek et al., 1996; van Heel et al., 1997). In the refinement phase (Penczek et al., 1992; 1994), an existing reconstruction (i.e., the density map obtained by a bootstrap reconstruction) is used to generate a library of even-spaced projections with which each of the experimental projections is compared to assign refined angles to it for the next round of reconstruction (Fig. 17).

The mid-90s marked the time when a methodology of 3D reconstruction in EM could first be discerned in outline, as best documented in various contributions to the Proceedings of the 15th Pfefferkorn Conference (1997). At about that time I compiled a book for the first time summarizing computational methods of single-particle 3D EM (Frank, 1996; re-edited in Frank, 2006).

From the time where the first refined, CTF-corrected reconstructions were obtained, in the mid-90s, more than 15 years had to go by before the "resolution revolution" brought us to the resolution, 2–4Å, where atomic

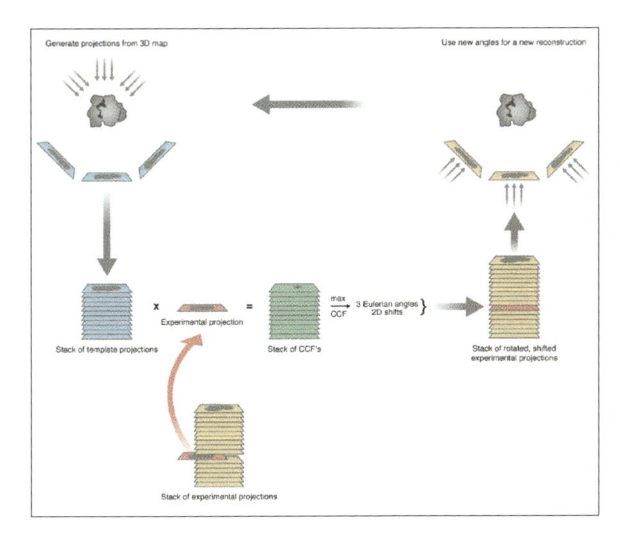

Figure 17. Iterative angular refinement scheme by projection matching, a scheme that underlies practically all cryo-EM reconstructions. (Reproduced with permission from Frank, 2011).

modeling becomes possible. The interpretation of the many low-resolution cryo-EM reconstructions obtained during that time was often dismissed as "blobology," a characterization that was unjust and unfair in most instances. In the following I would like to make this point by showing a few examples just from the area I'm most familiar with – the structural basis of protein biosynthesis. These examples demonstrate that well before it reached the present state of perfection, cryo-EM gave us important insights into pivotal processes of translation by the ribosome. This happened on the level of resolution that allowed the constellations and movements of entire domains to be described.

In the late 90s, one of my postdocs, *Rajendra Agrawal*, prepared a sample containing elongation factor G (EF–G) bound to the ribosome as it

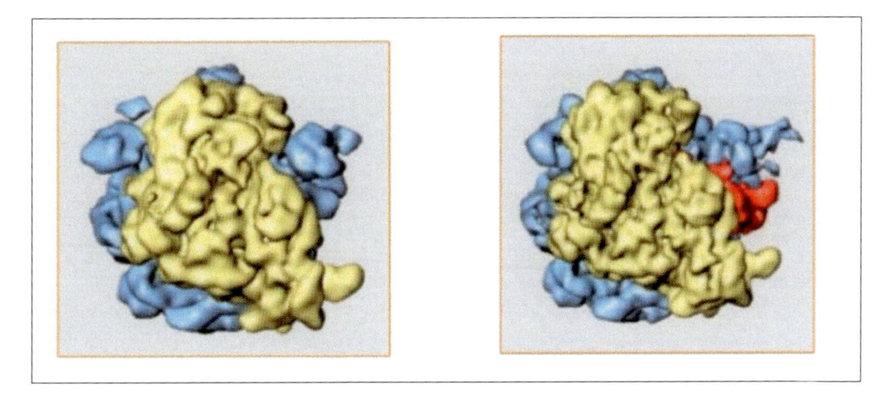

Figure 18. Ratchet-like motion of the E. coli ribosome during mRNA-tRNA translocation. Upon binding of EF-G, the small subunit (yellow) is seen to rotate relative to the large subunit (blue). (Reproduced with permission from Valle et al., 2003a).

catalyzes translocation of mRNA and tRNAs (Agrawal et al., 1998, 1999). He used a GTP analog to arrest the factor at the point where GTP hydrolysis is normally triggered. Comparison of the cryo-EM reconstruction with that of the unbound ribosome showed a dramatic change: the small subunit had rotated by seven degrees with respect to the large subunit (Frank and Agrawal, 2000; Valle et al., 2003a) (Fig. 20). This finding of the "ratchet-like" motion provided first clues on the mechanism of mRNA-tRNA translocation.

In 2002, *Mikel Valle*, another of my postdocs, found that during the decoding process, aminoacyl-tRNA (aa-tRNA) enters the ribosome in complex with the protein factor EF-Tu in a strongly distorted form, in the so-called A/T state (Valle et al., 2002; 2003b) (Fig. 19). In this case the antibiotic kirromycin was used to keep the factor from leaving the ribosome after GTP hydrolysis. This observation indicated that the tRNA acts as a molecular spring, apparently setting the threshold for discrimination between cognate and near-cognate codon-anticodon pairing (Yarus et al., 2003). A later study in my lab confirmed that the same mechanism holds for all three classes of tRNA (Li et al., 2008).

To give a third example, we teamed up in 2001 with *Jennifer Doudna*, then at Yale, to visualize the mRNA from hepatitis C virus in the process of hijacking the human ribosome. My postdoc *Christian Spahn*, working alongside *Jeff Kieft* from the Doudna team, discovered the way the so-called IRES element of the virus' mRNA engages the small subunit of the ribosome (Spahn et al., 2001).

Once density maps reached the sub-nanometer mark, it became possible to interpret them on the basis of existing structures and to build "quasi-" atomic models. Over the decade following the appearance of the ribosome X-ray structure, my group spearheaded three methods of such interpretations: real-space refinement in which entire domains were fitted

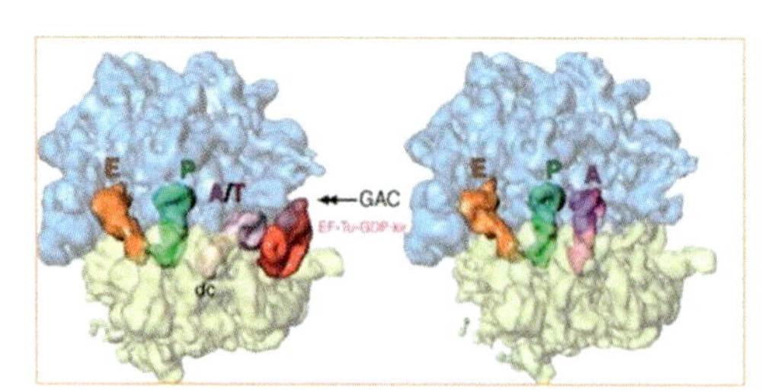

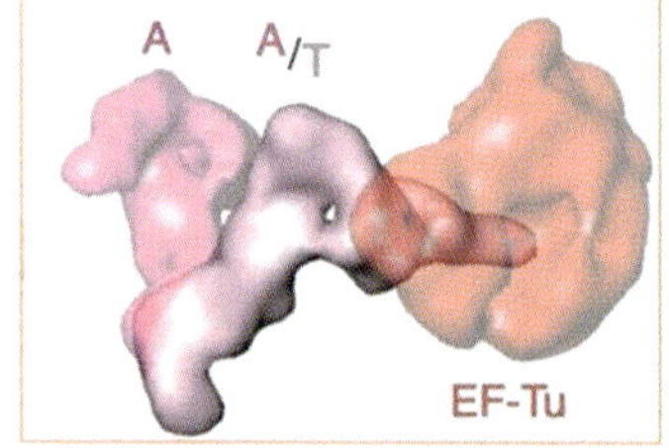

Figure 19. The tRNA captured in the process of entering the ribosome. As tRNA arrives in ternary complex with elongation factor Tu (EF-Tu) and GTP, it binds the ribosome in a strongly distorted form, first seen by cryo-EM using kirromycin. Left: Juxtaposition of ternary complex entering the ribosome, with tRNA in the distorted A/T state, and ribosome after departure of EF-Tu and accommodation of cognate tRNA. Right: tRNA conformations in A/A and A/T states. (Reproduced with permission from Valle et al., 2003b).

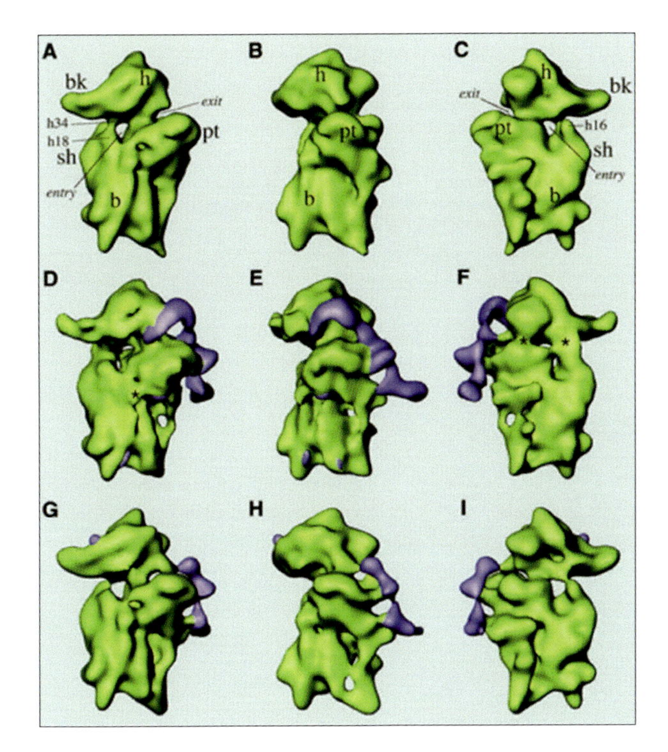

Figure 20. (A)–(C) 40S ribosomal subunit of HeLa cells (control) shown in different views; (D)–(F) 40S ribosomal subunit of HeLa cells with hepatitis C virus IRES bound; (G)–(I) same as (D)–(F) after removal of one domain. (Reproduced with permission from Spahn et al., 2001).

as rigid bodies (Gao et al., 2003; Gao and Frank, 2005), normal mode analysis (Tama et al., 2003), and *Molecular Dynamics Flexible Fitting,* or MDFF (Trabuco et al., 2007). More about the use of MDFF below.

MOVE TO COLUMBIA UNIVERSITY – STORY IN A SAMPLE AND RESOLUTION REVOLUTION

In 2008 I joined Columbia University and the faculties of two departments: Biochemistry and Molecular Biophysics, and Biological Sciences. What attracted me particularly were opportunities for many collaborations on the two vibrant campuses at the Medical Center and Morningside.

Two examples for MDFF fitting among many obtained in my lab fall into this next stage of my career: (i) a 6.7Å density map representing a snapshot of the decoding process (LeBarron et al., 2008) and its interpretation by *Elizabeth Villa* (Villa et al., 2009) (Fig. 21, left), and (ii) the atomic model of the ribosome from *T. brucei* built from a 5.5Å map by *Yaser Hashem* (Hashem et al., 2013) (Fig. 21, bottom).

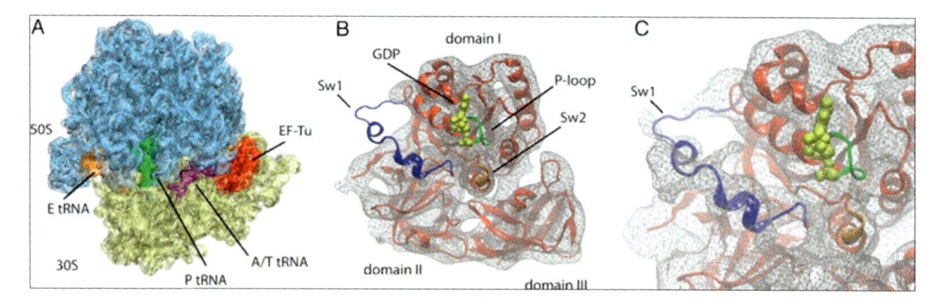

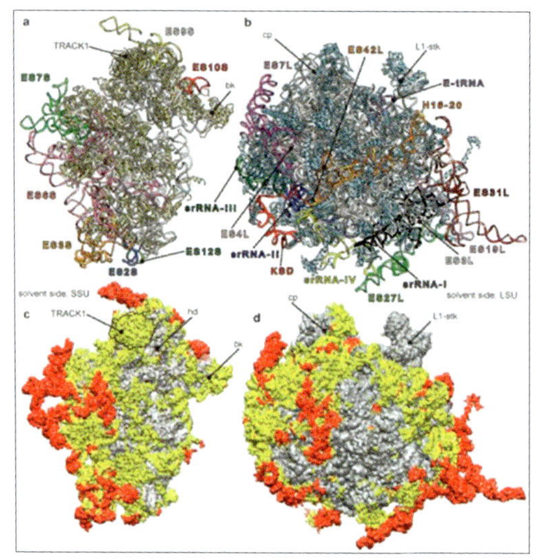

Figure 21. Top: the E. coli 70S ribosome bound with the aminoacyl-tRNA · EF-Tu · GDP ternary complex in the presence of the antibiotic kirromycin. A 6.7-Å cryo-EM density map was interpreted by atomic modeling with the aid of MDFF (Reproduced, with permission, from Villa et al., 2009).

Bottom: structure of the *T. brucei* ribosome obtained by flexible fitting of a 5.5-Å cryo-EM density map. Left: 40S subunit, right: 60S subunit. Expansion segments are painted red. (Reproduced, with permission, from Hashem et al., 2013a).

These two examples illustrate the potential of modeling using flexible fitting, but at the same time give a sense of the large effort that was required to obtain an atomic model prior to the advent of the direct electron detectors, in 2013.

Starting at my new place, I focused on a unique advantage of single-particle cryo-EM that is now utilized in many applications. Here, unlike in X-ray crystallography, molecules are unconstrained by crystal packing and thus are able to assume the full range of conformations present in solution. For a processive molecular machine such as the ribosome, unless it is stopped by a chemical intervention, many different states coexist (Schmeing and Ramakrishnan, 2009; Frank and Gonzalez, 2010). The full potential of single-particle cryo-EM to present an inventory of structures coexisting in a sample began to be realized with the introduction of maximum likelihood classification by Sjors Scheres, who worked as a student in the lab of *Jose-Maria Carazo* (Scheres et al., 2007) and later, as an independent principal investigator, at the Laboratory of Molecular Biology (LMB) of the Medical Research Council (MRC) in Cambridge (Scheres 2012). These algorithms, first introduced into single-particle processing by Fred Sigworth

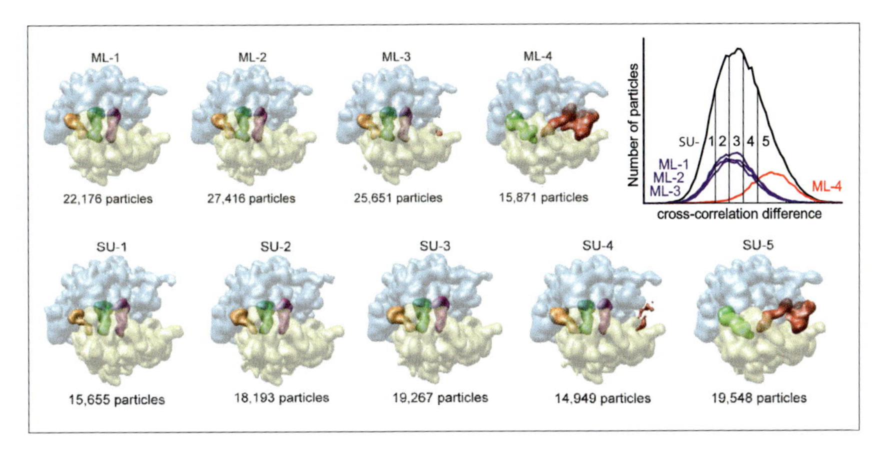

Figure 22. Maximum likelihood classification of ~90,000 images of the *E. coli* ribosome with and without EF–G bound. Upper row: maximum-likelihood classification into four classes. Lower row: supervised classification into five classes, using an EF–G-bound ribosome as a reference. (In this inferior method, classes ar defined and ordered by increasing similarity with the reference. Adapted, from Scheres et al., 2007).

(1998), perform the seemingly impossible task of disentangling changes of view angle from changes in structure (Sigworth, 2007). A large well-characterized ribosome data set we supplied was correctly (i.e., conforming with supervised classification, as the ground truth was unknown) classified in the initial study by Scheres et al. (Fig. 22). Working with this toolset, multiple structures are now routinely recovered from a single sample (Agirrezabala et al., 2012; Hashem et al., 2013b; Budkevich et al., 2014; des Georges et al., 2016; Loveland et al., 2017).

Four years after my move to Columbia University the field was revolutionized with the introduction of commercial direct electron detecting cameras already mentioned above. The fact that the poor quality of image recording presented a serious bottleneck in attempts to reach high resolution was implicit in the results of an earlier study of Richard Henderson (Henderson, 1995). Since 2013, as a consequence of the superior performance of the new cameras, a large number of cryo-EM maps have been obtained at near-atomic resolution (2–4 Å). Cryo-EM has become a mainstream technique of structural biology (Nogales, 2016).

Again resorting to the ribosome to make a general point, I show two examples for the "Story in a Sample" paradigm at near-atomic resolution. In one study, several states of the elongation cycle are "fished out" from images of ribosomes purified from a cell extract of *Plasmodium falciparum*, the malaria parasite (Sun et al., 2015) (Fig. 23 left). These states proved to correspond to those earlier identified by trapping the eukaryotic ribosome from yeast or mammals with GTP analogs. In another study by my group, part of the elongation cycle is visualized in a sample

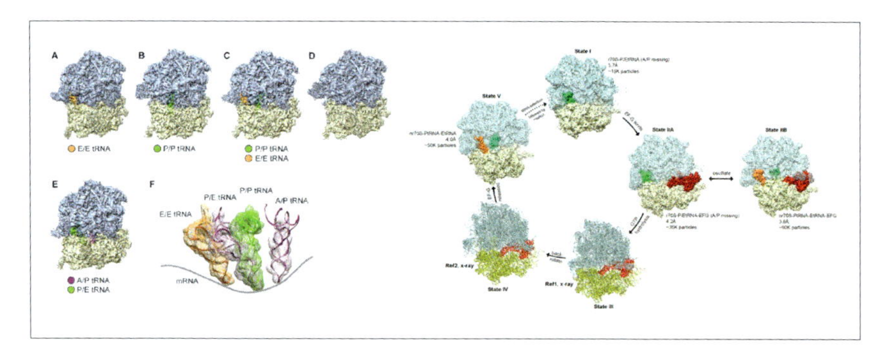

Figure 23. Two examples for determination of multiple structures from the same sample. Left: five 80S structures from Plasmodium falciparum ribosomes purified from a cell extract. Four of these (upper row) bear tRNAs and one (bottom) is empty. Inventory of all tRNA positions observed is on bottom right. (Reproduced, with permission, from Sun et al., 2015). Right: four E. coli 70S ribosome structures obtained from a sample where EF–G.GTP is present and bears a mutation H74A on the Switch 2 loop, slowing down GTP hydrolysis by a large factor (Li et al., 2015). The four structures are arranged along with two models in the order of the translation elongation cycle.

containing a mutant of EF–G (Li et al., 2015) (Fig. 23 right). Literally, cryo-EM is now able to tell us a story from a single sample of molecules in equilibrium, showing how the molecule changes its shape and binds or sheds ligands.

This potential for resolving dynamic changes of molecules in equilibrium is augmented by the development of time-resolved techniques that are able to trap short-lived states evolving in a non-equilibrium experiment (Berriman and Unwin, 1994; Chen and Frank, 2015; Chen et al., 2015; Fu et al., 2016; Lu et al., 2009; Shaikh et al., 2014). Ultimately, even the recovery of a continuum of structures reflecting the states of a biological molecule at work, and the mapping of its free-energy landscape are no longer distant goals (Dashti et al., 2014; Dashti et al., 2017; Frank and Ourmazd, 2016). This most recent development is made possible by our ability to collect large quantities of data, ensuring that even states encountered with low probability are represented in the ensemble.

Even though I have illustrated the progress achieved by using my favorite molecule, the ribosome (best resolution thus far at 2.5 Å – Liu et al., 2016; Fig. 24a), the range of applications in biology is virtually unlimited except for some lower bound on molecule size, and, of course, the necessity of having the molecule suspended in solution. My own recent collaborations with the groups of *Andrew Marks, Wayne Hendrickson, Alexander Sobolevsky,* and *Filippo Mancia* demonstrate the gain in knowledge achievable now for membrane-bound channels and receptors (Zalk et al., 2015; des Georges et al., 2016; Twomey et al., 2016; 2017a; 2017b; Chen et al., 2016) (Fig. 24 b,c).

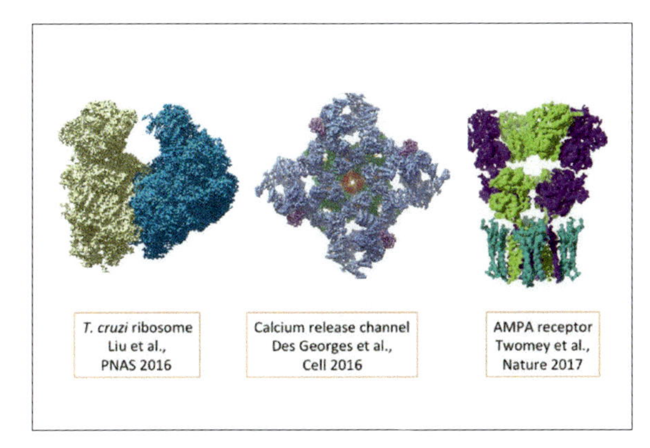

Figure 24. Three recent reconstructions at near-atomic resolution obtained in my lab in various collaborations. Left: *T. cruzi* ribosome. Liu et al., 2016)(reproduced with permission, from Frank, 2017); Center: Calcium release channel. (reproduced with permission, from des Georges et al., 2016); Right: AMPA receptor (reproduced, with permission, from Twomey et al., 2017b).

CONCLUSIONS

Structural Biology, as a field, is in the process of being remade as the relative contributions of X-ray crystallography, NMR and cryo-EM are being reevaluated and repositioned. This is a process that will go on for a number of years, whose dynamics are difficult to predict. Right now there are three points to be made: (i) As there is no need for crystals, a large gap in structural knowledge can be filled now, particularly concerning membrane-bound channels and receptors, but also large molecular assemblies with high flexibility. (ii) The lower size limit on molecules is rapidly receding as phase plates improve contrast and cameras promise to become even more powerful. The recent determination of the structure of hemoglobin (Koshouei et al., 2017), with 64 kD molecular weight, is a feat once unthinkable. (iii) The ability to recover multiple structures, or even a continuum of structures, from a single sample sets off cryo-EM sharply from X-ray crystallography, and promises to give us very detailed information on molecular mechanisms of ligand binding, allosteric switching, and gating events.

To conclude this account of a 40-year journey, I must confess that even though I was always a firm believer in the technique I dreamed up many years ago, I never thought I would see it come to fruition in this spectacular way – let alone that it would earn me a share of the highest prize coveted by scientists all over the world. It has been a truly fantastic journey overall.

Knowing that Human Medicine as a whole stands to benefit from the new technology is ultimately the best reward.

ACKNOWLEDGMENTS

In my account of more than four decades of work I have focused on contributions of my own lab. It goes without saying that as the field devel-

Figure 25. Albany reunion picture of then-current and former postdocs and colleagues, at the 1999 Gordon Conference for 3D Electron Microscopy. Upper row: Nicolas Boisset, Michael Radermacher, Joachim Frank, Carmen Mannella, Roland Beckmann, Martin Kessel, Pawel Penczek, Bruce McEwen, Jose-Maria Carazo. Lower row: Rasmus Schroeder, Montsertat Samso, Rajendra Agrawal, Bob Grassucci, and Holland Cheng.

oped many people have contributed to it worldwide, in areas reaching from sample preparation over instrument design and the development of computational tools, as recounted in many recent reviews (e.g., Nogales, 2016). As a community the cryo-EM crowd is unusual in its spirit of cooperation and mutual respect, as reflected in the cordial atmosphere at topical meetings such as the Gordon Conference on 3DEM and the so-called Hybrid Meetings at Lake Tahoe.

I should at this point give tribute to all the people who have made this development possible over the years. First of all, without the pivotal contributions and leadership of my co-laureates *Jacques Dubochet* and *Richard Henderson* the field would not have come together and prospered in this way. My thanks go to my students, postdocs, and collaborators who shared my vision and put in so much hard work to make it all happen (Fig. 25). *Bob Grassucci* has been with my group for more than 30 years in management and support of instruments of ever-increasing complexity.

My very special thanks goes to my wife *Carol Saginaw* who has steadily supported me for close to 40 years and never lost faith in me, and my family and non-scientist friends who have cheered me on at every turn.

Much of the earlier developments of the technique took place at the Wadsworth Center in Albany, which provided a sheltered, supportive environment with generous funding by the State of New York, particularly in the starting phase when it counted most. In my expression of gratitude to the Wadsworth Center I would like to single out *Carmen Mannella,* who supported me throughout those years as a colleague, collaborator and friend. My affiliation with the Department of Biomedical Sciences of the University at Albany since 1985 allowed me to foster academic interactions with the SUNY faculty at large. Since 2008, when I joined the faculties of Columbia's Department of Biochemistry and Molecular Biophysics and Department of Biological Sciences, I enjoyed the multifaceted intellectual environment fostered and nurtured by one of the country's greatest universities.

My work has been generously supported by the National Institute of General Medical Sciences of NIH almost without interruption since 1981. Funding for two inspiring Sabbatical stays in England and Germany was provided, respectively, by the Fogarty Foundation and the Humboldt Foundation. Over the years, I received sizeable awards from the National Science Foundation for several instrument purchases. Support by the Howard Hughes Medical Institute for almost 20 years has allowed me to build up a state-of-the-art, professionally supported cryo-EM facility first at the Wadsworth Center and later at Columbia University. It is important to point out in these times that Government support is absolutely necessary for science to prosper and, with it, the chances of success in fighting disease.

REFERENCES

Adrian, M., Dubochet, J., Lepault, J., and McDowall, A.W. (1984). "Cryo-electron microscopy of viruses," *Nature* **308**, 32–36.

Agirrezabala, X., Liao, H., Schreiner, E., Fu, J., Ortiz-Meoz, R.F., Schulten, K., Green, R., and Frank, J. (2012). "Structural characterization of mRNA-tRNA translocation intermediates," *Proc. Natl. Acad. Sci. USA* **109**, 6094–6099.

Agrawal, R.K., Penczek, P., Grassucci, R.A., and Frank, J. (1998). Visualization of elongation factor G on the *Escherichia coli* 70S ribosome: the mechanism of translocation. *Proc. Natl. Acad. Sci. USA* **95**, 6134–6138.

Agrawal, R.K., Heagle, A.B., Penczek, P., Grassucci, R.A., and Frank, J. (1999). EF-G-dependent GTP hydrolysis induces translocation accompanied by large conformational changes in the 70S ribosome. *Nat. Struct. Biol.* **6**, 643–647.

Ban, N., Freeborn, B., Nissen, P., Penczek, P., Grassucci, R.A., Sweet, R., Frank, J., Moore, P.B., and Steitz, T.A. (1998). "A 9 Å resolution X-ray crystallographic map of the large ribosomal subunit," *Cell* **93**, 1105–1115.

Ban, N., Nissen, P., Hansen, J., Moore, P.B., Steitz, T.A. (2000). "The complete atomic structure of the large ribosomal subunit at 2.4 Å resolution," *Science* **289**, 905–920.

Baumeister and Hahn (1975). "Relevance of of three-dimensional reconstruction of stain distributions for structural analysis of biomolecules," *Hoppe-Seyler's Z. Physiol. Chem.* **356**, 1313–1316.

Beer, M., Frank, J., Hanszen, K.J., Kellenberger, E., and Williams, R.C. (1974). "The possibilities and prospects of obtaining high-resolution information (below 30 Å) on biological material using the electron microscope." Some comments and reports inspired by an EMBO workshop held at Gais, Switzerland, October 1973. *Quart. Rev. Biophys.* **7**, 211–238.
Berriman, J. and Unwin, P.N.T. (1994). "Analysis of transient structures by cryo-microscopy combined with rapid mixing of spray droplets," *Ultramicroscopy* **56**, 241–252.
Budkevich, T.V., Giesebrecht, J. Behrmann, E., Loerke, J., Ramrath, D.J., Mielke, T., Ismer, J., Hildebrand, P.W., Tung, C.S., Nierhaus, K.H., Sanbonmatsu, K.Y. and Spahn, C.M. (2014). "Regulation of the mammalian elongation cycle by subunit rolling: a eukaryotic-specific ribosome rearrangement," *Cell* **158**, 121–131.
Chen, B., Kaledhonkar, S., Sun, M., Shen, B., Lu, Z., Barnard, D., Lu, T., Gonzalez, R.L., and Frank, J. (2015). "Structural dynamics of ribosome subunit association studied by mixing-spraying time-resolved cryo-EM," *Structure* **23**, 1097–1105.
Chen, B., and Frank, J. (2015). "Two promising future developments of cryo-EM: capturing short-lived states and mapping a continuum of states of a macromolecule," *Microscopy* **65**, 69–79.
Chen, Y., Clarke, O.B., Kim, J., Stowe, S., Kim, Y.-K., Assur, Z., Cavalier, M., Godoy-Ruiz, R., von Alpen, D.C., Manzini, C., Blaner, W.S., Frank, J., Quadro, L., Weber, D.J., Shapiro, L., Hendrickson, W.A., and Mancia, F. (2016). "Structure of the STRA6 receptor for retinol uptake," *Science* **353**. DOI: 10.1126/science.aad8266 .
Crowther, R.A., Amos, L.A., Finch, J.T., De Rosier, D.J., and Klug, A. (1970). "Three dimensional reconstructions of spherical viruses by Fourier synthesis from electron micrographs," *Nature* **226**, 421–425.
Crowther, R.A. (1971). "Procedures for three-dimensional reconstruction of spherical viruses by Fourier synthesis from electron micrographs.," *Phil. Trans. R. Soc. Lond.* B **261**, 221–230.
Dashti, A., Schwander, P., Langlois, R., Fung, R., Li, W., Hosseinizadeh, A., Liao, H.Y., Pallesen, J., Sharma, G., Stupina, V.A., Simon, A.E., Dinman, J., Frank, J., and Ourmazd, A. (2014). "Trajectories of the ribosome as a Brownian nanomachine," *Proc. Natl. Acad. Sci. USA* **111**, 17492–17497.
Dashti, A., Hail, D.B., Mashayekhi, G., Schwander, P., des Georges, A., Frank, J., and Ourmazd, A. (2017). "Conformational dynamics and energy landscapes of ligand binding in RyR1," bioRxiv. DOI: 10.1101/167080.
DeRosier, D.J. and Klug, A. (1968). "Reconstruction of 3-dimensional structures from electron micrographs," *Nature* **217**, 130–134.
Des Georges, A., Clarke, O.B., Zalk, R., Yuan, Q., Condon, K.J., Grassucci, R.A., Hendrickson, W.A., Marks, A.R., and Frank, J. (2016). "Structural basis for gating and activation of RyR1," *Cell* **167**, 145–157.
Dubochet, J., and McDowall, A. W. (1981). "Vitrification of pure water for electron microscopy," *J. Microsc.* **124**, 3–4.
Frank, J. (1969). Nachweis von Objektbewegungen im lichtoptischen Diffraktogramm von elektronenmikroskopischen Aufnahmen. *Optik* **30**, 171–180.
Frank (1970) Untersuchungen von elektronenmikroskopischen Aufnahmen mit hoher Auflösung mit Bilddifferenz- und Rekonstruktionsverfahren. [*Analysis of electron micrographs using image subtraction and reconstruction methods*]. Thesis. Re-printed by TUM University Press, Muenich 2019.
Frank, J. (1972). "A study on heavy/light atom discrimination in bright field electron microscopy using the computer," *Biophys J* **12**, 484–511.

Frank, J. (1973). "The envelope of electron microscopic transfer functions for partially coherent illumination," *Optik* **38**, 519–539

Frank, J. (1976). Determination of source size and energy spread from electron micrographs using the method of Young's fringes. *Optik* **44**, 379–391.

Frank, J. (1980). The role of correlation techniques in computer image processing. In *Computer Processing of Electron Microscope Images, Topics in Current Physics.* Vol. 13. P.W. Hawkes, editor. Springer, Berlin. 187–222.

Frank, J. (1984). "The role of multivariate image analysis in solving the architecture of the Limulus polyphemus hemocyanin molecule," *Ultramicroscopy* **13**, 153–164.

Frank, J. (1996). *Three-dimensional Electron Microscopy of Macromolecular Assemblies*, San Diego, Academic Press. (2nd edition: Frank, J. (2006), New York, Oxford U. Press).

Frank, J. (1998). "How the ribosome works," *American Scientist* **86**, 428–439.

Frank, J. (2011). "Visualization of molecular machines by electron microscopy," In: *Molecular Machines in Biology – Workshop of the Cell.* J. Frank, ed. Cambridge University Press, Cambridge, UK, pp. 20–37.

Frank, J. (2017). Advances in the field of single-particle cryo-electron microscopy over the last decade. *Nature Protocols* **12**, 209–212.

Frank, J., and Al-Ali, L. (1975). "Signal-to-noise ratio of electron micrographs obtained by cross correlation," *Nature* **256**, 376–379.

Frank, J., and Gonzalez, R.L. (2010). "Structure and dynamics of a processive Brownian motor: The translating ribosome," *Ann. Rev. Biochem.* **79**, 381–412.

Frank, J., and Ourmazd, A. (2016). "Continuous Changes in Structure Mapped by Manifold Embedding of Single-Particle Data in Cryo-EM," *Methods* **100**, 61–67.

Frank, J., Bußler, P., Langer, R., and Hoppe, W. (1970). Einige Erfahrungen mit der rechnerischen Analyse und Synthese von elektronenmikroskopischen Bildern höher Auflösung. [*Experience gained from digital analysis and synthesis of high-resolution electron micrographs*]. *Ber. Bunsenges. Phys. Chem.* **74**, 1105–1115.

Frank, J., Goldfarb, W., Eisenberg, D., and Baker, T.S. (1978). "Reconstruction of glutamine synthetase using computer averaging," *Ultramicroscopy* **3**, 283–290.

Frank, J., Verschoor, A., and Boublik, M. (1981a). "Computer averaging of electron micrographs of 40S ribosomal subunits," *Science* **214**, 1353–1355.

Frank, J., Shimkin, B., and Dowse, H. (1981b). "SPIDER – A modular software system for image processing," *Ultramicroscopy* **6**, 343–358.

Frank, J., Penczek, P., Grassucci, R., and Srivastava, S. (1991). Three-dimensional reconstruction of the 70S *Escherichia coli* ribosome in ice: The distribution of ribosomal RNA. *J. Cell Biol.* **115**, 597–605.

Frank, J., Zhu, J., Penczek, P., Li, Y., Srivastava, S., Verschoor, A., Radermacher, M., Grassucci, R., Lata, R.K., and Agrawal, R.K. (1995). "A model of protein synthesis based on cryo-electron microscopy of the E. coli ribosome," *Nature* **376**, 441–444.

Fu, Z., Kaledhonkar, S., Borg, A., Sun, M., Chen, B., Grassucci, R.A., Ehrenberg, M., and Frank, J. (2016). "Key intermediates in ribosome recycling visualized by time-resolved cryo-electron microscopy," *Structure* **24**, 2092–2101.

Glaeser, R.M. (1971). "Limitations to significant information in biological electron microscopy as a result of radiation damage," *J. Ultrastruct. Res.* **36**, 466–482.

Gao, H., and Frank, J. (2005). "Molding atomic structures into intermediate-resolution cryo-EM density maps of ribosomal complexes using real-space refinement," *Structure* **13**, 401–406.

Gao, H., Sengupta, J., Valle, M., Korostelev, A., Eswar, N., Stagg, S.M., Van Roey, P., Agrawal, R.K., Harvey, S.C., Sali, A., Chapman, M.S., and Frank, J. (2003). Study of the structural dynamics of the *E. coli* 70S ribosome using real-space refinement. *Cell* **113**, 789–801.

Goncharov, A.B., Vainshtein, B.K., Ryskin, A.I., and Vagin, A.A. (1987). "Three-dimensional reconstruction of arbitrarily oriented particles from their electron photomicrographs," *Sov. Phys. Crystallogr.* **32**, 504–509.

Harauz, G. and van Heel, M. (1986). "Exact filters for general geometry three-dimensional reconstruction," *Optik* **73**, 146–156.

Hashem, Y., des Georges, A., Fu, J., Buss, S.N., Jossinet, F., Jobe, A., Zhang, Q., Liao, H.Y., Grassucci, R.A., Bajaj, C., Westhof, E., Madison-Antenucci, S., and Frank, J. (2013a). "High-resolution cryo-electron microscopy structure of Trypanosoma brucei ribosome," *Nature* **494**, 385–389.

Hashem, Y., des Georges, A., Dhote, V., Langlois, R., Liao, H.L., Grassucci, R.A., Hellen, C.U.T., Pestova, T.V., and Frank, J. (2013b). "Structure of the mammalian ribosomal 43S preinitiation complex bound to the scanning factor DHX29," *Cell* **153**, 1108–1119.

Henderson, R. and Unwin, P.N.T. (1975). "Three-dimensional model of purple membrane obtained by electron microscopy," *Nature* **257**, 28–32.

Henderson, R. (1995). "The potential and limitations of neutrons, electrons and X-rays for atomic resolution microscopy of unstained biological molecules," *Q. Rev. Biophys.* **28**, 171–193.

Holmes, K. C. (2017). *Aaron Klug – A Long Way from Durban: A Biography.* Cambridge University Press.

Hoppe, W., Gassmann, J., Hunsmann, N., Schramm, H.J. and Sturm, M. (1974). "Three-dimensional reconstruction of individual negatively stained yeast fatty-acid synthetase molecules from tilt series in the electron microscope," *Hoppe Seylers Z. Physiol. Chem.* **355**, 1483–1487.

Kessel, M., Frank, J., and Goldfarb, W. (1980). "Averages of glutamine synthetase molecules as obtained with various stain and electron dose conditions," *J. Supramol Struct.* **14**, 405–422.

Khoshouei, M., Radjainia, M., Baumeister, W., and Danev, R. (2017). "Cryo-EM structure of haemoglobin at 3.2 Å determined with the Volta phase plate," *Nature Communications* **8**, doi:10.1038/ncomms16099

Lambert, O., Boisset, N., Penczek, P., Lamy, J., Taveau, J.C., Frank, J., and Lamy, J.N. (1994). "Quaternary structure of *Octopus vulgaris* hemocyanin. Three-dimensional reconstruction from frozen-hydrated specimens and intramolecular location of functional units Ove and Ovb," *J. Mol. Biol.* **238**, 75–87.

Langer, R., Frank, J., Feltynowski, A., and Hoppe, W. (1970). Anwendung des Bilddifferenzverfahrens auf die Untersuchung von Strukturänderungen dünner Kohlefolien bei Elektronenbestrahlung. [*Application of the image difference method to the analysis of structural changes of thin carbon films during electron exposure*]. *Ber Bunsenges Phys Chem* **74**, 1120–1126.

LeBarron, J., Grassucci, R.A., Shaikh, T.R., Baxter, W.T., Sengupta, J., and Frank, J. (2008). "Exploration of parameters in cryo-EM leading to an improved density map of the E. coli ribosome," *J. Struct. Biol.* **164**, 24–32.

Li, W., Agirrezabala, X., Lei, J., Bouakaz, L., Brunelle, J.L., Ortiz-Meoz, R.F., Green, R., Sanyal, S., Ehrenberg, M., and Frank, J. (2008). "Recognition of aminoacyl-tRNA: A common molecular mechanism revealed by cryo-EM," *EMBO J* **27**, 3322–3331.

Li, W., Liu, Z., Koripella, R.K., Langlois, R., Sanyal, S., and Frank, J. (2015). Activation of GTP hydrolysis in mRNA-tRNA translocation by Elongation Factor G. *Science Advances* **1** e1500169.

Liu, Z., Gutierrez-Vargas, C., Wei, J., Grassucci, R.A., Ramesh, M., Espina, N., Sun, M., Tutuncuoglu, B., Madison-Antenucci, S., Woolford, J.L., Tong, L., and Frank, J. (2016). "Structure and assembly model for the *Trypanosoma cruzi* 60S ribosomal subunit," *Proc. Natl. Acad. Sci. USA* **113**, 12174–12179.

Loveland, A.B., Demo, G., Grigorieff, N., and Korostev, A.A. (2017). "Ensemble cryo-EM elucidates the mechanism of translational fidelity," *Nature* **546**, 113–117.

Lu, T.R., Barnard, D., X., Mohamed, H., Yassin, A., Mannella, C.A., R.K., Lu, T.-M. and Wagenknecht, T. (2009). "Monolithic microfluidic mixing-spraying devices for time-resolved cryo-electron microscopy," *J. Struct. Biol.* **168**, 388–395.

Mossman, K. (2007). "Profile of Joachim Frank," *Proc. Natl. Acad. Sci. USA* **104**, 19668–19670.

Nogales, E. (2016). "The development of cryo-EM into a mainstream structural biology technique," *Nature Methods* **13**, 24–27.

Penczek, P., Radermacher, M., and Frank, J. (1992). "Three-dimensional reconstruction of single particles embedded in ice," *Ultramicroscopy* **40**, 33–53.

Penczek, P.A., Grassucci, R.A., and Frank, J. (1994). "The ribosome at improved resolution: new techniques for merging and orientation refinement in 3D cryo-electron microscopy of biological particles," Ultramicroscopy 53, 251–270.

Penczek, P.A., Zhu, J., and Frank, J. (1996). "A common-lines based method for determining orientations for N > 3 particle projections simultaneously," *Ultramicroscopy* **63**, 205–218.

Penczek, P.A., Zhu, J., Schröder, R., and Frank, J. (1997). "Three -dimensional reconstruction with contrast transfer compensation from defocus series," *Scanning Microscopy* **11**, 147–154.

Radermacher, M., and Frank, J. (1984). "Representation of three-dimensionally reconstructed objects in electron microscopy by surfaces of equal density," *J. Microsc.* **136**, 77–85.

Radermacher, M., Wagenknecht, T., Verschoor, A., and Frank, J. (1986). "A new 3-D reconstruction scheme applied to the 50S ribosomal subunit of E. coli," *J. Microsc.* **141**, RP1-2.

Radermacher, M., Wagenknecht, T., Verschoor, A., and Frank, J. (1987a). "Three-dimensional reconstruction from a single-exposure, random conical tilt series applied to the 50S ribosomal subunit of Escherichia coli," *J. Microsc.* **146**, 113–136.

Radermacher, M., Wagenknecht, T., Verschoor, A., and Frank, J. (1987b). "Three-dimensional structure of the large ribosomal subunit from Escherichia coli," *EMBO J.* **6**, 1107–1114.

Radermacher, M., Wagenknecht, T., Grassucci, R., Frank, J., Inui, M., Chadwick, C., and Fleischer, S. (1992). "Cryo-EM of the native structure of the calcium release channel/ryanodine receptor from sarcoplasmic reticulum," *Biophys. J.* **61**, 936–940.

Radermacher, M., Rao, V., Grassucci, R., Frank, J., Timerman, A.P., Fleischer, S., and Wagenknecht, T. (1994). "Cryo-electron microscopy and three-dimensional reconstruction of the calcium release channel/ryanodine receptor from skeletal muscle," *J Cell Biol* **127**, 411–423.

Saxton, W.O., and Frank, J. (1977). "Motif detection in quantum noise-limited electron micrographs by cross-correlation," *Ultramicroscopy* **2**, 219–227.

Saxton, W.O. and Baumeister, W. (1982). The correlation averaging of a regularly arranged bacterial cell envelope protein. *J. Microscopy* **127**, 127–138.
Scheres, S.H., Gao, H., Valle, M., Herman, G.T., Eggermont, P.P., Frank, J., and Carazo, J.M. (2007). "Disentangling conformational states of macromolecules in 3D-EM through likelihood optimization," *Nat Methods* **4**, 27–29.
Scheres, S.H. (2012). "A Bayesian view on cryo-EM structure determination," *J. Mol. Biol.* **415**, 406–418.
Scherzer, O. (1948). "The theoretical resolution limit of the electron microscope," *J. Appl. Phys.* **20**, 20–29 (1948).
Schmeing, T. and Ramakrishnan, V. (2009). "What recent ribosome structures have revealed about the mechanism of translation. *Nature* **461**, 1234–1242.
Shaikh, T.R., Yassin, A.S., Lu, Z., Barnard, D., Meng, X., Lu, T.M., Wagenknecht, T. and Agrawal, R.K. (2014). "Initial bridges between two ribosomal subunits are formed within 9.4 milliseconds, as studied by time-resolved cryo-EM," *Proc. Natl. Acad. Sci. USA* **111**, 9822–9827.
Sigworth F.J. (1998). "A maximum-likelihood approach to single-particle image refinement," *J. Struct. Biol.* **122**, 328–339.
Sigworth, F. J. (2007). "From cryo-EM, multiple protein structures in one shot," *Nature Methods* **4**, 20–21.
Spahn, C.M.T., Kieft, J.S., Grassucci, R.A., Penczek, P.A., Doudna, J.A., and Frank, J. (2001). "Hepatitis C Virus IRES RNA–induced changes in the conformation of the 40S ribosomal subunit," *Science* **291**, 1959–1962.
Steitz, T. (2009) "From the Structure and Funktion of the Ribosome to new Antibiotics", *Les Prix Nobel*, 179–204.
Sun, M., Li, W., Blomqvist, K., Das, S., Hashem, Y., Dvorin, J.D., and Frank, J. (2015). Dynamical features of the Plasmodium falciparum ribosome during translation. *Nucleic Acids Res.* **43**, 10515–10524.
Tama, F., Valle, M., Frank, J., and Brooks, C.L., 3rd (2003). "Dynamic reorganization of the functionally active ribosome explored by normal mode analysis and cryo-electron microscopy," *Proc Natl Acad Sci USA* **100**, 9319–9323.
Taylor, K.A. and Glaeser, R.M. (1974). "Electron diffraction from Frozen, hydrated protein crystals," *Science* **186**, 1036–1037.
Thon, F. (1966). "Zur Defokussierunsabhängigkeit des Phasenkontrastes bei der elektronenmikroskopischen Abbildung," Z. *Naturforsch.* **21a**, 476–478.
Trabuco, L.G., Villa, E., Mitra, K., Frank, J., and Schulten, K. (2008). "Flexible fitting of atomic structures into electron microscopy maps using molecular dynamics," *Structure* **16**, 673–683.
Twomey, E.C., Velshanskaya, M.V., Grassucci, R.A., Frank, J., and Sobolevsky, A.I. (2016). "Elucidation of AMPA receptor–stargazin complexes by cryo–electron microscopy," *Science* **353**, 83–86.
Twomey, E.C., Yelshanskaya, M.V., Grassucci, R.A., Frank, J., and Sobolevsky, A.I. (2017a). "Channel opening and gating mechanism in AMPA-subtype glutamate receptors," *Nature* **549**, 60–65.
Twomey, E.C., Yelshanskaya, M.V., Grassucci, R.A., Frank, J., and Sobolevsky, A.I. (2017b). "Structural bases of desensitization in AMPA receptor-auxiliary subunit complexes," *Neuron* **94**, 569–580.
Valle, M., Sengupta, J., Swami, N.K., Grassucci, R.A., Burkhardt, N., Nierhaus, K.H., Agrawal, R.K., and Frank, J. (2002). "Cryo-EM reveals an active role for aminoacyl-tRNA in the accommodation process," *EMBO J* **21**, 3557–3567.
Valle, M., Zavialov, A., Sengupta, J., Rawat, U., Ehrenberg, M., and Frank, J. (2003a). "Locking and unlocking of ribosomal motions," *Cell* **114**, 123–134.

Valle, M., Zavialov, A., Li, W., Stagg, S.M., Sengupta, J., Nielsen, R.C., Nissen, P., Harvey, S.C., Ehrenberg, M., and Frank, J. (2003b). "Incorporation of aminoacyl-tRNA into the ribosome as seen by cryo-electron microscopy," *Nat. Struct. Biol.* **10**, 899–906.

van Heel, M. and Frank, J. (1981). "Use of multivariate statistics in analysing the images of biological macromolecules," *Ultramicroscopy* **6**, 187–194.

van Heel, M., Keegstra, W., Schutter, W. and van Bruggen, E.F.J. (1982), in: *Life Chemistry Reports, Suppl. 1, The Structure and Function of Invertebrate Respiratory Proteins, EMBO Workshop*, Leeds, ed. E.J. Wood, pp. 69–73.

van Heel, M. (1987). "Angular reconstitution: a posteriori assignment of projection directions for 3D reconstruction," *Ultramicroscopy* **21**, 111–124.

van Heel, M., Orlova, E.V., Harauz, G., Stark, H., Dube, P., Zemlin, F. and Schatz, M. (1997). "Angular reconstitution in three-dimensional electron microscopy: Historical and theoretical aspects," *Scanning Microsc.* **11**, 195–210.

Villa, E., Sengupta, J., Trabuco, L.G., LeBarron, J., Baxter, W.T., Shaikh, T.R., Grassucci, R.A., Nissen, P., Ehrenberg, M., Schulten, K., Frank, J. (2009). "Ribosome-induced changes in elongation factor Tu conformation control GTP hydrolysis," *Proc. Natl. Acad. Sci. USA* **106**, 1063–1068.

Wade, R.H., and Frank, J. (1977). "Electron microscopic transfer functions for partially coherent axial illumination and chromatic defocus spread," *Optik* **49**, 81–92.

Yarus, M., Valle, M., and Frank, J. (2003). "A twisted tRNA intermediate sets the threshold for decoding," *RNA* **9**, 384–385.

Zalk, R., Clarke, O.B., des Georges, A., Grassucci, R.A., Reiken, S., Mancia, F., Hendrickson, W.A., Frank, J., and Marks, A.R. (2015). "Structure of a mammalian ryanodine receptor," *Nature* **517**, 44–49.

Zhu, J., Penczek, P.A., Schröder, R., and Frank, J. (1997). "Three-dimensional reconstruction with contrast transfer function correction from energy-filtered cryoelectron micrographs: procedure and application to the 70S Escherichia coli ribosome," *J. Struct. Biol.* **118**, 197–219.

Zingsheim, H.P., Neugebauer, D.C., Barrantes, F.J., and Frank, J. (1980). "Structural details of membrane-bound acetylcholine receptor from *Torpedo marmorata*," *Proc. Natl. Acad. Sci. USA* **77**, 952–956.

Richard Henderson. © Nobel Prize Outreach AB. Photo: A. Mahmoud

Richard Henderson

Biography

EARLY YEARS

I was born in Edinburgh, Scotland, on 19th July 1945. My mother Grace Goldie, after two weeks convalescence, took me back on the train to Berwick-upon-Tweed, England, to re-join my father John Henderson, who was a baker at Bryson's in Berwick. My mother was born in Edinburgh and my father in Tadcaster, North Yorkshire. They met in Edinburgh and started their married life in Berwick, where my father had been brought up. We lived in several rented flats in Berwick before moving when I was 3 years old to a council house across the river in Tweedmouth, where my father kept pigeons and then budgerigars. My mother persuaded Tweedmouth primary school to let me start early when I was four years old. I have no idea why she was so keen to get me out of the house. Just before my sixth birthday, my parents moved to the small rural village of Newcastleton, 2 miles north of the border between Scotland and England. My father was one of four bakers working in the local bakery, Oliver's, graduating from baking bread to cakes. I attended Newcastleton primary school for 5 years (aged 6 to 10). Towards the end of primary school, the four most academic pupils in our class of 19 (Figure 1) were selected to attend Hawick High School, which is about 20 miles north of Newcastleton and was at that time accessible by a 45-minute morning and evening journey by steam train. We would set off at 8am every morning and get back at 5pm each evening. The total travel time of over an hour per day meant we were able to complete any homework during the train journeys. At Hawick, the classes were "streamed" by academic ability, with the entry year having 13 classes of about 35 pupils per class. The four of us from Newcastleton were in the most academic A stream (see Figure 2) and thrilled to be taught Chaucer, Shakespeare, Latin and French as well as Science, Mathematics and my favourite Metalwork. I also remember being delighted to find that mathematics lessons were subdivided into algebra, geometry and trigonometry.

Figure 1. Primary school photo of an outing to Silloth, a seaside resort in Cumberland across the border in England, when I was 10 years old. The photo shows pupils from two classes, ours and the one from the year ahead. My three classmates Robert Davidson and Foster Harkness (rear left to right), and Maurice Carruthers (front right), who subsequently travelled by train to Hawick High School every day, are circled, with Richard Henderson in the centre of the front row.

My great aunt, who owned a corner shop in Edinburgh, sent me every week copies of all the children's comics that were published in the 1950s, so I did read but definitely not literature. Although my mother tried to persuade me to read when I was young, I did not succeed in completing any novel until the compulsory school English syllabus forced me to read Walter Scott's "Heart of Midlothian", which I managed by reading 12 pages each night. When my paternal grandmother died, she left me a set of Arthur Mee's "Children's Encyclopedia", which kept me occupied, especially on "Things to make and do". Most children left school aged 15 at that time, so in our fifth and sixth form examinations, then called Highers and Lowers in Scotland, the entry year of about 400 pupils had dropped to about 55. For my final two years, at age 15 in Hawick and 16 in Edinburgh, my parents received a £50 family allowance to encourage them to encourage me to continue into higher level education.

When I was 15, the bakery where my father worked in Newcastleton ran into financial troubles. My father resigned and took a new job in Edinburgh about 3 months before I would sit my first national school examinations in June. My mother and younger brother Ross moved with him to Edinburgh, but it was decided that it would be best if I stayed on for 3 months with our next-door neighbours in Newcastleton, the Zurbriggens, so that I could sit the examinations without having to move to a new school. For my final

Figure 2. Hawick High School class 1A photo 1956. Three people from Newcastleton, circled left to right, are Richard Henderson, Robert Davidson and Maurice Carruthers.

(6th) year of High School, I moved to Boroughmuir in Edinburgh, following a recommendation by John Low, the headmaster at Hawick.

During that final school year at Boroughmuir, we were all asked whether anyone would like to apply to Oxford or Cambridge University, for which extra lessons geared to their entrance exams would be given, but only one Latin and Greek scholar decided to apply, unsuccessfully. Everyone else decided to apply to Edinburgh University, so when I started a Physics degree course at Edinburgh, I had four or five classmates from Boroughmuir and a similar number from Hawick. Our fourth and final year in Physics at Edinburgh University had a class of 45, including 4 others from Boroughmuir and one other from Hawick. Our final year Physics class photo is shown in Figure 3. The large representation from Boroughmuir arose from the enthusiastic teaching of our Boroughmuir physics teacher Bill Cow, or "Bilko". Bill once played a recording to the 6th form physics class that he had made of a lecture by Dr Jack Dainty, Reader in Biophysics at Edinburgh University, in which he talked about his work on ion fluxes in *Nitella*, algae with giant cells. Bill's view was that biophysics was an important developing area in the future of physics. Although it did not make a deep impact on my thinking at the time, it is possible that my later decision to follow a career in biophysics derived from this initial exposure to Bill Cow's enthusiasm.

Figure 3. Final year Physics IV class photo in 1966, with those from Boroughmuir, David Hogg, Andrew White, Harry Dooley, Brian Renwick and Richard Henderson (left to right) circled in red. Brian Mitchell from Hawick (yellow), and Craig Mackay who also came to Cambridge and became an astronomer (blue) are also marked. Andrew and Brian were the co-owners of our first, very old car.

My mother, Grace, and both grandmothers were my strongest supporters; my father and grandfather were very busy working and often tired after working all day. My mother had to leave school at 14 to help earn money for her family (her father had been unemployed from 1930–1938 in the Depression), but had really wanted to continue in school, so my higher education allowed her vicariously to fulfil some of her aspirations. She was very supportive and delighted when I did well academically.

The following are some brief memories of the four schools I attended.

1. My first school was Tweedmouth West First School (at age 4–5), which was a 10-minute walk from our house. I can remember even then being fascinated by numbers, and less interested in literary topics. There was a strong emphasis on learning arithmetic skills, and I can remember reciting the "times tables" while walking to school. My paternal grandmother, Jinny Henderson, lived next door to the school, and a great aunt on my father's side a few doors farther down the same street, so occasionally I would visit my grandmother for lemonade and a biscuit on my way home.

2. When we moved to Scotland, I attended Newcastleton Primary School. Newcastleton is a small village midway between the Scottish and English towns of Hawick and Carlisle, each about 20 miles away in opposite directions. Its population then was about 800, but this has since dwindled to about 600.The two teachers I remember were Miss Russell and Mrs Fleming (when I was aged 6–10). During the summer when we moved, my mother realised that the Scottish schools were ahead of the English schools in their syllabus, so I was set some exercises by Miss Russell during the summer holidays, so that I could catch up. One difference was that my new classmates had all graduated to writing with "joined up" letters, whereas in Tweedmouth everyone was still writing using separate "printed" lettering. In those days also, the UK currency consisted of pounds, shillings and pence. So, when our class was set some problems involving the long-division of money, which my new schoolmates had already been taught, I remember having no trouble completing the task from first principles and getting the correct answer. However, I was told that my procedure was not correctly laid out and that I should not make up my own method. Towards the end of primary school, our class was subjected to a series of tests or qualifying exams that were used to decide on the type of secondary school education each pupil would be offered. Scotland had introduced universal free education up to age 15 in 1945, in a parallel reform to follow the 1944 Butler Education Act in England and Wales. I realised much later that the tests were the Scottish equivalent of the 11-plus exam, although, having started school earlier, I was only 10 years old at the time. The outcome of this testing was that 4 of us from a class of 19 were sent off to Hawick High School, leaving the remainder of the class to continue for another 3 years of secondary education in Newcastleton, which had a separate wing for older pupils.

3. At Hawick High school, when I was aged 11–15, the teachers I remember were "Jeemie" Allen, an excellent maths teacher and dahlia fancier, and Bill McLaren, who taught us for gym and rugby. One of my Hawick classmates, Myra Thomson, wrote to me after the 2017 Nobel Prizes were announced to remind me of the occasion when I was told off by Jeemie for simply writing down the answer to a maths question without bothering to write out the working. I remember being quite surprised when I was 14 to receive a school book prize, valued at 5 shillings, for being placed 3rd in maths, 3rd in science and 3rd equal in geography, having in earlier years always been nearer the bottom of the class. I chose a paperback book called "The Cockleshell Heroes" about a daring kayak raid to sink some German battleships in the French river port of La Rochelle during the second world war. This may have

led to my later enthusiasm for kayaking. Although I gradually worked my way up in maths and science at Hawick, my language skills were always poor, and I was the only one in the class to fail the "Lower" in French. In Hawick, there was a shop I visited in the school lunch hour along with one or two friends, which sold ex-WD (War Department) electronics, so we would acquire very inexpensive components and build our own valve radios.

4. At the end of my 5th year of secondary school, the headmaster at Hawick recommended three schools in Edinburgh for my final (6th) year of secondary school, namely Royal High, George Heriot's or Boroughmuir. The first two had (nominal) school fees whereas Boroughmuir was free, so we chose Boroughmuir Secondary School. My final year of school, at age 16, was spent at Boroughmuir, with Bill Cow, an excellent Physics teacher, and Dr Young our chemistry teacher, who had a Ph.D. and had worked on explosives during the war. Since we had already finished our science "Highers" examinations at the end of the 5th form, the 6th form science had no special curriculum. One difference between Hawick and Boroughmuir was that the sciences were taught in an integrated class by a single teacher at Hawick, whereas there were separate physics, chemistry and biology classes at Boroughmuir. Boroughmuir also had selective entry, so that the overall academic level and teaching standards were higher than at Hawick, which was an all-inclusive comprehensive school. A few of us expressed an interest in biology and studied some plant biology in our free periods. Dr Young also allowed us to choose any chemistry experiment we wanted to do on Friday afternoons, when we had a double period of Chemistry. One member of our class decided to take advantage of our teacher's wartime experiences to test a different explosive each week, so every Friday at around 3.30pm we all had to crouch down below the bench while that week's test explosive was ignited. Only once was the explosion strong enough to leave marks on the walls. My abysmal performance in French was rectified by having tuition in the 6th year in a class that had only two pupils. I had failed Lower French at Hawick and my classmate had failed Higher French at Boroughmuir. After an entire year of individual tuition, I scraped through the French exam with a bare pass and was thus able to meet the entrance requirements for the University of Edinburgh.

In my academic education, I thus benefitted greatly from the post-war education reforms, which opened new opportunities for working class children and brought in free secondary school education for all. At one stage aged 13, I tried to drop French in high school, but our headmaster called me into his office and told me that I should not do that if I wanted to keep open the option of going to University, because a language at O-level (Scottish Lower) was an entry requirement. I don't think I even knew of the existence of Universities at that point.

UNIVERSITY OF EDINBURGH (AGE 17–20)

I pursued a B.Sc. in "Natural Philosophy" which was the traditional name for Physics. This consisted of 4 years of physics, maths and mathematical physics. Peter Higgs was our mathematical physics lecturer in 1962–64, at just about the time he was writing his famous paper predicting what came to be known as the "Higgs boson". I was delighted finally to be allowed to focus entirely on the subjects I found most interesting.

During my undergraduate years, I took many jobs during the Christmas, Easter and Summer breaks, partly to earn enough money to pay for running a car, and partly to get direct experience of different working environments. With two school friends, we bought a car (£10 each) when we were 17. The car, a 15-year-old Morris 8 Series E, was very unreliable, so the three of us learned a lot about car engines, clutches, gearboxes, back-axles and half-shafts, since they all seemed to need frequent repair or replacement, mostly from scrapyards. Although only one of us had passed his driving test, he taught the other two. During the three summers from 1963 to 1965, I worked in the technical drawing office at the electrical engineering company Ferranti designing a slide projector, at the UK Atomic Energy Authority (UKAEA) at AWRE Aldermaston in a small group evaluating lithium-drifted germanium detectors for gamma rays, and with Dr John Muir in the Physics Department on a summer project using microwaves to analyse the dielectric properties of kaolinite clay, with support from a Carnegie Trust Vacation Scholarship. Each of these three positions exposed me to different cultures. In the company, it was very hierarchical: the man in charge of the drawing office did not cope well with the more competent students. At UKAEA, a science campus operating as part of the civil service, there were many brilliant scientists who were a pleasure to work alongside, but their research was part of bigger projects, so their enthusiasm was muted. In contrast, those carrying out research in the university were enthusiastic, highly motivated and clearly excited about their work. I therefore decided that an academic research career would be my best option.

DECISION TO CHOOSE BIOPHYSICS FOR GRADUATE RESEARCH

During my final year as an undergraduate, I spent a long time trying to decide which of the many exciting directions that physics was taking would be most interesting for me in a future research career. I can remember considering fusion research which promised to provide unlimited power generation, solid state physics which has transformed our lives through development of a multitude of semiconductor devices, high energy particle physics which has led to a deep understanding of nuclear structure, or astrophysics which has transformed our understanding of the universe from the big bang to black holes, neutron stars and gravitational waves. In the end, I decided that biophysics had great potential in bringing the power of physics to understand biological phenomena. One of the most important factors in making this choice was that I was keen to do individual hands-on research either myself or with one or two close colleagues, rather than to work as part of a large team, which would have been essential for some of the other directions. The final year physics exam in Edinburgh consisted of 6 papers on successive days on different topics ending with a final essay paper with a broader scope, and encouragement to be somewhat light-hearted. I wrote an essay on "Time", in which I explained that time consisted of the past, present and future. Since the past and present could simply be looked up in a history book or encyclopaedia, only the future was interesting. This then led into a consideration of where physics was heading, with a discussion of the above range of topics and ending up with the conclusion that biophysics had great potential and might offer rewarding opportunities for the individual.

Having decided on biophysics, the question was then where to go for a Ph.D. research project. Since, at the age of 20, I had no desire to do any more studying, attending lectures or sitting exams, this ruled out all American Universities, and on further investigation also ruled out Leeds (R.D. Preston, successor to Astbury) and Norwich, where Jack Dainty had moved to a Professorship, his promotion having been turned down by Edinburgh. Both those biophysics departments had compulsory M.Sc. degrees that took at least a year before they would allow students to pursue research for Ph.D. That left only King's College London, which I visited in November 1965. I talked with many people at Kings (Randall, Wilkins, Jean Hanson, Jack Lowy, Watson Fuller, Struther Arnott), and eventually wrote back after my return to Edinburgh asking whether I might be allowed to work on surface forces with Dr Anita Bailey. King's did not offer me a place immediately but said they would let me know next Spring. Having made my plan after a lot of investigation, I thought it might be tactful to go and tell our new Professor, Bill Cochran, who had arrived from Cambridge in 1964 and quickly built up an outstanding solid-state physics group, about my decision to go into biophysics. Without hesitation, he quickly advised me that I should write to his friend Max Perutz at the Medical Research Council (MRC) Laboratory of Molecular Biology in Cambridge (MRC-LMB). Somehow, in spite of all my efforts to talk to many other people in Edinburgh and to contact and visit other places around the UK, I had not managed to identify the MRC-LMB. This was primarily because most research in UK universities at that time was listed in the Science Research Council (SRC) Handbook. The MRC-LMB in Cambridge was listed only in the MRC Handbook, which was not available in the Physics Department. In contrast, King's College Biophysics was both a Biophysics Department in the University and an MRC Biophysics Unit, so was listed in both handbooks.

I therefore wrote to Max Perutz in January, received a reply in February and visited the MRC-LMB on a Saturday morning in March for the student Open Day, arriving on the overnight train from Edinburgh and returning on the same evening. There were about 20 students visiting, almost all from Cambridge. David Blow gave an informal talk about some of the research. I was also interviewed individually by Max Perutz and John Kendrew. Kendrew simply asked whether I had any questions. I said I was concerned that I had studied no biology or chemistry, only physics and maths, but he said I should not worry: I could easily pick up biology and chemistry as I went along. The laboratory was a hive of activity with more people at work on a Saturday morning at MRC-LMB than in other places I had visited midweek.

On my return to Edinburgh I therefore immediately wrote to say I would be very interested in becoming a Ph.D. student at MRC-LMB. Perutz wrote back two days later accepting me. That year two physics students started as Ph.D. students at MRC-LMB. Peter Gilbert who was a Physics undergraduate from Cambridge was the other. I had also learned from Cochran that another physics student from Edinburgh, Keith Moffat, had gone to MRC-LMB the year before to start a Ph.D. with Max Perutz, so I also wrote to Keith and asked him to tell me a bit more about Cambridge, especially the College system, about which I knew very little. Keith very kindly replied with a 4-page letter giving a thumbnail sketch of the positive and negative aspects of each college. Keith recommended Darwin and Corpus Christi, largely because Corpus had just opened new postgraduate accommodation in 1964 in the George Thomson Building. I spent a year living in the George Thomson building, and later became a fellow at Darwin and an Honorary Fellow at Corpus.

CAMBRIDGE (AGE 21–24)

I carried out research for my Ph.D. at the MRC Laboratory of Molecular Biology in Cambridge (MRC-LMB), with a thesis on "X-ray analysis of chymotrypsin: substrate and inhibitor binding". With David Blow, my supervisor, and Tom Steitz, who was a Jane Coffin Childs postdoctoral fellow at that time, we worked out the mechanism of action of this enzyme, which was the first serine protease to have its structure determined. It was also the third or fourth protein structure to be determined at atomic resolution, after myoglobin (1959) and lysozyme (1964). The structures of chymotrypsin, ribonuclease and carboxypeptidase were all determined in 1967. By 2018, there were atomic coordinates for 140,000 macromolecular structures deposited in the Protein Data Bank (PDB).

Before I started my postgraduate work in Cambridge, I applied in June 1966 to attend a 2-week Summer School in Molecular Biology and Biophysics in Oxford, which was held to mark the inauguration of the Laboratory of Molecular Biophysics in Oxford, under Professor David Phillips. My application had arrived after all the places had been filled but Max Perutz had also written to Phillips in support of my application. Consequently, I received another letter a few weeks later from Oxford to say that "due to a withdrawal" I could now be offered a place. Thus, after a week or two at MRC-LMB, I spent 2 weeks in Oxford listening to 50 superb lectures by 25 outstanding scientists, including David Phillips, Max Perutz, Fred Sanger, Maurice Wilkins, Aaron Klug and Mark Bretscher. The only lecture for which my notes were almost a blank, save for the word "supercilious", was by Sydney Brenner. Brenner was arrogant but also very clever, and shared the 2003 Nobel Prize in Physiology or Medicine for his work developing the nematode as a model organism. One of the highlights of life at MRC-LMB in the 1960s was the Saturday morning coffee meeting in the "Molecular Genetics" kitchen where Brenner would entertain six or ten weekend researchers with his wide-ranging, acerbic comments, phenomenal memory and ability to provide an integrated overview of all aspects of molecular biology as it was then.

Having learned only a limited amount of chemistry at school, since the syllabus for Scottish Higher Science at that time stopped with inorganic chemistry, with only the briefest mention of organic chemistry, I attended Cambridge University Part IA organic chemistry in which Peter Sykes was the most memorable lecturer, and also spent one term doing ten afternoon laboratory practicals in synthetic organic chemistry, the most memorable of which was the synthesis of methyl orange.

Towards the end of my Ph.D., I spent some time thinking about what to do next. I had realised by then that MRC-LMB was a superb laboratory with many truly outstanding scientists, and that its success depended on a very deep investigation and understanding of a few very narrow research topics. However, with my own very narrow training in physics and mathematics, I also realised that I would need a much broader grasp of a wider range of problems if I were to be able to choose a productive research topic and research direction, following the philosophy of Peter Medawar's description of scientific progress as "The Art of the Soluble". I therefore looked around for a postdoctoral opportunity where I would be able to get a much broader overview of the importance of different research areas across biology. I had been impressed by two scientists at Yale. One was Fred Richards (1925–2009) for his work on the structure and mechanism of ribonuclease, where his broad knowledge and insight had allowed him to provide the definitive explanation of the mechanism of action of the enzyme ribonuclease; although his structural work lagged significantly behind that of another US group at Buffalo led by David Harker, who had determined a better structure for ribonuclease, Richards was much more successful at explaining its importance and in relating structure to mechanism. The second impressive person at Yale was Jui Wang (1921–2016) in the Chemistry Department, who had proposed a hydrolytic mechanism for chymotrypsin that was appealing and showed deep insight into the chemistry of catalysis. I therefore wrote to both enquiring about the possibility of postdoctoral work. Wang replied immediately offering me a place, whereas Richards' reply, although equally encouraging, did not come for 6 weeks: he was a keen yachtsman and was away sailing. I therefore decided to join Jui Wang and wrote two postdoctoral fellowship applications to the Helen Hay Whitney Foundation (HHWF) and the Jane Coffin Childs (JCC) Memorial Fund. Maclyn McCarty, the chairman of HHWF came to interview Jonathan Greer and me in Cambridge, and we were both subsequently offered HHWF fellowships. Since the HHWF stipend was higher than that of any other postdoctoral fellowship at that time, I accepted their offer, withdrew from JCC, and went to Yale accompanied by my wife Penny and our new-born daughter Jennifer, arriving in New Haven on 20th June 1970. It was Jennifer's first birthday on the day we arrived at Yale.

POSTDOCTORAL (AGE 25–27)

I thus ended up as a Helen Hay Whitney Foundation Postdoctoral Fellow at Yale University, in the group of Prof. Jui Wang for 2 years in the Chemistry Department, then spent my third year with Prof. Fred Richards in the Department of Molecular Biophysics and Biochemistry, with a bench in Tom Steitz' lab. I tried to work on voltage-gated sodium channels in nerve and muscle membranes with the goal of determining the structure, but after 2 years decided that this goal was premature because the methods were inadequate, so then decided to tackle a simpler membrane protein, which was the light-driven proton pump bacteriorhodopsin, which had just been discovered by Walther Stoeckenius in 1971.

When I first arrived at Yale, based on my postdoctoral fellowship application and my own thinking at that time, I had planned to embark on two projects. The first was to label a peptide substrate of chymotrypsin with ^{13}C at its carbonyl carbon and to carry out ^{13}C Nuclear Magnetic Resonance (NMR) analysis to explore the chemical environment in the active site. The second was to choose another enzyme, perhaps slightly more interesting than chymotrypsin, and to purify, crystallise and solve its structure. I had written to David Blow to tell him about my plans and he replied to say that he thought Bob Shulman, then at Bell Labs in Newark, New Jersey might be already trying the ^{13}C experiment, so he recommended that I should contact him. I did contact him and was invited to give a seminar at Bell Labs in 1970, where I met Shulman and Dinshaw Patel, his NMR right-hand man. After my seminar, in which I explained how I planned to go about obtaining an equilibrium concentration of the ^{13}C-labelled enzyme-substrate complex using a high concentration of the substrate leaving group to increase the level of the desired structure by mass action, Shulman said it was a very good idea and they would do it. Since he had much better NMR facilities at Bell Labs than at Yale, I agreed to leave it to them and indeed George Robillard carried out and published the experiment, which provided some interesting insights.

I discussed my second proposed project with Wang. This was to purify and crystallise another enzyme. For this, I had selected NADP reductase from spinach, and purified it with one of Wang's Ph.D. students, Jim Keirns. We succeeded in producing small pale-yellow crystals of spinach NADP-reductase before discovering that Martha Ludwig, by then at Ann Arbor, Michigan was already making progress on it. After some discussion with Wang, he explained that there were thousands of enzymes and, since I was still a young postdoc, it would be much better to pick a new longer-term problem that would come to fruition in 20 years, rather than aiming to take a small incremental step in a topical field. After a day or two, I realised that this was very good advice, and abandoned my initial plans to work on enzyme mechanisms and enzyme structure.

At that time, there was a lot of enthusiasm to understand the structure of membrane and membrane proteins, so I asked myself what was the most interesting membrane protein and chose voltage-gated sodium channels. Voltage-gated sodium and potassium channels had been at the heart of the 1963 Nobel Prize winning work of Alan Hodgkin and Andrew Huxley, and for someone with a physics background these were very attractive research targets. In addition, Wang had published some theoretical speculations about the mechanism of ion channels, and was keen for someone to look for microwave emissions from ion channels in nerve membranes as they opened and closed during the action potential. We therefore ordered some microwave equipment, and I began synthesizing some small molecules which I hypothesized might bind to and block sodium channel currents in nerves. After synthesizing 3 or 4 compounds, I then spent a week or two looking around at Yale to find someone who could measure nerve action potentials, ending up by making contact and collaborating with Murdoch Ritchie, who was chairman of the Pharmacology Department on the Medical School campus on the other side of New Haven. Although all my compounds had absolutely zero effect on nerve impulse propagation, this initial contact nevertheless led to a fruitful 3-year collaboration with Murdoch Ritchie and two others based in Ritchie's laboratory, namely David Colquhoun, who was a sabbatical visitor from London, and Gary Strichartz, another postdoctoral fellow.

Our experiments on ion channels that eventually produced some useful insights began with the idea of producing radiolabelled tetrodotoxin using a tritium gas electrical discharge. Prof. Martin Saunders, also in the Yale Chemistry Department, had a moribund basement laboratory that was full of spider webs, but also housed a fume cupboard, vacuum equipment and 10 Curies of pure tritiated water, T_2O. After arranging to have the basement lab reactivated, I managed to produce some tritium-labelled tetrodotoxin, purify it using flat-bed electrophoresis, and in collaboration with Ritchie demonstrate specific tetrodotoxin binding to nerves and nerve membranes from a variety of sources. We published a number of interesting papers, but my early steps to extract and purify these voltage-gated sodium channels (VGSCs) were disappointing, because the channels extracted using either Triton X-100 or deoxycholate detergent, were quite unstable with a lifetime of a few minutes at room temperature or a day at 4°C. I guessed that it might take another 30 years to solve the stability problem so decided to abandon working on VGSCs and choose instead a simpler membrane protein, with the criterion that it should be stable after detergent solubilisation and available in reasonable quantity. The microwave experiments also failed.

In June 1971, I had attended a meeting in San Francisco of the American Society for Biochemistry and Molecular Biology (ASBMB) and heard a wonderful talk by Walther Stoeckenius describing his discovery of the purple membrane from *Halobacterium halobium* (subsequently renamed as *Halobacterium salinarum*) and his finding with Dieter Oesterhelt and Allen Blaurock that it was composed of a two-dimensional crystalline array of a single membrane protein to which the chromophore retinal was bound via a Schiff base in a 1:1 stoichiometry, responsible for its characteristic purple colour. After following the work that Stoeckenius and his colleagues Allen Blaurock and Glen King were doing during the next year or two to try to elucidate the structure of the purple membrane, I felt they were heading completely in the wrong direction. Therefore, in early 1973, I decided it would be an opportune time to try one or two new ideas to solve the structure of bacteriorhodopsin, the single protein in purple membrane. Bacteriorhodopsin fitted perfectly the criteria of being stable and available in large amounts. By chance, Don Engelman, then Assistant Professor at Yale, had worked with Stoeckenius as a postdoc a couple of years earlier and knew him well, so Don and I phoned up Stoeckenius and asked whether he could send us a culture of *H. halobium*, which he kindly did. Neither of my initial ideas, either to use heavy atom derivatives to determine the phases of the powder pattern rings by multiple isomorphous replacement, or to solubilise bacteriorhodopsin and crystallise it in three dimensions for X-ray crystallography, worked out but these were the two approaches I was pursuing when my 3-year postdoctoral fellowship at Yale came to an end.

Penny and I made many friends at Yale and have kept in touch with them over the subsequent decades. During our years in New Haven, Penny gave birth to our second child Elizabeth, born in January 1971, but she had hydroencephalus at birth and died just over seven months later. Our third child, Alastair, was born on 22 March 1973, a few months before our planned return to the U.K.

BACK TO CAMBRIDGE – 1973 UNTIL NOW

Phase I – bacteriorhodopsin at low resolution

We returned to Cambridge on 20th June 1973 to the MRC-LMB on a 5-year appointment, exactly three years after our departure in 1970 (on an American J-1 visa which had a 3-year limit). Jennifer was now four years old and Alastair three months. During my ultimately fruitless efforts to make any progress in analysing the structure of the purple membrane and bacteriorhodopsin using X-ray powder pattern phasing or three-dimensional crystallisation, I was impressed by a talk that Nigel Unwin gave in October 1973 in the annual MRC-LMB symposium. He spoke about electron microscopy (EM) using a phase plate made from a single thread of spider web silk coated with gold. He was clearly thinking that his images of tobacco mosaic virus contained features that represented the protein structure as well as the negative stain he was using to embed the structure. After his talk, we discussed using EM to study the structure of bacteriorhodopsin in its natural two-dimensional crystalline form without using any heavy metal stain. We worked together very productively for about 18 months, ending up with a low-resolution, three-dimensional structure of the first membrane protein, determined by a novel method. The structure showed seven well-resolved trans-membrane α-helices oriented almost perpendicular to the membrane plane, with the implication that this α-helical architecture might be found in other membrane proteins, as indeed it has. After that early success, I switched my efforts from X-ray diffraction to elec-

Figure 4. Nigel Unwin and Richard Henderson (left to right) sitting on the steps of the original entrance to the MRC Laboratory of Molecular Biology in the 1990s.

tron diffraction, and eventually to electron microscopy. Nigel switched from working on viruses using negative stain to working on membrane proteins and two-dimensional or helical crystals without using heavy metals, so our 1973–1975 collaboration had a profound impact on the direction of both of our future scientific careers. We were jointly awarded the 1999 Gregori Aminoff Prize of the Royal Swedish Academy of Sciences. A photograph of us taken around that time is shown in Figure 4.

Phase II – bacteriorhodopsin at high resolution

After spending about 7 or 8 years unsuccessfully trying to extend the resolution of our bacteriorhodopsin map from low resolution (7 Å) to high resolution (3 Å), where we expected to resolve the chemistry of the structure (i.e. to see the amino acid side chains and understand the mechanism), I eventually concluded that the methods that our group had been trying (model building, molecular replacement and heavy atom derivatives) were simply not powerful enough, and that we would have to embrace the necessity of recording high resolution electron cryomicroscopy (cryoEM) images. We did this from 1984 until 1990, by visiting and collaborating with a number of other laboratories (EMBL Heidelberg with Jacques Dubochet and Jean Lepault; Fritz Haber Institute Berlin with Fritz Zemlin and Elmar Zeitler; and Berkeley with Ken

Downing and Bob Glaeser) as well as trying to improve our in-house EM capabilities in Cambridge. Eventually, the problems of high-resolution cryomicroscopy imaging and computer-based image processing were largely solved and we obtained a high-resolution map in 1990, into which we were able to build a nearly complete atomic model of bacteriorhodopsin. In the end, it was only the second membrane protein structure to be determined at high resolution. The first was Hartmut Michel's bacterial reaction centre membrane protein complex from *R. viridis* solved by crystallisation and X-ray crystallography, for which he shared the 1988 Chemistry Nobel Prize with Hans Deisenhofer and Robert Huber. After 1990, Sriram Subramaniam and I did some trapping of intermediates to help work out the mechanism of the bacteriorhodopsin light-driven protein pump. Our bacteriorhodopsin work was essentially completed by 1999.

On a more personal note during this period, Penny and I arranged an amicable divorce in 1988, Jennifer married Richard Morris in 1993, I married Jade Li in 1995, and Alastair married Laura Williams in 1999. The resulting clan including 6 grandchildren all came to Stockholm on 10th December 2017 (Figure 5).

Phase III – single particle cryoEM

From 1995, following publication of a review I wrote on "The potential and limitations of neutrons, electrons and X-rays for atomic resolution microscopy of unstained biological molecules", I was convinced that the future of cryoEM would involve imaging of single particles embedded in vitreous ice, a specimen preparation method that had been developed by Jacques Dubochet's group in the 1980s. This "single particle cryoEM" method was potentially very powerful because it did not require the protein of interest to be crystallised nor did it require the use of crystallographic methods, which was what I had worked on exclusively until that point using X-ray and electron diffraction. Single particle electron microscopy had started with the image processing methods of Joachim Frank and Marin van Heel, as well as Owen Saxton and Wolfgang Baumeister, prior to the development of Dubochet's plunge-freeze method of producing thin films of amorphous ice, but it was the combination of the two that promised to be particularly powerful.

We worked from 1995 until 2013 to analyse the problems and barriers to making progress, and gradually understood and solved them. The most important were the need for brighter sources, better vacuums, more stable cold stages and better detectors. The electron microscope companies, under pressure from users, addressed the first three. Our group in Cambridge, collaborating with another group at Rutherford-Appleton-Laboratory (RAL) near Oxford, worked out how to improve the detectors, and other people developed better computer programs to take advantage of some of the features of the new detectors (Steve Ludtke, Niko Grigorieff, and Sjors Scheres). As a result, almost overnight in early 2013, everyone started to obtain maps with much higher resolution. This was termed the "Resolution Revolution" by Werner Kühlbrandt (*Science* (2014) **343**, 1443). From that point on, there was great enthusiasm from the entire structural biology community and a wider adoption of single particle cryoEM methods, so that now in 2018 cryoEM has become the prime method for many structural biology problems.

OTHER CONTRIBUTIONS

I was Joint Head of the Structural Studies Division at MRC-LMB (1986–1999), Director of MRC-LMB (1996–2006) and a member of the Medical Research Council, which is the governing Board of MRC (2008–2014). Previous Directors were Max Perutz (1962–1979), Sydney Brenner (1979–1986) and Aaron Klug (1986–1996). The current Director is Hugh Pelham (2006–2018). All previous directors at MRC-LMB were also Nobel Prize winners. Probably the most significant achievement that we made during my Directorship was to advocate in 1999 the construction of a new building to house the MRC Laboratory of Molecular Biology in the 21st century, to carefully time the initiative to request funds for the construction to coincide with an upswing in the economic cycle around 2003, and to negotiate the subsequent hurdles for land acquisition and planning permission. This resulted in a superb new 30,000 square metre building, which opened in 2013, is attuned to the needs of modern molecular biology, and has the flexibility to have space reconfigured to parallel the changing needs of research. During my time as Director, Hugh Pelham as Deputy Director took a deep interest in the design and layout of the new building and carried the early planning through to completion during the first half of his Directorship. At the same time, Dr Megan Davies, the Assistant Director from 1996 and Head of the MRC Centre in Cambridge, ensured that our relationships with MRC in London and the local biomedical community in Cambridge were strengthened. An initial invitation soon after I started my Directorship to have dinner with Dr Keith Peters, the Regius Professor of Physik (an old term for Medicine) and Head of the Clinical School, developed into an annual strategic tête-à-tête that helped to keep the interests of MRC-LMB aligned with those of the Clinical School and the NHS Addenbrooke's Hospital Trust.

Figure 5. Clan gathering in Stockholm. Left to right: Joshua Morris, Jessica Morris, Grace Morris, Rosie Morris, Rachel Henderson, Jade Li, Alastair Henderson, Richard Henderson, Jennifer Morris, Richard Morris, Tom Henderson, Laura Henderson.

NOBEL CEREMONY

My wife, Jade Li, and I spent a wonderful 10 days in Stockholm for the awards of the 2017 Nobel Prizes, with the good fortune to be able to hear the Physics lectures about how gravitational waves were observed for the first time, to meet the Physiology or Medicine laureates who had worked out the molecular basis of circadian rhythms, and also the literature and economics laureates, Kazuo Ishiguro and Richard Thaler. A photograph of our family taken on the stage of Stockholm Concert Hall immediately after the awards is shown in Figure 5.

From Electron Crystallography to Single Particle CryoEM

Nobel Lecture, December 8, 2017 by
Richard Henderson
MRC Laboratory of Molecular Biology, Cambridge, UK.

INTRODUCTION AND BACKGROUND IN X-RAY CRYSTALLOGRAPHY

After completing an undergraduate physics degree at Edinburgh University in 1966, and deciding to pursue Ph.D. research in biophysics, I had the good fortune to consult Professor Bill Cochran who suggested I write to Max Perutz, at that time head of the recently opened Medical Research Council Laboratory of Molecular Biology (MRC-LMB) in Cambridge. Perutz offered me a 3-year MRC Scholarship to work with David Blow on the proteolytic enzyme chymotrypsin. I arrived just as the chymotrypsin group was calculating a 3-dimensional (3D) Fourier map using two heavy-atom derivatives for phasing. Unfortunately, that first map was only partly interpretable, with electron density for only 10 of the 241 amino-acid residues recognisable, and since Brian Matthews was just leaving for a new postdoctoral position at the NIH, I was invited to join the "chymotrypsin team", in which Paul Sigler was the only other scientist, to help determine the structure. After about 6 months' work collecting data for a third heavy-atom derivative, the next 3D Fourier map proved to be fully interpretable, so I found myself soon after my arrival in Cambridge transformed into a trained X-ray crystallographer and co-author of a paper (Matthews et al, 1967) describing the 3D structure of chymotrypsin. At the end of that first year, I then embarked on my thesis research into substrate and inhibitor binding to chymotrypsin, working initially alongside and then in collaboration with Tom Steitz, who had arrived as a postdoctoral fellow that summer. By 1969 we had obtained a number of informative 3D difference Fourier maps that allowed us to understand substrate and inhibitor binding to chymotrypsin and to explain the hydrolytic mechanism (Steitz et al, 1969; Henderson, 1970).

My transition from X-ray crystallographer to electron crystallographer followed indirectly from my postdoctoral experiences at Yale, where I had decided to work on membrane protein structure and had tried to tackle voltage-gated sodium channels (VGSCs) from garfish olfactory nerves. I had found that the VGSCs, assayed by a tritiated-tetrodotoxin ligand-binding assay were unstable after solubilisation in detergent (Henderson & Wang, 1972), so had switched to working on the small, stable and abundant membrane protein bacteriorhodopsin that had been discovered by Walther Stoeckenius and his collaborators in the purple membrane fraction from *H. halobium* (Oesterhelt & Stoeckenius, 1971; Blaurock & Stoeckenius, 1971).

BACTERIORHODOPSIN AT 7 Å, THEN 3.5 Å, REFINEMENT & KINETICS

Following my return to the MRC-LMB, I gave a talk in the annual laboratory symposium in October 1973 about my ideas for trying to solve the structure of bacteriorhodopsin. Since bacteriorhodopsin had been shown (Blaurock & Stoeckenius, 1971) to consist of well-ordered two-dimensional (2D) crystals in the membranes of *H. halobium*, I had two ideas. One was to use X-ray powder diffraction of these native membranes with multiple heavy atom derivatives to phase and resolve the problem of overlapping reflections. The other was to make 3D crystals from detergent-solubilised monomeric bacteriorhodopsin. Neither of these ideas worked out, but in the same symposium I heard an impressive talk by Nigel Unwin about his work to record high-quality electron microscope images of negatively stained tobacco mosaic virus (TMV) using a phase plate that he had constructed from a single thread of spider web silk coated with gold. Afterwards, we discussed the possibility of recording images and electron diffraction patterns from 2D crystals of bacteriorhodopsin without using negative stain. A very productive 18-month collaboration ensued, culminating in the determination of the 7 Å 3D structure of bacteriorhodopsin (Unwin & Henderson, 1975; Henderson & Unwin, 1975), shown in Figure 1 & Figure 2, determined using electron diffraction and electron microscopy of 2D crystals of bacteriorhodopsin at room temperature embedded in a thin film of glucose. Nigel and I wondered why this electron crystallographic method had produced a 3D density

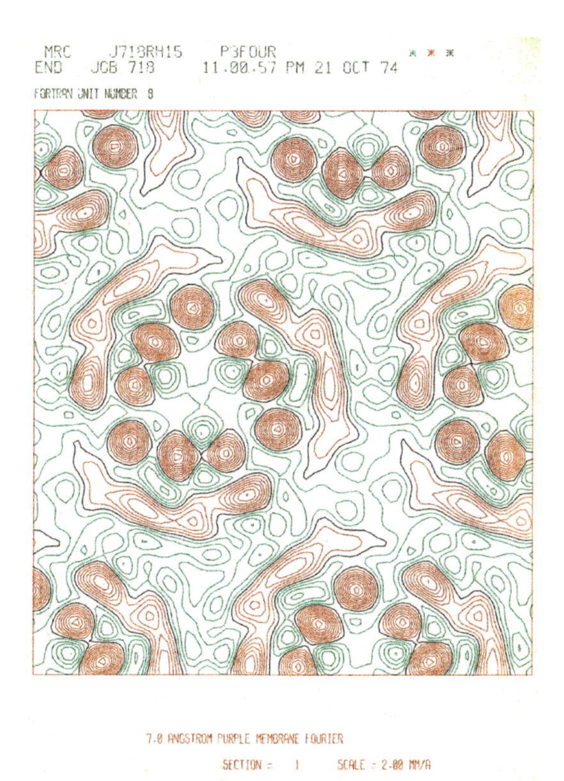

Figure 1. The first projection structure at 7 Å resolution of the purple membrane calculated in October 1974 using 36 reflections obtained by room-temperature electron diffraction and imaging of glucose-embedded 2D crystals of bacteriorhodopsin (Unwin & Henderson, 1975).

map at only 7 Å resolution, when there was nothing about the approach that intrinsically limited the resolution. We thought that the recording of images on film might be a limiting factor and spent time investigating different photographic emulsions. We also thought that the film scanners that were available in the 1970s for digitising the images might be degrading the information and spent time building and improving film scanners. This produced only fairly small improvements.

At that stage, having come into structural biology through X-ray diffraction in which all the phases of the Fourier components, as observed through Bragg diffraction from the crystal lattice, had to be determined indirectly, I also thought that electron diffraction was intrinsically more promising than electron microscopy because the elegant simplicity of recording electron diffraction patterns compared favourably with the multiple difficulties of recording good images. We therefore spent several years trying to extend the resolution of the bacteriorhodopsin structure using a number of diffraction-based approaches. Figure 3 summarises the different ideas we tried. Tom Ceska tried to make heavy atom derivatives (Ceska & Henderson, 1990). Joyce Baldwin and Michael Rossmann tried molecular replacement (Tsygannik & Baldwin, 1987; Rossmann & Hender-

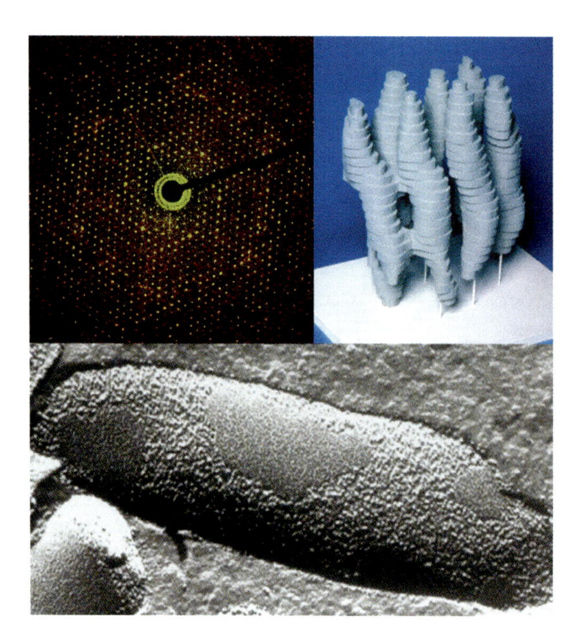

Figure 2. The structure of bacteriorhodopsin at 7 Å resolution in 3D from 18 images and 15 diffraction patterns. The collage shows (a) freeze-fracture picture from Walther Stoeckenius, (b) electron diffraction pattern obtained much later using a phosphor/fibre-optics/CCD camera, (c) the 1975 balsawood model of a single bacteriorhodopsin molecule (Henderson & Unwin, 1975).

son, 1982). David Agard tried to extend the phases using a multi-parameter model building approach (unpublished). Although all of these approaches gave hints of success that were encouraging at times, none of them were powerful enough to give phases that resulted in convincing maps that were interpretable much beyond the resolution obtained in 1975. It was not until Tzyy-Wen Jeng and Wah Chiu demonstrated, in a collaboration with Fritz Zemlin (Jeng et al, 1984), that images showing clearly visible diffraction spots at 3.9 Å resolution could be obtained from thin 3D crystals of rattlesnake venom crotoxin using an electron microscope in Berlin with a liquid-helium superconducting objective lens, that I became convinced electron cryomicroscopy could produce high quality images. We therefore embarked, as a last resort, on using electron cryomicroscopy for high-resolution phase determination (see Figure 4). In earlier years, Bob Glaeser's group had shown that freezing thin 3D crystals of catalase could produce good electron diffraction patterns and images (Taylor & Glaeser, 1974; 1976) and that there was a benefit in terms of reduced radiation damage (Glaeser, 1971), but I had been unconvinced by earlier attempts to show that electron cryomicroscope images of purple membrane contained high-resolution information (Hayward & Stroud, 1981).

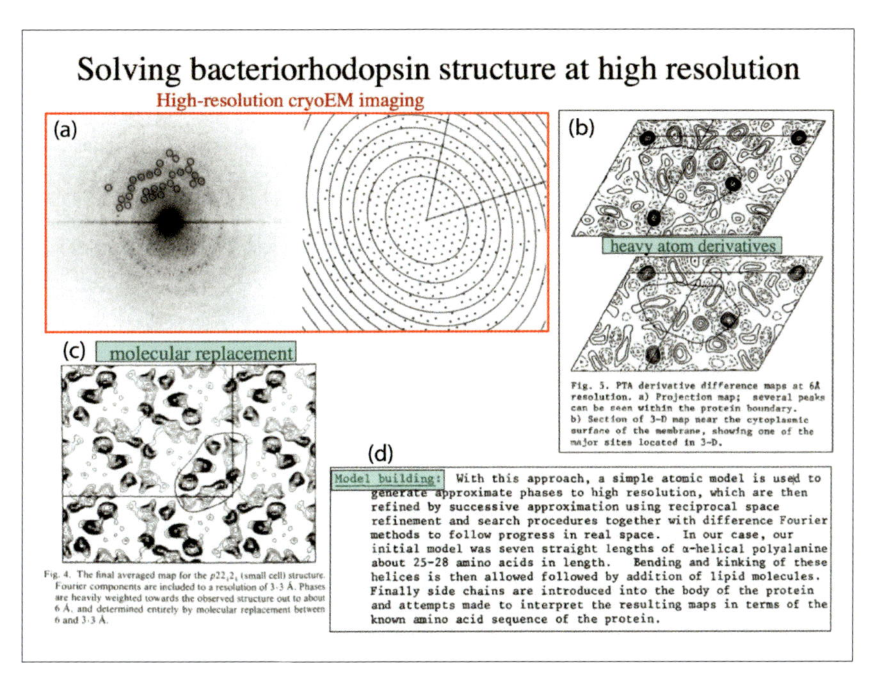

Figure 3. Overview of methods used in the early 1980s to try to solve the structure of bacteriorhodopsin at high resolution. (a) an optical diffraction pattern of a high-resolution projection image from the cryomicroscope in Berlin with the detectable spots encircled alongside the computed transform with the same Fourier components that were detected after computer processing from Henderson et al (1986). (b) difference Fourier maps of a heavy atom derivative, in this case phosphotungstate (PTA) from Ceska & Henderson (1990). (c) result of an attempt to extend the phases by molecular average and phase refinement from Rossmann & Henderson (1982). (d) attempts to bootstrap the phases to high resolution by using a model consisting of a bundle of 7 α-helices, from unpublished work by David Agard. Although each method did produce improvements in the 3D maps, only method (a) was powerful enough to solve the high-resolution structure.

The path from 7 Å resolution to 3.5 Å and atomic model

- Cooling specimen to liq. N_2 or liq. He temperature reduces the effects of radiation damage and gives 4- to 5-fold increase in diffraction
- Very few electron microscopes were stable enough in 1980s to achieve imaging with 3.5 Å resolution using cold stages
- Collaborations with and travelling to three different labs were essential:

  Lepault/Dubochet at EMBL

  Zemlin/Beckmann/Zeitler at Fritz-Haber-Institute in Berlin

  Downing/Glaeser at Berkeley
- Beam tilt was a key feature that required computational correction
- Correcting for defocus gradient when tilting specimen was another technical challenge
- Finally, 70 images allowed a map to be calculated, adequate to build an atomic model
- Refinement by Niko Grigorieff + increase to 100 images with 30 more from Ken Downing
- Yoshi Fujiyoshi independently determined the structure with an improved map
- All subsequent X-ray structures used the cryoEM coordinates for molecular replacement

Figure 4. Summary of the key steps in the path from 7 Å to 3.5 Å resolution.

The change of emphasis from diffraction to imaging proved to be very challenging. I began with a visit to Jacques Dubochet's laboratory at the European Molecular Biology Laboratory (EMBL) in Heidelberg in 1984, working with Jean Lepault to record images on their hybrid Zeiss/Siemens microscope with the same design of superconducting liquid-helium objective lens as on the Berlin microscope. We spent a week with that home-constructed microscope, which turned out to be very unreliable. Fortunately, we managed to obtain just one image that showed diffraction beyond 4 Å resolution, although because of the difficulty of alignment and the short mean time between failures, that image had over 5000 Å of astigmatism. We did not pursue further imaging at EMBL. Nevertheless, that was the first image that allowed us to begin developing procedures for the computer-based processing of high-resolution images from 2D crystals of unstained membrane proteins. After my visit to EMBL Heidelberg, Elmar Zeitler invited me to the Fritz-Haber Institute of the Max-Planck-Society in Berlin where the superconducting lens was installed on an old Siemens 100 keV electron microscope with a conventional tungsten electron source. My first visit to Berlin, which initiated a decade-long collaboration with Fritz Zemlin and Erich Beckmann, proved even less productive than the visit to EMBL. No good images were obtained at all, but Fritz Zemlin was able to use the problems we encountered as justification to initiate a programme of improvements in the reliability of the microscope, which they called Suleika, so that by 1986, we had obtained a reasonable number of high resolution images of bacteriorhodopsin in projection. Finally, during his sabbatical visit to MRC-LMB in 1984, Bob Glaeser had suggested that Ken Downing should record some cryoEM images from purple membranes on their JEOL 100B at Berkeley. As a result, we also had an image from Ken Downing on a third electron cryomicroscope that also showed diffraction beyond 4 Å resolution.

After extensive computer processing of these early "high-resolution" projection images of 2D crystals of bacteriorhodopsin, the diffraction peaks at and beyond 4 Å resolution were clearly visible well above the noise level, just as they had been on Wah Chiu's crotoxin images two years earlier, yet when we looked for consistency, the phases from different images were in total disagreement. The phases were essentially random numbers beyond about 6 Å resolution. In the end, the explanation was that we were not taking into consideration the beam tilt arising from inaccurate alignment of the illumination along the optical axis of the microscope. By reading the literature, especially publications in *Ultramicroscopy*, I found two papers. One was entitled "The importance of beam alignment ... in high resolution electron microscopy" (Smith et al, 1983). The other was by our collaborator Fritz Zemlin! (Zemlin, 1979). Both explained how beam-tilt misalignment perturbed the high-resolution phases with an error that was proportional to the cubed power of resolution. As soon as a beam-tilt correction factor, consisting of two extra parameters, was added to our computer programs, all the observations immediately clicked into perfect agreement and we were able to publish a comprehensive paper (Henderson et al, 1986) describing the projection structure of bacteriorhodopsin at 3.5 Å resolution, which in passing showed that previous efforts at determination of the projection structure, including our own, had all been incorrect. The final hurdle was to develop a method to correct for the gradient of defocus due to the height difference across images of tilted and highly tilted specimens, which was needed to extend the method into three dimensions. We called this the tilt-transfer function (TTF) correction (Henderson & Baldwin, 1986). It also proved much harder to obtain high quality images from tilted specimens than from untilted specimens because beam-induced charging and physical motion caused image blurring and thus greater loss of information in the vertical direction than in the plane parallel to membranes. This beam-induced image blurring problem on highly tilted specimens was helped by spotscan imaging (Bullough & Henderson, 1987; Downing 1988), especially when coupled with the improved coherence from the field emission source on the Berkeley JEOL microscope. In parallel with visits to and collaborations with Berlin and Berkeley, we also tried to develop a better side-entry cold stage at MRC-LMB in Cambridge (Henderson et al, 1991), so the eventual high-resolution 3D map of bacteriorhodopsin (Henderson et al, 1990), which allowed us to build an atomic model for most of the amino acids in the structure (Figure 5), contained a small number of images from Cambridge that supplemented the bulk of the data from Berlin and Berkeley.

Later on, Werner Kühlbrandt used the same methods, in collaboration with Yoshi Fujiyoshi, to determine the structure of the light-harvesting complex LHC-II from green plants (Kühlbrandt et al, 1994), and Ken Downing, Eva Nogales and Sharon Wolf determined the atomic structure of the αβ-tubulin dimer from 2D "zinc sheet" crystals (Nogales et al, 1998). The difficulties and limitations encountered during the bacteriorhodopsin work made it clear that the development of cryoEM would need substantial improvements in the microscope technology, especially more stable cold stages, higher vacuums, and brighter field emission sources. Also, at that time we thought higher acceleration voltage was needed to improve the electron optics of the column. These improvements were all developed slowly over the next 10 years, laying the foundations for other types of cryoEM including work with single particles.

After the 1990 publication describing the first atomic model of bacteriorhodopsin, we worked on trapping the intermediates in the light-driven

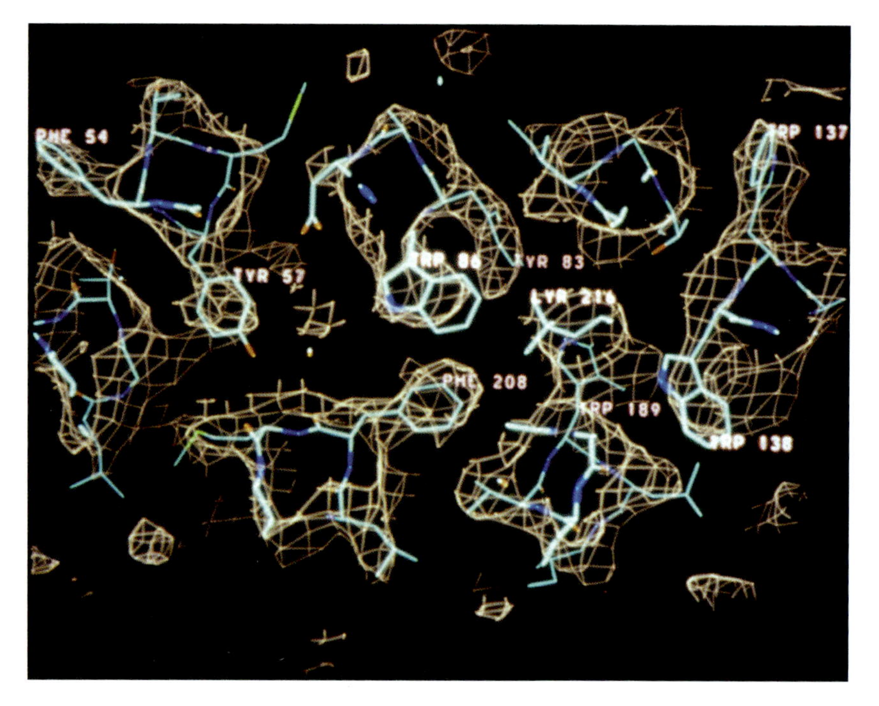

Figure 5. A slice through the central region of the 3.5 Å resolution 3D map with the corresponding atomic model superimposed, showing side chains of phenylalanine, tyrosine and tryptophan residues as well as part of the β-ionone of the chromophore retinal, which was the highest density feature in the map, from Henderson et al (1990).

photocycle (Subramaniam et al, 1993; 1999; Subramaniam & Henderson, 2000) and on the crystallographic refinement of the atomic model after the addition of a few more images from tilted specimens (Grigorieff et al, 1996). Although our work had come to its natural conclusion, the structure of bacteriorhodopsin continued to be improved both by electron microscopy (Kimura et al, 1997) and by X-ray crystallography once 3D crystals that diffracted well without twinning were obtained (PebayPeyroula et al, 1997; Lücke et al, 1999). There are now well over 100 sets of bacteriorhodopsin coordinates deposited in the Protein Data Bank (PDB).

DUBOCHET PLUNGE-FREEZE METHOD

In 1978, John Kendrew persuaded Jacques Dubochet to join the EMBL in Heidelberg to develop electron cryomicroscopy and to investigate the properties of frozen water with the goal of the determination of biological structures using cryoEM. In a series of seminal papers in the early 1980s, Dubochet and his colleagues worked out the conditions necessary to produce hexagonal ice, cubic ice and amorphous ice and how to interconvert them (Dubochet et al, 1982a; 1982b; 1984). This led to the development of their

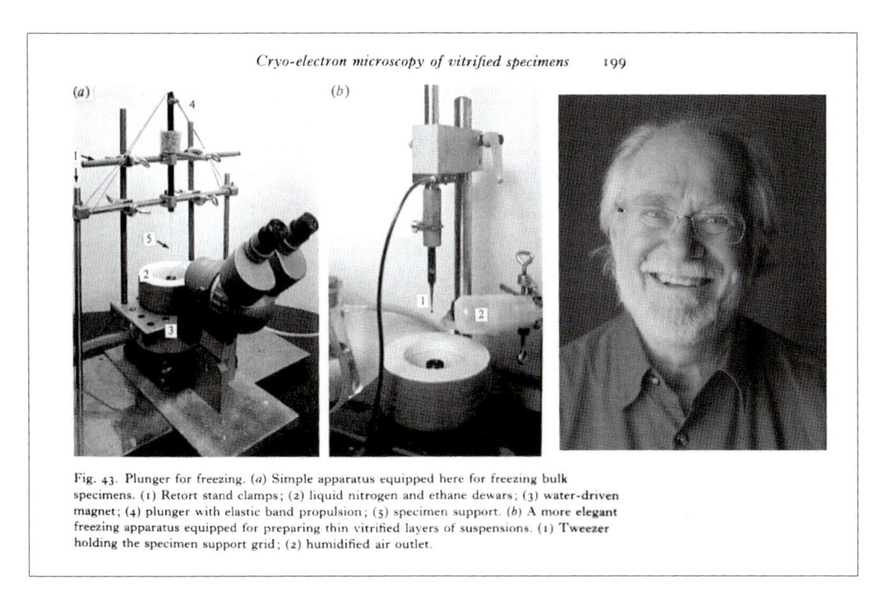
Cryo-electron microscopy of vitrified specimens 199

Fig. 43. Plunger for freezing. (*a*) Simple apparatus equipped here for freezing bulk specimens. (1) Retort stand clamps; (2) liquid nitrogen and ethane dewars; (3) water-driven magnet; (4) plunger with elastic band propulsion; (5) specimen support. (*b*) A more elegant freezing apparatus equipped for preparing thin vitrified layers of suspensions. (1) Tweezer holding the specimen support grid; (2) humidified air outlet.

Figure 6. Early apparatus developed for plunge-freezing by the group of Jacques Dubochet at EMBL, from Dubochet et al (1988). A recent photograph of Dubochet is also shown.

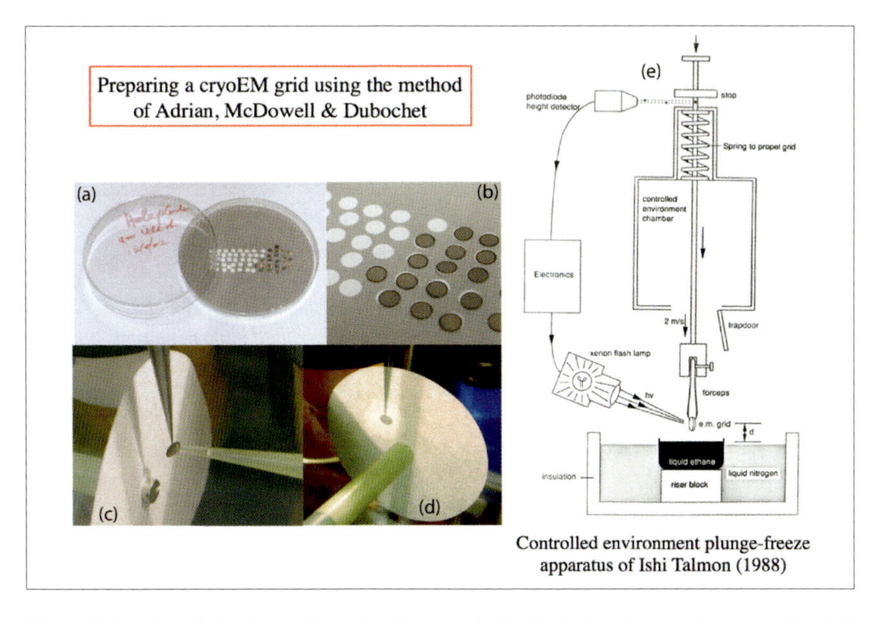

Figure 7. A series of photographs and a diagram to illustrate the plunge-freeze method of Dubochet. (a) Petri dish with about 60 grids coated with holey carbon made by Claudio Villa at MRC-LMB in 2002. (b) at higher magnification. (c) single grid to which a 3ml droplet is being applied. (d) blotting procedure. (e) schematic diagram of the plunge-freeze apparatus we used for trapping bacteriorhodopsin intermediates in 1992. The apparatus was a more sophisticated version of early EMBL devices with a controlled environment developed in Haifa by Talmon's group (Bellare et al, 1988), to which we added a time-resolved xenon flash unit that was kindly donated by Nigel Unwin.

plunge-freeze method for preparing a thin film of vitreous ice in which the biological structures of interest were suspended (Adrian et al, 1984; Dubochet et al, 1988). A photograph of their early apparatus is shown in Figure 6 and a collage explaining the principle is shown in Figure 7. This method, which consists of applying a drop of solution to an electron microscope grid, then blotting with filter paper for a few seconds to form a thin film, followed by plunging the grid into liquid ethane at liquid nitrogen temperature, is essentially the same method that most people still use 35 years later. The procedure together with many beautiful cryoEM images was explained in a comprehensive review (Dubochet et al, 1988).

EARLY SINGLE PARTICLE IMAGE ANALYSIS

Joachim Frank was the earliest to appreciate that structural information could be extracted from noisy electron microscope images of single particles (Frank, 1975; Frank & Al-Ali, 1975). With Marin van Heel, he introduced a powerful method, called multivariate statistical analysis, for extracting averages representing the typical, noise-free appearances of the different image subpopulations found in a stack of individual images (van Heel & Frank, 1981; Frank & van Heel, 1982). This early single particle work on classification of projection images of negatively stained biological structures became more powerful when the transition was made from 2D into 3D with the introduction of angular reconstitution by van Heel (1987) and the Random Conical Tilt (RCT) method by Radermacher et al. (1987). These methods allowed 3D structures to be obtained for the first time from single particle images of non-symmetrical structures. When these single particle methods were then applied to cryoEM images of specimens made using the Dubochet plunge-freeze method, the first single particle 3D structures of the ribosome were obtained (Frank et al, 1991), initially at low resolution, and then gradually improving (Gabashvili et al, 2000).

SINGLE PARTICLE CRYOEM – BLOBOLOGY IN THE EARLY DAYS

By the early 1980s, the steady progress in electron microscopy, electron cryomicrosopy, and calculation of 3D structures from EM images of all sorts of specimens led Wah Chiu and Nigel Unwin to propose a new Gordon Research Conference (GRC) theme, which they called "Three-dimensional electron microscopy of macromolecules", abbreviated to 3DEM. The first conference photograph is shown in Figure 8, with Nigel and Wah in the front row, surrounded by many others already mentioned above. The topic was timely and the 3DEM GRC has grown in size and frequency over the years.

Figure 8. Group photograph from the first Gordon Research Conference on "Three Dimensional Electron Microscopy of Macromolecules" in 1985, with Wah Chiu and Nigel Unwin as chairman and vice chairman, circled in yellow. Also shown, circled in red are the three 2017 Chemistry Nobel laureates sitting or standing as close as possible to the organisers. In the back row, circled in blue are Bob Glaeser, Ken Taylor and Ken Downing, who also had key roles in the early development of cryoEM.

Around 1987, the X-ray crystallographers had also started to explore freezing 3D crystals to liquid nitrogen temperature, which had been applied already to crystals of small organic molecules (Hope & Nichols, 1981). Håkon Hope spent a year working with Ada Yonath's group and managed to obtain much better diffraction patterns of ribosome 3D crystals than could be obtained without freezing (Hope et al, 1989). At that time, the intensity of X-ray sources at synchrotrons was not sufficient to observe any fading of the diffraction patterns from frozen crystals due to X-ray radiation damage but it was clear from a comparison of the amount of energy deposited by electron and X-ray irradiation that this was simply due to the relatively weak X-ray beams available then (Henderson, 1990). As a consequence of my interest in the importance of radiation damage in electron microscopy, I was invited to give a talk at a meeting in Grenoble to discuss the possibility of building an X-ray microscopy beam line at the planned European Synchroton Radiation Facility (ESRF). To my surprise, many of those present did not know about the mechanisms and consequences of radiation damage, so I decided to write a review comparing radiation damage by electrons with that from X-rays. While I was writing the review, a copy of "Neutron News" arrived with a centre-page pull-out supplement listing the nuclear reactions and cross-sections for the interaction of neutrons with all the isotopes of all the elements (Sears, 1992). This allowed a calculation of the ratio of elastic to inelastic cross-sections and the resulting energy deposited during neutron illumination to be added to the review. The result was a broad review describing the potential and limitations of neutrons, electrons and X-rays for high-resolution imaging of biological macromolecules. The conclusion was that electrons produced the least damage per useful elastically scattered event, by a factor of 3 less than neutrons and a factor of 1000 less than X-rays. The review went on to estimate the minimum molecular weight of a macromolecular assembly and the approximate number of single particle images that would be required to determine the atomic structure by single particle cryoEM without resorting to crystallisation either in 2D or 3D (Henderson, 1995).

Since by then it was also becoming clear that it was just as difficult to make well-ordered 2D crystals as it was to make well-ordered 3D crystals of membrane proteins, I decided to switch the efforts of our group from electron crystallography to single particle cryoEM. This brought with it a number of new requirements. Since all of the information, both amplitudes and phases, would now come from the images, the electron diffraction patterns could no longer compensate for poor quality images. The microscopes would need more stable stages, better vacuums and the much brighter sources that were provided by field emission electron guns. Consequently, at MRC-LMB we purchased a Hitachi HF-2000 and this was used by Bettina Böttcher in Tony Crowther's group to obtain the first single particle structure with sub-nanometre resolution. The structure of the icosahedral assembly of hepatitis B core protein reached 7.4 Å resolution (Böttcher et al, 1997), and revealed the presence of a bundle of 4 α-helices protruding from the surface, which could be interpreted in terms of the amino acid sequence. Bettina used over 6000 particles in her work, which was still several orders of magnitude greater than the theory (Henderson, 1995) suggested and had a B-factor of about 500 $Å^2$. The B-factor, also called temperature factor or Debye-Waller factor, is an excellent way to describe how the power in Fourier components fades with resolution. The very high B-factor of 500 $Å^2$ in this case realistically ruled out being able to go to higher resolution without understanding the origin of the loss of contrast that limited the resolution. Many other structures of icosahedral viruses and helically ordered assemblies had been studied by cryoEM, but few reached 10 Å resolution by then. This was the era when cryoEM was termed "blobology" because the resolution of all the maps, except those from 2D crystals, merely revealed blobs of density for individual protein domains, which was insufficient to resolve the path of the polypeptide or the chemistry of the amino acids. A summary of the historical progress of cryoEM for HepB is shown in Figure 9: atomic resolution was not reached until 2013 when Hong Zhou's group reached 3.5 Å by collecting thousands of images on film. A summary of the

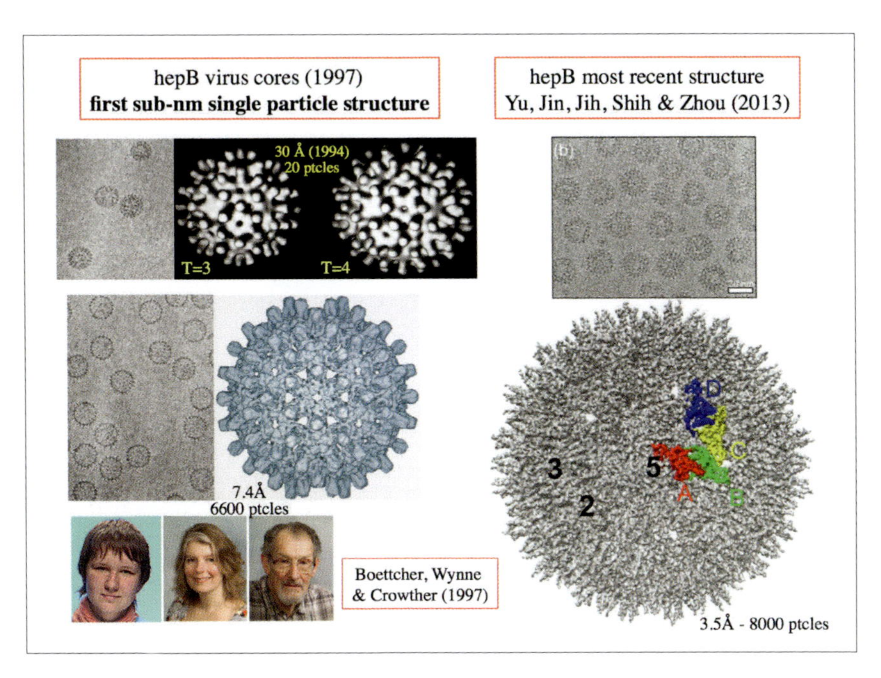

Figure 9. Three stages in the progress of cryoEM studies of the hepatitis B virus cores. The top left panel shows the first cryoEM 3D maps at 30 Å resolution obtained in 1994 from samples brought by Nikolai Kiselev from Paul Pumpens in Riga, Latvia. The sample had a mixture of T=3 and T=4 particles, but both 3D maps show similar protrusions, from Crowther et al (1994). The bottom left panels show a cryoEM image and 7.4 Å structure from the work of Böttcher et al (1997), which was the first sub-nm single particle cryoEM structure. It was calculated using ~6400 images of T=4 particles from an improved preparation. It showed that each protrusion consisted of a bundle of 4 α-helices. Finally, the panels on the right show a more recent cryoEM image and the 3.5Å resolution 3D structure from the work of Yu et al. (2013) in which they used many more images on a better microscope, but still using film as the recording medium. Similar resolutions can now be obtained using the new detectors with substantially fewer particles.

state-of-the-art of cryoEM in 2001 is shown in Figure 10, which includes a panel showing the structure of the *E. coli* 70S ribosome at 11.5 Å resolution (Gabashvili et al, 2000).

Niko Grigorieff had joined our group initially to work on the refinement of the bacteriorhodopsin structure. After completing this work (Grigorieff & Henderson, 1995; 1996; Grigorieff et al. 1995; 1996), he wrote a new single particle program called Frealign, which he used to determine at 22 Å resolution the first cryoEM structure of mitochondrial Complex I (Grigorieff, 1998), chosen to be an interesting structure without any internal symmetry. Grigorieff wrote Frealign specifically to treat all the electron optical parameters, such as defocus and astigmatism that were required for high-resolution single particle cryoEM, with the aim of being complementary to earlier 3DEM program suites such as Imagic (van Heel et al, 1996) and Spider (Frank et al, 1996).

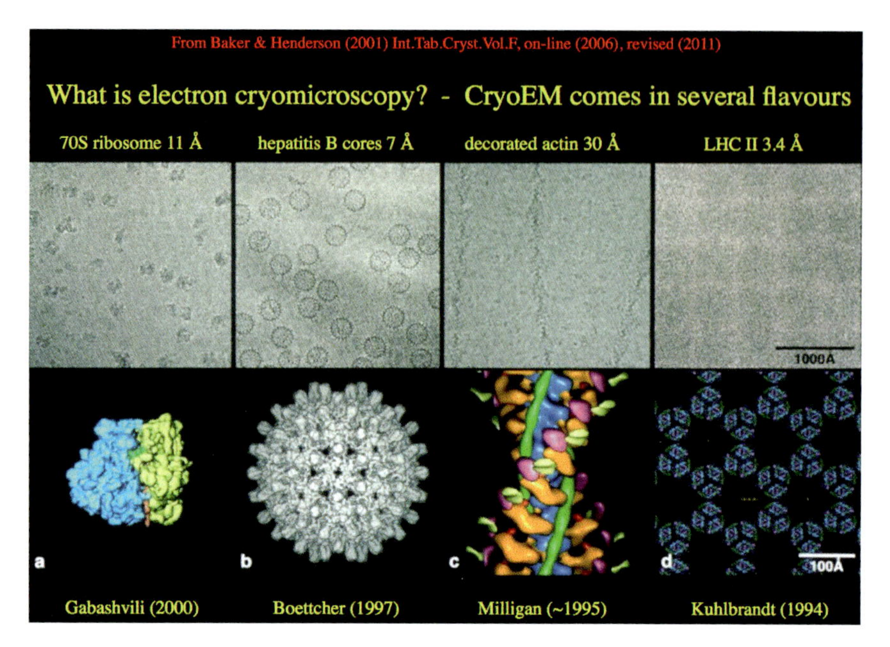

Figure 10. Reproduced from Baker & Henderson, 2001; 2012. Examples of macromolecules studied by cryoEM and 3D image reconstruction and the resulting 3D structures (bottom row) after cryoEM analysis dating from around the year 2000. All micrographs (top row) are displayed at ~170,000x magnification and all models at ~1,200,000x magnification. (a) A single particle without symmetry. The micrograph shows 70S E. coli ribosomes complexed with mRNA and fMet-tRNA. The surface-shaded density map, made by averaging 73,000 ribosome images from 287 micrographs, has a resolution of 11.5 Å. The 50S and 30S subunits and the tRNA are coloured blue, yellow and green, respectively. The identity of many of the protein and RNA components were known and some RNA double helices were clearly recognisable by their major and minor grooves (e.g. helix 44 is shown in red). Courtesy of J. Frank, using data from Gabashvili et al. (2000). (b) A single particle with symmetry. The micrograph shows hepatitis B virus cores. The 3D reconstruction, at a resolution of 7.4 Å, was computed from ~6400 particle images taken from 34 micrographs. From Böttcher et al (1997). (c) A helical filament. The micrograph shows actin filaments decorated with myosin S1 heads containing the essential light chain. The 3D reconstruction, at a resolution of 30–35 Å, is a composite in which the differently coloured parts are derived from a series of difference maps that were superimposed on F-actin. The components include: F-actin (blue), myosin heavy-chain motor domain (orange), essential light chain (purple), regulatory light chain (yellow), tropomyosin (green) and myosin motor domain N-terminal beta-barrel (magenta). Courtesy of A. Lin, M. Whittaker & R. Milligan (Scripps Research Institute, La Jolla). (d) A 2D crystal: light-harvesting complex LHCII at 3.4 Å resolution (Kühlbrandt et al, 1994). The model shows the protein backbone and the arrangement of chromophores in a number of trimeric subunits in the crystal lattice. In this example, image contrast is too low to see any hint of the structure without image processing. Courtesy of W. Kühlbrandt (Max-Planck-Institute for Biophysics, Frankfurt).

At around this time, Peter Rosenthal joined our group and recorded single particle cryoEM images of several interesting structures (e.g. Rosenthal et al, 2003). Sriram Subramaniam and Jacqueline Milne had also arrived on a sabbatical visit that for a variety of reasons was extended to 3 years. Jacqueline calculated a 3D structure of the pyruvate dehydrogenase complex (Milne et al, 2002) from images of single particles embedded in amorphous ice, using Marin van Heel's Imagic package (van Heel et al, 1996) to get started and Niko Grigorieff's Frealign (Grigorieff, 1998) for higher resolution refinement, with samples supplied by Gonzalo Domingo in Richard Perham's group in the Biochemistry Department.

An initial project by Peter Rosenthal to develop a semi-automatic procedure for determination of the absolute hand of a single particle cryoEM structure by using tilt pair images developed into a much broader publication (Rosenthal & Henderson, 2003), which allowed us to propose a theoretical framework to describe and understand the results of single particle cryoEM studies, and to propose the "tilt-pair validation procedure". The tools he developed helped to explain why the resolution was limited in 3D cryoEM studies at that time and to suggest what would be needed to do better. The basic idea was that the predicted resolution-dependence of the electron scattering, at resolutions beyond about 10 Å, could be described by four factors – the electron-scattering form-factors for individual-atoms, Wilson statistics, image blurring and errors in the determination of the orientation parameters of the particles in the analysis. To a first approximation, a single B-factor could explain the observations both in theory and for a set of experimental data that consisted of 3600 single particle images of pyruvate dehydrogenase, with a B-factor of ~1000 Å^2. This high B-factor limited the resolution to 8.7 Å. Rosenthal also introduced a novel plot of the natural logarithm of the number of particles required to achieve different resolutions versus the reciprocal of resolution squared. In a light-hearted way, we referred to this as a universal resolution calculator, or "Rosenthal plot". It showed graphically that the most important factor to achieving higher resolution single particle cryoEM structures was to acquire better images in which the higher resolution Fourier components were recorded with less blurring and therefore less contrast loss. Better images would have lower intrinsic B-factors and would allow more accurate orientation determination leading to lower computational blurring, and lower overall B-factors. This image quality problem was essentially the same problem that had been identified in earlier publications (Henderson & Glaeser, 1985; Henderson, 1992), and provided the rationale for increased efforts to develop better electron detectors.

DEVELOPMENT OF DIRECT ELECTRON DETECTORS, IMPACT ON SINGLE PARTICLE CRYOEM, AND THE "RESOLUTION REVOLUTION"

During the 1990s, Wasi Faruqi who had worked earlier on the development of X-ray detectors with Hugh Huxley at MRC-LMB, switched his emphasis to the development of better detectors for electrons. Before that and in practice right up until 2012, photographic film had been the best medium for recording electron images, but suffered from the fact that the images were not immediately available since the film had to be developed, fixed, washed, dried and digitised on a film scanner. In addition, the emulsions had to be desiccated for weeks before use, otherwise the microscope vacuum would be compromised and the residual water vapour in the column would rapidly build up as a contaminating layer of ice on the cryo-specimens. After various projects to develop electron detectors based on phosphor/fibre-optics/CCD (Faruqi et al, 1995; 1999) or on the Medipix series of hybrid pixel detectors (Faruqi et al, 2003; 2005), Wasi identified work by Renato Turchetta at the Rutherford Appleton Laboratory (RAL) near Oxford on monolithic active pixel sensors (MAPS) using CMOS (complementary metal oxide semiconductor) technology. Turchetta had been investigating the use of these detectors for charged particle detection (Caccia et al, 1999) and had brought the Startracker CMOS detector to MRC-LMB in 2002. Tests immediately showed excellent sig-

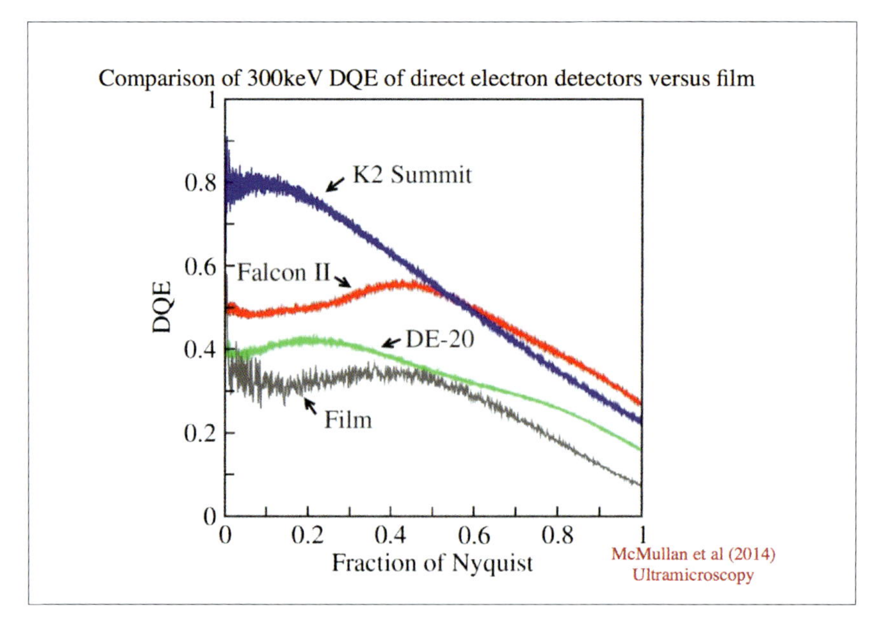

Figure 11. Comparison of performance of three direct electron detectors, reproduced from McMullan et al (2014). The DQE is measured as a function of spatial frequency for the DE-20 (green), Falcon-II (red) and K2 Summit (blue). The corresponding DQE of photographic film is shown in black.

nal-to-noise ratio for detection of 120 keV electrons, but also showed that the electron beam rapidly damaged the pixels. A long project to improve the radiation hardness and to optimise the detective quantum efficiency (DQE) then ensued, until publications in 2009 (McMullan et al, 2009a; 2009b; 2009c) showed that this type of direct detection device (DDD) would exceed the performance of film when used for imaging of high energy electrons (preferably 300 keV or higher), provided the sensors were backthinned.

The work with the RAL group eventually led to the commercial development of the Falcon detectors by FEI, now Thermo Fisher Scientific. In parallel, work by Gatan to develop the K2 CMOS detector, based on earlier work by Peter Denes, and by Direct Electron to develop the DE-12 detector, based on earlier work by Kleinfelder, Xuong and Ellisman, both produced similar CMOS cameras. A comparison of these three detectors with film is shown in Figure 11. All three had improved detective quantum efficiency (DQE) compared with film, but only the K2 detector at that time had fast enough read-out to allow implementation of an electron counting mode in which the analogue images were processed to replace the stochastic signal from individual electron events with an equal signal that represented the true nature of the image, and which gave the K2 a

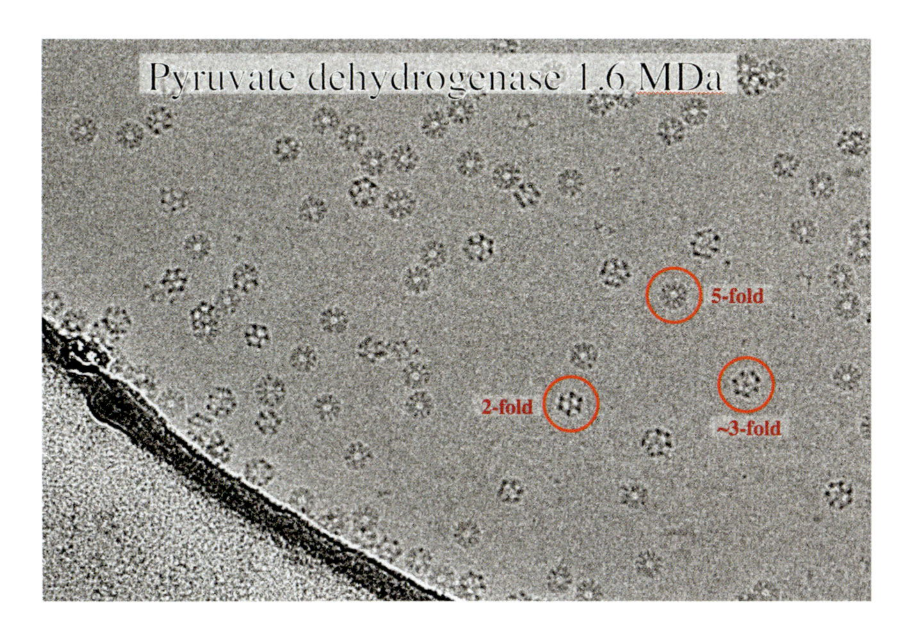

Figure 12. An example of a state-of-the-art cryoEM image with excellent signal-to-noise ratio. It shows each macromolecule very clearly with obvious orientations. This specimen was plunge-frozen by Peter Rosenthal in 2001 using a sample of pyruvate dehydrogenase from Richard Perham's group. The image was recorded in 2015 by Vinothkumar on a Falcon-II detector in integrating mode. The specimen, kept under liquid nitrogen for 14 years, still has perfectly amorphous ice.

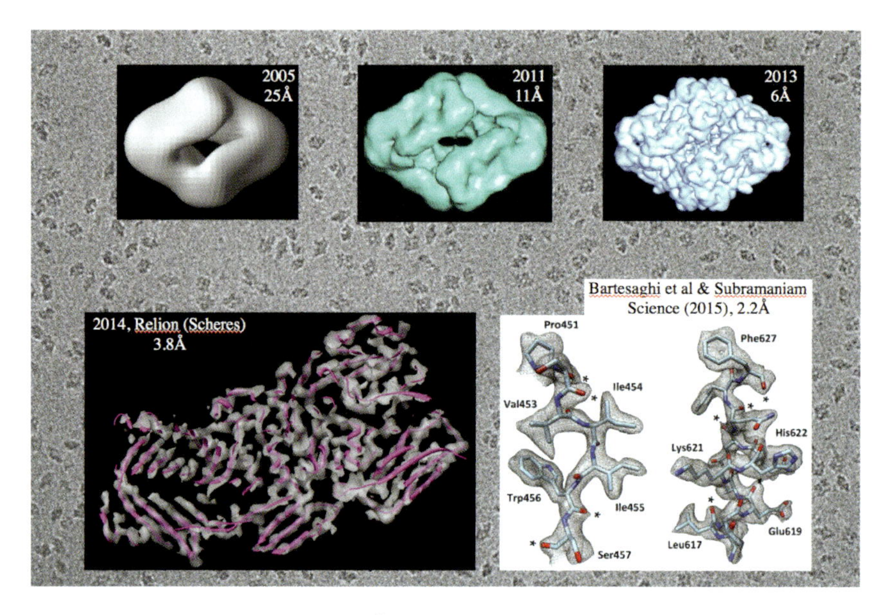

Figure 13. Progress of work on *E. coli* β-galactosidase, which was selected as a challenging test object in 1997. The underlying image shows a field of view recorded in 2013 on one of the direct electron detectors. The five superimposed panels show how technical progress has greatly improved the resolution during the last 20 years. The top left panel shows an attempt to obtain a low-resolution 3D structure from the earliest images recorded on film: we did not believe this structure even though it is roughly right. The top centre panel shows a medium resolution structure using images recorded on photographic film with 80 keV electrons. This was the first structure we proved was correct because it passed the tilt-pair validation test (Rosenthal & Henderson, 2003; Henderson et al, 2011). The top right structure was obtained using a direct electron detector that first became available in 2013 and immediately showed higher resolution. The use of Relion (bottom left) improved the resolution further to 3.8 Å. Finally, the work of Bartesaghi et al (2015), using higher magnification, produced a superb map at 2.2 Å resolution.

higher DQE at low resolution. In addition, the rolling shutter read-out mode for these CMOS sensors allowed the recording of the images as dose-fractionated exposure series, or "movies", which allowed subsequent computer-based correction for beam-induced specimen motion to be carried out (Brilot et al, 2012; Campbell et al, 2012; Bai et al, 2013; Li et al, 2013; Scheres, 2014; Vinothkumar et al, 2014a; Rubinstein & Brubaker, 2015; Grant & Grigorieff, 2015). The combination of increased DQE and specimen motion correction greatly improved the quality of the images, and this alongside the development of improved computer image processing algorithms such as in Relion (Scheres, 2012) resulted in a quantum leap in the resolution of single particle cryoEM structures (e.g. Liao et al, 2013; Amunts et al, 2014; Allegretti et al, 2014). This advance was characterised by the term "Resolution Revolution" (Kühlbrandt, 2014) and has proved to be an apt description. An example of the clarity seen in these new images is shown in Figure 12.

There have been many superb structures determined by single particle cryoEM at near-atomic resolution during the last 5 years; I briefly mention three with which I had an early involvement, and which were subsequently pursued to higher resolution by younger colleagues. Two of these are shown in Figure 13 and Figure 14. The first (Figure 13) is β-galactosidase from *E. coli*, which had reached a resolution of only 11 Å when images had been recorded at 80 keV on film (Henderson et al, 2011), but rapidly went to 6 Å (Vinothkumar et al, 2014a), then 3.8 Å using Relion, and finally 2.2 Å through the work of Bartesaghi et al (2015), when higher quality images recorded with the new DDD cameras were obtained. The second structure (Figure 14) is that of mitochondrial Complex I from the work of Judy Hirst's group first at 5 Å resolution (Vinothkumar et al, 2014b) and then at 4.2 Å (Zhu et al, 2016), which allowed them to identify and build atomic models of all 45 polypeptides that make up the structure of this large macromolecular complex. Finally, the work of John Rubinstein on F_1F_O-ATPases also shows the impact of the resolution revolution on a structure that was only barely tractable 15 years ago – the 30 Å resolution obtained in 2003 (Rubinstein et al, 2003) has recently reached 3.7 Å (Zhou et al, 2015; Guo et al, 2017).

OUTSTANDING PROBLEMS

Two kinds of analysis are very revealing when applied to current cryoEM structural determinations. One of these is the Rosenthal plot, the slope of which can reveal the underlying B-factor that describes the behaviour of the data and limits the resolution of the final density map, in a robust way that does not require estimation of the modulation transfer function (MTF) of the detector. The best recent single particle cryoEM structure determinations show B-factors of 90–100 $Å^2$, which is an enormous improvement over the values of 500–1000 $Å^2$ that were obtained a decade earlier (Böttcher et al, 1997; Rosenthal & Henderson, 2003; Rubinstein et al, 2003). The other useful plot shows the information content against the frame number or electron dose from dose-fractionated "movies", as shown in Figure 15, reproduced from Henderson (2015). This second type of plot shows that the first few frames of the movies, which should have the least radiation damage, are actually much worse than those with exposures in the range 5–10 el/$Å^2$. They have less information, characterised by higher B-factors than later frames. These two diagnostic tools show that the images being acquired using present state-of-the-art approaches still fall significantly short of what would be expected in perfect images limited only by radiation damage and no other factor. The first frame of such a perfect image should show the highest contrast and the lowest B-factor, with a gradual increase in disorder in subsequent frames

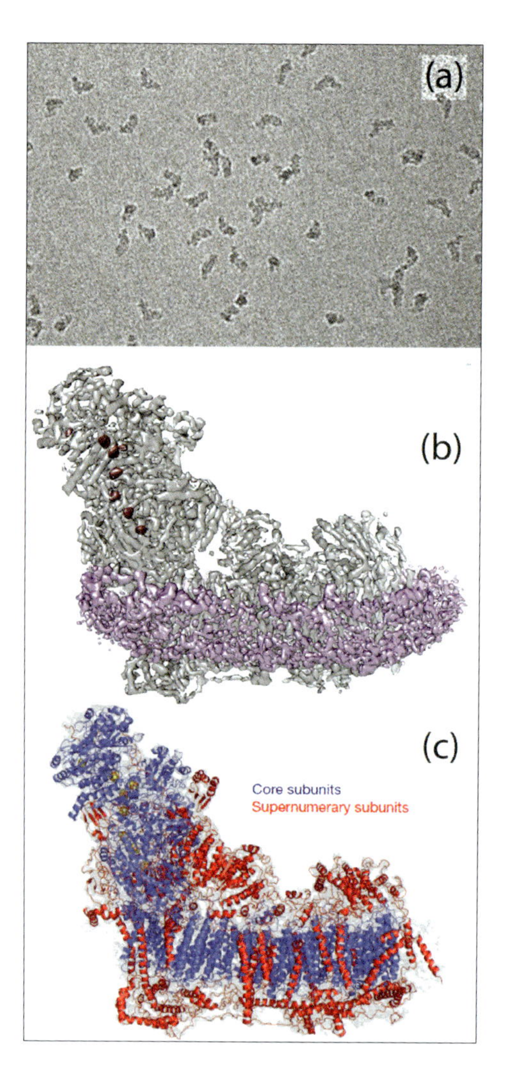

Figure 14. Mitochondrial Complex I. (a) cryoEM image. (b) 3D density map at three contour levels. (c) 3D model showing all 45 protein subunits. From Vinothkumar et al (2014b) and Zhu et al, (2016).

as radiation damage causes slightly different changes in the structure of the different molecules in the dataset. In addition, it is well-known that radiation damage causes mass loss due to the release of volatile radiation products (Müller & Engel, 2001). The dose-dependence of the intercept (C_f) parameter in the particle polishing procedure of Relion (Scheres, 2014) invariably shows a reduction of ~20% during the exposure, which is believed to be due to mass loss. To help understand the origin of the phenomena seen in these two types of plot, some recent publications allow the relative importance of several possible contributory factors to be estimated.

At liquid nitrogen temperature, hydrogen and oxygen are lost from both surfaces of the thin film of ice during electron irradiation, but the conse-

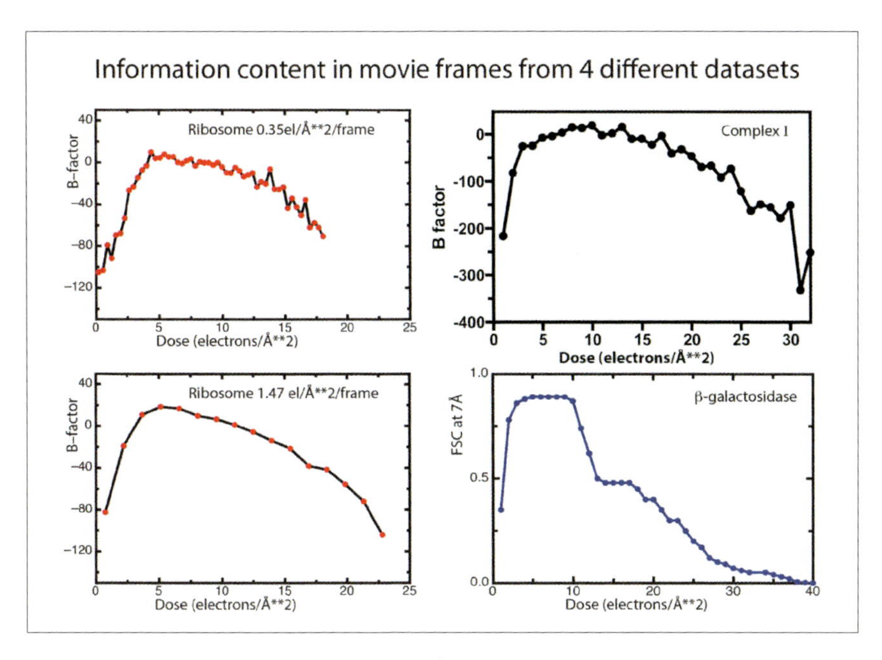

Figure 15. Plot of the B-factors or signal at 7 Å resolution in typical movie sequences, reproduced from Henderson (2015) with permission.

quence of radiolysis deeper inside the specimen simply consists of recombined water molecules whose positions have moved. McMullan et al (2015) showed that the water molecules in pure amorphous ice move during electron irradiation by about 1 Å after an electron dose of 1 el/Å^2 using 300 keV electrons. After a typical exposure of 25 el/Å^2, the average water molecule has therefore moved by ~5 Å. This pseudo-Brownian motion of the water molecules pushes around the embedded macromolecular assemblies being studied, but fortunately this causes only a small movement of the macromolecule, such as 0.5 Å for a ribosome and slightly more for smaller structures (McMullan et al, 2015). The resulting blurring of the images adds a small uniform B-factor to summed images but cannot explain the poor contrast and high B-factor in the first frames of the movies.

Recent work by Russo and Henderson has also allowed the impact of charge build-up during irradiation due to the "Berriman effect" (Brink et al, 1998), and of charge fluctuations in the thin layers of amorphous, non-conductive ice, often termed the "beeswarm effect", to be estimated. Both of these were measured to be finite but small. The Berriman effect was found to have an impact only during the very earliest part of the first frame of an exposure series and reached an equilibrium during the rest of the exposure (Russo & Henderson, 2018a). The beeswarm effect also produces a measurable perturbation in the images, which is manifested as a small decrease in the amplitude of the envelope function that is detectable only at very high defocus values, well outside the range normally used for single particle cryoEM images (Russo & Henderson, 2018b).

The clear conclusion is that physical (i.e. mechanical) motion is the principal remaining factor that is causing the quality of the best current images to fall short of that expected in theory, which should be limited only by statistical disorder due to radiation damage. There are two possible causes of this physical motion and resultant image blurring. One is the beam-induced relaxation of stresses frozen into the specimen at the point of plunge-freezing, due to the different linear coefficients of expansion of protein (positive, since proteins shrink on freezing) and water (negative, since it expands on freezing), as well as the support, which constrains both. The second is a consequence of covalent bond breakage after radiation damage to the protein or nucleic acid in the macromolecular assembly. Bond breakage causes covalent bonds of length ~1.5 Å to increase to ~3.5 Å producing radiolytic fragments that are separated by van der Waals distances. Depending on their size, these radiolytic fragments will then be either trapped causing an increase in internal pressure or will diffuse away and evaporate creating a cavity and a decrease in internal pressure. Unless the contributions of trapped and released radiolytic fragments are exactly balanced, this will cause beam-induced local specimen motion, with resulting image blurring, of just the kind observed. We could say that we have a diagnosis but not a cure for the outstanding problem of beam-induced image blurring.

FUTURE

What will be the consequence of successfully eliminating or ameliorating the remaining problems of specimen motion and image blurring? Such an advance might be achieved as a consequence of improvements in specimen supports (e.g. Russo & Passmore, 2014), or of improvements in imaging protocols (e.g. Berriman & Rosenthal, 2012), or of improvements in computer-based motion correction of the dose-fractionated movies (e.g. Zheng et al, 2017). We do not yet have an accurate estimate of the slope of the intrinsic increase in disorder as a function of radiation dose, which is also likely to depend to some extent on the composition of the specimen. For example, it is known (Glaeser, 1971) that nucleic acid bases are on average more radiation resistant than amino acids, due to conjugation in the ring structures of the bases, and that the aromatic side chains of phenylalanine, tyrosine and tryptophan are more radiation resistant than other amino acids. However, if we estimate that the disorder in average protein structures increases by a B-factor equivalent of 6 $Å^2$ for each additional exposure to a dose of 1 el/$Å^2$ using 300 keV electrons, then the slope of the plot of information content versus electron exposure will look similar to the asymptotic tail at higher doses in the plots of Figure 15. After 25 el/Å^2, the B-factor of that frame would then be 150 Å^2, and the average B-factor over the first 10 el/Å^2 of a dose-fractionated movie would be 30 Å^2. The resulting overall decrease from the current state-of-the-art B-factor (B_0) in single particle cryoEM of about 90 Å^2 about 30 Å^2 (B_n) would translate into a 30-fold reduction in the number of images required to reach 3 Å resolution [*viz.* $\exp((B_0 - B_n)/2d^2)$], or alternatively an increase in resolution from 3.0 Å to ~1.7 Å [*viz.* $\sim d(B_n/B_0)^{1/2}$], or the ability to resolve an increased number of multiple states by 3D classification (Scheres et al, 2007) using the same number of images. It is safe to conclude that single particle cryoEM has a promising future.

ACKNOWLEDGEMENTS

Figure 16 shows a gallery of those who have contributed to the work described in this article. I am grateful to them all for being such wonderful collaborators and colleagues.

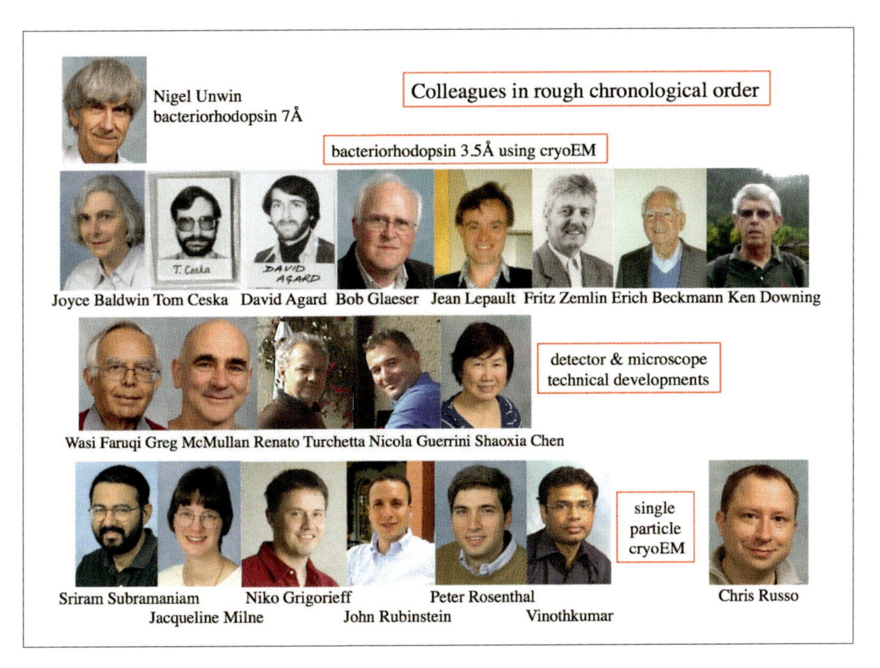

Figure 16. Colleagues who have contributed to the work described in this lecture. Top: Nigel Unwin. Second row: Joyce, Baldwin, Tom Ceska, David Agard, Jean Lepault, Fritz Zemlin, Erich Beckmann, Bob Glaeser, Ken Downing. Third row: Wasi Faruqi, Greg McMullan, Renato Turchetta, Nicola Guerrini, Shaoxia Chen. Bottom: Sriram Subramaniam, Jacqueline Milne, Niko Grigorieff, John Rubinstein, Peter Rosenthal, Kutti Ragunath Vinothkumar, Chris Russo.

REFERENCES

Adrian, M., Dubochet, J., Lepault, J. & McDowall, A. W. (1984). *Nature* **308**, 32–36.
Allegretti, M., Mills, D. J., McMullan, G., Kühlbrandt, W. & Vonck, J. (2014) *eLife*.01963.
Amunts, A., Brown, A., Bai, X.-C., Llacer, J. L., Hussain, T., Emsley, P., Long, F., Murshudov, G., Scheres, S. H. W. & Ramakrishnan, V. (2014) *Science* **343**, 1485–1489.
Bai, X.-C., Fernandez, I. S., McMullan, G. & Scheres, S. H. W. (2013). *eLife*.00461.
Baker, T. S. & Henderson, R. (2001). *In: International Tables for Crystallography, Vol. F, "Crystallography of Biological Macromolecules", Chapter 19.6 pp. 451–463; 473–479. (Ed. Rossmann, M.G. & Arnold, E.). Dordrecht: Kluwer Academic Publishers.*
Baker, T. S. & Henderson, R. (2012). *In: International Tables for Crystallography, Vol. F (second edition), "Crystallography of Biological Macromolecules", Chapter 19.6 pp. 593–614. (Ed. Rossmann, M.G. & Arnold, E.). Wiley.*
Bartesaghi, A., Merk, A., Banerjee, S., Matthies, D., Wu, X. W., Milne, J. L. S. & Subramaniam, S. (2015). *Science* **348**, 1147–1151.
Bellare, J. R., Davis, H. T., Scriven, L. E. & Talmon, Y. (1988). *J Electron Micr Tech* **10**, 87–111.
Berriman, J. A. & Rosenthal, P. B. (2012). *Ultramicroscopy* **116**, 106–114.
Blaurock, A. E. & Stoeckenius, W. (1971). *Nature-New Biol* **233**, 152–155.
Böttcher, B., Wynne, S. A. & Crowther, R. A. (1997). *Nature* **386**, 88–91.
Brilot, A. F., Chen, J. Z., Cheng, A. C., Pan, J. H., Harrison, S. C., Potter, C. S., Carragher, B., Henderson, R. & Grigorieff, N. (2012). *J Struct Biol* **177**, 630–637.
Brink, J., Sherman, M. B., Berriman, J. & Chiu, W. (1998). *Ultramicroscopy* **72**, 41–52.
Bullough, P. & Henderson, R. (1987). *Ultramicroscopy* **21**, 223–229.
Caccia, M., Campagnolo, R., Meroni, C., Kucewicz, W., Deptuch, G., Zalewska, A. & Turchetta, R. (1999). *arXiv*:hep-ex/ 9910019v1.
Campbell, M. G., Cheng, A. C., Brilot, A. F., Moeller, A., Lyumkis, D., Veesler, D., Pan, J. H., Harrison, S. C., Potter, C. S., Carragher, B. & Grigorieff, N. (2012). *Structure* **20**, 1823–1828.
Ceska, T. A. & Henderson, R. (1990). *J Mol Biol* **213**, 539–560.
Crowther, R. A., Kiselev, N. A., Böttcher, B., Berriman, J. A., Borisova, G. P., Ose, V. & Pumpens, P. (1994). *Cell* **77**, 943–950.
Downing, K. H. (1988). *Ultramicroscopy* **24**, 387–398.
Dubochet, J., Adrian, M., Chang, J. J., Homo, J. C., Lepault, J., McDowall, A. W. & Schultz, P. (1988). *Q Rev Biophys* **21**, 129–228.
Dubochet, J., Adrian, M., Teixeira, J., Alba, C. M., Kadiyala, R. K., Macfarlane, D. R. & Angell, C. A. (1984). *J Phys Chem* **88**, 6727–6732.
Dubochet, J., Chang, J. J., Freeman, R., Lepault, J. & McDowall, A. W. (1982a). *Ultramicroscopy* **10**, 55–61.
Dubochet, J., Lepault, J., Freeman, R., Berriman, J. A. & Homo, J. C. (1982b). J *Microsc* **128**, 219–237.
Faruqi, A. R., Andrews, H. N. & Henderson, R. (1995). *Nucl Instrum Meth* **A 367**, 408–412.
Faruqi, A. R., Cattermole, D. M., Henderson, R., Mikulec, B. & Raeburn, C. (2003). *Ultramicroscopy* **94**, 263–276.
Faruqi, A. R., Henderson, R., Pryddetch, M., Allport, P. & Evans, A. (2005). *Nucl Instrum Meth* **A 546**, 170–175.

Faruqi, A. R., Henderson, R. & Subramaniam, S. (1999). *Ultramicroscopy* **75**, 235–250.
Frank, J. (1975). *Ultramicroscopy* **1**, 159–162.
Frank, J. & Alali, L. (1975). *Nature* **256**, 376–379.
Frank, J., Penczek, P., Grassucci, R. & Srivastava, S. (1991). *J Cell Biol* **115**, 597–605.
Frank, J., Radermacher, M., Penczek, P., Zhu, J., Li, Y. H., Ladjadj, M. & Leith, A. (1996). *J Struct Biol* **116**, 190–199.
Frank, J. & vanHeel, M. (1982). *J Mol Biol* **161**, 134–137.
Gabashvili, I. S., Agrawal, R. K., Spahn, C. M. T., Grassucci, R. A., Svergun, D. I., Frank, J. & Penczek, P. (2000). *Cell* **100**, 537–549.
Glaeser, R. M. (1971). *J Ultrastruct Res* **36**, 466–482.
Grant, T. & Grigorieff, N. (2015). *eLife*.06980
Grigorieff, N. (1998). *J Mol Biol* **277**, 1033–1046.
Grigorieff, N., Beckmann, E. & Zemlin, F. (1995). *J Mol Biol* **254**, 404–415.
Grigorieff, N., Ceska, T. A., Downing, K. H., Baldwin, J. M. & Henderson, R. (1996). *J Mol Biol* **259**, 393–421.
Grigorieff, N. & Henderson, R. (1995). *Ultramicroscopy* **60**, 295–309.
Grigorieff, N. & Henderson, R. (1996). *Ultramicroscopy* **65**, 101–107.
Guo, H., Bueler, S. A. & Rubinstein, J. L. (2017). *Science* **358**, 936–940.
Hayward, S. B. & Stroud, R. M. (1981). *J Mol Biol* **151**, 491–517.
Henderson, R. (1970). *J Mol Biol* **54**, 341–354.
Henderson, R. (1990). *P Roy Soc B-Biol Sci* **241**, 6–8.
Henderson, R. (1992). *Ultramicroscopy* **46**, 1–18.
Henderson, R. (1995). *Q Rev Biophys* **28**, 171–193.
Henderson, R. (2015). *Arch Biochem Biophys* **581**, 19–24.
Henderson, R. & Baldwin, J. M. (1986). *44th Annu. Proc. EMSA* (Bailey, G. W., ed.), pp. 6– 9, San Francisco Press.
Henderson, R., Baldwin, J. M., Ceska, T. A., Zemlin, F., Beckmann, E. & Downing, K. H. (1990). *J Mol Biol* **213**, 899–929.
Henderson, R., Baldwin, J. M., Downing, K. H., Lepault, J. & Zemlin, F. (1986). *Ultramicroscopy* **19**, 147–178.
Henderson, R., Chen, S. X., Chen, J. Z., Grigorieff, N., Passmore, L. A., Ciccarelli, L., Rubinstein, J. L., Crowther, R. A., Stewart, P. L. & Rosenthal, P. B. (2011). *J Mol Biol* **413**, 1028–1046.
Henderson, R. & Glaeser, R. M. (1985). *Ultramicroscopy* **16**, 139–150.
Henderson, R., Raeburn, C. & Vigers, G. (1991). *Ultramicroscopy* **35**, 45–53.
Henderson, R. & Unwin, P. N. T. (1975). *Nature* **257**, 28–32.
Henderson, R. & Wang, J. H. (1972). *Biochemistry* **11**, 4565–4569.
Hope, H., Frolow, F., Vonbohlen, K., Makowski, I., Kratky, C., Halfon, Y., Danz, H., Webster, P., Bartels, K. S., Wittmann, H. G. & Yonath, A. (1989). *Acta Crystallogr* **B 45**, 190–199.
Hope, H. & Nichols, B. G. (1981). *Acta Crystallogr* **B 37**, 158–161.
Jeng, T. W., Chiu, W., Zemlin, F. & Zeitler, E. (1984). *J Mol Biol* **175**, 93–97.
Kimura, Y., Vassylyev, D. G., Miyazawa, A., Kidera, A., Matsushima, M., Mitsuoka, K., Murata, K., Hirai, T. & Fujiyoshi, Y. (1997). *Nature* **389**, 206–211.
Kühlbrandt, W. (2014). *Science* **343**, 1443–1444.
Kühlbrandt, W., Wang, D. N. & Fujiyoshi, Y. (1994). *Nature* **367**, 614–621.
Li, X. M., Mooney, P., Zheng, S., Booth, C. R., Braunfeld, M. B., Gubbens, S., Agard, D. A. & Cheng, Y. F. (2013). *Nat Methods* **10**, 584–590.
Liao, M. F., Cao, E. H., Julius, D. & Cheng, Y. F. (2013). *Nature* **504**, 107–112.

Lücke, H., Schobert, B., Richter, H. T., Cartailler, J. P. & Lanyi, J. K. (1999). *J Mol Biol* **291**, 899–911.
Matthews, B. W., Sigler, P. B., Henderson, R. & Blow, D. M. (1967). *Nature* **214**, 652–656.
McMullan, G., Chen, S., Henderson, R. & Faruqi, A. R. (2009a). *Ultramicroscopy* **109**, 1126–1143.
McMullan, G., Faruqi, A. R., Henderson, R., Guerrini, N., Turchetta, R., Jacobs, A. & van Hoften, G. (2009b). *Ultramicroscopy* **109**, 1144–1147.
McMullan, G., Clark, A. T., Turchetta, R. & Faruqi, A. R. (2009c). *Ultramicroscopy* **109**, 1411–1416.
McMullan, G., Faruqi, A. R., Clare, D. & Henderson, R. (2014). *Ultramicroscopy* **147**, 156–163.
McMullan, G., Vinothkumar, K. R. & Henderson, R. (2015). *Ultramicroscopy* **158**, 26–32.
Milne, J. L. S., Shi, D., Rosenthal, P. B., Sunshine, J. S., Domingo, G. J., Wu, X. W., Brooks, B. R., Perham, R. N., Henderson, R. & Subramaniam, S. (2002). *Embo J* **21**, 5587–5598.
Müller, S. A. & Engel, A. (2001). *Micron* **32**, 21–31.
Nogales, E., Wolf, S. G. & Downing, K. H. (1998). *Nature* **391**, 199–203.
Oesterhelt, D. & Stoeckenius, W. (1971). *Nature-New Biol* **233**, 149–152.
Pebay-Peyroula, E., Rummel, G., Rosenbusch, J. P. & Landau, E. M. (1997). *Science* **277**, 1676–1681.
Radermacher, M., Wagenknecht, T., Verschoor, A. & Frank, J. (1987). *J Microsc-Oxford* **146**, 113–136.
Rosenthal, P. B. & Henderson, R. (2003). *J Mol Biol* **333**, 721–745.
Rosenthal, P. B., Waddington, L. J. & Hudson, P. J. (2003). *J Mol Biol* **334**, 721–731.
Rossmann, M. G. & Henderson, R. (1982). *Acta Crystallogr* **A 38**, 13–20.
Rubinstein, J. L. & Brubaker, M. A. (2015). *J Struct Biol* **192**, 188–195.
Rubinstein, J. L., Walker, J. E. & Henderson, R. (2003). *Embo J* **22**, 6182–6192.
Russo, C. J. & Henderson, R. (2018a). *Ultramicroscopy* **187**, 43–49.
Russo, C. J. & Henderson, R. (2018b). *Ultramicroscopy* **187**, 56–63.
Russo, C. J. & Passmore, L. A. (2014). *Science* **346**, 1377–1380.
Scheres, S. H. W. (2012). *J Struct Biol* **180**, 519–530.
Scheres, S. H. W. (2014). *eLife*.03665
Scheres, S. H. W., Gao, H. X., Valle, M., Herman, G. T., Eggermont, P. P. B., Frank, J. & Carazo, J. M. (2007). *Nat Methods* **4**, 27–29.
Sears, V. F. (1992). *Neutron News* **3**(3), 26–37.
Smith, D. J., Saxton, W. O., O'Keefe, M. A., Wood, G. J. & Stobbs, W. M. (1983). *Ultramicroscopy* **11**, 263–281.
Steitz, T. A., Henderson, R. & Blow, D. M. (1969). *J Mol Biol* **46**, 337–348.
Subramaniam, S., Gerstein, M., Oesterhelt, D. & Henderson, R. (1993). *Embo J* **12**, 1–8.
Subramaniam, S. & Henderson, R. (2000). *Nature* **406**, 653–657.
Subramaniam, S., Lindahl, M., Bullough, P., Faruqi, A. R., Tittor, J., Oesterhelt, D., Brown, L., Lanyi, J. & Henderson, R. (1999). *J Mol Biol* **287**, 145–161.
Taylor, K. A. & Glaeser, R. M. (1974). *Science* **186**, 1036–1037.
Taylor, K. A. & Glaeser, R. M. (1976). *J Ultrastruct Res* **55**, 448–456.
Tsygannik, I. N. & Baldwin, J. M. (1987). *Eur Biophys J* **14**, 263–272.
Unwin, P. N. T. & Henderson, R. (1975). *J Mol Biol* **94**, 425–440.
van Heel, M. (1987). *Ultramicroscopy* **21**, 111–123.
van Heel, M. & Frank, J. (1981). *Ultramicroscopy* **6**, 187–194.

van Heel, M., Harauz, G., Orlova, E. V., Schmidt, R. & Schatz, M. (1996). *J Struct Biol* **116**, 17–24.
Vinothkumar, K. R., McMullan, G. & Henderson, R. (2014a). *Structure* **22**, 621–627.
Vinothkumar, K. R., Zhu, J. P. & Hirst, J. (2014b). *Nature* **515**, 80–84.
Yu, X. K., Jin, L., Jih, J., Shih, C. H. & Zhou, Z. H. (2013). *Plos One* 0069729.
Zemlin, F. (1979). *Ultramicroscopy* **4**, 241–245.
Zheng, S. Q., Palovcak, E., Armache, J.-P., Verba, K. A., Cheng, Y. & Agard, D. A. (2017) *Nat Methods* **14**, 331–332.
Zhou, A. N., Rohou, A., Schep, D. G., Bason, J. V., Montgomery, M. G., Walker, J. E., Grigorieff, N. & Rubinstein, J. L. (2015). *eLife*.10180.
Zhu, J. P., Vinothkumar, K. R. & Hirst, J. (2016). *Nature* **536**, 354–360.

Chemistry 2018

one half to

Frances H. Arnold

"for the directed evolution of enzymes"

the other half jointly to

George P. Smith and Sir Gregory P. Winter

"for the phage display of peptides and antibodies"

The Nobel Prize in Chemistry, 2018

Presentation speech by Professor Sara Snogerup Linse, Member of the Royal Swedish Academy of Sciences; Member of the Nobel Committee for Chemistry, 10 December 2018.

Your Majesties, Your Royal Highnesses, Esteemed Nobel Laureates, Ladies and Gentlemen,

We humans believe we know everything. Sometimes it may be more fruitful to acknowledge our incompetence and trust the superior performance of Mother Nature. An element of chance can work wonders in the lab. Just like in everyday life, the level of success may be higher if one allows oneself the luxury of good luck.

Evolution has over billions of years adapted and refined the chemistry of life. This allows various organisms to co-exist and thrive in all sorts of possible and seemingly impossible environments.

In nature, evolution has no plan. Changes that make an organism better adapted to its environment increase the chances for survival and new improvement can be added over generations to come. Frances Arnold had a defined plan when she set up directed evolution in her laboratory. She wanted to make a greener chemical industry and produce biofuels in a sustainable way.

Frances Arnold harnessed the power of evolution and made it a versatile tool for improving nature's own catalysts, enzymes. She also made it possible to create new biocatalysts that can speed up reactions not seen in nature. Time after time, through smart combinations of knowledge and chance, she created enzymes that are for the greatest benefit of humankind.

The proteins in nature can also make wonders by interacting and intriguing. George Smith developed a method to create very large collections of similar proteins and then using a molecular bait he could fish out the members of the collection that were most strongly attracted to the bait. He constructed his method so that every protein carries with it a recipe for its own production. This feature makes it easy to make new copies of the best proteins and engage them in a tighter competition for the bait.

Gregory Winter sharpened the fishing tools for the development of pharmaceuticals. Antibodies are large and complex molecules, but Winter chose to work with a small fragment that carries all the variation seen among natural antibodies. The result became a powerful method for deriving new antibodies for diagnostics and treatment, antibodies that with high precision can adhere, strongly and persistently, to other molecules or cell surfaces to facilitate or interfere with their duties.

Frances Arnold, George Smith and Gregory Winter:

Your work has led to the development of enzymes for a greener chemistry and antibodies that save lives. That is a truly great achievement. On behalf of the Royal Swedish Academy of Sciences I wish to convey to you our warmest congratulations. May I now ask you to step forward and receive your Nobel Prizes from the hands of His Majesty the King.

Frances H. Arnold. © Nobel Prize Outreach AB. Photo: A. Mahmoud

Frances Hamilton Arnold

Biography

I WAS BORN IN EAST PITTSBURGH, Pennsylvania, on July 25, 1956, misshapen after my mother's twenty-four-hour labor and with no hair. My father called me his 'Swan' (until that nickname was swapped for 'Vampira' when I was a teenager). William Howard Arnold and Josephine Inman Routheau, both twenty-five, already had two-and-a-half-year old Bill when I came along. I was followed thirteen months later by Edward (a sweet 'surprise'), then by David, and finally by Thomas when I was twelve and old enough to take care of a baby by myself.

This crowd of boys, which I learned to navigate, was usually organized by Bill, named after my father and after my grandfather, General William Howard Arnold, who had served in the U.S. Army in the Pacific theater during WWII, commanded the U.S. forces in Austria after the war, and retired as commander of the 5th U.S. Army. We were part of an extended Catholic family, many of whom to this day gather at summer cottages in Macatawa, on the east shore of Lake Michigan. The women in the family ran the show, ably filling in to organize the troops at home. For years we shared my grandparents' turn-of-the-century cottage "Stack Arms" with various cousins, until my father built his own place in 1965. My grandfather, the powerful general, died of a broken heart only weeks after his beloved and even more commanding wife, Elizabeth Welsh Mullen, succumbed to breast cancer in 1976.

Macatawa, Michigan was paradise because we could run around freely, sometimes in packs, and sometimes alone. My mother nearly had a heart attack one day upon seeing my tricycle abandoned at the end of the dock. I was found underneath the dock, digging up crayfish. A couple of summers later, I had to be persuaded not to launch the raft I built to take me to Chicago, ninety miles across the lake. I learned to sail, and to respect and use the forces of nature, on Lake Michigan. Without television or internet, we enjoyed books, bicycles, and friends. I read every single issue of the 1950s *Readers' Digest* from the stack next to the *Analog Science Fic-*

tion and Fact magazines my father adored. I was especially entranced by the reports of severed limbs being reattached in miraculous surgeries. I envisaged myself following an early idol, Dr. Christiaan Barnard, who performed the first human heart transplant in 1967. In a single summer I tore through every medical book available in the local Holland Public Library. But I abandoned the idea of being a transplant surgeon when I found out that the mere sight of blood made me nauseous.

Summers were heaven, but the school years in Pittsburgh were another story. No one knew what to do with a smart little girl in the 1960s. Attempts to keep me busy included music lessons (piano and violin), any number of sewing and art projects, ice skating – which involved walking more than a mile in freezing weather to the rink – and Saturday catechism classes in Wilkinsburg, again on foot, snow, rain, or shine. I spent as much time as I could outdoors, digging under stones for salamanders, and also finding and collecting used soda bottles which I could turn into the local drug store for two cents. Three bottles = one fudgesicle. I led my younger brother on adventures that involved exploring large drainpipes; and when that was not possible – it seemed like we were in an ice age then, with regular school closures for four feet of snow – we watched *I Love Lucy* re-runs on the round TV set and played infinite variations of war games.

My father, an experimental physicist who received his PhD from Princeton in 1955 at the age of 24, spent 1954 doing experiments on top of Mt. Evans in Colorado. My mother was not thrilled to be in such a remote site (elevation 10,700 ft), and they moved down-mountain to Idaho Springs (7,500 ft) shortly before my brother Bill was due. Both my parents contracted polio there and spent time in iron lungs, cared for by my fearless maternal grandmother, Josephine Routheau. Upon graduation, my

Figure 1. Frances, 1961.

Figure 2. War games. Eddy, Frances and Bill, 1961.

father set aside academics to work in the nascent nuclear industry. The Westinghouse Electric Corporation in Pittsburgh was going to provide "electricity too cheap to meter," and my father helped design the pressurized water reactor technology needed to make that dream come true.

In the 1960s, my father was away from home much of the time. He seemed to be in Nevada a lot; later I learned it was for nuclear testing. He would bring us silver dollars from the casinos to which he had to accompany the big bosses on their gambling sprees. When at home, he loved to build houses or model airplanes, listen to classical music, read, and work on his coin and stamp collections. I thought he was the smartest person in the world because he knew all the answers, could explain how everything worked, and could fix nearly anything. To spend time with him, Eddy and I vied for the duplicates of the stamp collections. We divided up the world: I got the British colonies; Eddy took the rest of the world. From stamp collecting I learned geography, and learned that geographical boundaries, governments, and even languages, changed over time. I learned that empires crumbled, and former colonies gained independence.

My brilliant success in elementary school got me into typing class. At the age of ten, far ahead of my classmates, I spent most of my time drawing pictures, making little paper people for my friends, and perfecting my mirror writing (being left-handed made it easy for me to write backwards and impress my friends when they held my coded messages up to the bathroom mirror). My parents somehow convinced Edgewood Elementary School to allow me to take some classes at the high school next door. One of my favorite extra classes was typing, although I was not very good at it. I would perch on two telephone books in order to reach the typewriter while the high school students laughed at the tiny 5th grader with legs dangling from the elevated seat. I still have letters I typed to my father during class. I also took mechanical drawing, which was also challenging, but taught me the important skill of looking at and describing objects from different perspectives.

By age 13 I was pretty much fed up with classroom learning. It was 1969, and Baltimore, the city we lived in at that time, was burning. I was not invited back for 9th grade to the private girls' school that my mother had worked very hard to get me admitted to, and which I hated anyway. Instead of attending school, I began hitchhiking to anti-war protests in Washington D.C. I spent only half of 9th grade at the large public, inner city high school before my father's job took us back to Pittsburgh. The move back North just made my trips to the protests longer.

At Allderdice, an excellent public school in the Squirrel Hill neighborhood of Pittsburgh, I picked up a few words of Yiddish as my second foreign language (after French) and delighted in a whole new set of descriptions far more colorful than the morose ones of Catholic culture. As a teenager who needed to understand the world, but who lacked the power to navigate it, I distanced myself from my classmates and parents. I lived on my own in a terrible, run-down, and bug-infested third-floor apartment in a gritty neighborhood, working at various jobs to pay my rent and bills while dreaming of a future that would free me from the limitations of being young and female in the 1960s and early 1970s. My jobs included selling seeds (yes, I earned the bicycle at age 10), lunch counter waitress (age 14), pizza parlor helper (15), department store clerk (16), receptionist (16), cocktail waitress (17, I told them I was 22 and no one ever checked, as fewer young people had driver's licenses then), waitress in Pittsburgh's famous jazz club Walt Harper's Attic, and finally taxi driver (age 18). By the time I left for college, I had become adept at maneuvering a massive 1960s Yellow Cab up and down the terrifyingly steep hills and pot-holed streets of Pittsburgh. Those streets were narrower than my cab, but my customers would insist I could get through, and they were (usually) right. Without GPS, I constructed maps in my head and still benefit from the good sense of direction that I developed. Driving a taxi was hard work. Long days (sometimes more than 10 hours) netted 20 or 25 dollars, but in just a few weeks I worked my way up from the filthy, banged up cabs the old dispatchers gave to new drivers, to clean, newer (but still just as wide) cabs that elicited better tips. The fun had to end, however, because I was on my way to college.

In 1974 I somehow managed to convince the admissions officers at Princeton University to admit me. I thought it was my convincing essay, or perhaps the fact that I was a very rare female candidate applying in engineering, but it probably also did not hurt that my father had received his PhD in physics there and knew the Dean of Engineering well. I started in 1974, when the first women were graduating, since Princeton only began accepting them in 1969. There were probably about 15% women in my class, and far fewer in Mechanical and Aerospace Engineering. But being the only female was nothing new to me, and I stayed in MAE because there was no good reason to switch to something else. I was busy absorbing as much knowledge and as many new ideas as I could: Italian language, economics, socialist theory, Russian language and literature, art history, and plenty of math and physics. My lack of interest in chemistry was reflected in my freshman grade, and I did not progress in that field at that time.

At Princeton, I continued driving taxis, worked at the library, assembled electronic equipment, and cleaned the house of philosopher Thomas S. Kuhn. I needed the money to support my addiction to Laker Airways, which made it possible to fly to London for $99 if you were willing to line up at the ticket office in Manhattan at 4 am and fly out around midnight the same day. During my last two years at Princeton, I spent every break in London or Italy or Paris.

Travel opened a fascinating world of different cultures and especially cuisines (I love people's creativity with food and was delighted to learn that daily food could be completely delicious). After my junior year in high school (1973), my maternal grandparents took my brother Bill and me to Europe for The Grand Tour. They had been visiting Europe for many years, traveling to their favorite little towns in Austria, Germany, France, and Italy. We spent two full months on the road, never staying more than two nights in any one place. Everyone seemed to know and love my grandparents, Col. Edward and Josephine Routheau. "Mama" and "Baba" taught me everything I needed to know to enjoy life with little money. Their secret: a bottle of wine, a fresh baguette, and a bit of pâté on a sunny, grassy knoll by the side of a country road in southern France.

Eager to return to Europe and experience it on my own, I took time off after my sophomore year at Princeton to work in Madrid and Milan in 1976–1977. During this time, I never spoke English, and discovered whole new cultures and friends. My Italian boyfriend and I motorcycled all over northern Italy, and in the summer, we went all the way to Istanbul and back on his 1956 Moto Guzzi 500, the classic bike of the Italian carabinieri. We traveled to the Cinque Terre, before there was a paved road, and camped or slept wherever a farmer would have us. With my guitar I shared the songs of Bob Dylan and Italy's equivalent, Francesco Guccini, with anyone who would listen.

During my last two years at Princeton, with renewed interest in completing my degree and finding something meaningful to do, my former lackadaisical attitude to coursework changed. I loved the upper level classes and found that, with a little bit of effort, I had genuine talent for math and engineering; in 1979 I earned my degree magna cum laude in Mechanical and Aerospace Engineering. The energy crisis of the 1970s and mentors at Princeton whose passion for connecting science and benefits to society sparked what became a lifelong interest in alternative energy.

After graduating, I donned my backpack once again and, with about $2/day saved for expenses, traveled from Ecuador to São Paulo, Brazil, for an internship on solar energy projects with Professor José Goldemberg, who later became Brazil's Minister for the Environment and the 'father' of its ethanol fuels program. It took six weeks over the Inca trail by bus to make my way from Guayaquil, Ecuador, to Santa Cruz, Bolivia. The trip from Lima to Ayacucho took more than 36 hours on a slow, steep climb up a rocky path, shared with a goat covered with fleas. We stopped every hour, it seemed, so the federal police, looking for Shining Path members, could empty the bus, puzzle over my passport, and hours later send us on our way. I loved Peru, but not the regular bouts of food poisoning. That summer I perfected an ability to sleep anywhere and bolstered my immune system. During my time in Brazil, I picked up a bit of Portuguese and a taste for beans and rice, served for lunch every single day in the cafeteria, and especially for the traditional feijoada, a fantastic, rich stew of black beans and every part of the pig, served on Sundays at noon because it required the rest of the day to digest.

With a degree in mechanical engineering and the Carter administration's emphasis on clean, renewable energy sources, I took my first 'real' job (1979–1980) at a new national laboratory, the Solar Energy Research Institute (now NREL) in Golden, Colorado. My duties in Frank Kreith's Heat Transfer Group were primarily to develop new passive solar heating and cooling technologies; I also helped write position papers for the United Nations on solar energy in the developing world. Outside the office, I was learning how to ride an off-road motorcycle and improve my skiing, which I had tried for the first time while living in Italy. In exchange for free rent, I lived on a horse property and cared for the animals when the owner was away. I took up classical guitar to combat the onslaught of country western music blaring from every radio station.

1981–1985, UNIVERSITY OF CALIFORNIA, BERKELEY

With the election of Ronald Reagan as President of the United States, the future for passive solar heating and cooling seemed somewhat limited. I'd never been to California, but at the end of 1980 I packed my few belongings into my 1971 red Volkswagen Super Beetle and headed west to start graduate studies at the University of California, Berkeley. The chemical engineers there had decided to take a risk on a mechanical engineer who also happened to be a woman, and I was accepted into the PhD program, beginning in January 1981.

Although my first desire was to work on cellulosic biofuels, interest in that technology had waned: automobiles ballooned again to giant proportions, and the oil embargoes were forgotten. We also forgot how to care for the planet. Funding for alternative energy projects became scarce; the professor I had come to work for retired; and I had to change direction. Professor Harvey Blanch, Australian by birth and recently recruited to Berkeley from the University of Delaware, however, was ready to support a whole new industry on the horizon, the biotechnology industry. A revolution was taking place in California and Boston – new companies with names like 'Genentech' and 'Amgen' were looking for engineers to scale up their processes for making protein therapeutics using recombinant DNA technology. Someone would have to produce

Figure 3. At University of California, Berkeley, College of Chemistry c. 1984.

Figure 4. Hiking in the Sierras, c. 1981.

and purify the recombinant proteins that promised to change the face of medicine.

Thus, I took on research in bioseparations, studying affinity chromatography, and developing and validating mathematical models of chromatographic separations. I also developed an appreciation for the challenges of working with proteins: everything was designed around keeping the proteins happy. This was not easy, since proteins are only marginally stable and, it seemed, would denature at the slightest provocation. Furthermore, most protein therapeutics involved highly complex, post-translationally modified structures that are easily rendered useless when manufactured, purified, or stored under the wrong conditions. Process engineers had little experience with proteins, or with biochemistry for that matter, and the standard chemical engineering separation processes were ill-suited to protein preservation.

Graduate school, like college, was another feast of learning, but this time it was organic chemistry, biochemistry, immunology, enzymology, advanced mathematics, and of course the entire undergraduate and graduate chemical engineering curriculum. Organic chemistry made sense to me – making molecules was like building a puzzle. After taking it for credit, I happily audited multiple organic chemistry courses as the official note-taker for the student-run Black Lightning note service (beloved by students who dreaded waking up in time for the 8 am class). The superb biochemists Jack Kirsch and Judith Klinman introduced me to the remarkable catalytic capabilities of enzymes from a quantitative, physical chemistry perspective that I especially appreciated, and Allan Wilson introduced me to molecular evolution of protein sequences. I soaked up new knowledge, taking science and math courses just for the fun of it, something I continued to do decades later.

It was not until my last year as a graduate student that it occurred to me to try my hand at being a professor. I played bridge and went backpacking with some of the young professors in the UC Berkeley College of Chemistry, but had little idea of what a professor actually did for a living, other than teach a course or two. My PhD advisor, Harvey Blanch, however, also started companies and consulted for industry, as did some of the biochemistry professors, and this multifaceted activity made the academic enterprise much more interesting to me. I wanted a connection to the 'real world', but also the independence I had not experienced in my various previous industrial and national laboratory positions. I therefore decided to apply for academic positions. The year 1984 was a very good time to do so: U.S. universities were waking up to the fact that while more and more women were interested in science and engineering, there were essentially no women on engineering faculties. At that time, chemical engineers did not do postdoctoral research before starting their academic careers, and I was offered positions at a number of very good places, including MIT. In 1985 I accepted a position at the University of Minnesota, which enjoyed the #1 ranking in chemical engineering, but I also parlayed my many job offers into funding for a one-year postdoc at UC Berkeley with biophysical chemist Ignacio Tinoco, to learn spectroscopic methods of characterizing biomolecules that I thought I would use in my future laboratory. I was 29.

Around that time, I met Jay Bailey, a world-renowned biochemical engineering professor at Caltech, a small, private institute in southern California that I knew very little about. Its chemical engineering faculty was tiny, and its reputation in engineering was for very 'academic' research. Caltech's PhDs tended to become professors rather than industry leaders. Since the University of Minnesota could not absorb someone of Bailey's stature, I applied to Caltech for an assistant professor position. Jay and I married in 1987, in Macatawa, surrounded by friends and (lots of) family.

DIRECTED EVOLUTION AT CALTECH 1986–2003

I moved to Caltech in mid-1986, where I was given a temporary position as a postdoctoral researcher (my official title was 'Visiting Associate', as I was already on the rolls as an assistant professor at the University of Minnesota). I was pleased to have a bench in Jack Richards's laboratory, where I would learn how to engineer a protein's sequence, a technology I wanted for my own research. Richards had recently developed 'cassette mutagenesis', one of the first site-directed mutagenesis methods for engineering proteins. My first foray into molecular biology and genetic engineering was to make a couple of mutated cytochrome *c* proteins for a collaboration with Harry Gray's group, who wanted to use them to probe biological electron transfer. Everything was difficult in 1986: synthesizing oligonucleotides, DNA sequencing, cloning, and working with restriction enzymes were all problematic, while my experience with trouble-shooting cloning experiments was very limited. But I persevered, and when my first mutant sequence was confirmed, I was a proud protein engineer.

Auspiciously for me, two Caltech chemical engineering professors moved elsewhere, thus opening a junior position in this new 'bio' part of chemical engineering; an offer was extended to me to join the faculty, and I started as Assistant Professor of Chemical Engineering at the California Institute of Technology in January 1987.

Jay Bailey's was the first, or at least one of the first, chemical engineering laboratories to use molecular biology methods to approach problems in industrial biotechnology. With his own educational and research background limited to mathematical modeling of chemical reaction systems, Bailey ably demonstrated how to use fearless graduate students to bring new techniques into the laboratory. He attracted some of the brightest students from all over the world to Caltech, and I gratefully recruited from that stellar pool to get my own protein-oriented group started. Over the years, I would fine-tune my own ability to remind graduate students and postdocs that they could learn and do anything. The new Arnold lab was going to make sure that protein engineering would become part of chemical engineering, just as Bailey and others were doing for metabolic engineering. These early efforts to genetically engineer biological systems became important and industrially relevant foundations for what is now known as 'synthetic biology'.

The problem was that no one really knew how to engineer useful proteins. Proteins, especially enzymes, are fascinating and do many things that people find useful, from monitoring blood glucose levels to taking stains off clothes. But many of the problems that chemical engineers (and others) faced with using proteins for industrial applications came from their inability to function under non-natural conditions. Proteins often performed poorly outside their natural environments, and the engineer had to develop Rube Goldberg mechanisms to purify, store, and use them. However, with the advent of technologies for engineering protein sequences, and therefore their properties, it became possible for the first time in the 1980s to consider engineering the protein itself according to the process engineer's or the industrial biotechnologist's specifications. My group was going to engineer protein sequences to make them behave in a process or application, rather than design the process or application around the protein.

I also wanted to show that proteins could be engineered for unusual but useful properties that would open up whole new applications. Alex Klibanov of MIT had surprised the world in the 1980s by showing that enzymes could work when suspended in dry, nonpolar solvents. Dissolving enzymes in high concentrations of polar solvents, however, immediately obliterated activity, even if it could be shown that they retained their folded structures. The prevailing view at that time was that proteins could not exhibit highly non-natural properties, such as the ability to function in organic solvents. The argument seemed to be that because Nature never did it, it could not be done. But in fact, that was precisely why it could be done, and why it might even be easy to do so. I therefore took on the challenge of engineering enzymes that would catalyze their reactions when dissolved in polar organic solvents, but no one knew how to alter their sequences for this purpose. My feeble attempts at 'rational design' of enzymes for organic solvents were failures, as were most experiments aimed at improving proteins at that time. While it was easy to diminish or even destroy an enzyme's function, there were very few reports of making better enzymes. And the process was difficult – it required having the enzyme's crystal structure, of which there were very few, and then understanding the protein's structure and function sufficiently well to identify not just the sites of useful mutations, but which amino acids ought to be placed there.

In the 1980s, some labs were starting to engineer nucleic acids, peptides and even proteins using phage display and other methods to make huge libraries of biomolecules, which they then sorted using binding assays or genetic selections to find the useful sequences. Appreciating protein complexity as well as the combinatorial explosion in sequence possibilities that comes with targeting multiple sites for mutation and the low frequency of beneficial mutations that could be expected, I developed an alternative and highly general approach suitable for the problems I was interested in, engineering better enzymes. Adopting a simple, newly-developed method for making mutations randomly in a specific gene, the polymerase chain reaction under error-prone conditions, my students and I made libraries of bacteria having genes randomly mutated at just one or two sites and screened them for the properties we wanted using rapid assays in petri dishes or 96-well plate readers. Then we would take the genes for the best proteins and repeat the process to accumulate benefits, evolving the proteins step-wise until we achieved the functional goals.

To my delight, beneficial and surprising mutations appeared in our laboratory-evolved enzymes. Beneficial mutations were not so rare that we could not find them with a carefully controlled screen, and we could accumulate them for further improvements. When we mapped the mutations to the protein structure, we were surprised to see that they often happened on the enzyme surface, where protein chemists at the time argued that mutational effects were mostly neutral. Activating mutations also appeared far from the enzyme active sites, where no one could explain their effects, much less predict them *a priori*. By 1990 we were finally on our way to engineering useful enzymes, using evolution as our guide. Another golden moment was the arrival of my first son, James Howard Bailey, in April 1990. I was 34 years old, untenured, overworked, but had a beautiful baby boy, was full of energy, and knew exactly where I needed to go.

Not everyone agreed with my approach, however. The protein engineering field, with its main roots in biochemistry, was very much focused on 'rational design'. Protein chemists argued they could predict beneficial mutations and sequences using structure-guided approaches and even computational methods. Some of my Caltech chemistry colleagues, dismissive of the contemporaneous popularity of 'combinatorial chemistry', which involved synthesizing and screening large molecular libraries to find drug leads, looked askance at my random mutagenesis efforts as intellectual laziness. I was undeterred, however, because I had an approach that worked.

While I was finally on the right track to engineering proteins, life outside the lab was going off the rails. My marriage was failing, and Jay moved to Switzerland, to become a professor at the ETH in Zurich. Caltech stepped in to help me financially overcome the difficulty of living on my own with my one-year-old baby in a Pasadena house I could not afford, for which I will always be grateful. Caltech Provost Paul Jennings and President Thomas Everhart also helped me through a difficult tenure process and demonstrated to me what real leadership means: a real leader has a moral compass and sometimes has to make decisions that go against the wishes of powerful people. Tenured, I could do what I loved best, and that was engineering enzymes by directed evolution. I dropped all other projects related to protein-metal recognition, a much more 'standard' part of my research that came out of my physical chemistry training, to focus exclusively on evolving enzymes.

In 1992, I met Andrew Lange, a brilliant and charismatic young cosmologist, at the annual meeting of the Packard Fellows at the Monterey Bay Aquarium. We had both received Packard fellowships in 1989, but had somehow never met. It was as close as one can come to love at first sight. All throughout 1993, while we tried to find positions in the same place, Andrew would fly down from Oakland to Pasadena on Thursdays after teaching his freshman physics class and then fly back up in time to teach his Tuesday morning class. He wanted a family, and with me he had an instant one. James adored him. Andrew, an experimental physicist, encouraged James's natural curiosity of the mechanical world with hands-on deconstruction of everything one could find in the Caltech dumpsters. There were a lot of oscilloscopes in the trash in those days, and I still have cathode ray tubes scavenged from various deconstruction projects. When Caltech came through with a professor position for

Andrew in 1994, he moved down from Berkeley. Some of the physics professors genuinely thought they were responsible for recruiting the most promising young cosmologist in the country to Caltech, but that gold star belonged to me. Very soon thereafter we were overjoyed to welcome William Andrew Lange (1995) and Joseph Inman Lange (1997) to our family.

We struggled to raise three little boys while also establishing our careers. Andrew traveled for his experiments and team meetings to Antarctica and to other distant places for weeks at a time. He was a devoted and loving father, but loved his science just as much. He longed to be with his team during the deployments of experiments. His 1998 BOOMERANG experiment, contained in a balloon that circumnavigated the South Pole and collected photons from the early universe, was a huge success and cemented his name in astrophysics and cosmology. He was Caltech's golden boy and was said to be on track to a Nobel Prize.

I could not have been more proud, or more exhausted. Life would have been impossible without our dear "Mama", Carmen, who gave my boys her abundant love, along with homespun treatments for minor ailments, delicious food, and practical tips for navigating the vast Los Angeles public transit system. During the 1990s I attended local conferences, sometimes very pregnant, but traveled relatively little. My own work, while not nearly as visible as Andrew's, was going well, and talented scientists from disparate disciplines came to Caltech to discuss directed evolution with me and my students. During that time, I was pushing enzymes into entirely new directions. The field was expanding rapidly, and directed evolution methods were becoming widely adopted. I took on a few industrial challenges with Proctor & Gamble, Degussa, and Dow Chemical that would allow me to demonstrate the power of directed evolution with 'real' problems, not just model systems. I am still grateful to their scientists for introducing me to interesting and challenging problems and for sharing their deep experience with what makes an enzyme useful – or not. I am especially grateful that they invested time and money in a brand-new and still largely unproven technology spearheaded by a young, female engineer. Importantly, Pim Stemmer, who independently published a directed enzyme evolution paper a year after my 1993 paper, visited Caltech for a few weeks in 1996. We planned a new start-up company, Maxygen, to commercialize our shared vision of using evolution to create virtually any protein or gene. Maxygen negotiated a license to all Caltech directed evolution intellectual property, and I served as a founding scientific advisor.

I was elected to the National Academy of Engineering in 2000, at the same age of forty-three as my father was when he was elected. When I came on stage at the induction ceremony in Washington D.C., my father stood up and shouted, to the delight of his many friends in the audience,

"That's my daughter!" I believe we are still the only father-daughter member pair in the NAE.

2003–2004: A SABBATICAL AROUND THE WORLD

Jay Bailey died of colon cancer at age 58 when our son James was eleven years old. Family life at home with Andrew, who was prone to debilitating depression, became increasingly strained. I reasoned that a trip around the world would give all of us much to learn about and pull us together; with my four gentlemen we could share adventures, such as those of my youth, and also quiet times. To his enormous credit, Andrew agreed to this plan. I chose two of our sabbatical destinations, Australia and South Africa, where I had friends but no real work to do. Andrew chose Cardiff, where he had real collaborators and a chance to get some science accomplished.

We arrived in Alice Springs for the Australian winter of 2003 for our first adventure, with Aboriginal people in the Red Center. Our sons immediately went feral. By the third night in Oz, we checked into our million-star hotel, sleeping on the red dirt in swags well-designed to keep snakes and spiders out. We spent the first two magical weeks of our sabbatical with the Southern Cross overhead, the smell of campfire in our nostrils, and our heads filled with the walkabout stories and legends of the native people. My two littlest boys dug for honey ants and grubs and played with the local children, climbing in to join their family groups piled onto old mattresses spread out on the desert floor. We settled into an eight-week stay at Swinburne Tech, in Melbourne, and Joe and William went to kindergarten and first grade in the Hawthorne public school. I secured for James a two-month stay at a fine boys' boarding school, Scotch College, nearby. Every weekend the family would go visit gold mines and farm stays.

It was winter again when we came back to Caltech for a short stay, and then headed to Africa for the next stage of our voyage, which included

Figure 5. Joe and Australian friend, 2003.

Figure 6. Family on a five-day camel trek across the Sinai Desert, 2004.

Figure 7. William and camel.

Egypt, South Africa, Namibia, and Madagascar, followed by the United Kingdom. It was a fairy-tale sabbatical, and the best year of my life. We were all happy, healthy, and delighted to watch our sons soak up the adventures.

My research group did exceptionally well during this time. I realized they could do more on their own, and that they had developed real leadership skills. From then on, I have given group members as much freedom as I can to pursue their own ideas and mentor others.

2005–2010: DARK TIMES

I returned from this magical year at the end of 2004 to find out that I had breast cancer which had spread to my lymph nodes. I underwent two surgeries and 1.5 years of debilitating chemotherapy and radiation. I took up yoga for physical and mental health, and somehow managed to work every day, from which I derived both pleasure and purpose.

Our science started to focus on a problem I had long been interested

Figure 8. William and Joe with friends in Namibia, 2004.

Figure 9. Joe with Himba friends in Namibia, 2004.

in, alternative energy. Oil prices were steadily climbing. Since 2000 we had been engineering cytochrome P450s to oxidize alkanes, one goal being to make recombinant organisms that could convert gaseous alkanes to liquid fuels. In 2005, with Matt Peters and Peter Meinhold, and funding from a prominent venture capitalist, we started what soon became Gevo, Inc., one of the first biofuels start-ups in the new 'synthetic biology' space. Undergoing intense treatment for breast cancer, however, I was in no condition to spend much time on that project, and Matt and Peter ran the circus. Gevo, still in business today, makes renewable jet fuel starting from biomass using an engineered yeast and chemistry.

On January 22, 2010 Andrew Lange committed suicide, which shocked the world and left behind grieving family members, friends, students, and colleagues. We had not lived together for more than two years, but I now had to pick up the pieces of our family and cut some sort of path for our three shattered sons, then ages 17, 13, and 11. That year was a blur. Members of my research group continued to take care of each other, and they and Caltech were my rock. My friends helped me throughout. I continuously remind myself that no one is guaranteed an easy life, but we can make it easier for others.

2011–PRESENT: A NEW STAGE

In good health, and perhaps just a bit wiser than before, I made a conscious decision to take risks again, in my professional life as well as my personal one. I traveled with my sons, encouraged their own adventures that took them to distant lands, and made many new friends outside my usual circles. I continued favored activities like scuba diving, and hiking to my historic one-room cabin in the San Gabriel Mountains. And for the first time I accepted invitations to give scientific talks to general audiences. I discovered that both scientific and general audiences respond warmly to story-telling and efforts to convey the big picture; they want to be reminded of the wonder and power of evolution, which we see all around us, and to think that science can lead to a better future. I like to end my talks with an open and exciting future full of questions to answer, rather than closing the box on a problem.

Perhaps most important, I felt free, even compelled, to pursue new and more challenging problems with my science. I had always wanted to make enzymes do new chemistry, and catalyze reactions not known in the biological world. I posed to several good chemists in my group the specific question: can you get a cytochrome P450 to catalyze reactions using nitrogen rather than oxygen? They rose to the challenge, and we engineered the first 'nitrene transferase' and 'carbene transferase' enzymes in 2012. Building on this realization that nature is poised for all sorts of new

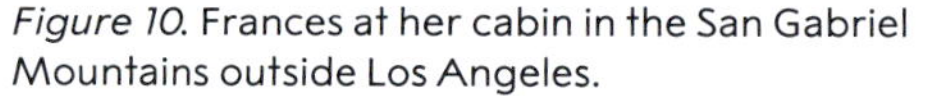

Figure 10. Frances at her cabin in the San Gabriel Mountains outside Los Angeles.

Figure 11. William, James, Frances, Joseph in the Peruvian Amazon, 2016.

capabilities, we have been exploring a whole new world of enzyme chemistry.

Today my laboratory feels very much like it did in the exciting days of the 1990s: intense, with a palpable sense of discovery and the knowledge that we are laying the foundations of how molecules will be made in the future, using genetically-encoded biological systems that include enzymes engineered to do chemistry first invented by human beings. I am grateful to experience that excitement and focus for a second time, again shared with a tremendously talented group of young people.

My work with directed enzyme evolution was recognized by the Charles Stark Draper Prize in 2011, the highest honor an engineer can receive in the United States. I was the first and remain to this day the only woman to win this award, which has been given by the U.S. National

Figure 12. Enjoying chicha with William, James and Joseph, Peru, 2016.

Academy of Engineering since 1989. My two youngest sons and I were welcomed by President Barack Obama at the White House in 2013, when I received the U.S. National Medal of Technology and Innovation. (James was serving in the U.S. Army in Afghanistan and could not join us.) And in 2016 I received the Millennium Technology Prize, again the first (and only) woman to be so honored. I did not set out to be the first female engineer to break into this rarefied territory, but I was one of the first to be given the chance to show what she could do. Only the ninth woman to be hired on the Caltech faculty, I am the first female Nobel Laureate there. Many brilliant women have joined science and engineering faculties in my lifetime, and I predict that many more of the highest recognitions of women's scientific contributions are coming.

CONCLUDING REMARKS

I am sure the reader will note that I have not commented directly on the contributions of the many students, postdocs, and colleagues I have worked with and drawn inspiration from. Some were called out specifically in my Nobel Lecture, but I have not been able to thank everyone who contributed to the conception and wide application of directed enzyme evolution. I would like now to thank my mentors and those I have tried to mentor, for I learned much from you. I also want to thank Ben and Donna Rosen for all that they have done and continue to do for Caltech; it has been my great honor to direct the Donna and Benjamin M. Rosen Bioengineering Center for the last six years.

My father, William Howard Arnold, died in 2015. I miss him very much. He would have been very proud and would have especially loved the Stockholm festivities. My dear middle son, William Andrew Lange, died in 2016 at the age of 20; his short life was enriched by caring for monkeys in South Africa, for children in Kenya and India, and for his friends. Both Williams are still very much in my heart. I'd like to think that Andrew, too, would have been happy for me. My son Joseph, and my son James and his wife Alanna, and my stepson Sean Bailey came to Stockholm together with nearly sixty friends, family members, and former students to celebrate my Prize. I am very grateful for all that I have and for all the people, and animals, who have enriched my life.

Innovation by Evolution: Bringing New Chemistry to Life

Nobel Lecture, December 8, 2018 by
Frances H. Arnold
California Institute of Technology, USA.

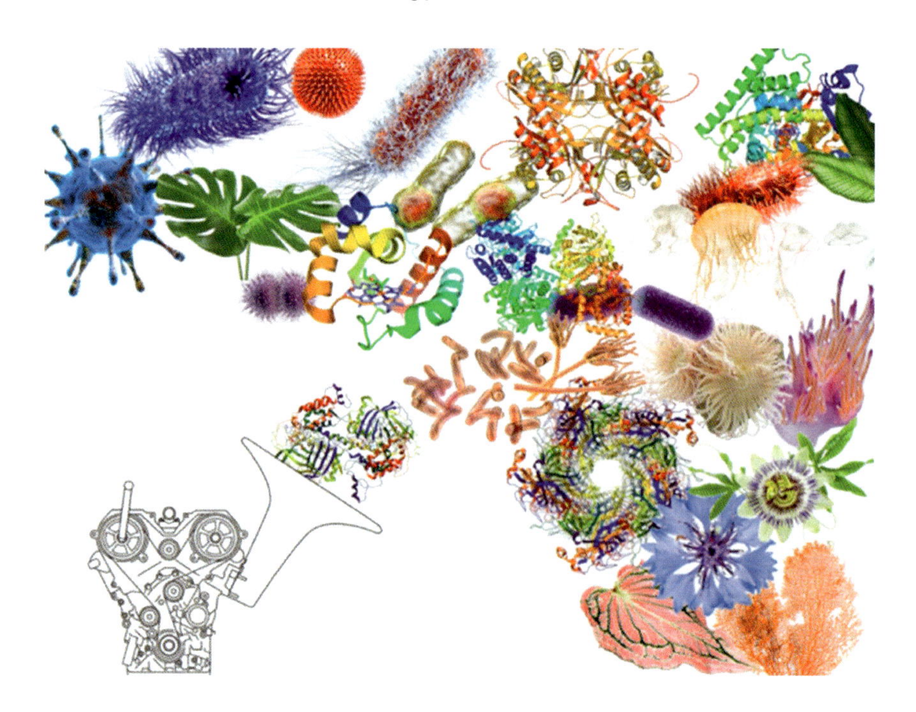

I AM A CHEMIST AND ENGINEER who looks upon the living world with the deepest admiration. Nature, herself a brilliant chemist and by far the best engineer of all time, invented life that has flourished for billions of years under an astonishing range of conditions. I am among the many inspired by the beauty and remarkable capabilities of living systems, the breathtaking range of chemical transformations they have invented, the complexity and myriad roles of the products. I am in awe of the exquisite specificity and efficiency with which Nature assembles these products from simple, abundant, and renewable starting materials.

Where does this chemistry come from? It derives from enzymes, the DNA-encoded protein catalysts that make life possible, molecular machines that perform chemistry no human has matched or mastered.

EVOLUTION, A GRAND DIVERSITY-GENERATING MACHINE

Equally awe-inspiring is the *process* by which Nature created these enzyme catalysts and in fact everything else in the biological world. The process is evolution, the grand diversity-generating machine that created all life on earth, starting more than three billion years ago. Responsible for adaptation, optimization, and innovation in the living world, evolution executes a simple algorithm of diversification and natural selection, an algorithm that works at all levels of complexity from single protein molecules to whole ecosystems. No comparably powerful design process exists in the world of human engineering.

I wanted to engineer Nature's enzymes to make ones tailored to, and uniquely suited for, human purposes. For close to five thousand years we have made use of microbial enzymes to brew beer and leaven bread. Once the protein catalysts were identified and isolated, many more diverse applications were devised. Today, enzymes are used to diagnose and treat disease, reduce farm waste, enhance textiles and other materials, synthesize industrial and pharmaceutical chemicals, and empower our laundry detergents. But so much more could be achieved if we understood how to build new ones.

Early protein engineers struggled mightily with this goal. In those days (the 1980s), we did not know enough about how a DNA sequence encodes enzyme function to design enzymes for human applications. Unfortunately, this is still true: today we can for all practical purposes read, write, and edit any sequence of DNA, but we cannot *compose* it. The code of life is a symphony, guiding intricate and beautiful parts performed by an untold number of players and instruments. Maybe we can cut and paste pieces from nature's compositions, but we do not know how to write the bars for a single enzymic passage. However, evolution does.

EXPLORING THE UNIVERSE OF *POSSIBLE* PROTEINS

Some researchers think of the protein universe as the set of all proteins that Nature has devised. But these proteins, relevant to biology, are an infinitesimal fraction of the *possible* proteins. The universe of *possible* proteins, my universe, contains solutions to many of humanity's greatest needs: there we will find cures for disease, solutions to energy crises and a warming world, food and clean water for a growing population, and ways to arrest the miseries of aging. I wanted to explore this universe to find those proteins that will serve humanity.

But how does one discover a useful protein in the infinitude of possible proteins, a set larger by many orders of magnitude than all the particles in the universe? In his fascinating short story, the *Library of Babel,* Jorge Luis Borges describes a collection comprising all possible books assembled from an alphabet of letters [1]. Most texts in Borges' library are gibberish, and his despairing librarians, for all their lifelong efforts, cannot locate a single meaningful sentence, much less a complete story.

Similarly, most possible protein sequences encode nothing we would recognize as meaningful. Unlike Borges' librarians, however, I am entirely surrounded by proteins with meaningful stories. They are everywhere and can literally be scraped from the bottom of my shoe, captured from the air I breathe, or extracted from a database. These are the products of billions of years of work performed by mutation and natural selection. And evolution continues to create new ones from these rare functional sequences that were themselves discovered by evolution. Thus, I decided to start my exploration by using this gift from evolution, the existing functional proteins.

There are thousands of ways to make one change in the amino acid sequence of a protein. There are millions of ways to modify it by two changes, and so on – the numbers grow so rapidly that making a single copy of each protein altered by only 1% of its sequence would require the weight of the world in materials. And the vast majority of these modified sequences are neither usable nor useful. The challenge therefore is to discover protein sequences that provide new benefits and deliver novel improvements on a thrifty scale of weeks, rather than millennia or eons, and with the help of one graduate student rather than that of an army. To outperform Nature, I needed a strategy that sidesteps the despair of the Babel librarians.

John Maynard Smith helped answer this challenge for me in a beautiful paper published in 1970 [2]. Consider an ordered space in which any protein sequence is surrounded by neighbors that have a single mutation. For evolution to work, he reasoned, there must exist functional proteins adjacent to one another in this space. Although most sequences do not encode functional proteins, evolution will work even if just a few meaningful proteins lie nearby. Given low levels of random mutation, the filter of natural selection can find those sequences that retain function. In fact, many of today's proteins are the products of a few billion years of mostly such gradual change. Many of these mutations are neutral and change little, but others can be deleterious. Natural selection picks the wheat from the chaff and guides mutating proteins along continuously functional paths through the vast space of sequences mostly devoid of function.

But by using evolution I want to make *better* proteins, proteins that serve my purposes. Thus, directed protein evolution becomes a search on a new fitness landscape, where fitness is performance and is defined by the artificial selection I impose. This is a landscape whose structure we knew very little about in the 1980s. Evolution on a rugged landscape is difficult, as mutation propels sequences into crevasses of non-function. However, latching onto Maynard Smith's argument that proteins evolve on a landscape smooth in at least some of its many dimensions, I reasoned that directed evolution could find and follow continuous paths leading to higher fitness [3].

A PROCESS FOR EVOLVING PROTEINS IN THE LABORATORY

Science, like all human endeavors, is evolutionary. We progress by adding to and recombining what is present. Important developments in the 1980s and 1990s influenced my thinking. Manfred Eigen speculated on *in vitro* molecular evolution [4], and Gerry Joyce was selecting RNA 'enzymes' that could cleave DNA from pools of billions, perhaps trillions, of mutated sequences [5]. Error-prone PCR (polymerase chain reaction) [5, 6] became a useful tool for random mutagenesis of genes. Jim Wells [7] and others demonstrated that beneficial mutations in proteins could be accumulated. Stuart Kauffman quantified evolutionary trajectories on model fitness landscapes [8], and the philosopher Daniel Dennett supplied the conceptual framework that helped me convey the power of evolution to others [9].

Protein genotypes and phenotypes do not coexist in one molecule, as they do for RNA, and protein fitness landscapes differ fundamentally from those of RNA. Thus, directed protein evolution would require different strategies and experimental tools. To devise a directed evolution strategy suitable for enzymes, I started with the fundamental rule: "You get what you screen for."

We were generating enzymes of interest in recombinant microorganisms by inserting genetic material that we could mutate in the test tube. We used common microbes like *Escherichia coli* or yeast to produce 'libraries' of mutant enzymes to test for desired functions. Since we were making enzymes for human applications, we rejected microbial growth or survival selections favored by microbiologists and geneticists. While those approaches enable a straightforward search through thousands and even millions of variants in one experiment, they do not meet our criteria of affording function in novel environments, over-expression in a production host, compatibility with new substrates, specific product formation, and so on. Thus, we turned to good old-fashioned analytical chemistry to develop reproducible, reliable screens that reported what mattered to us.

To measure what mattered, we were limited to monitoring the few thousand protein variants we could express and array in readily available 96-well plates or on a petri dish. Therefore, we could only search deeply those sequences one or two mutations away from the starting protein. Given that such a small change in sequence would be expected to generate only small improvements in function, we would have to deploy reproducible screening assays capable of finding those rare and only slightly improved protein progeny. A desirable mutation might yield only a two-fold increase in catalytic activity or a few degrees' step up in melting temperature. To achieve significant changes, we would have to multiply those benefits over successive generations.

This strategy works well when re-optimizing enzymes for new tasks. While a natural enzyme generally performs well in its biological job, it is often less enthusiastic about doing a new job and initially works poorly (Figure 1). New demands change the fitness landscape, often knocking a protein down from a position that was painstakingly acquired through the work of natural evolution. Sequential rounds of random mutation and screening for improved performance, however, can accumulate the beneficial mutations needed to climb to a new peak.

To illustrate, in the late 1980s my research group started to re-engineer a protease, subtilisin E, to perform its hydrolytic reaction under unusual and non-natural conditions. We chose to have the enzyme function in high concentrations of a polar organic solvent (dimethylformamide, DMF) that causes wild-type subtilisin E to lose most of its activity. We used random mutagenesis and screening to recover activity lost by adding low concentrations of DMF, combining the beneficial mutations [10]. Emboldened by these results, Keqin Chen performed iterations of random mutagenesis and screening for activity in increasing concentrations of the organic solvent and evolved an enzyme that performed as well in 60% DMF as its wild-type parent did in the absence of DMF, a 256-fold increase in activity [11].

Strikingly, this enzyme adapted rapidly to a challenge it presumably had not encountered during its evolution. Furthermore, the mutations that led to the improved performance were unexpected. We could not explain how mutations located on loops surrounding the enzyme's active site enhanced activity in high concentrations of organic solvent, much

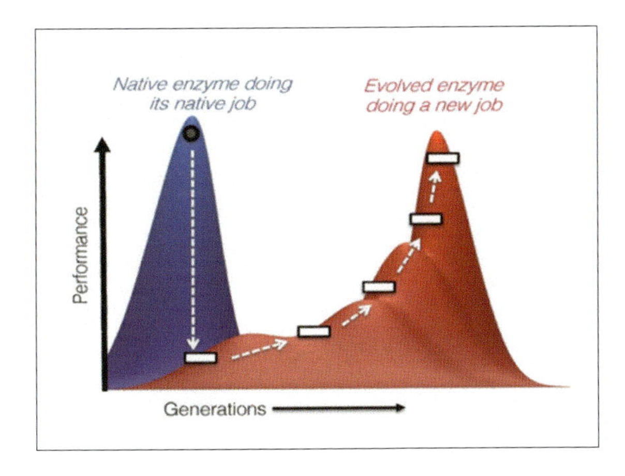

Figure 1. An enzyme whose function is optimized for its native job generally performs poorly in a new role. Directed evolution through rounds of mutation and screening can discover changes in sequence that improve performance, climbing a new fitness peak.

less plan them in a rational approach to engineering an enzyme with this new capability. But we had a *process* that gave the right result, even if that result would require much more reverse engineering to understand fully.

I met Pim Stemmer at a workshop organized by Stuart Kauffman at the Santa Fe Institute in 1995, not long after his landmark "DNA shuffling" paper was published [12]. Pim introduced sex – recombination – as a search strategy for protein evolution and called his method molecular breeding, a description I now often use to explain what I do. At Maxygen, the company he started in 1997 that licensed our technologies, and where I served on the founding Science Advisory Board, Pim's vision was grand: he wanted to evolve viruses, metabolic pathways, plant traits, and human therapeutics. My focus was entirely on enzymes and getting useful results quickly.

Those results ensued. A few examples of how enzymes could be evolved to accept challenging, non-natural substrates [13] or function at high temperatures [14, 15], work of intrepid lab members Jeffrey Moore, Huimin Zhao, and Lori Giver, convinced many researchers, especially those in industry where deadlines were tight and interest in understanding why individual mutations were beneficial lagged behind the need for the enzyme. Directed evolution offered a reliable optimization algorithm: find a starting enzyme, develop a moderate-throughput assay, and turn the crank.

The methods we developed and demonstrated in the 1990s were adopted rapidly. That decade saw the explosive rise of directed evolution in industrial and academic laboratories around the world, especially those of Andy Ellington, Manfred Reetz, Uwe Bornscheuer, George Georgiou, Romas Kazlauskas, and Don Hilvert, who introduced many novel concepts and improvements. In its original and in many modified forms, directed evolution produces new gene editing tools, therapeutic enzymes, and enzymes for diagnostics, DNA sequencing and synthesis, imaging, agriculture, textiles, cleaning aids, and much more. Some of those developments are detailed in excellent reviews [e.g., 16,17].

What important lessons did we learn about enzymes from the early directed evolution experiments? First and foremost, we learned that enzymes can adapt to new challenges. It often takes only a few mutations for an enzyme to acquire the targeted trait. We had not known when we started out how many generations would be needed to obtain useful changes in function. Nature, after all, takes a circuitous route to achieving new properties, combining neutral or even negative mutations with beneficial ones. Those paths can involve hundreds of changes. Our approach, however, collected only adaptive mutations that yield steep changes in function. Useful traits could emerge in less than ten, or even five, generations.

We also learned that much remains to be done before we can reliably design good enzymes. Beneficial mutations found by directed evolution are often far from the site of catalysis. Even today we struggle to explain their effects, and are unable to predict them reliably or easily. Nevertheless, practitioners now enjoy a dependable process for improving enzymes that does not require us to understand their structures, folding, or catalytic mechanisms.

EVOLUTION OF ENZYMES FOR REACTIONS INVENTED BY CHEMISTS

What fascinates me today is the evolution of new enzymes. I wish to go beyond optimizing biological functions that are already known, and instead bring to life whole new chemistries. But how can one create enzymes that catalyze reactions invented by chemists? One cannot go to a biochemical database to find enzyme sequences annotated for such transformations. For many years, in fact, creating new chemistry by directed evolution seemed to me an insurmountable challenge. Enzymes position functional groups in exquisite arrangements to bind substrates and stabilize reaction transition states. For a long time I could not see how my conservative directed evolution strategy of accumulating one or two beneficial mutations per generation would create entirely new enzyme active sites. Unless, of course, an active site is already largely there...

When innovating, Nature does not invent new active sites *de novo*. Rather, to support the fight for survival or to move into a new niche, emerging enzymes exploit existing catalytic mechanisms and machineries [18]. The biological world is replete with proteins whose chemical capabilities extend well beyond the functions for which they are selected at any given time. These "promiscuous" activities can become advantageous, such as when a new food source becomes available, and provide the basis for evolution of a new enzyme that gives a fitness advantage to its host [19]. Promiscuous functions can also be useful for human applications [20]. If a new catalytic activity is already present, even at a low level, our conservative process of accumulating beneficial mutations can mold it into a new enzyme. Dan Tawfik, in particular, compellingly demonstrated how known promiscuous enzyme activities, sometimes relics of their own ancestral origins, may be evolvable in the laboratory [21]. Directed evolution can innovate when *the innovation is already present.* For several years we have capitalized on this realization to create whole families of enzymes that catalyze reactions previously unknown in biology [22].

To explain this process, we turn to the cytochrome P450 enzyme family. Nature draws on P450's highly reactive iron-oxo Compound I (and other intermediates) to perform varied reactions that presumably evolved from the promiscuous functions of ancestral P450s (Figure 2). Today, the cytochrome P450 family has members that can transfer an oxygen atom to organic molecules to make specific hydroxylated compounds or epoxides, oxidize heteroatoms, nitrate aromatics, and much more. The biological world shaped these enzymes using the diversity-generating machine of evolution, and hundreds of thousands of their sequences are stored in databases.

This magnificent biological diversity now drives laboratory innovations. With insights and inspiration from chemistry, directed evolution can take us where biology has never gone. For instance, if a P450 can

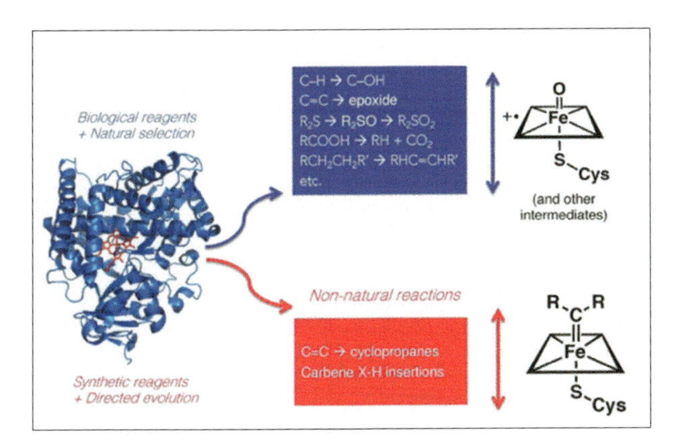

Figure 2. Expanding the scope of P450 chemistry. The cytochrome P450 family, whose members were presumably created by gene duplication and natural selection of promiscuous functions, comprises enzymes that use reactive oxygen intermediates to catalyze a wide range of reactions. We reasoned that we could expand the scope of P450 chemistry by using synthetic carbene and nitrene precursors to drive formation of new reactive intermediates. Directed evolution would be used to mold the enzyme, controlling and enhancing new-to-nature activities.

transfer reactive oxygen species to substrates, perhaps it can also be directed to transfer reactive nitrogen or carbon species and assemble molecules using efficient strategies hitherto unused by biology. To Pedro Coelho and Eric Brustad, members of my lab in 2012, the P450 enzyme's reactive Compound I intermediate resembled an iron carbenoid, long used by chemists to transfer carbenes to carbon-carbon double bonds in alkenes, or participate in X-H insertion reactions to form new heteroatom-carbon bonds (Figure 2). These reactions, unknown in biology, are possible in chemistry when humans supply synthetic diazo carbene precursors and transition metal catalysts modeled after heme cofactors.

We thought perhaps by offering man-made carbene precursors to heme proteins we could discover promiscuous 'carbene transferase' activities [23]. If so, we might use directed evolution to draw out and improve such biologically irrelevant but synthetically interesting capabilities.

An early achievement for this approach was alkene cyclopropanation, a transformation well known in transition metal catalysis but unknown in biology. Inspired by early reports of heme mimics catalyzing carbene transfer to alkenes in organic solvents, we discovered that iron-heme proteins do indeed promote cyclopropanation when provided with diazo carbene precursors and a suitable alkene substrate, in water. Furthermore, mutations altered both the activity and the selectivity of product formation so that enzymes produced individual cyclopropane stereoisomers [23].

This new reaction has many practical applications. For example, directed evolution resulted in a highly efficient enzyme for efficient production of the chiral cis-cyclopropane precursor to the antidepressant medication levomilnacipran [24]. We and Rudi Fasan have since engineered a variety of heme proteins to synthesize other pharmaceutical precursors [25,26]. Because alkene cyclopropanation proceeds in whole *Escherichia coli* cells that express the evolved enzyme, as well as in cell lysate, preparing the catalyst is as simple as growing bacteria.

At the same time, we also discovered that some engineered cytochromes P450 performed nitrene chemistry, generating the iron-nitrenoid from a synthetic azide nitrene precursor and directing the nitrene to C-H bonds for C-H amination (Figure 2) [27]. Earlier chemical research stimulated these experiments as well. In 1985, Gellman and coworkers reported that supplying an iminoiodane nitrene precursor to a rabbit liver cytochrome P450 led to three turnovers of intramolecular C-H amination [28]. We and Rudi Fasan re-discovered this promiscuous nitrene transfer activity at more or less the same time, almost thirty years later [27, 29]. More enzymes catalyzing abiological nitrene transfer reactions followed, brought about by a combination of chemical insight for reaction discovery and directed evolution to improve nascent activities [30].

Since a goal of this lecture is to introduce foundational concepts and

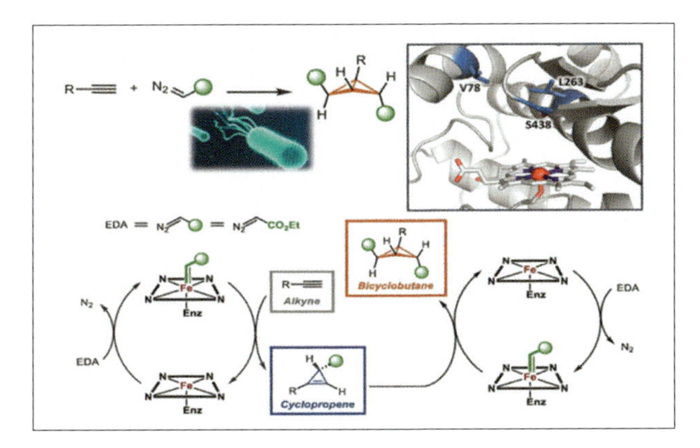

Figure 3. Enzyme-catalyzed bicyclobutanation through two carbene transfers to an alkyne, catalyzed by a serine-ligated variant of a cytochrome P450 [31]. The enzyme must generate the reactive carbene and transfer it to the alkyne substrate and then do it a second time to the cyclopropene intermediate in order to generate the bicyclobutane. Evolved enzymes make a single stereoisomer of these highly strained rings, which indicates a well-defined orientation of the substrates in the newly-evolved active site.

explain how we came to them, rather than to review research results, I will mention just one recent example of making products that chemists find very challenging: highly strained rings. Producing bicyclobutanes by two carbene transfers to an alkyne is a transformation not known in biology. It is rare in the world of human chemistry, and was never reported to be catalyzed using iron. Kai Chen first evolved an engineered, serine-ligated cytochrome P450 to transfer a carbene to an alkyne with perfect selectivity and make single stereoisomers of cyclopropenes. These carbocycles, whose ring strain is greater than 50 kcal/mol, are highly challenging to synthesize stereoselectively, but the enzyme does it with ease. Using the appropriate alkyne, Kai Chen also coaxed the enzyme to transfer a second carbene, cyclopropanating the double bond of the cyclopropene in the protected enzyme active site to make bicyclobutanes having >60 kcal/mol of ring strain (Figure 3). Following directed evolution, the enzymes delivered single stereoisomers of highly strained cyclopropenes or bicyclobutanes with turnovers in the thousands [31].

When supplied with alkynes and carbene precursors, *E. coli* expressing these new enzymes churn out cyclopropenes and bicyclobutanes. For sugar and a few growth-promoting trace elements, these living catalysts perform their chemistry in water (buffer), at room temperature. We hope their remarkable selectivities, low cost, and ability to use Earth-abundant iron to make strained rings that are otherwise difficult to obtain will open new applications for these fascinating products.

BRINGING NEW BONDS TO BIOLOGY

For a final glimpse into the exciting future chemistry that laboratory-evolved enzymes will enable, I will describe how we can now create biocatalytic machinery to make *bonds unknown in biology*. Silicon is the second most abundant element in the Earth's crust. Despite their ubiquity, carbon-silicon bonds are non-existent in the biological world. Yet laboratories make them, and lots of them. The room in which you are reading contains caulks, sealants, earphones, hair gels, and many more products whose carbon-silicon bonds are human-made.

In 2016, Jennifer Kan and her team discovered that heme proteins can catalyze carbene insertion into Si-H bonds to make various organosilicon products [32]. We are particularly fond of a marvelous little cytochrome *c* from *Rhodothermus marinus*, isolated from a hot saltwater pool in Iceland and now in the Protein Data Bank. This cytochrome c holds onto its heme through a covalent attachment; it is a manageable 124 amino acids long, its three-dimensional structure is known, and it is hyper-stable. With a melting temperature above 100° C, it can even be boiled and not lose its fold or metal cofactor. Although its biological function is electron transfer, it also happens to catalyze Si-H insertion: *Rma* cyt *c* inserts the carbene derived from methyl ethyldiazoacetate into dimethylphenylsilane with 40 turnovers and 97% enantiomeric excess (ee).

Those who know cytochromes c may find this puzzling. The iron in a cytochrome *c* is coordinatively saturated, i.e., four equatorial nitrogen ligands come from the porphyrin ring, while in *Rma* cyt *c* a methionine and a histidine provide the two axial ligands. Hence it is reasonable to ask where the reactive carbene forms, and how this protein binds the silane substrate. In fact, measuring the volume of the active site by rolling a computer ball over the crystal structure delivers an answer of zero.

Yet nature cares nothing for our calculations. This protein catalyzes its Si-C bond-forming reaction almost as effectively as the best human-invented catalysts for a similar reaction (which, by the way, use precious metals rather than readily-available iron), and it evolves. Just three generations of mutations directed to residues in the active site and screening for higher activity generated an enzyme now 15 times more active than human-invented catalysts. The new enzyme makes single enantiomers of its products and has a good substrate scope, producing new organosilicon compounds from at least 20 different silanes with hundreds to thousands of turnovers and >99% ee [32]. Once again, the enzyme is fully genetically encoded, so that bacteria expressing the gene form new silicon-carbon bonds, perhaps for the first time ever in a living system. Of course, we cannot know for sure that it is the first time, given that so much chemistry of the biological world remains unexplored.

A wonderful feature of engineering by evolution is that solutions come first; an understanding of the solutions may or may not come later. The evolved protein can be studied biochemically in an effort to discern how its new features emerged. In the case of the Si-H insertion enzyme derived from *Rma* cytochrome *c*, the x-ray crystal structure showed that the three activity-enhancing mutations changed the structure of a loop over the iron. What was the methionine axial ligand in the wild-type protein became an aspartic acid that no longer ligated the iron and instead pointed out to the solvent. A flip in the configuration of the loop created a binding pocket and also made the loop more dynamic so that it accommodates both the carbene and a range of silane substrates. Rusty Lewis was even able to trap the reactive carbene in the evolved enzyme and observe its orientation in the protein crystal at high resolution. Structural and spectroscopic studies combined with computational models allowed us to begin to explain how the new catalytic activity arose [33]. That does not mean, however, that we could *predict* the mutations that produced these effects.

Why stop at silicon? Another element of interest is boron, richly represented in the deserts not far from my southern California home. As with silicon, a wide range of organoboron compounds have been created in laboratories, but carbon-boron bonds have never been found in the biological world. Biology uses boron in the form of borates incorporated into natural products, most likely without the help of specific enzymes. Similarly, silicates are widely found in plants and marine animals such as diatoms.

To make carbon-boron bonds biologically, we again turned to our favorite cytochrome *c*. Jennifer Kan and Xiongyi Huang uncovered some activity for forming organoboron compounds by carbene insertion in the B-H bonds of water-stable borane adducts. Directed evolution created an enzyme that catalyzed 400 times more turnovers than the best small-molecule catalysts reported for similar transformations. Again, the enzyme is fully genetically encoded and carries out this new function inside living bacteria [34].

Carbon-X bonds known in biology include mostly C-H, C-N, C-O, C-S

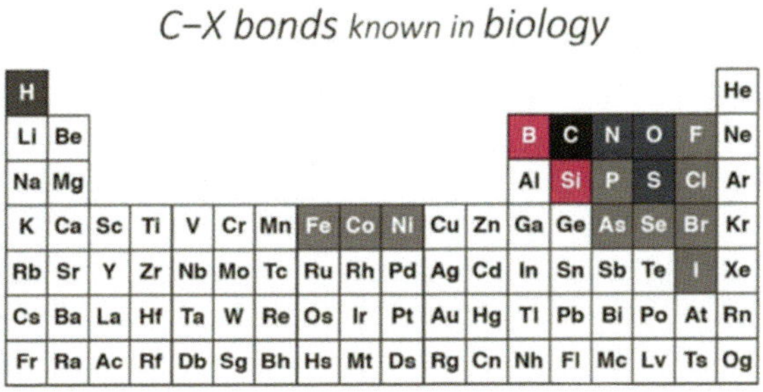

Figure 4. New 'carbene transferases' made by directed evolution have added C-Si and C-B bonds to biology's DNA-encoded synthetic repertoire.

bonds, some bonds to halogens, and a few to P, As, Se, and some metals (Figure 4). This leaves vast swaths of the periodic table untouched. With help from directed evolution and inspiration from the transition metal catalysis literature, two whole new elements have been added to biology's C-X bond repertoire. This ability can be exploited to bring boron and silicon into life and into products that can be manufactured in engineered microorganisms. The enzymes can also be used in synthetic chemistry, where their exquisite and tunable selectivities will enable low-cost and easy preparation of organoboron and organosilicon products.

FINAL THOUGHTS ABOUT THE FUTURE OF DIRECTED EVOLUTION

Life – the biological world – is the greatest chemist, and evolution is her design process. In fact, the internet of *living* things has been crowd-sourcing problem solving this way for more than three billion years. Evolution can circumvent our profound ignorance of how sequence encodes function, and may allow us to find new solutions to human problems. I have described how we are now moving into a future of readily-tunable catalysts that perform challenging reactions using Earth-abundant materials and producing minimal waste. Biological systems are good models for sustainable chemistry that uses abundant, renewable resources and recycles a good fraction of its products. I dream of the day that much of our chemistry becomes genetically encodable, and microorganisms and plants are our programmable factories.

I am continually amazed at the ease with which evolution innovates. With the power of evolution realized for engineering, we can look at diverse products of natural evolution in an entirely new way. Instead of asking what enzymes do in the natural world, we can now ask, "What *might* they do?" Enzymes will perform chemistry in more ways than we could have imagined, especially when we use evolution to unleash their latent potential. A treasure trove of new enzymes awaits discovery for carrying out chemistry that we could not even contemplate just a few years ago.

Existing diversity provides the fuel for these innovations; both natural and directed evolution uses this diversity to solve challenges, exploit opportunities, and evade catastrophe. As countless examples from the natural world attest, the alternative to diversity is extinction.

I thank the Royal Swedish Academy of Sciences, my family, friends, colleagues, Sabine Brinkmann-Chen, Cheryl Nakashima, and especially my current and former students. I am grateful to the California Institute of Technology, where I was inspired to explore the unknown for new understanding and found the team with which to do it.

REFERENCES

1. Borges, J.L. (1941) The Library of Babel. Editorial Sur.
2. Maynard-Smith, J. (1970) Natural Selection and the Concept of a Protein Space. *Nature,* **225**, 563–564.
3. Arnold, F.H. (2011). The Library of Maynard Smith: My Search for Meaning in the Protein Universe. *Microbe, ASM News,* **6**, 316–318.
4. Eigen, M., Gardiner, W. (1984) Evolutionary Molecular Engineering Based on RNA Replication. *Pure. Appl. Chem.,* **56**, 967–978.
5. Beaudry, A.A., Joyce, G.F. (1992) Directed Evolution of an RNA Enzyme. *Science,* **257**, 635–641.
6. Leung, D.W., Chen, E., Goeddel, D.V. (1989) A Method for Random Mutagenesis of a Defined DNA Segment using a Modified Polymerase Chain Reaction. *Technique,* **1,** 11–15.
7. Wells, J.A. (1990) Additivity of Mutational Effects in Proteins. *Biochemistry,* **29**, 8509–8517.
8. Kauffman, S.A. (1993) The Origins of Order: Self-Organization and Selection in Evolution. Oxford University Press.
9. Dennett, D.C. (1995) Darwin's Dangerous Idea: Evolution and the Meaning of Life. New York: Simon & Schuster.
10. Chen, K., Arnold, F.H. (1991) Enzyme Engineering for Nonaqueous Solvents: Random Mutagenesis to Enhance Activity of Subtilisin E in Polar Organic Media. *Biotechnology* **9**, 1073–1077.
11. Chen, K., Arnold, F.H. (1993) Tuning the Activity of an Enzyme for Unusual Environments: Sequential Random Mutagenesis of Subtilisin E for Catalysis in Dimethylformamide. *Proceedings of the National Academy of Sciences USA,* **90**, 5618–5622.
12. Stemmer, W.P.C. (1994) Rapid Evolution of a Protein *in vitro* by DNA Shuffling. *Nature,* **370**, 389–391.
13. Moore, J.C., Arnold, F.H. (1996) Directed Evolution of a para-Nitrobenzyl Esterase for Aqueous-Organic Solvents. *Nature Biotechnology,* **14**, 458–467.
14. Giver, L., Gershenson, A. Freskgard, P.O., Arnold, F.H. (1998) Directed Evolution of a Thermostable Esterase. *Proceedings of the National Academy of Sciences USA* **95**, 12809–12813.
15. Zhao, H., Arnold, F.H. (1999) Directed Evolution Converts Subtilisin E into a Functional Equivalent of Thermitase. *Protein Engineering,* **12**, 47–53.
16. Bornscheuer U.T., Huisman G.W., Kazlauskas, R.J., Lutz, S., Moore, J.C., Robins, K. (2012) Engineering the Third Wave of Bocatalysis. *Nature,* **485**, 185–194.
17. Zeymer, C., Hilvert, D. (2018) Directed Evolution of Protein Catalysts. *Annual Review of Biochemistry,* **87**, 131–157.
18. Babbitt, P.C., Gerlt, J.A. (2000) New Functions from Old Scaffolds: How Nature Reengineers Enzymes for New Functions. *Advances in Protein Chemistry,* **55**, 1–28.
19. O'Brien, P.J., Herschlag, D. (1999) Catalytic Promiscuity and the Evolution of New Enzymatic Activities. *Cell Chemical Biology,* **6**, R91–R105.
20. Kazlauskas, R.J. (2005) Enhancing Catalytic Promiscuity for Biocatalysis. *Current Opinion in Chemical Biology,* **9**, 195–201.
21. Khersonsky, O., Tawfik, D.S. (2010) A Mechanistic and Evolutionary Perspective. *Annual Review of Biochemistry,* **79**, 471–505.
22. Arnold, F.H. (2018) Directed Evolution: Bringing New Chemistry to Life. *Angewandte Chemie International Edition,* **57**, 4143–4148.

23. Coelho, P.S., Brustad, E.M., Kannan, A., Arnold, F.H. (2013) Olefin Cyclopropanation via Carbene Transfer Catalyzed by Engineered Cytochrome P450 Enzymes. *Science*, **339**, 307–310.
24. Wang, Z.J., Renata, H., Peck, N.E., Farwell, C.C., Coelho, P.S., Arnold, F.H. (2014) Improved Cyclopropanation Activity of Histidine-Ligated Cytochrome P450 Enables the Enantioselective Formal Synthesis of Levomilnacipran. *Angewandte Chemie International Edition* **53**, 6810–6813.
25. Hernandez, K.E., Renata, H., Lewis, R.D., Kan, S.B.J., Zhang, C., Forte, J., Rozzell, D., McIntosh, J.A., Arnold, F.H. (2016) Highly Stereoselective Biocatalytic Synthesis of Key Cyclopropane Intermediate to Ticagrelor. *ACS Catalysis,* 6, 7810–7813.
26. Bajaj, P., Sreenilayam, G., Tyagi, V., Fasan, R. (2016) Gram-Scale Synthesis of Chiral Cyclopropane-Containing Drugs and Drug Precursors with Engineered Myoglobin Catalysts Featuring Complementary Stereoselectivity. *Angewandte Chemie International Edition*, **55**, 16110–16114.
27. McIntosh, J.A., Coelho, P.S., Farwell, C.C., Wang, Z.J., Lewis, J.C., Brown, T.R., Arnold, F.H. (2013) Enantioselective Intramolecular C–H Amination Catalyzed by Engineered Cytochrome P450 Enzymes *in vitro* and *in vivo*. *Angewandte Chemie International Edition*, **52**, 9309–9312.
28. Svastits, E.W., Dawson, J.H., Breslow, R., Gellman, S.H. (1985) Functionalized Nitrogen Atom Transfer Catalyzed by Cytochrome P-450. *Journal of the American Chemical Society,* **107**, 6427–6428.
29. Singh, R., Bordeaux, M., Fasan, R. (2014) P450-Catalyzed Intramolecular sp3 C–H Amination with Arylsulfonyl Azide Substrates. *ACS Catalysis*, **4**, 546–552.
30. Zhang, R.K., Huang, X., Arnold, F.H. (2018) Selective C-H Bond Functionalization with Engineered Heme Proteins: New Tools to Generate Complexity. *Current Opinion in Chemical Biology*, **49**, 67–75.
31. Chen, K., Huang, X., Kan, S.B.J., Zhang, R.K., Arnold, F.H. (2018) Enzymatic Construction of Highly Strained Carbocyles. *Science*, **360**, 71–75.
32. Kan, S.B.J., Lewis, R.D., Chen, K., Arnold, F.H. (2016) Directed Evolution of Cytochrome c for Carbon–Silicon Bond Formation: Bringing Silicon to Life. *Science*, **354**, 1048–1051.
33. Lewis, R.D., Garcia-Borràs, M., Chalkley, M.J., Buller, A.R., Houk, K.N., Kan, S.B.J., Arnold, F.H. (2018) Catalytic Iron-Carbene Intermediate Revealed in a Cytochrome c Carbene Transferase. *Proceedings of the National Academy of Sciences USA*. **115**, 7308-7313.
34. Kan, S.B.J., Huang, X., Gumulya, Y., Chen, K., Arnold, F.H. (2017) Genetically Programmed Chiral Organoborane Synthesis. *Nature*, **552**, 132–136.

George P. Smith. © Nobel Prize Outreach AB. Photo: A. Mahmoud

George P. Smith

Biography

EARLY LIFE

I was born March 10, 1941 to Albert Mark Smith II (March 2, 1908 to February 3, 1978) and Jessie Patton Biggs Smith (September 14, 1909 to June 14, 2000). My brother Mark (A. Mark Smith III) was born December 29, 1942 and my sister Helen (now Helen Boyd) was born June 22, 1947.

My father graduated from West Point (the U.S. Military Academy) in 1930, but left the army for civilian life immediately afterward. He had steady employment during the Depression, and he and my mother, who married in 1936, lived in relative prosperity in those years.

The Japanese attack on Pearl Harbor occurred nine months after my birth. My father immediately re-joined the army and continued as a career officer until his retirement in 1965. This meant that our family moved very frequently while Mark, Helen, and I were growing up. By the time I graduated from ninth grade in 1955, I'd attended an extraordinary 11 or 12 schools. Most of the schools were on army bases, where all the kids moved as frequently as we did. That meant that the new kid in school was rapidly assimilated into society, without having to spend a painful year or two as outsider. I learned how to make new friends rapidly, but not how to maintain long-term friendships – a pattern that continues to some degree to the present day.

Mostly my father was posted up and down the East Coast, but during the Korean War he served in Japan from 1951 to 1954. The rest of the family joined him there in April 1952, just as the U.S. occupation was coming to an end.

MY CAREER AS HERPETOLOGIST

In the summer of 1949, our family went on vacation for a few weeks at Penobscot Bay, Maine. The place was overrun with snakes. I ran across two snakes eating the same frog or toad from opposite ends (I don't know

Figure 1. Our family sitting under the stone lion at the Toshogu shrine, Ueno Park, Tokyo, September 7, 1952. I'm the 11-year old dreamer on the right.

Figure 2. My friend and I have just caught two elegant rat snakes. Washington Heights U.S. military housing complex, Tokyo, summer 1952.

how it turned out). I caught a beautiful green snake and carried it triumphantly by the tail into the parlor, where my mother was hosting a tea party for some proper local ladies. There were some "eeks," and I was ordered back outside. My herpetology career had begun.

HAM RADIO

I caught the radio bug while we were still in Japan, and continued through high school. I got my novice license (KN4OWA) around 1956, and built a small rig, which included a classy straight telegraph key I got off an army surplus telegraph set. I never got very proficient in Morse code, however, and dropped out after high school. By that time, I'd learned quite a bit about circuitry, and not a little calculus and analytical geometry (without learning their names), from obsessive study of the Radio Amateur's Handbook. That's knowledge that has turned out to be of use in the sequel.

HIGH SCHOOL

My last three years of schooling were at Philips Academy Andover, a private boarding school in Massachusetts founded in 1778. Both President

George Herbert Walker Bush and his son President George Walker Bush graduated from Andover. So did 2018 Economics laureate Bill Nordhaus, a year behind me; he was a fellow member of the Radio Club, though I don't remember him from then.

Education at Andover was rigorous in a patrician sort of way. Traditional skills like English composition were prominent in the curriculum, while I remember instruction in math and science to be rather old-fashioned. I was attracted to biology, whose head, Harper Follansbee, appointed in 1940, did nothing to discourage my herpetological ambitions. There was a magnificent indigo snake in one of the buildings. An ambitious young biology teacher, John Kimball, was appointed in 1956, and modernized the biology curriculum over the years. I barely remember him from my years at Andover, but have come to appreciate his great contribution to biology education, especially the free online textbook http://www.biology-pages.info/, which he continues to maintain to this day, and which I made liberal use of in my own teaching career.

Apart from Latin, French, and English, which I'll discuss in the next section, my most memorable course at Andover was senior American history with Fred Allis. Memorable because Mr. Allis assumed a level of sophistication in the analysis of Supreme Court decisions, the economic theories of (I seem to remember) Joseph Schumpeter, etc., that was utterly beyond my abilities. Some of my classmates seemed to thrive on this stuff. I vaguely remember that a few even argued with Mr. Allis! Not me. Long, miserable evenings in the library earned me a (no doubt undeserved) gentleman's C.

Altogether, though, I graduated from Andover well prepared for college – provided that I didn't major in history.

LANGUAGE

Three language classes at Andover were more memorable than American history: Latin, French, and senior English.

My Latin teacher was Frank Benton, who was appointed in 1918 and retired when I graduated in 1958. He looked upon our attempts to "construe" (translate out loud in class) our assigned Latin texts with curmudgeonly good humor. Wisely eschewing Cicero for the pimply teens in his third year Latin class, he assigned an up-to-date subject instead: medieval Latin. I hold our reader, Helen Waddell's *A Book of Medieval Latin for Schools*, first published in 1931, before me as I write. The verse has understandable meter and clear rhymes that even a teen could appreciate. As I took up choral singing in college and afterward, thus regularly encountering medieval Latin, I came to appreciate Mr. Benton's class even more.

My French teacher was James Grew, appointed in 1935. Mr. Grew had adopted l*a méthode directe.* On the first day of French I, he briefly explained the rules of the class in English. That was the last English we were to hear from him for many months. He'd talk to us in French, ask us questions in French, sometime right in our faces, and eventually we'd start responding in French – broken French at first, of course, but gradually more and more idiomatic French. It was as if we were toddlers first learning our mother tongue – which is pretty much the logic of *la méthode directe*. French is the only foreign language in which I attained anything approximating fluency (mais maintenant j'ai presque tout oublié).

My senior English teacher was Dudley Fitts, appointed in 1941, who was in addition a prominent and well-respected poet, literary critic, and translator of classical Latin and Greek literature. What I remember most in his class was Chaucer, whose Middle English language we learned to understand and recite with some facility. I can't say that senior English in general or Chaucer in particular had the same specific influence on my life as did French and Latin. But it did come in handy a few years ago, when our dinner group gathered for a meal whose theme was Spring. Someone had laboriously typed out the first 18 lines of *Canterbury Tales,* and asked if anyone would like to try reading it during the meal. "Sure," I said, standing up and ignoring the typescript. I rendered those first 18 lines from memory, more than half a century after high school graduation, with (I flatter myself) dramatic flair and a creditable Middle English accent.

My engagement with language – both my own and foreign – has persisted to the present day. I continued to study French during my year in England as an exchange student between high school and college, and I took German and Spanish in college. I have learned to recognize roots in those and other languages, including some in Hebrew and Arabic. The language instinct has declined dramatically with age, of course; language is largely an academic enterprise now.

A YEAR IN ENGLAND

Before going to college, I spent the 1958–1959 academic year as an exchange student at Wellington School, a combination day and boarding school in Wellington, Somerset, England. I was considered a Sixth Former (senior), and was assigned to one of the school's houses (I don't remember which one). By then I knew I was going to be a biology major in college, but I decided to take English, French, and History as my A-level subjects. Surprisingly considering what I wrote above about American history at Andover, I did OK in history: I think the History Master, whose name I believe was Victor Finn, was impressed by my writing ability, not my historical acumen. Evidently, I did pretty well in French too, because I have a copy of the *Oxford Companion to English Literature* with an inscription saying that it was the 1959 Modern Languages Prize.

An incident at Wellington School stands out as a shameful embarrassment in retrospect. Two representatives of South Africa's apartheid government were invited to address a school assembly. For half an hour or so, one of them explained calmly and patiently how reasonable South Africa's apartheid policy was, then asked for questions. I stood up and asked a respectful question – thank God I don't remember what – in my Yank accent, which brought forth an amused comment from the speaker before he answered. After assembly, the History Master (again, I think Victor Finn) took some of us Sixth Formers aside and vented his fury that such an outrage had been allowed to occur in the school. I don't remember him using the word "racism," but I understand now that that was what he was denouncing. I felt mildly embarrassed at the time, but now I squirm at how unthinkingly I accepted such racist ideology as if legitimate.

HAVERFORD COLLEGE

My choice of Haverford College, a small Quaker school outside Philadelphia, counts as a sort of rebellion from my Andover background: Andover in those days was still largely a feeder school for Harvard, Yale, and Princeton.

I came just too late (September 1959) for any hope of a career in herpetology. Emmett "Dixie" Dunn, a prominent herpetologist, had been chair of the Biology Department, but he died in 1956. His replacement, Ariel Loewy, was even then (at age 34 in 1959) a prophet-like visionary on campus, whose soft-spoken charisma is still very much in evidence (along with a hacking cough) in an interview he gave in 1996 (Loewy, 1996). Loewy was born to a prosperous Jewish family in Bucharest, but the family fled to England in 1936 and then to Canada in 1941. He went to the University of Cambridge in England for a research fellowship in 1952–1953, and was thus eyewitness to Watson's and Crick's elucidation of the structure of DNA and the birth of the molecular genetics revolution. He accepted a position in Haverford's Biology Department in 1953, and by the time I arrived had persuaded the college to let him utterly transform the curriculum, abandoning broad coverage of traditional subdisciplines in order to concentrate on cell and molecular biology. This was a unique educational choice at the time. It was also a terrific piece of luck for me: molecular biology was a far better fit to my abilities than herpetology, its math-like style congruent with the mathematical habit of mind I came to discover in myself at Haverford.

Loewy managed to recruit two ambitious young scientists who shared his curricular vision in full measure, and his charisma in some measure as well: Irv Finger and Mel Santer, who along with Loewy taught most of the biology courses I took.

In 2009 I reconnected with Santer, and told him about an incident in one of his courses that significantly influenced my scientific life. In one of his tests, he asked us (as I remember) to reconstruct the logic of an experiment by Howard Dintzis that had been published the year before (Dintzis, 1961), and that was immediately recognized as an important contribution to molecular biology. Dintzis showed that polypeptides are synthesized from amino terminus to carboxy terminus, by adding amino acids sequentially to the carboxy terminal end of the growing chain of amino acids. I was very pleased with my answer: I thought I'd caught on to the rationale of a complex experiment without a teacher's help – an unusual or perhaps unique experience in my intellectual life to date. As I remember, Santer didn't agree: he gave me a low mark on the question. Santer didn't remember any of this in 2009, of course. Nevertheless, the "Dintzis experiment" (as it came to be called) reappeared in my life on several occasions, including in graduate school as I'll explain below. This incident is emblematic of the many ways teachers can have lasting influences on their students without realizing it at the time.

A core component of Loewy's curricular reform was the senior research tutorial, in which every senior biology major was required to complete a research project with one of the faculty. My senior year (1962–1963) tutor was Meg Mathies, who served as temporary replacement for Santer during his sabbatical leave. Mathies was an immunologist who had just earned her Ph.D. under Abram Stavitsky at Case Western Reserve University; it was her influence that steered much of my subsequent research career toward molecular immunology.

My project was nothing if not ambitious: to show that the antigen-binding specificity of an antibody was determined by the messenger RNAs (mRNAs) that encode it. That seems like an obvious truth today, but it wasn't then. The alternative *template theory,* which I'll explain later in this essay, implied that the amino acid sequences of an antibody's polypeptides (thus the nucleotide sequences of the corresponding mRNAs) did not suffice to explain its antigen specificity. My plan was to immunize a rabbit with a phage (phage T4, a virus that infects bacterial cells), extract RNA from the spleen (a primary immune organ) as the rabbit responded to the immunization, add that RNA to the cell-free protein synthesis system reported the year before (Nirenberg and Matthaei, 1961), and determine whether the resulting protein included antibodies that could *neutralize* T4 phages (that is, block their ability to infect bacterial cells). The experiment was hopelessly naïve. I failed to demonstrate any T4-neutralizing antibodies; it was Edgar Haber, who turned out to be my Ph.D. mentor a few years later, who defeated the template theory, as I'll explain below. Still, a few things did work: the cell-free system did synthesize protein when artificial RNA was added in large amounts; T4 infection could be quantified using a plaque assay; the immunized rabbit did produce antibodies with strong T4-neutralizing activity as measured using the plaque assay. I used T4 to immunize rabbits and mice at several points in my later research career, and phage neutralization assays were key in my first phage-display publication (Smith, 1985).

I took a course called (I think) College Mathematics with a master teacher named Cletus Oakley. It explored a handful of carefully selected areas of mathematical analysis. I found this a most congenial kind of thinking, and went on to take the calculus sequence in college and a number of other courses afterward. Elementary math has been a significant component of my science throughout my career. Experimental results can often be presented in a more perspicuous, compelling, and workmanlike manner using simple mathematics. I can't claim expertise in any area of the discipline, and I'm not an amateur do-it-yourself purist either. I cheerfully outsource challenging integrations or differential equations to online tables or utilities, or to colleagues who are actual mathematicians. My philosophical interest in probability theory, first encountered in Oakley's class, as the fundamental rules of empirical reasoning (see below) arises from my engagement with math.

Haverford was still a Quaker school when I attended, and I admired the Quaker attitudes and commitment to social justice that I encountered in college. Many of my college friends had come to Haverford from Quaker schools. I participated in a number of Quaker weekend work camps in the inner city of Philadelphia. All this led to a short-lived engagement with Christianity (which had had little importance in my earlier life) during my college years, but religion dropped back out of my life for good after graduation. It has been Quakerism's social attitudes and commitment to justice, not its devotional dimension, that has stuck with me.

SINGING

During one of the summers in my college years, I got a tech job at the Marine Biological Laboratory in Woods Hole, on Cape Cod in Massachusetts. The college students like me spent many evening and weekend hours socializing, and I joined several other young people in an informal madrigal group. I have been singing in choruses ever since, with some interruptions. Years later, in 1978 in Columbia, Missouri, I was a founding member of the Ad Hoc Singers (a name we made up in haste before our first gig in a nursing home), which has evolved through several name-changes to the Columbia Chorale today. I'm still singing in the Chorale, progressing (regressing?) from second bass to second tenor as my low notes disappeared with age.

Engagement with choral singing meshes with my interest in language.

I've written a number of musical essays for the Columbia Chorale, and I've posted an essay on *Jerusalem of Gold* on the Mondoweiss website. It is the language of the lyrics and how it relates to the musical diction that is my chief concern in these essays.

BRIEFLY A HIGH SCHOOL TEACHER

I didn't graduate from Haverford with an academic career in mind. Instead, my experience with the Quaker inner-city weekend work camps inspired me to be a high school teacher in neighborhoods like the ones I'd encountered. I took a few basic education courses in my senior year, and enrolled in the Master of Arts in Teaching (MAT) program at Temple University that summer. A semester of practice teaching in an inner-city school made me change my mind. I told myself that it was a teacher's personal relations with young students, not engagement with their developing intellects, that was of prime importance, and I had little talent in that area. A more honest assessment might be my realization that teaching was going to be really hard work, and that maybe I should go for an easier life.

In any case, I resolved to apply to graduate school and took a temporary technician job with Martin Nemer at the Institute for Cancer Research (now the Fox Chase Cancer Center) in Philadelphia. Nothing could have locked in my new career choice more effectively than my brief experience there. Every afternoon there was a tea at which the whole institute, including many eminent scientists, were invited to socialize. As I remember, one of the regular participants, a man who made time for lowly techs like me, was Irwin Rose (he was awarded the Nobel Prize in Chemistry forty years later for the discovery of ubiquitin). This was heady stuff for a beginner!

GRADUATE SCHOOL AT HARVARD MEDICAL SCHOOL AND THE MASSACHUSETTS GENERAL HOSPITAL

In Fall 1964, I was admitted, along with about a dozen other students, to a PhD program at Harvard Medical School called (I think) the Biomedical Sciences (BMS) program. During our first year, we took an integrated basic medical sciences curriculum taught by a large team of professors, after which we would choose a dissertation mentor in any of a large selection of departments. BMS was a concentrated and remarkably comprehensive survey of modern biomedical research by master teachers. We students formed our own little society of friends, though in keeping with the army brat pattern of friendship I mentioned earlier I haven't kept up with them.

I chose a cellular immunologist in the Department of Bacteriology and Immunology, Hugh McDevitt, as my dissertation advisor. I already had a project in mind: to use the "Dintzis experiment" that I had learned about in Santer's class at Haverford (see above) to determine whether or not the variable (V) and constant (C) parts of an immunoglobulin polypeptide, which had been defined by the first immunoglobulin amino acid sequence results (Hilschmann and Craig, 1965), were synthesized as separate polypeptide chains that were subsequently joined together. (We now know, but didn't then, that the V and C portions are joined in multiple steps at the chromosomal DNA level.) I didn't get to embark on this project, however, since McDevitt was recruited to Stanford and I switched to the lab of Ed Haber (already mentioned above) at the Massachusetts General Hospital (MGH). "My" Dintzis experiment was published two years later by other immunologists, who showed that the V and C portions are synthesized as a single polypeptide (Knopf et al., 1967; Lennox et al., 1967). I wasn't disappointed to be "scooped"; indeed, that established molecular immunologists would consider "my" experiment worth doing seemed to be a personal validation rather than a setback.

Haber, at age 34, was already chief of the Cardiac Unit at MGH (Editors, 1998) and an accomplished protein chemist. He had been born February 1, 1932, to a German Jewish family who escaped to Palestine in 1933 and moved to the U.S. in 1939. He had done his postdoc in the lab of Chris Anfinsen at the National Institutes of Health. Anfinsen was to receive the Nobel Prize in Chemistry in 1972 for showing that a protein's primary structure (i.e., the amino acid sequence of its polypeptide) sufficed to specify its final three-dimensional ("folded") structure, thus its biological activity. Haber had used Anfinsen's methods to demonstrate the same principle in the case of antibodies (Haber, 1964). This put paid to the template theory (to be explained below) that had been the target of my failed senior tutorial project at Haverford.

Haber had a dreary basement lab in a dreary wing of the dreary (but storied) MGH. Sequencing immunoglobulins was a core experimental approach to molecular immunology at the time, and the Edman degradation was the flagship of sequencing technology. Machines called "sequenators" for automating the Edman degradation had been recently introduced. My plan was to develop a new kind of sequenator in which the protein to be sequenced would be exposed to the Edman reagents in the gas phase rather than dissolved in liquids. This innovation could have the important benefit of applying to small peptides that would be washed away by liquids. I had spectacular success at the beginning, but those successes could never be repeated, and the project was ultimately abandoned without any resulting publications.

The foregoing experimental failure wasn't the end of my graduate career, for I had meantime been engaged in a theoretical study of the problem of "antibody diversity" (as we called it at the time). This problem arose in its starkest form from the work of Karl Landsteiner, who had shown in the early 20th century that animals were able to mount a specific antibody response to almost any synthetic organic chemical he tested (Landsteiner, 1962). “It is not reasonable that an animal produces predetermined antibodies against thousands of such synthetic substances,” Fritz Breinl and Felix Haurowitz had argued (Breinl and Haurowitz, 1930).

Breinl and Haurowitz’s solution to this conundrum, as elaborated later by Linus Pauling (Pauling, 1940), was *the template theory*. They proposed that a very limited number of generic antibody polypeptides could suffice for an unlimited number of different antibodies with distinct antigen-binding specificities. Antigen specificity arose (so went the theory) when these generic polypeptides first wrapped around the antigen as a template, then disengaged from the template as an antibody molecule with an antigen-specific binding site molded into its three-dimensional architecture. Antigen templates thus acted catalytically, each template molecule serving to stamp antigen-specificity into many antibody molecules. It’s the template theory that had been the target of my senior tutorial project at Haverford, and that Haber’s work had refuted (Haber, 1964). The great diversity of antibodies was now understood to reflect a corresponding great diversity of antibody-coding genes. And as the amino acid sequences of antibodies accumulated with increasing use of sequenators and other technical advances, it was clear that that diversity was concentrated in the V regions of their polypeptide chains.

My theoretical study focused on the evolution of V genes (the genes encoding V regions). Walter Fitch at the University of Wisconsin and I applied the computerized phylogenetic reconstruction algorithms he and Emanuel Margoliash had just developed (Fitch and Margoliash, 1967) to the increasing number of V region amino acid sequences that were becoming available. This is was the study at the core of my dissertation and a review by me, Leroy Hood, and Fitch (Smith et al., 1971). Two years later I extended these studies in a full-length monograph (Smith, 1973c). I used these analyses in a critical assessment of two competing explanations of V gene diversity: the somatic mutation and germline theories. The somatic mutation theory supposed that the observed diversity of V region sequences arises through somatic mutation in V genes in individual B-cell clones during the lifetime of each individual animal or person (B cells are the lymphocytes that produce antibodies). According to this theory, the number of V region sequences far exceeds the number of germline V genes. The germline theory, in contrast, supposes that each possible V region an individual can express is directly encoded by a corresponding germline V gene. This theory required a large, though not infinite, number of V genes to account for the large number of different V region sequences. I argued that the evidence favored the germline theory, though not conclusively. Many predictions of, and arguments for, that theory have been confirmed by subsequent findings; in particular, there are indeed large numbers of V genes for some families of V regions. Nevertheless, the central claim of the germline theory has turned out to be completely wrong: extensive somatic mutation in individual B-cell clones is clearly the fundamental explanation for V region diversity.

My dissertation was finished in December 1969 with some help from my father, who used the drafting set he still had from West Point to draw several of the most complicated figures. The dissertation posed a problem for my committee: it reported no experimental results, my attempt at developing a new sequenator having failed. This was, to say the least, unusual, but I passed anyway, and got my degree at the beginning of 1970. Not for the last time in my career did I receive special treatment.

Haber wouldn't put his name on any of these publications, but he was a very engaged mentor. In particular, he took care that I had a chance to interact with many other leaders in molecular immunology, an advantage that greatly enhanced my career.

WAR RESISTANCE

Between the Gulf of Tonkin resolution in August 1964 and the March on the Pentagon in October 1967, I became progressively more politically active against the war in Vietnam. Although I was too old for the military draft, I turned in my draft card to my local Selective Service board as a symbolic act of resistance. As was usual in those days, I was promptly drafted into the army, but refused induction very publicly in the presence of dozens of inductees in early 1968. This could well have resulted in a prison sentence, which would undoubtedly have greatly altered the course of my life. But as it turned out, I wasn't imprisoned and suffered no other consequence of my act. The reason may be that the military was coming to realize that its policy of drafting resisters was backfiring. Refusing the draft was a highly visible form of resistance that was being effectively used to recruit young people to the anti-war cause.

POSTDOC WITH OLIVER SMITHIES AT THE UNIVERSITY OF WISCONSIN

I first encountered Oliver Smithies at the 1967 Cold Spring Harbor Symposium on Antibodies, a landmark meeting at which major issues that were engaging immunologists at the time, including theories of antibody diversity, were vigorously discussed. Smithies presented a creative new theory of diversity, in which V-gene diversity was generated by extensive somatic recombination rather than somatic mutation. Francis Crick was effusive about Smithies's hypothesis; he castigated the audience for not hailing the solution to the puzzle they had been seeking so long. A less modest person might have been tempted to gloat in triumph at praise from such high quarter. Not Smithies. He responded respectfully to the several participants who objected strenuously to his theory. I was more impressed with his integrity than Crick was with his theory.

I came to Smithies's lab at the beginning of 1970. My main experimental project there (and completed at the University of Missouri) was sequencing the immunoglobulin light chains from a mouse myeloma tumor that was peculiar in secreting two light chains (Rose et al., 1977; Smith, 1973a, 1978b). One of them consisted of a signal sequence joined directly to the C region. We interpreted this as aberrant expression of an unjoined C gene, but the two light chains were later shown to be alternative splicing products of the primary RNA transcript of a single aberrantly joined light chain locus (Sikder et al., 1985).

Smithies was the one who in 1955 had introduced gel electrophoresis into the armamentarium of biological research in the form of starch gel electrophoresis, which was a regular component of our work in his lab. I'm one of the few living scientists who remembers the art of pouring, loading, running, and staining a starch gel. I say "art" advisedly, because success required mastery of skills that were exceedingly difficult to systematize in a manual: recognizing exactly the right "blurbling" sound as you boiled up the starch in a flask over an open flame, slapping a spatula just so onto the petroleum jelly covering the gel in order to peel it off in one piece after the run was over, etc. When Smithies turned his attention to DNA in collaboration with Fred Blattner toward the end of my postdoc, he devised an absolutely characteristic style of agarose gel electrophoresis. The molten gel was poured directly onto the benchtop, with a comb held in place by lumps of clay. After the DNA samples were loaded into the wells, warm petroleum jelly was poured over the surface to prevent evaporation (as in starch gel electrophoresis), and electrode trays were connected to the ends with paper wicks. After the run, the petroleum jelly was removed with a spatula (again, as for starch gels) and the gel was lifted off the benchtop and stained with ethidium bromide – the final step being the only one that would be recognizable to today's practitioners. Smithies was indeed an inveterate improviser. Otto Hiller, who made starch gel electrophoresis apparatuses in his shop in Madison, was a fellow tinkerer and Smithies's regular Saturday afternoon companion, as he explains in his Nobel biography (Smithies, 2007). I admired but never emulated the improvisational style I witnessed in his lab.

The Genetics Building in Madison, where Smithies's lab resided, was also home to two doyens of population genetics: Sewall Wright (December 21, 1889 to March 3, 1988), one of the founders of the field; and James F. Crow (January 18, 1916 to January 4, 2012), a master teacher and author of "Crow's Notes," a concise paperback guide for undergraduate genetics students, as well as coauthor with Motoo Kimura of the best-selling population genetics text *An Introduction to Population Genetics Theory*. Crow's influence added a new dimension to my theoretical studies, which were continuing in Smithies's lab. The germline theory of antibody diversity that I favored presumed that multiple germline V genes formed long tandem arrays in chromosomal DNA, as was already known for ribosomal RNA genes. Such arrays would be subject to occasional unequal crossover events in the germline, leading to repeated small contractions and expansions of the arrays over evolutionary time. Tandem genes subject to these expansions and contractions could be looked on as a population subject to many of the same principles of population genetics as a population of interbreeding organisms. This idea was included in my monograph on antibodies (Smith, 1973c) and presented at the 1973 Cold Spring Harbor Symposium (Smith, 1973b). My first publication from the University of Missouri was a lead article in Science arguing that any long stretch of chromosomal DNA not subject to natural selection would tend to turn into a tandem array, thus explaining the abundance of repetitive DNA with no obvious function in the genomes of many organisms (Smith, 1976).

UNIVERSITY OF MISSOURI AND COLUMBIA

In 1975 I was recruited to the newly-established Division of Biological Sciences at "Mizzou" by Abe Eisenstark, its first director. Eisenstark's remit was a little like Loewy's at Haverford: to modernize biological research and teaching on campus. I became part of a small coterie of young cell and molecular biologists who together constituted a favored establishment that was resented by the "Old Turks," my name for a few older faculty who (with some justification) felt devalued in Eisenstark's *Risorgimento*.

In my first few years at Mizzou I continued my experimental and theoretical study of antibody diversity, including the amino acid sequencing project I described above. My first doctoral student was Jamie Scott, who undertook to count the number of V genes in the mouse λ family. A few dozen different V region sequences in this family had been published, but their diversity was severely limited. According to the somatic mutation theory, these V regions could have arisen from only two V genes; the germline theory, in contrast, would require dozens. Counting the mouse λ V genes thus seemed a promising way to decide definitively between the somatic mutation and germline theories. Using two independent approaches, high-precision mathematical analysis of hybridization kinetics and denaturing gradient gel electrophoresis (Fischer and Lerman, 1980), Scott demonstrated that two genes accounted for all the V regions in the family, thus refuting the germline theory in this case (Scott et al., 1985). Her trip to Leonard Lerman's lab in Albany to learn denaturing gradient gel electrophoresis, a lead pig with radioactive hybridization probe in her luggage, and the exciting results that ensued, are still fond memories. Scott's article was never accepted for publication except as an abstract. One reason for this injustice is that the reviewers and editors believed, without justification, that the problem of antibody diversity had already been settled in favor of somatic mutation – an example of the impatient inattention to detail that is an unpleasant side-effect of the hurried pace of modern science. Scott returned to my lab as a postdoc after medical school, and played a key role in the development of that technology (Smith, 2018). She and her husband Felix Breden have been friends of my family ever since.

A few years after coming to Mizzou, I embarked on an ill-starred developmental biology project with the roundworm *Caenorhabditis elegans*, a model organism that my Biological Sciences colleague Don Riddle had brought from Sydney Brenner's lab in Cambridge, England. My interest in filamentous phage biology arose from this project (Bauer and Smith, 1988; Crissman and Smith, 1984; Nelson et al., 1981; Smith, 1988; Zacher et al., 1980).

In August 1983, my wife Margie Sable started graduate school in the School of Public Health at the University of North Carolina. This was an opportunity for me to continue research in a prominent filamentous phage lab, Bob Webster's at Duke University, only 10 miles away from Margie's school. The first phage display experiment (Smith, 1985) was started in Bob's lab at the end of my sabbatical year there.

The contribution of four coworkers to phage display was highlighted in my Nobel Prize lecture (Smith, 2018), and won't be repeated here: Steve Parmley, Robert Davis, Jamie Scott (mentioned above), and Jinan Yu. But a number of other coworkers have undertaken phage-display projects that lay outside the scope of the lecture. Prominent among them were Valery Petrenko and Leslie Matthews.

Petrenko arrived as a visiting professor from Russia in 1993. The late professor Richard Perham of Cambridge University had run across his phage-display work in Novosibirsk (Minenkova et al., 1993a; Petrenko et al., 1991), and mediated an invitation to a Banbury Center conference in April 1992 (Minenkova et al., 1993b). Petrenko introduced an entirely new line of research in my lab: fashioning innovative "new materials" by engineering the major coat protein of filamentous virions (Petrenko and Smith, 2000; Petrenko et al., 1996; Petrenko et al., 2002); he has continued this research after moving to Auburn University as a professor in 2000. Petrenko and I have also collaborated on reviews and more conventional phage-display projects (Kouzmitcheva et al., 2001; Petrenko and Smith, 2005; Smith and Petrenko, 1997; Smith et al., 1998). Petrenko and his wife Natasha Petrenko have become friends of my family.

Leslie Matthews came to the lab as a doctoral student and stayed on as a postdoc. She and another postdoc, Melissa Nevils, were the lead scientists in the "epitope discovery" (ED) project that was my lab's main research initiative for about five years starting in 1999. ED's goal was to use phage display as a new gateway to discovery of promising synthetic vaccine candidates, especially for difficult diseases like malaria. After an auspicious start (Matthews et al., 2002), the project suffered a severe setback when Matthews's and Nevils's demonstration project with a malaria-like model, babesiosis of cattle, was not accepted for publication. Despite multiple attempts, I was unable to secure funding for applying the ED concept to human malaria.

My closest colleague in the Division of Biological Sciences has been Miriam Golomb. She is a gifted science writer who was one of the chief architects of the university's Campus Writing Program – an endeavor to which she recruited Matthews, also a gifted writer. I myself taught many writing intensive courses. Golomb embraces an inspiring approach to undergraduate teaching that emphasizes strong student engagement, including regular extended conferences with students and lab courses with hands-on experiments that she's constantly updating – recently adding a yeast CRISPR module, for instance.

For the final six years of my university career, my chief endeavor was a teaching initiative called Mathematics in Life Sciences (MLS), directed by math professor Dix Pettey. The program included a campus learning community called a freshman interest group for which I was faculty co-facilitator; and an alternative beginning biology lab that integrated elementary mathematics more intimately into the curriculum. Golomb and I developed the first MLS lab in Fall 2009, and I took charge of the Fall lab for the succeeding five years. The program was funded by the National Science Foundation for its first five years. Two of its lab modules have been published in a biology education journal (Smith, 2017; Smith et al., 2015).

I have had a long-standing philosophical and technical interest in probability theory as the fundamental guidebook for making rational scientific judgments about the world in light of the available evidence. This viewpoint goes by the name Bayesian statistics or philosophy because of the central role of Bayes's theorem as the rule for updating our assessment of contending theories in the light of new evidence. Bayesian principles were included in a presentation I gave at a university genetics symposium in 1978 (Smith, 1978a), and in an article I co-authored with Hans Lehrach and his coworkers in Germany using Bitnet to exchange drafts (Michiels et al., 1987). I continue to talk and write informally about the subject.

My brother Mark and sister-in-law Lois Honeycutt also live in Columbia, where they're history professors at the University of Missouri. My wife Margie was director of the School of Social Work until her retirement in 2016. Nepotism evidently isn't a thing of the past at Mizzou.

Margie is Jewish, and since our marriage (next section) I've become increasingly engaged in Jewish culture. I'm not technically a Jew, however; that's because for a non-believer like me, religious conversion would be dishonest. As I learned more and more about Zionism, I came to understand it as a great ongoing injustice against the Palestinian people, as well as a threat to Israeli Jews and to the wider Jewish society that I value as one of my most important adoptive communities. I helped organize Mid-Missourians for Justice in Palestine, a community organization that supports the global boycott, divestment, and sanctions (BDS) campaign against Israel until it ends its regime of subjugation and dispossession.

TANKSUIT VERY MUCH

I met Margie in October 1979 during a faculty/staff swim at the university's natatorium. She was wearing a green tank suit, which lives on in our basement as an affectionate memento of our encounter. Soon after our romance began, she sent me a card with the title of this section as the printed message inside. We married October 10, 1981.

Our older son Alex Sable-Smith was born July 15, 1985 – the result of many plane trips between Columbia and Chapel Hill, where Margic had stayed on to continue her doctorate in public health after my sabbatical leave ended. I was the only solo male student in a child-birthing class in Columbia. Alex's birth was a personal watershed, as fatherhood and family life in general grew to occupy at least as central a place in life as science. Alex is now a family physician, working in hospice and palliative medicine at the Veterans Administration Hospital in Palo Alto, California. Our younger son Bram Sable-Smith was born April 1, 1988. He's now a freelance journalist in Madison, Wisconsin, where he lives with his wife Emma Brown.

Figure 3. Margie and I welcome guests to our wedding party in American Legion Hall 202, October 10, 1981. It was a square dance, with no alcohol but plenty of pies. You can see the band in the background.

Figure 4. Family get-together, Christmas Day 2015. Seated: my brother Mark, Jesse Boyd (son of Matthew and Helen), our son Alex, Margie, my brother-law Matthew Boyd, and my sister Helen Boyd. Standing: Mark's wife Lois Honeycutt, our son Bram, me, and Aubrey Smith (Mark and Lois's son). Missing: Mark and Lois's other son Derek, and Matthew and Helen's daughter Casey.

Figure 5. Our family celebrates immediately after the Nobel Prize Ceremony, December 10, 2018. From left: Emma Brown (Bram's wife), our son Bram, Margie, me, and our son Alex.

REFERENCES

Bauer, M., and Smith, G.P. (1988). Filamentous phage morphogenetic signal sequence and orientation of DNA in the virion and gene-V protein complex. *Virology* **167**, 166–175.

Breinl, F., and Haurowitz, F. (1930). Chemische Untersuchungen des Präzipitates aus Hämoglobin und Antihämoglobin-Serum und Bemerkungen über die Natur der Antikörper. Hoppe-Seylers *Zeitschrift für physiologische Chemie* **192**, 45.

Crissman, J.W., and Smith, G.P. (1984). Gene-III protein of filamentous phages: evidence for a carboxyl-terminal domain with a role in morphogenesis. *Virology* **132**, 445–455.

Dintzis, H.M. (1961). Assembly of the peptide chains of hemoglobin. *Proceedings of the National Academy of Sciences of the United States of America* **47**, 247–261.

Editors (1998). In memoriam (for Edgar Haber). *Bioconjug Chem* **9**, 3.

Fischer, S.G., and Lerman, L.S. (1980). Separation of random fragments of DNA according to properties of their sequences. *Proceedings of the National Academy of Sciences of the United States of America* **77**, 4420–4424.

Fitch, W.M., and Margoliash, E. (1967). Construction of phylogenetic trees. *Science* **155**, 279–284.

Haber, E. (1964). Recovery of Antigenic Specificity after Denaturation and Complete Reduction of Disulfides in a Papain Fragment of Antibody. *Proceedings of the National Academy of Sciences of the United States of America* **52**, 1099–1106.

Hilschmann, N., and Craig, L.C. (1965). Amino acid sequence studies with Bence-Jones proteins. *Proceedings of the National Academy of Sciences of the United States of America* **53**, 1403–1409.

Knopf, P.M., Parkhouse, R.M., and Lennox, E.S. (1967). Biosynthetic units of an immunoglobulin heavy chain. *Proceedings of the National Academy of Sciences of the United States of America* **58**, 2288–2295.

Kouzmitcheva, G.A., Petrenko, V.A., and Smith, G.P. (2001). Identifying diagnostic peptides for lyme disease through epitope discovery. *Clinical and Diagnostic Laboratory Immunology* **8**, 150–160.

Landsteiner, K. (1962). *The Specificity of Serological Reactions,* Revised Edition (New York, Dover Publications, Inc.).

Lennox, E.S., Knopf, P.M., Munro, A.J., and Parkhouse, R.M. (1967). A search for biosynthetic subunits of ight and heavy chains of immunoglobulins. *Cold Spring Harb Symp Quant Biol* **32**, 249–254.

Loewy, A.G. (1996). Michael Freeman interview with Ariel Loewy (http://hdl.handle.net/10066/1633).

Matthews, L.J., Davis, R., and Smith, G.P. (2002). Immunogenically fit subunit vaccine components via epitope discovery from natural peptide libraries. *J Immunol* **169**, 837–846.

Michiels, F., Craig, A.G., Zehetner, G., Smith, G.P., and Lehrach, H. (1987). Molecular approaches to genome analysis: A strategy for the construction of ordered overlapping clone libraries. *Comput Appl Biosci* **3**, 203–210.

Minenkova, O.O., Il'ichev, A.A., Kishchenko, G.P., Il'icheva, T.N., Khripin Iu, L., Oreshkova, S.F., and Petrenko, V.A. (1993a). [Preparation of a specific immunogen based on bacteriophage M13]. *Molekuliarnaia Biologiia* **27**, 561–568.

Minenkova, O.O., Ilyichev, A.A., Kishchenko, G.P., and Petrenko, V.A. (1993b). Design of specific immunogens using filamentous phage as the carrier. *Gene* **128**, 85–88.

Nelson, F.K., Friedman, S.M., and Smith, G.P. (1981). Filamentous phage DNA cloning vectors: a noninfective mutant with a nonpolar deletion in gene III. *Virology* **108**, 338–350.

Nirenberg, M.W., and Matthaei, J.H. (1961). The dependence of cell-free protein synthesis in E. coli upon naturally occurring or synthetic polyribonucleotides. *Proceedings of the National Academy of Sciences of the United States of America* **47**, 1588–1602.

Pauling, L. (1940). A theory of the structure and process of formation of antibodies. *J American Chem Society* **62**, 2643–2657.

Petrenko, V.A., Kuzmicheva, G.A., Tatkov, S.I., Karpenko, L.I., and Ilyichev, A.A. (1991). Mutagenic properties of oligonucleotides with modified sugar-phosphate backbone. *Nucleic Acids Symp Ser*, 213–214.

Petrenko, V.A., and Smith, G.P. (2000). Phages from landscape libraries as substitute antibodies. *Protein Engineering* **13**, 589–592.

Petrenko, V.A., and Smith, G.P. (2005). Vectors and modes of display. In *Phage Display in Biotechnology and Drug Discovery*, S.S. Siddhu, ed. (New York, CRC Press), pp. 63–110.

Petrenko, V.A., Smith, G.P., Gong, X., and Quinn, T. (1996). A library of organic landscapes on filamentous phage. *Protein Engineering* **9**, 797–801.

Petrenko, V.A., Smith, G.P., Mazooji, M.M., and Quinn, T. (2002). Alpha-helically constrained phage display library. *Protein Engineering* **15**, 943–950.

Rose, S.M., Kuehl, W.M., and Smith, G.P. (1977). Cloned MPC 11 myeloma cells express two kappa genes: A gene for a complete light chain and a gene for a constant region polypeptide. *Cell* **12**, 453–462.

Scott, J.K., Smith, G.P., Lerman, L.S., and Gerlach, J. (1985). Cataloging germline immunoglobulin V-lambda genes by direct analysis of cellular DNA. Paper presented at: *Advances in Gene Technology, Vol II: The Molecular Biology of the Immune System* (Cambridge University Press).

Sikder, S.K., Kabat, E.A., and Morrison, S.L. (1985). Alternative splicing patterns in an aberrantly rearranged immunoglobulin kappa-light-chain gene. *Proceedings of the National Academy of Sciences of the United States of America* **82**, 4045–4049.

Smith, G.P. (1973a). Mouse immunoglobulin kappa chain MPC 11: Extra amino-terminal residues. *Science* **181**, 941–943.

Smith, G.P. (1973b). Unequal crossover and the evolution of multigene families. *Cold Spring Harb Symp Quant Biol* **38**, 507–513.

Smith, G.P. (1973c). *The Variation and Adaptive Expression of Antibodies* (Cambridge, MA, Harvard University Press).

Smith, G.P. (1976). Unequal crossover and the evolution of repeated DNA sequences. *Science* **191**, 528–535.

Smith, G.P. (1978a). Non-Darwinian evolution and the beard of life. *Stadler Symp* **10**, 105–110.

Smith, G.P. (1978b). Sequence of the full-length immunoglobulin kappa-chain of mouse myeloma MPC 11. *The Biochemical journal* **171**, 337–347.

Smith, G.P. (1985). Filamentous fusion phage: Novel expression vectors that display cloned antigens on the virion surface. *Science* **228**, 1315–1317.

Smith, G.P. (1988). Filamentous phage assembly: Morphogenetically defective mutants that do not kill the host. *Virology* **167**, 156–165.

Smith, G.P. (2017). Understanding Reversible Molecular Binding. *The American Biology Teacher* **79**, 746–752.

Smith, G.P. (2018). Nobel lecture in Chemistry, 2018: Phage display: Simple evolution in a Petri dish (https://www.nobelprize.org/prizes/chemistry/2018/smith/lecture/) (Stockholm, Sweden, Nobel Foundation).
Smith, G.P., Golomb, M., Billstein, S.K., and Montgomery Smith, S. (2015). The Luria-Delbrück Fluctuation Test as a Classroom Investigation in Darwinian Evolution. *The American Biology Teacher* **77,** 614–619.
Smith, G.P., Hood, L., and Fitch, W.M. (1971). Antibody diversity. *Annual Review of Biochemistry* **40**, 969–1021.
Smith, G.P., and Petrenko, V.A. (1997). Phage Display. *Chemical Reviews* **97**, 391–410.
Smith, G.P., Petrenko, V.A., and Matthews, L.J. (1998). Cross-linked filamentous phage as an affinity matrix. *J Immunol Methods* **215**, 151–161.
Smithies, O. (2007). Biographical (https://www.nobelprize.org/prizes/medicine/2007/smithies/biographical/) (Stockholm, Sweden, Nobel Foundation).
Zacher, A.N.I., Stock, C.A., Golden, J.W.I., and Smith, G.P. (1980). A new filamentous phage cloning vector: fd-tet. *Gene*, 127–140.

Phage Display: Simple Evolution in a Petri Dish

Nobel Lecture, December 8, 2018 by
George P. Smith
University of Missouri, Columbia, USA.

INTRODUCTION

In 1997, Valery Petrenko and I published a perspective on phage display in *Chemical Reviews* (Smith and Petrenko, 1997) that adumbrated the over-arching "directed evolution" theme of the 2018 Nobel Prize in Chemistry. "Imagine, then, the applied chemist, not as designer of molecules with a particular purpose, but rather as custodian of a highly diverse population of chemicals evolving *in vitro* as if they were organisms subject to natural selection," we wrote. "A chemical's 'fitness' in this artificial biosphere would be imposed by the custodian for his or her own ends. For instance, the population might be culled periodically of individuals who fail to bind tightly to some biological receptor; the population would then evolve toward specific ligands for that receptor ... Progress toward the custodian's chosen goal would in a sense be 'automatic': once appropriate selection conditions are devised, no plan for how the system is to meet the demands of selection need be specified. And if the chemical population is sufficiently diverse, perhaps this 'blind' process will outperform rational design. The custodian may not comprehend, even in retrospect, how the products of selection work, just as biologists have only the sketchiest understanding of how a fruitfly functions."

Unlike the 1997 perspective, the present essay, like my oral presentation (Smith, 2018), will not attempt to position phage display within grand themes in the history of scientific ideas. I will try instead to reconstruct the story of the phage-display idea as I personally experienced it. As you will see, it is not a tale of heroic flashes of insight. Rather, it is a case study in how a scientific advance emerges gradually in incremental steps within overlapping global scientific communities. To foster science, to enjoy its many material and cultural benefits, I believe our society must sustain whole scientific communities like the ones from which phage display emerged, not try to identify a small cadre of individuals who are somehow specially destined to make breakthrough discoveries or innovations – or to be awarded Nobel prizes.

A PHAGE PRIMER

Phages are viruses that infect bacteria. They are particularly convenient research subjects, both because they can be propagated to enormous numbers inexpensively in microbiological cultures, and because the virus particles, called *virions*, are sturdy structures that can be readily freed of most impurities and manipulated in the laboratory.

The virion has an outer shell, called the *capsid*, composed of a geometric array of phage coat proteins. Inside the capsid is the phage chromosome, whose genes encode the phage proteins (including the coat proteins). The capsid serves not only to protect the chromosome, but also to mediate the infection process, whereby the virion attaches to an uninfected bacterium and the phage chromosome is transferred into the cell's cytoplasm. Once inside the cell, the phage genes reprogram the cell's machinery to make progeny virions, which are ultimately released from the cell and can go on to infect hundreds of uninfected cells to initiate another round of infection.

Different classes of phages infect different bacterial species and have diverse virion sizes and shapes. Filamentous phages, the particular class chosen for phage display, infect the enteric (gut) bacterium *Escherichia* coli. The virions are long and thin, as seen in the electron micrograph in Figure 1; wildtype virions are about 1 μm long and 6 nm across. The capsid consists of a long tubular array of thousands of major coat protein subunits. One end of the tube is capped by 5 copies each of two minor coat proteins; the opposite end is capped by 5 copies each of two other minor coat proteins.

One of the minor coat proteins, pIII, is the subject of the work described in this essay. Its surface-exposed portion is sometimes visible as irregular knobs at one tip of the virion, as explained in the next section. Inside the capsid is the phage DNA chromosome. At the right in Figure 1 is a schematic representation of the virion. The phage chromosome is represented by the black links inside. Coat protein gene III is

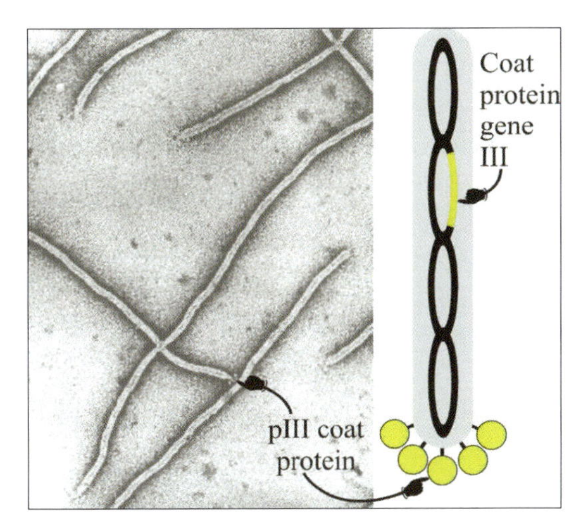

Figure 1. Electron micrograph of filamentous virions (left) and the schematic representation that will be used in this essay (right). The black links represent the phage chromosome; the only phage gene represented, by the yellow segment, is gene III, encoding phage coat protein pIII. Five copies of pIII cap one tip of the virion, with surface-exposed portions that are visible as irregular knobs at the tip of one particle in the electron micrograph (left-hand pointer), and that are represented by yellow circles in the schematic (right-hand pointer).

represented as a yellow segment in the chromosome; the other phage genes aren't represented. Similarly, only the surface-exposed part of pIII is explicitly represented as 5 yellow circles at one tip of the virion.

The infection process is extraordinarily efficient; when virions are mixed with a large excess of bacterial cells, the number of successfully infected cells is typically at least half the number of virions (infectivity ≥ 50 percent). It's also possible to artificially introduce "naked" phage DNA (stripped of its protein capsid) into bacterial cells, a procedure called *transfection*. Artificial introduction of the phage chromosome into the cell via transfection initiates the same infection program as does natural infection. Transfection is extremely inefficient (transfectivity many orders of magnitude less than 1 percent), but that inefficiency can be compensated by using huge numbers of input DNA molecules. Transfection is a staple of recombinant DNA technology, which involves manipulating naked DNA molecules in vitro before introducing them into cells.

FILAMENTOUS PHAGE DISPLAY IS BORN

From July 1983 to August 1984, I went on sabbatical to Bob Webster's lab in the Department of Biochemistry at Duke University. He was an accomplished investigator of filamentous phage biology, while I was a newcomer to the field. Much of my phage work before (Crissman and Smith, 1984; Nelson et al., 1981) and during the sabbatical was focused on coat protein pIII. The five copies of this protein were known to include an N-terminal domain that is exposed on the surface at one tip of the virion, and that is sometimes visible in electron micrographs as irregular knobs loosely connected to the bulk of the virion by very thin (usually invisible) threads (Armstrong et al.,

1981; Gray et al., 1981); the pointers in Figure 1 point to such knobs in the micrograph and in the accompanying schematic diagram.

It occurred to me that it might be possible to genetically fuse all or part of a foreign protein to the exposed parts of pIII without greatly impairing pIII's function in the phage infection cycle. If so, the foreign amino acids would be displayed at the tip of the virion, where they'd be accessible to macromolecules such as antibodies and receptors. This was hardly a great leap of imagination. Genetically fusing all or parts of two genes to make a recombinant gene encoding a recombinant "fusion protein" was a commonplace of recombinant DNA technology at the time (and today); choosing a phage coat protein as one of the fusion partners did not seem to stray very far from a well-traveled path.

Paul Modrich's lab down the hall from Webster's lab had three resources I could use to pursue my project: a purified protein called *Eco*RI, which he was studying at the time; the gene encoding the protein; and antibody against the protein. Figure 2 shows schematically how I used the *Eco*RI gene to produce the first filamentous phage-display construct. First, a DNA fragment from the *Eco*RI gene was inserted into the gene-III segment of naked phage DNA *in vitro*, resulting in a modified phage DNA with a recombinant gene III encoding an *Eco*RI-pIII fusion protein; foreign DNA fragments inserted into a chromosome in this way are called *inserts*. Second, the modified DNA was transfected into bacterial cells, which were then cultured in the hope that they would release modified virions bearing the fusion protein at one tip. The cells indeed released virions, which turned out to be infective – the first validation of my plan. I will call the modified virions "test phages."

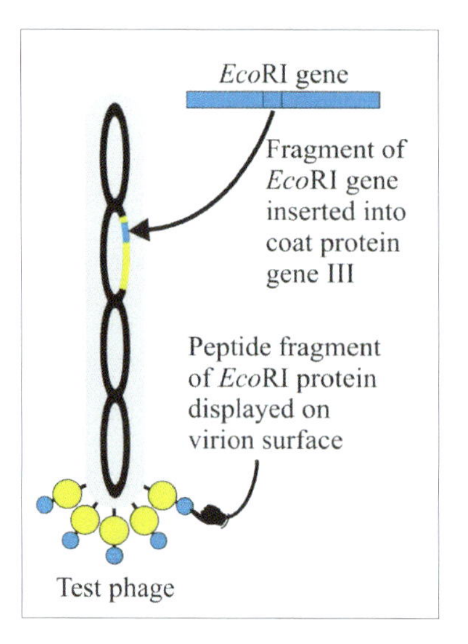

Figure 2. Construction of test phage. See text for details.

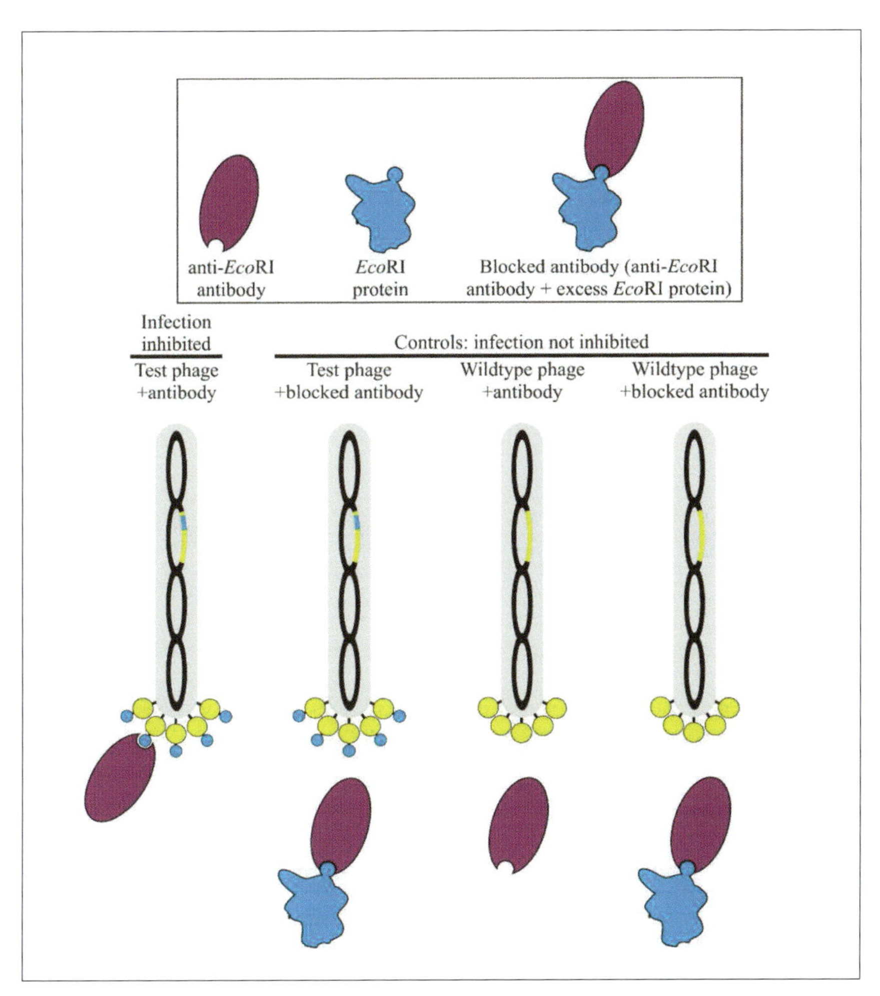

Figure 3. Proposed demonstration that EcoRI peptide was displayed on surface of test phage. See text for details.

According to my scheme, a peptide fragment of the *Eco*RI protein – the amino acids encoded by the insert – should be exposed at one tip of the virion surface; that fragment is symbolized by the blue circles in Figure 2. From now on in this essay, I'll refer to the surface-exposed amino acids encoded by a foreign DNA insert generically as a "peptide," regardless of the number of amino acids, and whether or not they constitute all or part of a natural protein. Figure 3 above shows schematically how I used Modrich's anti-*Eco*RI antibody and *Eco*RI protein to test this supposition. If the test phages were mixed with a large excess of the anti-*Eco*RI antibody (purple ovals), the antibody might well bind to the surface-exposed *Eco*RI peptide (blue circles), and thereby impair test-phage infectivity;

while the antibody should have no such effect on the infectivity of wild-type control phages. However, if the antibody was pre-treated with a large excess of the *Eco*RI protein (irregular blue shape) before being mixed with the phages, the protein would occupy the antibody's binding sites (indentations in the purple ovals), thus blocking its ability to impair test-phage infectivity. Figure 4 graphs the results of this experiment. As anticipated in Figure 3, mixing the antibody with the test phages caused a dramatic time-dependent decline in infectivity, while the antibody had no effect on the infectivity of wildtype control phages. Moreover, if the antibody was pre-reacted with excess EcoRI protein, it no longer affected infectivity of either the test phages or the wildtype control phages. These results showed that the peptide encoded by the insert was indeed exposed on the outside of the virion, where it was accessible to the antibody.

The results summarized in this section were reported in my first phage-display article, in which I called constructs like the test phage "filamentous fusion phage" (Smith, 1985).

REPLACING SCREENING WITH AFFINITY SELECTION

When I submitted my first phage-display publication, I had a specific application in mind: to replace laborious screening of small conventional phage expression libraries with easy affinity selection from enormous phage-display libraries. Explaining the difference between screening and selection will serve to illuminate the essential technological advance that phage display represents.

The signature conventional phage expression vector is λgt11 (Young and Davis, 1983), another phage that infects *E. coli* bacteria. Its chromo-

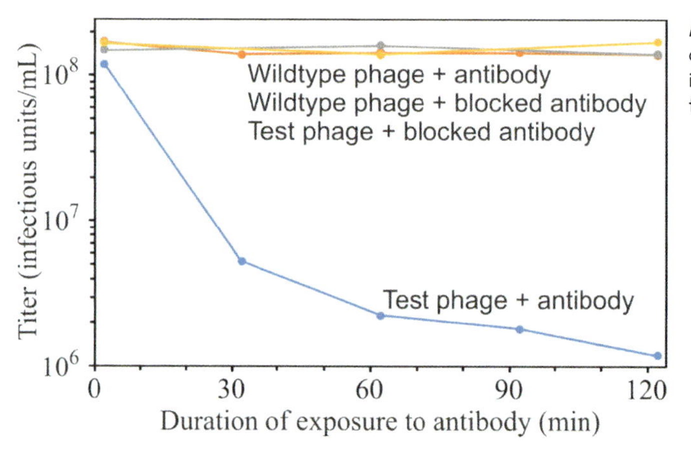

Figure 4. Results of test proposed in Figure 3. See text for details.

some has an extra, non-phage gene into which foreign DNA can be inserted. The resulting recombinant gene is a genetic fusion of the phage's extra gene and the foreign DNA insert. This recombinant gene is *expressed* in λgt11-infected cells: that is, the infected cells synthesize the fusion protein encoded by the recombinant gene. Phage-display constructs like the test phage in Figure 2 are also expression vectors, but there is a crucial difference: the fusion protein expressed in λgt11-infected cells accumulates in those cells but does not become part of the progeny virions they release.

When a limited number of λgt11 virions are mixed with a vast excess of bacteria and spread evenly over the surface of nutrient agar in a Petri dish, they form *plaques* after the dish has been incubated overnight at 37°C. This procedure is called *plating* the virions (or phages), "plate" here being a synonym for Petri dish. The plaques are visible as small circular zones of clearing in an opaque *lawn* of thick bacterial growth. Each clear zone stems from a single phage-infected cell, which releases hundreds of progeny virions after an hour or so, the cell itself being lysed in the process; the progeny virions immediately go on to infect hundreds of bacterial cells in the immediate neighborhood on the agar surface. A few rounds of infection, each increasing the number of lysed cells by a factor of hundreds, suffice to create a zone of cell lysis large enough to be visible as a clear plaque in the thick bacterial lawn. In its ability to form visible plaques, λgt11 resembles its wildtype ancestor λ as well as countless other classes of phage, including filamentous phages (Figure 5). All the virions in a plaque stem from a single infected cell, thus from a single progenitor virion; they therefore constitute a phage *clone.*

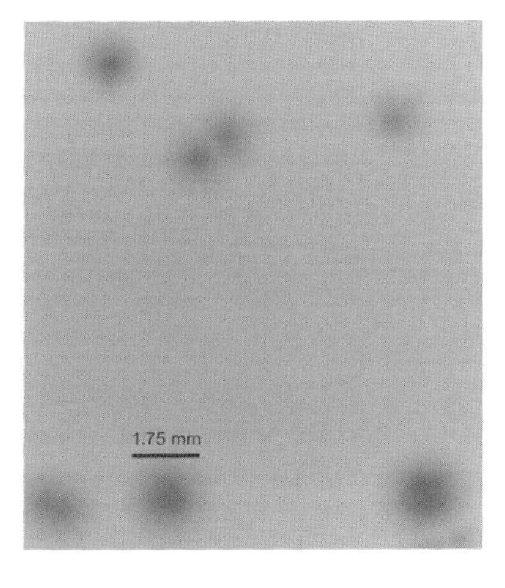

Figure 5. Plaques formed by wildtype filamentous phages. All the virions in one plaque constitute a clone. Unlike λ phages, filamentous phages don't kill the host cell, but they do slow down cell growth; their plaques are due to slowed growth, not cell lysis.

If a large collection of diverse DNAs – for example, DNAs representing all or most of the protein-encoding genes in an organism – are inserted *en masse* into λgt11 phage DNA, the result is a highly diverse collection of phage chromosomes called an *expression library*. The members of the library are identical apart from the foreign DNA insert portions of their recombinant genes. If the DNA library is transfected *en masse* into bacterial cells, and those cells are then shaken overnight at 37°C in a culture flask, the virions released by the transfected (and subsequently infected) cells are also collectively called a library. At the virion stage, a library's members are clones, not individual DNAs; each clone is represented by hundreds of thousands or millions of identical virions stemming from a single successfully transfected cell. A very large λgt11 library would comprise up to 10 million different clones, representing up to 10 million different recombinant genes with different foreign DNA inserts.

"Library" is in a sense a misnomer for such a collection of 10 million phage "books," each represented by hundreds of thousands of identical phage copies. There is no analog of a card catalog, and all copies of all "books" are jumbled together indiscriminately in a single container. Nevertheless, if a researcher has available an antibody specific for a particular protein of interest, that antibody can be used as a *probe* to *screen* the library for rare phage clones expressing fusion proteins whose foreign fusion partner derives from the protein of interest. Those rare phage clones are the *targets* of the screening.

Screening a λgt11 expression library is demanding of both labor and technical skill (Mierendorf et al., 1987). First, library virions are plated on a large Petri dish at very high density (about 50,000 plaques per 150-mm dish); the resulting plaques are exceedingly crowded, unlike the plaques in Figure 5. Second, once plaques have begun to develop, a special kind of membrane disc is carefully laid over the agar surface and plaques are allowed develop for a few more hours; during this development, fusion protein molecules from each plaque become irreversibly immobilized on the immediately apposing surface of membrane. Third, the membrane and the agar surface are marked so that they can be realigned after being separated, and the membrane is carefully lifted off the agar surface so as not to disturb the plaques; at this stage, the membrane is called a *plaque lift*; the Petri dish is stored temporarily in the refrigerator. Fourth, the plaque lift is incubated with the antibody probe, allowing the antibody to bind specifically to fusion proteins from target plaques; unbound antibody molecules are washed off the lift, which is treated in several steps with reagents that allow the bound antibodies to be detected as small purple spots on the lift surface; each of these spots corresponds to a presumptive target plaque on the stored Petri dish. Fifth, the lift is realigned with the Petri dish and used to localize a small circular area of the agar containing the presumptive target plaque (plus a few dozen non-target plaques). This circular area is excised as an agar plug, which is immersed in sterile medium and shaken vigorously to allow virions from the plaques to diffuse into the liquid. Sixth, the virions in the liquid are plated at much lower density – a hundred or so plaques per Petri dish – and the previous four steps are repeated, this time allowing presumptive target virions to be prepared in clonally pure form. The cloned virions are the gateway to further research on the target gene such as DNA sequencing. The scale of a screening project depends on the abundance of the target clones in the original library; a large project might involve a million original plaques on 20 Petri dishes, and even that may be too few for success.

Affinity selection from a phage-display library is in a sense the reverse of screening a conventional phage expression library with an antibody probe. In screening, the antibody probe in solution is used to bind specifically to hundreds of thousands of immobilized fusion proteins on the plaque lift. In affinity selection, which will be detailed in the next section, it's the antibody probe (or other specific binding protein) that's immobilized, and the billions of fusion proteins that are in solution. The exceedingly rare target fusion proteins bind specifically to the immobilized probe, and are thus captured on the immobilizing surface while all the unbound fusion proteins – the unwanted fusion proteins that aren't targets – are washed away. The captured target fusion proteins are then released from the immobilized probe and analyzed. The target fusion proteins are obtained in far too small yields for direct analysis. But that doesn't matter, because each target fusion protein is a coat protein that's attached to the virion whose recombinant coat-protein gene encodes it. In other words, the immobilized probe doesn't capture free-standing target fusion proteins; it captures whole infectious virions displaying the target fusion proteins on the virion surface. Those virions can be propagated to any desired scale simply by infecting fresh bacterial cells.

It's the physical linkage between each fusion protein and the infectious virion that encodes it that is the essential advance of phage display. It's that linkage that permits a vast increase in the number of phage clones that can be effectively surveyed. It's not necessary to prepare clones as separate (if very crowded) plaques on Petri dishes – a requirement that severely limits the number that can be screened. A probe immobilized on a solid support, such as the surface of a small empty plastic Petri dish, can readily be exposed to, say, ten trillion virions representing 100 billion clones. Even if only a handful of those 100 billion starting clones display a target fusion protein, a few rounds of affinity selection should suffice to clone them. Moreover, each round of affinity selection is far easier and less technically demanding than the screening procedure outlined above.

Although the physical linkage explained in the previous paragraph continues to be understood as the essence of phage display, my vision of the technology's field of application now seems very parochial indeed. Not until 1988 were my horizons greatly expanded, as I'll describe below.

AFFINITY SELECTION

As I returned to Missouri at the end of my sabbatical in August 1984, I resolved to refocus my research program on the phage-display idea. Three practical issues needed to be addressed: development of an efficient phage-display vector – one that would make it easy to insert foreign DNAs into the coat protein gene without unduly impairing phage function; development of an effective procedure for affinity selection; and development of methods for creating huge phage-display libraries, so that the potential of affinity selection could be fully exploited. The first two tasks were the dissertation project taken on by Steve Parmley, a graduate student who came to my lab at that time; his results were published three years later (Parmley and Smith, 1988). The fUSE vector system he introduced continues to be used today, especially vector fUSE5 (GeneBank Accession AF218364). The subject of this section is the affinity selection process introduced in my 1985 article and greatly improved in Parmley's dissertation project.

Affinity selection was closely modeled on "panning" (Wysocki and Sato, 1978), a staple of cellular immunology research of which I was well aware at the time (I didn't think to acknowledge this debt in any of my previous publications). In immunological panning, it is cells rather than virions that are captured by immobilized antibodies, which bind specifically to antigens on the surface of the cells. At first, we ourselves used the term "panning" (or "biopanning" if the super-strong biotin-streptavidin interaction was involved) before settling on the more suitably generic term affinity selection.

Parmley's affinity selection experiments closely followed the cellular immunology model. Thus the capturing molecules were antibodies, which were immobilized by non-specific adsorption to the polystyrene surface of Petri dishes or microplates. Since then, many different types of molecules, which we call generically *selectors*, immobilized on a variety of substrates by various types of bonding, have served to capture target virions.

Figure 6 below diagrams schematically the steps in a round of affinity selection. *Step 1*: Selector is immobilized by attachment to an immobilizing substrate, such as the surface of a polystyrene Petri dish or microplate well. Stringency can be adjusted by changing the density of immobilized selector on the substrate surface, as explained later in this section. *Step 2*: Input virions are added, so that virions whose displayed peptides bind the

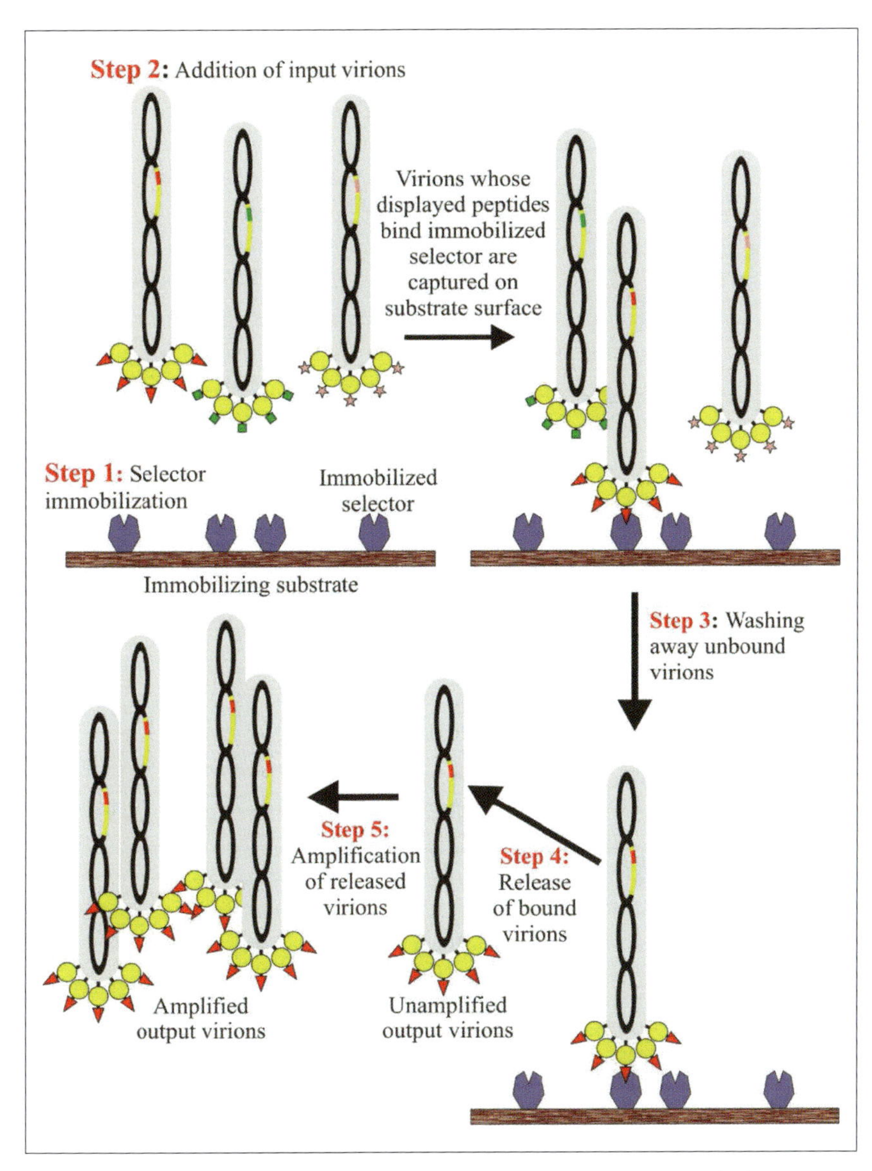

Figure 6. Schematic diagram of affinity selection. Steps will be detailed in the text.

receptor are captured on the substrate surface. The input to the first round of affinity selection is the starting unselected library; this input might consist of up to 10–100 trillion virions representing up to 100 billion clones. The inputs to each subsequent round of selection are the amplified output virions (Step 5 below) from the previous round of selection. *Step 3*: The substrate surface is thoroughly washed to remove virions that are not captured by immobilized selector. *Step 4*: Captured virions that remain after washing are released by weakening the bond between the selector and the displayed peptide; alternatively (not shown; see next paragraph), they may be released by cleaving the segment of amino acids linking the displayed peptide to the rest of the virion. The released virions are the unamplified output of selection. *Step 5*: The unamplified output virions are "amplified" by infecting fresh bacterial cells and shaking the cells overnight in culture medium. Each virion in an unamplified output gives rise to millions or billions of identical progeny virions in the amplified output. Amplified output virions are the input virions for the next round of affinity selection. Virions from either unamplified or amplified outputs can be cloned so that individual displayed peptides can be sequenced (by sequencing the DNA inserts that encode them) and analyzed in countless other ways.

Releasing captured virions from the immobilized selector is a key step in affinity selection (step 4 in Figure 6). At first this was accomplished by elution in acid, which filamentous phages tolerate without losing infectivity. That was another methodological choice arising from my immunological background: acidic elution buffer had long been used in immunoaffinity chromatography to release antibodies from immobilized antigens or antigens from immobilized antibodies. Acid greatly weakens the interaction between antibodies and most antigens, generally without irreversibly inactivating either binding partner. However, many practitioners have come to appreciate the advantages of releasing phages using the intestinal protease trypsin to cleave a bond between the displayed insert-encoded peptide and the remainder of the virion, as summarized by my graduate student Will Thomas and me in a critical review (Thomas and Smith, 2010). Filamentous virions have been known since the 1960s to be extremely resistant to trypsin and other intestinal proteases (Salivar et al., 1964), as expected for a phage that infects enteric bacteria like *E. coli.* Thomas and I added the trypsin-release phage-display vector f3TR1 (GeneBank Accession HM355479) to the fUSE family.

Affinity selection is not perfectly discriminatory; the yield of a binding clone might be only 10,000 times higher than the yield of each non-binding clone. If binding clones comprise fewer than one in a million clones in the input virions, therefore, they may well not be represented in the small sample of a hundred or so cloned output virions that are typically subject to further analysis. Consequently, one or two additional rounds of affinity selection are normally carried out, the input virions for each round being amplified output virions from the previous round. Only then are individual cloned output virions analyzed. More recently, next-gen sequencing (NGS) makes it possible to sequence the DNA inserts of many millions of phages, allowing identification of particularly abundant clones in early-round outputs, even when they comprise far fewer than 1 percent of the total ('t Hoen et al., 2012).

Quite apart from imperfect discrimination, any practical affinity-selection strategy must confront the possibility of "target-unrelated phages" (TUPs) – that is, phage clones that are favored for reasons other than affinity of their displayed peptide for the immobilized selector (unlike elsewhere in this essay, the "target" referred to in the definition of TUP is a synonym for the selector rather than displayed peptides that bind the selector). Thomas and I, along with our colleague Miriam Golomb, have published a critical review of the TUP problem (Thomas et al., 2010). The ability to distinguish TUPs from genuine selector-binding clones is an additional attraction of NGS as the read-out for affinity selections ('t Hoen et al., 2012).

Stringency is a key parameter of affinity-selection strategy. It refers to the degree to which selection favors virions displaying high-affinity peptides over virions displaying low-affinity peptides. Stringency can be increased in several ways, such as decreasing the density of selectors on the immobilizing substrate (Step 1 in Figure 6). High stringency must almost always be purchased at the cost of reduced *yield* – the percent of binding virions in the input that are successfully recovered in the output. In the first round of affinity selection, yield is all-important: in a very large input library, each clone (thus each displayed peptide) is represented by a limited number of virions in the input, and any clone that is lost in this round cannot be recovered in subsequent rounds. In the output of the first round of selection, thus in the input to the second round, the number of clones represented has been reduced by many orders of magnitude compared to the unselected library, and each clone is therefore represented by orders of magnitude more virions. Only in the second and subsequent rounds, therefore, does it make sense to increase stringency in order to favor virions displaying high-affinity peptides or proteins.

1988 – A YEAR OF DISCOVERY

The title of this section is from a preface I wrote to a 1996 phage-display laboratory manual (Smith, 1996). There I described three encounters – two personal and one with a published article – that greatly expanded my vision for the phage-display idea in 1988.

The first encounter occurred in January, when I visited Vidal de la Cruz and Tom McCutchan in the Laboratory of Parasitic Diseases at the National Institutes of Health in Bethesda, Maryland. Following my 1985 article, they had constructed phages displaying antigens from the malaria parasite *Plasmodium falciparum* (de la Cruz et al., 1988). In the course our conversation, McCutchan casually remarked "you know, you could use phage to out-Geysen Geysen." Mario Geysen, I learned, had developed methods for chemically synthesizing hundreds of short peptides simultaneously, and a clever plan for using this technology to delineate the peptide *epitope* recognized by an antibody, without the need for advance knowledge of the antibody's specificity (Geysen et al., 1986a, b); an epitope is the part of an antigen that makes close contact with an antibody's binding site. Geysen couldn't accomplish this task by testing all possible short peptides individually: even examining all possible hexapeptides would have required him to synthesize and test 64 million compounds. Instead, he tested 400 peptide mixtures, each mixture having specified amino acids at two positions and an equal mixture of all the amino acids at the other positions. A position with a mixture of amino acids is said to be *degenerate*, and a peptide mixture with degenerate positions is called a degenerate peptide. The degenerate peptide showing the best binding then serves as the starting-point for synthesis of a second series of 400 new degenerate peptides, this time with the optimal amino acids at the first two non-degenerate positions and 400 combinations of specified amino acids at two additional positions (thus reducing the number of degenerate positions by two). In this way, he could in theory progressively narrow the antibody's binding specificity to a single peptide sequence.

It was obvious to all three of us – de la Cruz, McCutchan, and me – that phage display might well allow 64 million (or many more) peptides to be surveyed individually for affinity for an antibody – not by testing them one by one, but by using immobilized antibody to select virions displaying binding peptides from a phage-display library displaying tens of millions (or many more) random peptides altogether. Furthermore, synthetic DNA inserts encoding tens of millions (or many more) random peptides could be purchased cheaply from chemical DNA synthesis companies. The companies wouldn't have to synthesize them one by one. Instead, they'd synthesize them all at once in the form of a degenerate oligonucleotide. Just as the synthetic degenerate peptides in the previous paragraph had mixtures of amino acids at various positions in the peptides, so the degenerate oligonucleotide would have a mixture of nucleotides at each position in the peptide coding sequence. Such degenerate oligonucleotides were already in widespread use in molecular biology.

Construction of a random peptide library would be easy. We'd start with a synthetic DNA insert in which each three-position codon would have the degenerate sequence NNK, where N stands for an equal mixture of all four nucleotides and K stands for an equal mixture of G and T. An NNK degenerate codon is thus an equal mixture of 32 codons, including the TAG stop (nonsense) codon and 31 sense codons that together include codons for all 20 amino acids. An insert with six such codons would comprise 32^6 = 1.074 billion sequences altogether, including 31^6 = 888 million sequences encoding all possible 64 million six-amino-acid peptides without stop codons. Random peptide libraries, which we called "epitope libraries" at the time, seemed a much more promising application of phage display than libraries of natural proteins or protein fragments. The epitope library concept was included at the last minute in the 1988 article (Parmley and Smith, 1988); and soon became the main focus of my research program, as I'll described in the next section.

The second encounter, in October 1988, was with an article reporting the construction of functional single-chain antibodies (Bird et al., 1988). I was electrified by this report. If 240-amino acid single-chain antibodies could be displayed on the virion surface (and the prospects seemed reasonably good), immobilized antigens could be used to affinity-select specific antibodies from synthetic phage-antibody libraries, completely bypassing the natural immune system. I soon learned that the same vision occurred independently to a number of other investigators. Indeed, Robert Ladner of Genex Corporation had fully articulated the phage-antibody concept in a patent application a year and a half earlier, though he chose phage λ rather than filamentous phage as the prospective display vector (Ladner et al., 1987). In any case, it is others, including my co-laureate Greg Winter and his colleagues, who have succeeded in bringing this vision to a markedly successful reality.

The third encounter, in December 1988, occurred in the office of Jim Larrick at Genelabs, a biotech company in Redwood City, California. "Drugs," he said cryptically after brief greetings. Larrick's message was that using cellular receptors or other medically significant biomolecules as immobilized selectors might open up a new gateway to drug discovery. This was a dramatic expansion of my horizons, which had hitherto been largely limited to immunological concerns. That's when we came to use the generic term "selector," and when selectors other than antibodies and antigens came to figure prominently in our plans and publications.

AFFINITY SELECTION FROM RANDOM PEPTIDE LIBRARIES

Our first attempts at constructing a random peptide library were undertaken in the 1988–1989 academic year by Shannon Flynn as part of his undergraduate research project. He ran up against a severe bottleneck: inability to transfect bacterial cells with naked DNA in sufficient numbers to put the phage-display concept to the test. Transfection was a necessary step in library construction, since inserting foreign DNA into the phage coat protein gene requires manipulation of naked phage DNA. At the end of 1988, Bill Dower and his colleagues reported a practical method for very large-scale transfection of bacterial cells (Dower et al., 1988), and work on large phage-display libraries began in earnest. That's when two key colleagues came to my lab: Robert Davis and Jamie Scott.

Davis served as my lab manager and chief technician for over two decades, reducing to practice several of the large-scale routines in the lab. One of the most important routines was a radioactive DNA sequencing strategy tailored specifically to the short inserts encoding random peptides (Haas and Smith, 1993). Davis was able to sequence almost 800 phage clones per week in this way, with time left over to take on other tasks as well. We estimate that he generated well over 1 million bases of sequence altogether before we finally started outsourcing sequencing to the University's DNA core facility – at considerable increase in cost.

Meanwhile, Scott was a postdoc, who, with Davis's expert technical help, constructed vector fUSE5 and used it to create our first large random peptide library: f3-6mer (GeneBank Accession AF246446), which has 200 million clones and displays random 6-amino acid peptides (6mers). With the aid of our colleague Hannah Alexander, who had come to Missouri from Richard Lerner's lab at the Scripps Research Institute, we obtained from that lab samples of two monoclonal antibodies specific for a known peptide epitope (Fieser et al., 1987). The sequences of the 6mer peptides Scott affinity-selected from the library closely matched the antibodies' known epitope, even though knowledge of that epitope had in no way influenced construction of the library or conduct of the experiment. This was a dramatic validation of the phage-display idea, and the essential breakthrough recognized by my Nobel Prize award. Scott's results were published in *Science* in July 1990 (Scott and Smith, 1990), simultaneously with analogous results from two other labs (Cwirla et al., 1990; Devlin et al., 1990).

IN VITRO EVOLUTION OF S-PEPTIDE ANTAGONISTS BY AFFINITY MATURATION

In 1992, David Schultz and John Ladbury brought the ribonuclease S protein/S peptide system to my lab (Smith et al., 1993). In the 1950s, Fred Richards at Yale (Steitz, 2009) found that partial digestion of bovine pancreatic ribonuclease A with the protease subtilisin cleaves the ribonuclease polypeptide chain at a single position, between amino acids 20 and 21. The cleaved protein retains its enzymatic activity, but when the two fragments – the 20 amino acid S peptide and the 104 amino acid S protein – are separated, as shown in Figure 7, neither one is enzymatically active. When the two fragments are mixed, however, S-peptide binds to S-protein, and enzymatic activity is restored.

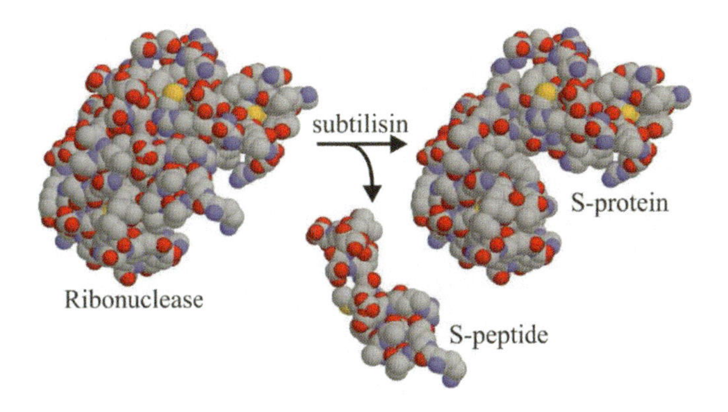

Figure 7. Cleavage of ribonuclease with subtilisin and separation of S-protein and S-peptide moieties. S-protein served as the model "receptor" and S-peptide as its natural ligand (e.g., "hormone") in Jinan Yu's affinity maturation study. Space-filling CPK models created from X-ray crystallographic coordinates in Protein Data Bank accession 1KF5.

For several years, S-protein was our selector molecule of choice in further development of phage-display technology. We presented it as a model "receptor," and S-peptide as its cognate natural ligand (e.g., "hormone"). The "physiological effect" that ensued when the "hormone" bound its "receptor" was restoration of enzyme activity. Peptides affinity-selected from a random peptide library by immobilized S-protein selector would be candidate "drugs" that agonize or antagonize the S-peptide "hormone," or in some other way modulate physiological activity. "In short," we wrote (Smith et al., 1993), "the S-protein/S-peptide model preserves the essential features of a pharmacologically significant receptor/ligand couple but is convenient for methodological development since both components can be purchased or prepared inexpensively in large quantities."

My graduate student Jinan Yu used this model system in an extensive study of *affinity maturation*, a strategy for *in vitro* evolution of high-affinity ligands from random peptide libraries (Yu and Smith, 1996). Her starting library, f3-15mer (GenBank Accession AF246445), generously provided by Japanese researchers (Nishi et al., 1996), has 250 million clones, and presumably displays the same number of different 15 amino acid random peptides (15mers). Those 15mers are an exceedingly sparse sampling of all 3.3×10^{19} possible 15mers. A library's "champion" – the displayed 15mer with the highest affinity for the S-protein selector – would almost certainly have much lower affinity than the best possible 15mer.

Affinity maturation is a strategy to overcome extreme sparseness in input random peptide libraries. It entails introducing random mutations into the DNA inserts during the amplification step (step 5 in Figure 6) of each round of selection except the last, while gradually increasing stringency in successive rounds of selection. Mutation creates a "clan" of closely-related mutant peptides from each peptide in the unamplified output, thus strategically increasing the number of peptides represented in the input to the next round of selection. Meanwhile, increasing stringency gradually, rather than immediately after the first round, opens up the possibility of discovering a "dark horse" peptide: a peptide with higher affinity than any peptide that arises through mutation of the initial champion.

Addition of ongoing mutation sharpens the parallelism between affinity selection of phage-displayed peptides and natural selection of organisms in the living world. The resulting peptide ligands can justifiably be said to have "evolved" from their peptide ancestors in the starting unselected library.

Affinity maturation is a term borrowed from immunology. In the course of the natural antibody response to an antigen, the affinity of circulating antibodies for the antigen improves. This improvement stems from hypermutation specifically targeted to the antibody genes in each B-cell clone that responds to the antigen, along with gradually increasing strength of selection in favor of clones with high-affinity antibodies as the concentration of antigen in the body steadily declines. To me, as to many other practitioners familiar with basic immunology, the extension from B-cell clones to phage-display clones was obvious. Unsurprisingly, this was especially true of researchers such as my co-laureate Greg Winter and his colleagues who were affinity selecting antibodies from phage-antibody libraries (Low et al., 1996).

Affinity maturation substantially increases the labor investment in an affinity selection project. Winter reminds us in his Nobel lecture that the need for affinity maturation (with its attendant cost in labor) stems from the limited number of clones in the initial library (Winter, 2018). He and his colleagues have now constructed phage-antibody libraries that are so large that high-affinity antibodies can be directly affinity-selected with many antigens without the need for affinity maturation.

Yu's affinity maturation experiments did not reveal any dark horse peptides, but they did reveal mutants of the library's initial champion with substantially higher affinity for the S-protein selector than the champion itself (Figure 8). All the affinity-selected peptides could be aligned with the natural ligand, S-peptide, such that the four amino acids in S-peptide that make close contact with S-protein in the S-peptide/S-protein complex are matched in the selected peptides. This is not unexpected, of course, but it's important to understand that the experiment might well have turned out differently: it might have revealed peptides that bind the S-protein selector in an entirely different way than does the natural ligand, and that don't have any discernible sequence similarity to that natural ligand.

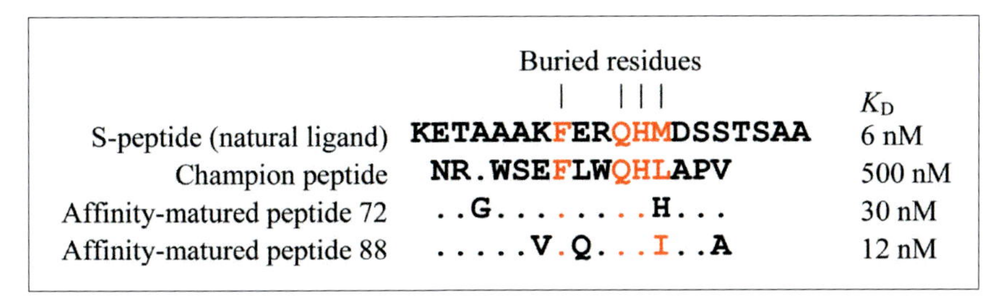

	Buried residues	K_D
S-peptide (natural ligand)	KETAAAKFERQHMDSSTSAA	6 nM
Champion peptide	NR.WSEFLWQHLAPV	500 nM
Affinity-matured peptide 72	..G........H...	30 nM
Affinity-matured peptide 88	.....V.Q...I..A	12 nM

Figure 8. Amino acid sequences using one-letter abbreviations for S-protein-binding peptides from Jinan Yu's affinity-maturation study, aligned with S-protein's natural ligand S-peptide. The "champion" peptide was affinity-selected from the f3-15mer library without mutagenesis; peptides 72 and 88, which are presumably mutant descendants of the champion peptide, were obtain through affinity maturation, as described in the text. Amino acids in peptides 72 and 88 that match those in the champion peptide are rendered as dots (·). The four amino acids in S-peptide that make close contact with S-protein in intact ribonuclease are colored red, as are the corresponding letters or dots in the affinity-selected peptides. L (leucine) and I (isoleucine) are considered matches to the M (methionine) in S-peptide, since S-peptide variants with those substitutions are still able to bind S-protein. Approximate dissociation equilibrium constants K_D for binding to S-protein are given for each peptide; these constants are inversely related to affinity.

The ligands Yu discovered have no known practical worth apart from their heuristic value. But ligands for other, pharmacologically significant selectors have been affinity-selected from the same library, including in the lab of my long-time University of Missouri colleague Sue Deutscher (Glinsky et al., 2000; Kumar et al., 2007; Landon and Deutscher, 2003; Landon et al., 2004a; Landon et al., 2004b; Larimer and Deutscher, 2014; Peletskaya et al., 1997; Peng et al., 2017; Soendergaard et al., 2014a, b; Zou et al., 2004). Indeed, is there any limit to the number of different ligands, some perhaps of great practical worth, that might evolve from the very same library by imposing different selection regimes, without the need for deep prior knowledge? Members of the phage-display community have constructed many additional libraries of innovative design, from which many valuable ligands have been selected, including the therapeutic antibodies described in Winter's Nobel lecture (Winter, 2018).

A COMMUNITY ACHIEVEMENT

The phage-display concept evolved gradually, and continues to evolve, in a phage-display community of which I'm only one of numerous members. At each step along my own lab's evolutionary path, an idea or experimental result that I and my coworkers had access to by virtue of our membership in overlapping global scientific communities – phage biology, immunology, molecular biology, evolutionary biology, others – brought us within sight of some incremental advance. The same is surely true of the other members of the community. At no point – certainly not when my first article was published (Smith, 1985) – was the significance of phage display revealed in anything like full clarity, as I have detailed in my own case for the year 1988. We did experience *eureka* moments of exultation, such as when Scott and I saw the results of our first large-scale affinity selections (Scott and Smith, 1990). But evolution of the phage-display concept certainly didn't stop there. Indeed, its qualification as a discovery or improvement meeting the standard in Alfred Nobel's Testament has been the collective achievement of the phage-display community, who have applied and modified it in countless innovative ways, very few of which I or any other individual member could possibly have foreseen. It is accordingly on behalf of that community that I have accepted this great honor.

REFERENCES

't Hoen, P.A., Jirka, S.M., Ten Broeke, B.R., Schultes, E.A., Aguilera, B., Pang, K.H., Heemskerk, H., Aartsma-Rus, A., van Ommen, G.J., and den Dunnen, J.T. (2012). Phage display screening without repetitious selection rounds. *Analytical Biochemistry* **421**, 622–631.

Armstrong, J., Perham, R.N., and Walker, J.E. (1981). Domain structure of bacteriophage fd adsorption protein. *FEBS Lett* **135**, 167–172.

Bird, R.E., Hardman, K.D., Jacobson, J.W., Johnson, S., Kaufman, B.M., Lee, S.M., Lee, T., Pope, S.H., Riordan, G.S., and Whitlow, M. (1988). Single-chain antigen-binding proteins [published erratum appears in *Science* 1989 Apr 28; **244** (4903): 409]. Science **242**, 423–426.

Crissman, J.W., and Smith, G.P. (1984). Gene-III protein of filamentous phages: evidence for a carboxyl-terminal domain with a role in morphogenesis. *Virology* **132**, 445–455.

Cwirla, S.E., Peters, E.A., Barrett, R.W., and Dower, W.J. (1990). Peptides on phage: A vast library of peptides for identifying ligands. *Proceedings of the National Academy of Sciences of the United States of America* **87**, 6378–6382.

de la Cruz, V.F., Lal, A.A., and McCutchan, T.F. (1988). Immunogenicity and epitope mapping of foreign sequences via genetically engineered filamentous phage. *Journal of Biological Chemistry* **263**, 4318–4322.

Devlin, J.J., Panganiban, L.C., and Devlin, P.E. (1990). Random peptide libraries: a source of specific protein binding molecules. *Science* **249**, 404–406.

Dower, W.J., Miller, J.F., and Ragsdale, C.W. (1988). High efficiency transformation of E. coli by high voltage electroporation. *Nucleic Acids Research* **16**, 6127–6145.

Fieser, T.M., Tainer, J.A., Geysen, H.M., Houghten, R.A., and Lerner, R.A. (1987). Influence of protein flexibility and peptide conformation on reactivity of monoclonal anti-peptide antibodies with a protein alpha-helix. *Proceedings of the National Academy of Sciences of the United States of America* **84**, 8568–8572.

Geysen, H.M., Rodda, S.J., and Mason, T.J. (1986a). The delineation of peptides able to mimic assembled epitopes. *Ciba Foundation Symposium* **119**, 130–149.

Geysen, H.M., Rodda, S.J., and Mason, T.J. (1986b). A priori delineation of a peptide which mimics a discontinuous antigenic determinant. *Molecular Immunology* **23**, 709–715.

Glinsky, V.V., Huflejt, M.E., Glinsky, G.V., Deutscher, S.L., and Quinn, T.P. (2000). Effects of Thomsen-Friedenreich antigen-specific peptide P-30 on beta-galactoside-mediated homotypic aggregation and adhesion to the endothelium of MDA-MB-435 human breast carcinoma cells. *Cancer Res* **60**, 2584–2588.

Gray, C.W., Brown, R.S., and Marvin, D.A. (1981). Adsorption complex of filamentous fd virus. *Journal of Molecular Biology* **146**, 621–627.

Haas, S.J., and Smith, G.P. (1993). Rapid sequencing of viral DNA from filamentous bacteriophage. *Biotechniques* **15**, 422–424, 426–428, 431.

Kumar, S.R., Quinn, T.P., and Deutscher, S.L. (2007). Evaluation of an 111In-radiolabeled peptide as a targeting and imaging agent for ErbB-2 receptor expressing breast carcinomas. *Clin Cancer Res* **13**, 6070–6079.

Ladner, R.C., Glick, J.L., and Bird, R.E. (1987). Method for the preparation of binding molecules (https://patentimages.storage.googleapis.com/9a/70/56/d588700bb3fc66/WO1988006630A1.pdf)

Landon, L.A., and Deutscher, S.L. (2003). Combinatorial discovery of tumor targeting peptides using phage display. Journal of cellular biochemistry 90, 509–517.

Landon, L.A., Harden, W., Illy, C., and Deutscher, S.L. (2004a). High-throughput fluorescence spectroscopic analysis of affinity of peptides displayed on bacteriophage. *Anal Biochem* **331**, 60–67.

Landon, L.A., Zou, J., and Deutscher, S.L. (2004b). Is phage display technology on target for developing peptide-based cancer drugs? *Current drug discovery technologies* **1**, 113–132.

Larimer, B.M., and Deutscher, S.L. (2014). Development of a peptide by phage display for SPECT imaging of resistance-susceptible breast cancer. *Am J Nucl Med Mol Imaging* **4**, 435–447.

Low, N.M., Holliger, P.H., and Winter, G. (1996). Mimicking somatic hypermutation: Affinity maturation of antibodies displayed on bacteriophage using a bacterial mutator strain. *Journal of Molecular Biology* **260**, 359–368.

Mierendorf, R.C., Percy, C., and Young, R.A. (1987). Gene isolation by screening lambda gt11 libraries with antibodies. *Methods in Enzymology* **152**, 458–469.

Nelson, F.K., Friedman, S.M., and Smith, G.P. (1981). Filamentous phage DNA cloning vectors: a noninfective mutant with a nonpolar deletion in gene III. *Virology* **108**, 338–350.

Nishi, T., Budde, R.J., McMurray, J.S., Obeyesekere, N.U., Safdar, N., Levin, V.A., and Saya, H. (1996). Tight-binding inhibitory sequences against pp60(c-src) identified using a random 15-amino-acid peptide library. *FEBS Lett* **399**, 237–240.

Parmley, S.F., and Smith, G.P. (1988). Antibody-selectable filamentous fd phage vectors: Affinity purification of target genes. *Gene* **73**, 305–318.

Peletskaya, E.N., Glinsky, V.V., Glinsky, G.V., Deutscher, S.L., and Quinn, T.P. (1997). Characterization of peptides that bind the tumor-associated Thomsen-Friedenreich antigen selected from bacteriophage display libraries. *J Mol Biol* **270**, 374–384.

Peng, Y., Prater, A.R., and Deutscher, S.L. (2017). Targeting aggressive prostate cancer-associated CD44v6 using phage display selected peptides. *Oncotarget* **8**, 86747–86768.

Salivar, W.O., Tzagoloff, H., and Pratt, D. (1964). Some physical-chemical and biological properties of the rod-shaped coliphage M13. *Virology* **24**, 359–371.

Scott, J.K., and Smith, G.P. (1990). Searching for peptide ligands with an epitope library. *Science* **249**, 386–390.
Smith, G.P. (1985). Filamentous fusion phage: novel expression vectors that display cloned antigens on the virion surface. *Science* **228**, 1315–1317.
Smith, G.P. (1996). 1988 – A year of discovery. *In Phage Display of Peptides and Proteins: A Laboratory Manual*, B.K. Kay, J. Winter, and J. McCafferty, eds. (New York, Academic Press), pp. xvii–xix.
Smith, G.P. (2018). Nobel lecture in Chemistry, 2018: Phage display: simple evolution in a Petri dish (https://www.nobelprize.org/prizes/chemistry/2018/smith/lecture/) (Stockholm, Sweden, Nobel Foundation).
Smith, G.P., and Petrenko, V.A. (1997). Phage Display. *Chemical Reviews* **97**, 391–410.
Smith, G.P., Schultz, D.A., and Ladbury, J.E. (1993). A ribonuclease S-peptide antagonist discovered with a bacteriophage display library. *Gene* **128**, 37–42.
Soendergaard, M., Newton-Northup, J.R., and Deutscher, S.L. (2014a). In vitro high throughput phage display selection of ovarian cancer avid phage clones for near-infrared optical imaging. *Combinatorial chemistry & high throughput screening* **17**, 859–867.
Soendergaard, M., Newton-Northup, J.R., and Deutscher, S.L. (2014b). In vivo phage display selection of an ovarian cancer targeting peptide for SPECT/CT imaging. *Am J Nucl Med Mol Imaging* **4**, 561–570.
Steitz, T.A. (2009). Retrospective. Frederic M. Richards (1925–2009). *Science* **323**, 1181.
Thomas, W., and Smith, G. (2010). The case for trypsin release of affinity-selected phages. *Biotechniques* **49**, 651–654.
Thomas, W.D., Golomb, M., and Smith, G.P. (2010). Corruption of phage display libraries by target-unrelated clones: Diagnosis and countermeasures. *Anal Biochem* **407**, 237–240.
Winter, G.P. (2018). Nobel Lecture in Chemistry, 2018: Harnessing evolution to make medicines (https://www.nobelprize.org/prizes/chemistry/2018/winter/lecture/) (Stockholm, Sweden, Nobel Foundation).
Wysocki, L.J., and Sato, V.L. (1978). "Panning" for lymphocytes: A method for cell selection. *Proceedings of the National Academy of Sciences of the United States of America* **75**, 2844–2848.
Young, R.A., and Davis, R.W. (1983). Efficient isolation of genes by using antibody probes. *Proceedings of the National Academy of Sciences of the United States of America* **80**, 1194–1198.
Yu, J., and Smith, G. (1996). Affinity maturation of phage-displayed peptide ligands. *Methods in Enzymology* **267**, 3–27.
Zou, J., Dickerson, M.T., Owen, N.K., Landon, L.A., and Deutscher, S.L. (2004). Biodistribution of filamentous phage peptide libraries in mice. *Mol Biol Rep* **31**, 121–129.

Sir Gregory P. Winter. © Nobel Prize Outreach AB. Photo: A. Mahmoud

Sir Gregory P. Winter

Biography

SUMMARY

I was born (14 April 1951) in Leicester, England but spent most of my childhood in the Gold Coast (later Ghana). The family returned to England in 1964, settling in Newcastle-upon-Tyne. There I went to the Royal Grammar School, which developed my interests in chemistry and biology and set me on a path to Cambridge University (Trinity College), a BA in Natural Sciences (1970–1973) and PhD in protein chemistry (1973–1976).

Most of my research career (1973–2012) was based at the Medical Research Council's (MRC) Laboratory of Molecular Biology (LMB), Cambridge. I undertook my PhD and postdoctoral work at the LMB. I was appointed to the MRC's scientific staff at the LMB in 1981, and became Head of the Division of Protein and Nucleic Acid Chemistry (1994–2008) and Deputy Director of the LMB (2006–2011). I was also Deputy Director of the MRC Centre for Protein Engineering (1990–2010), and held Fellowships at Trinity College, Cambridge (1976–1980; 1990–2012; 2019–). I was appointed Master of Trinity from 2012–2019.

My research developed from my interests in the chemistry and structure of proteins and nucleic acids. After a period (1973–1982) sequencing proteins and nucleic acids, I developed genetic engineering as a tool to analyse protein functions, applying it to a study of the binding and catalytic mechanism of an enzyme (1982–1988). I also developed genetic engineering strategies for the design (1984–1988) and evolution (1988–1997) of antibody pharmaceuticals suitable for treatment of cancer and immune disorders, helping to spearhead the development of antibodies as a new class of powerful biologics. Subsequently I have contributed to the development of bicyclic peptides and their conjugates as medicines. In the course of my academic research, I have worked with industry as a consultant, and also as a founder and director of two start-up companies based on antibodies, Cambridge Antibody Technology Ltd (founded in 1989, listed on the London Stock Exchange in 1997, sold to AstraZeneca

Figure 1. MRC Laboratory of Molecular Biology, 1970. Credit MRC Laboratory of Molecular Biology.

in 2006), Domantis Ltd (founded in 2000, sold to GlaxoSmithKlein in 2007), and one start-up company based on peptides, Bicycle Therapeutics Ltd (founded in 2009) and listed on Nasdaq in 2019.

ENZYMES

I started my PhD studies at the MRC Laboratory of Molecular Biology, Cambridge (LMB) in the autumn of 1973, funded by an MRC studentship and supervised by Brian Hartley, a protein chemist. With LMB colleagues

Figure 2. Brian Hartley in 1969.
© Jens Birktoft/MRC Laboratory of Molecular Biology.

David Blow (protein crystallographer) and Alan Fersht (enzyme kineticist), Brian aimed to compare the amino acid sequences, crystallographic structures and catalytic mechanisms of an ancient family of enzymes, the aminoacyl tRNA synthetases, and to gain insights into the early evolution of proteins. For my PhD, I was supposed to determine and compare the sequences of the tryptophanyl (TrpTS) and tyrosyl tRNA (TyrTS) synthetases by protein chemical methods, in the hope of identifying a sequence motif involved in catalysis. During my PhD studies, Brian was appointed to the Professorship of Biochemistry at Imperial College, and I had to follow him to London in the autumn of 1975 and help set up the protein chemical facilities. Nevertheless, I managed to complete the sequence of the TrpTS by the summer of 1976, the work forming the core of my Cambridge PhD thesis and a successful application for a Junior Research Fellowship at Trinity College.

In the meantime, David Blow had obtained good crystals of the TyrTS and needed the amino acid sequence to solve its three-dimensional structure. I postponed my return to Cambridge for over a year, trying to complete the sequence, but without success. Nevertheless, I was able to provide large segments of TyrTS sequence to David and to compare it with the TrpTS sequence. This revealed a cysteine (Cys 35) in the N-terminal region of both enzymes, a region identified by David as part of the tyrosyl adenylate binding site and hinting that Cys35 might have a role in catalysis.

After hearing a lecture from Fred Sanger on his new DNA sequencing methods, I concluded it would be easier to deduce the amino acid sequence of a protein from the DNA sequence of the corresponding gene than by direct chemical sequencing of the protein. Accordingly, I decided to learn DNA sequencing, and taking advantage of the salary provided by my Trinity Fellowship, applied to join Fred back at the LMB. In turn he passed on my application to his colleague George Brownlee.

George was planning to sequence the genome of influenza virus to help understand the mechanisms underlying influenza virus epidemics and pandemics. He had also taken on a PhD student, Stanley (Stan) Fields. We first had to grow the virus in eggs, isolate the viral RNA and then make cDNA copies. Instead of cloning each of the individual genome segments, we used an M13 shotgun strategy to sequence all the segments together, working closely with Fred's group. During this work George was appointed to the Professorship of Chemical Pathology in Oxford, but Stan and I were allowed to stay at the LMB to complete the sequence. By 1981, we had completed the sequence, and Stan left for a postdoctoral position in the USA.

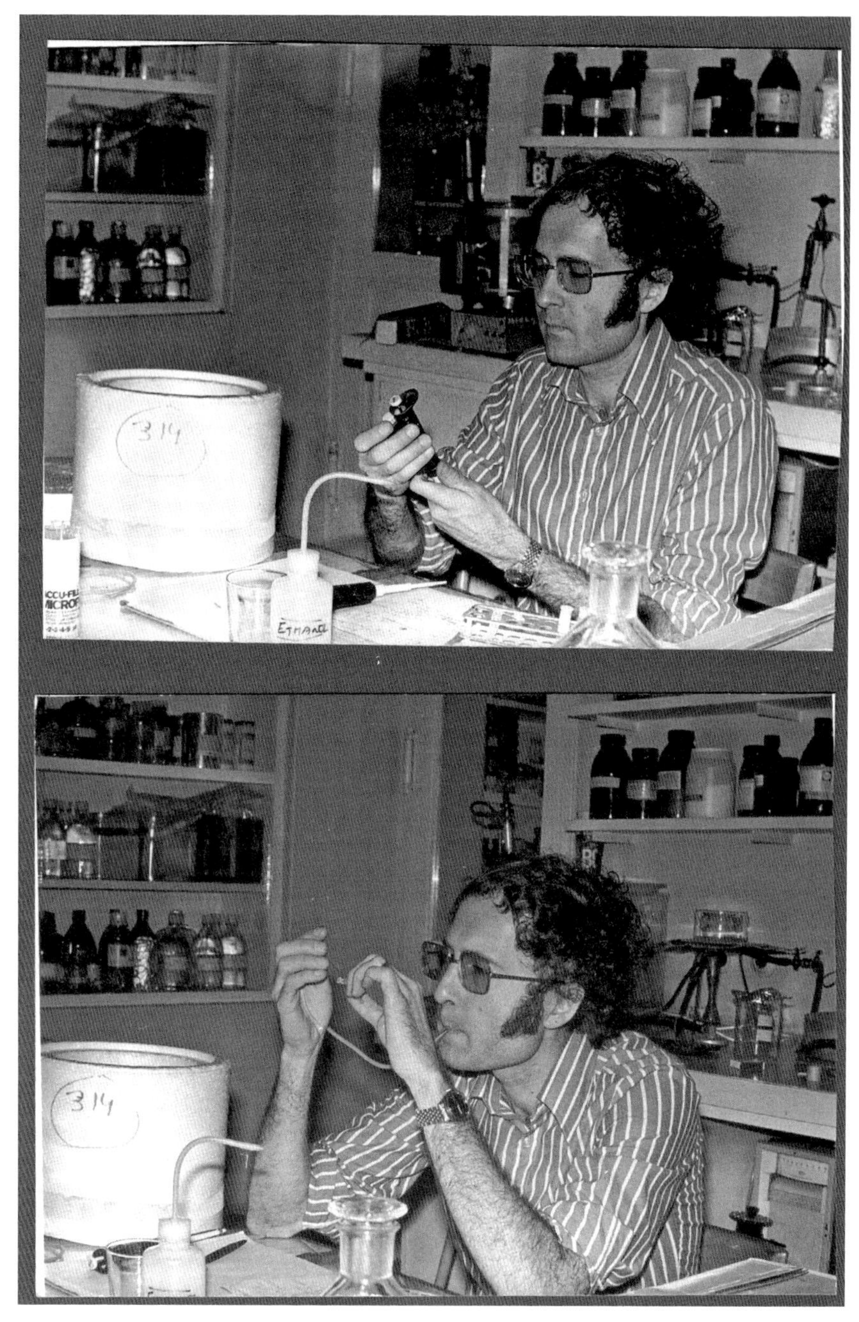

Figure 3. Gregory Winter in c.1980 demonstrating two methods for pipetting small volumes. Credit MRC Laboratory of Molecular Biology.

Although the sequence of the influenza genome was now available as a framework for further understanding of the epidemiology and biology of the virus, I couldn't shake off a longing to return to a world of proteins, structure and mechanism. Now thoroughly familiar with recombinant DNA technology, I saw huge potential in the application of this technology to the structure-function studies of proteins. Although my Trinity salary expired in the autumn of 1981, Fred Sanger offered me a short-term MRC position to explore these ideas.

Fortunately, an opportunity soon materialised: a former colleague at Imperial College, David Barker, had cloned the gene for TyrTS, and joined me for a couple of months in Cambridge to learn DNA sequencing and to help sequence the gene. The gene was soon sequenced, and David Blow (now at Imperial College) provided with the sequence information necessary to complete his model of the three-dimensional structure of TyrTS. David Barker also sequenced the methionyl tRNA synthetase (MetTS), revealing a constellation of residues (Cys35...His 45... His 48) conserved with the TyrTS, again implicating Cys35 in the catalytic mechanism.

To obtain experimental evidence for the role of Cys35, I decided to mutate the Cys35 to a serine residue in the cloned TyrTS gene, and to express and characterise the mutant enzyme. For this purpose, I used oligonucleotide-directed mutagenesis, travelling to Michael Smith's laboratory in Vancouver for six weeks to learn the latest methods from his postdoctoral worker Mark Zoller.

On my return from Canada, I collaborated with Alan Fersht (now at Imperial College) to understand the effects of this and other mutations on the enzyme mechanism. We confirmed that Cys35 was a key catalytic residue, but also found that the Ser35 enzyme was weakly active. In due course, by exploring other mutations around the active site, we established that the catalytic mechanism of the enzyme involved the preferential stabilisation of the tyrosyl adenylate transition state by multiple non-covalent bonds (including bonds from a mobile loop involving His 45).

As we improved our methods for site directed mutagenesis and synthesis of oligonucleotides, we started to undertake larger scale mapping projects. In particular we mapped the path of the tRNA across the enzyme, and later, by the same approach, mapped the binding site for complement C1q in antibodies. With site-directed mutagenesis established as a powerful analytical tool for studies of protein structure and function, I started on the next phase of my career.

ANTIBODIES

In October 1983 Fred Sanger retired and Cesar Milstein took over as Head of Division. Cesar's passion was antibodies – particularly the mechanisms underpinning their diversity. He agreed to my tenure but suggested that I use site directed mutagenesis to study the structure and function of antibodies. In January 1984, I was attacked on the way to work, and my shoulder dislocated, leaving my right arm temporarily paralysed. No longer able to work at the bench, and as a distraction from the pain caused by the nerve damage, I immersed myself into a virtual world of protein structures using the LMB's Evans and Sutherland PS300 computer graphics system. I was particularly interested in the architecture of binding sites. In the light of Cesar Milstein's suggestion, I started to look carefully at antibodies.

At that time, only the three-dimensional structures of human and mouse myeloma proteins were known. Human myeloma proteins are a species of antibody produced in myeloma patients and for which we do not know the cognate antigen. Nevertheless, it was supposed that the antigen binding sites were located in regions of hypervariable sequence in the loops of the variable (V) domains; indeed Elvin Kabat had called the hypervariable regions "complementarity determining regions" (CDRs), and the regions outside the CDRs he called Framework Regions (FRs).

Figure 4. Cesar Milstein and Fred Sanger, mid-1980s. Credit MRC Laboratory of Molecular Biology.

Inspection of the antibody architecture suggested to me that if Kabat was right, the antigen-binding activity of one antibody might be transferred to another by CDR transplant. This offered the prospect of endowing human myeloma proteins with the binding activities of rodent monoclonal antibodies, and so creating "humanised" [see Note 1] antibodies with predetermined binding activities for treatment of human disease, particularly non-infectious diseases such as cancer. I was aware that rodent monoclonal antibodies were seen as "foreign" in patients, provoking a blocking immune response, and anticipated that "humanised" antibodies (up to 95% human), might provoke a lesser response.

Accordingly, I designed a synthetic gene encoding an antibody variable domain in which the CDRs from a mouse monoclonal antibody with known binding activity were stitched into the framework regions of a human myeloma protein. Cesar Milstein allocated a post for a research officer, and was joined by Peter Jones, who acted as my right hand man until my retirement. We set about creating the gene by the chemical synthesis and assembly of oligonucleotides and produced the humanised antibody in myeloma cells using expression vectors kindly provided by Michael Neuberger (LMB). This took us some 18 months of work, and to our relief we found that the humanised antibody bound to the same target as the original mouse antibody, and with similar binding affinity.

We followed up with two more humanised antibodies; in these cases, we had to engineer mutations into the packing contacts between FR and CDR residues to fully restore binding affinities. One of these antibodies, a rat monoclonal antibody against a lymphocyte marker, was humanised in collaboration with Herman Waldmann's group (Cambridge Department of Pathology). This antibody had been chosen for its therapeutic potential and within months, the humanised antibody was used to destroy a large tumour mass from the spleens of two patients with non-Hodgkins lymphoma.

On the advice of Cesar Milstein, I had filed a patent on the humanising technology, but then found myself embroiled in discussions with my employers, the MRC, in formulating the best licensing strategy. Fortunately, the MRC finally agreed to adopt a largely non-exclusive licensing policy, similar to that used for the licensing of the Cohen Boyer patents on recombinant DNA technology, with a small upfront payment and a low royalty. Later the MRC's Collaborative Centre for Industry at Mill Hill offered a service to industry for humanising antibodies. These strategies encouraged the uptake of the humanising technology by companies, and led to the development of several therapeutic antibodies, marketed mainly for treatment of non-infectious diseases such as cancer and immune inflammatory disorders.

In the course of our work in humanising mouse hybridomas, we developed a set of PCR primers to amplify and clone the genes encoding the antibody variable domains from hybridomas. As explained in more detail in the Nobel Lecture, this technical advance also opened up the prospect of making human antibodies from V-genes harvested from human lymphocyte populations. However, we soon became aware of competition: the Scripps Research Institute and the biotechnology company Stratagene appeared to be working together along similar lines. We had yet to develop a screening method of sufficient power and realised that with our limited resources we would be outgunned.

COMPANIES

At that time the MRC was unable to offer more resources, and so I sought external collaborations with industry. We had filed patent applications on our work, but industry saw the ideas as too "blue sky". I therefore mused about starting my own biotechnology company and using it as a vehicle to develop the screening methodology. After a lecture at Amersham International in the summer of 1989, one of the Amersham employees, David Chiswell, offered to help me set up such a company. Around this time, I also had a visit from an Australian friend, Dr Geoffrey Grigg, who had already founded his own company (Peptech) in Sydney. He loved the "blue sky" ideas, and offered to help with funding, but we first had to

Figure 5. Geoffrey Grigg (photo from 2004). Credit Gregory Winter.

Figure 6. Caroline Winter (GW daughter) and Margaret Thatcher at the opening of the MRC Centre for Protein Engineering in 1991. Credit MRC Laboratory of Molecular Biology.

bring together all the various interested parties. After two months we had a deal: the MRC agreed to license the patents to a new company, Cambridge Antibody Technology (CAT), in return for an equity stake and a product royalty; Peptech agreed to provide CAT with a draw-down loan convertible to equity; David Chiswell agreed to become the Managing Director of the company and set up the laboratory facilities; and I agreed to split my time between my academic studies and the company for several years. The deal also included the provision that CAT would fund an employee to work in my group at the LMB to explore strategies for mass screening of antibodies, including the use of filamentous bacteriophage.

In early 1990 John McCafferty (formerly Amersham and now a CAT employee) joined my group. He soon discovered that antibody fragments could be displayed on filamentous bacteriophage and that binders could be enriched by factors of one thousand-fold in each round of affinity selection. More good news followed – in the autumn of the same year the MRC offered me further space and posts in a refurbished building adjacent to the LMB. This was the MRC's new Centre for Protein Engineering (CPE), a species of research institute originating from a government initiative to encourage academic and industry collaborations. Alan Fersht (now back in Cambridge at the Chemistry Department of Cambridge University) was appointed as Director of the CPE, and I became the Deputy Director.

Figure 7. Gregory Winter with Peter Jones early 1990s in the Centre for Protein Engineering. Credit MRC Laboratory of Molecular Biology.

The antibody work flourished in the CPE, and the explicit link with industry facilitated our collaborations with CAT. We isolated "binders" from human phage antibody libraries, explored methods for mutating and selecting antibodies with improved affinities, and showed that high affinity human antibodies could be isolated directly from very large libraries. The CPE became a hub of antibody expertise, with projects to develop new antibody formats (including diabodies and later single domains), and to clone all the human germline antibody segments which we used as building blocks for synthetic human antibodies. Most importantly we isolated human antibody fragments against human self-antigens from the antibody libraries.

In CAT the scope for rapid expansion was more limited due to the limited cash reserves and the difficulties of finding further investors. Fortunately, a collaborative contract with Knoll Pharmaceuticals (later acquired by Abbott Laboratories) was successful and led to the development of the antibody adalimumab (Humira) against the inflammatory mediator TNFalpha. This later became the first human therapeutic antibody to be approved for marketing by the US Food and Drug Administration (2002) and the world's top selling pharmaceutical drug. By 1996 the company had started to prepare for an initial public offering (IPO) on the London Stock Exchange and was floated in 1997. However, the preparations for the IPO brought out tensions between investors and split the

Board. As most of the antibody technology was by now well established, I stepped down from the Board and left CAT before the IPO.

I was also interested in developing some antibody technology that had not been established in CAT. In 1989, at the LMB we had isolated single antibody variable (VH) domains with excellent binding activities, but prone to aggregate. With improved properties, these domains had potential as small protein domains for topical applications, or as building blocks for bispecific antibodies. As CAT decided not to develop this technology, I helped to establish a new company, Domantis (originally named Diversys) to do so, with the MRC taking an equity stake. Ian Tomlinson, a former PhD student and group leader at the LMB, was a scientific co-founder, and as with CAT, Geoffrey Grigg and Peptech Ltd played a key role as seed investors. Domantis was founded in 2000, the company focusing on the selection of aggregation-resistant domains against potential pharmaceutical targets. The company established several research partnerships with pharmaceutical companies and in 2007 was acquired by GlaxoSmithKline, shortly after AstraZeneca acquired CAT. By 2018, the MRC had received more than £1 bn in royalties, sales of shares and other commercial payments in respect of the therapeutic antibody technologies we had created at the LMB, CPE, CAT and Domantis.

My next enterprise emerged from work with peptides. I had always been interested how proteins evolved and was much taken by a suggestion that proteins had evolved through stitching together multiple peptide segments by RNA splicing, a process termed exon shuffling by Walter Gilbert. We tried to mimic this process by randomly shuffling together peptide segments, displaying the combinatorial libraries on phage, and using proteolysis to select for those that folded. We found that the folded peptides were stabilised by forming multimers (dimers and/or tetramers), and/or by incorporation of a prosthetic group (heme). This led us to consider making small proteins by folding random peptide libraries around a prosthetic core, and to a postdoctoral worker, Christian Heinis, stapling peptide libraries to the core through three cysteine residues. These libraries, comprising highly constrained bicyclic peptides, proved to be a source of high affinity ligands against a range of protein targets. Indeed, we came to think of the bicyclic peptides as small antibody mimics, in which the β-sheet protein framework had been replaced a chemical framework. The "bicycles" were expected to have some advantages (and disadvantages) over antibodies; unlike antibodies they could be chemically synthesised, and on injection would penetrate deep into tissues.

The technology seemed ripe for development, and a start-up company the best vehicle to deliver this. Christian would soon depart for an academic post at the EPFL Lausanne. One of my previous seed-phase partners, Geoffrey Grigg had passed away, and Peptech had merged to form

Arana Therapeutics (swallowed in turn by Cephalon, then Teva). Fortunately, one of my former postdoctoral workers, Regina Hodits, was now with the Atlas Ventures and liked the technology. In 2009 she helped set up the company Bicycle Therapeutics, bringing in other venture capital partners and John Tite (formerly GSK) as CEO/CSO. The company industrialised the selection and synthesis of bicycles, and later with Kevin Lee (formerly Pfizer) as CEO, pressed ahead with the development of bicycles and bicycle conjugates for use in oncology. Bicycle Therapeutics was listed on the Nasdaq stock exchange in 2019.

INSTITUTIONS

Gradually I found myself becoming more interested in the possible applications of my work and more detached from the academic focus of the LMB. However, the LMB was my scientific home and where I had done most of my scientific research. I had worked my passage from PhD Student, to Postdoctoral Worker, to Group Leader, to Head of Division and finally Deputy Director. In the course of my career Brian Hartley, George Brownlee, Fred Sanger and Cesar Milstein had acted as scientific mentors; Alan Fersht, Michael Neuberger and Terence Rabbitts had proved excellent collaborators, and Hugh Pelham, Richard Henderson, Aaron

Figure 8. Alan Fersht in early 1990s in the Centre for Protein Engineering. Credit MRC Laboratory of Molecular Biology.

Klug and Sydney Brenner had been supportive as LMB Directors. The LMB was a wonderful place, acting as a magnet for brilliant PhD students, postdoctoral workers and technical staff – all too numerous to mention here [for names see Note 2].

In 2013, the LMB moved from a cramped 1960s building on one side of the Cambridge Biomedical campus into a large and superb new building on the other side. The case for funding the new building had been fortified by the revenues generated for the MRC and HM Treasury by the antibody technologies and the expectation that further revenues might follow from such opportunistic translation of curiosity-driven research. However, throughout my time, the LMB itself had limited space, resources or appetite for translational work. To see my work applied I had to "privatise" the more translational aspects outside the LMB, whether through the CPE, CAT, Domantis or Bicycle Therapeutics. The move to the new building weakened the spell that had bound me to the LMB.

When in 2012 I was offered the Mastership of Trinity College, I accepted it – I was very grateful to the College (and its Fellows) for their catalytic role in my life. Brian Hartley had been my undergraduate Director of Studies at College, and had persuaded me, through his infectious enthusiasm, to build a research career in molecular biology. David Blow, also a Fellow of the College, had taken me as a summer student to build a display model of the enzyme trypsin, igniting my interest in the structure and mechanism of proteins. Another Trinity man, Michael Neuberger, had provided me with the expression vectors that got me started in the antibody world; Michael had also sent me one of his Trinity undergraduates, Ian Tomlinson (later Domantis) as a PhD student.

As a Fellow of the College, I already had some idea of the changes in store for me as Master. I was installed in the Master's Lodge, a Tudor palace, and expected to "exercise a general superintendence over the affairs of the College." My main role was to maintain the good order of the College by attending or presiding over its meetings and activities, making speeches and representing the College externally. However almost all powers were reserved to the College Council, which generally aimed to reach a wide consensus before agreeing any change, delegating matters of substance for deliberation by sub-committees, which might only meet once or twice a year. The pace of change was glacial, but the effect of small tweaks here and there seemed to work, and the College continued to perform well academically and in its investments.

I did nevertheless preside over some changes, including the development of better relations with the College's alumni and the provision of more space and a higher profile for early stage companies at the College-owned Cambridge Science Park. I also continued with my involvement with Bicycle Therapeutics and some other biotechnology companies and started to advise venture funds in biotechnology investments. At the time of writing, as I come to the end of my tenure as Master, I expect that some combination of science, medicine and start-up companies will underpin my future activities.

Note 1. In our early publications, we referred to such antibodies as "reshaped" human antibodies: the field later adopted the alternative term of "humanised" mouse antibodies, and which I have used here.

Note 2. I would however like to acknowledge and thank the following who worked with me at the LMB and CPE, with apologies to those inadvertently omitted: David Barker, Hugues Bedouelle, Elise Bernard, Tim Bonnert, George Brownlee, Yvonne Bruggeman, Marianne Bruggeman, Jackie Bye, Paul Carter, David Chiswell, Cyrus Chothia, Daniel Christ, Peter Christensen, Tim Clackson, Graham Cook, Simon Corbett, Jonathan Cox, Stephanie de Bono, Ruud de Wildt, Paul Dear, John Doorbar, Alexander Duncan, M. James Embleton, Kristoffer Famm, Alan Fersht, Stanley Fields, Mariangela Figini, John Finch, Riccarda Finnern, Igor Fisch, Nicolas Fischer, Kevin FitzGerald, Jefferson Foote, Stefan Freund, Jean-Pol Frippiat, Michael Gait, Ermanno Gherardi, Steffen Goletz, Barbara Gorick, Guy Gorochov, Heather Griffin, Andrew Griffiths, Detlef Gussow, Oliver Hartley, Brian Hartley, Robert Hawkins, Christian Heinis, Regina Hodits, Philipp Holliger, Hennie Hoogenboom, Peter Hudson, Nevin Hughes-Jones, Olga Ignatovich, Leo James, Philip Jennings, Laurent Jespers, Jean-Luc Jestin, Kevin Johnson, Peter Jones, Abraham Karpas, Perry Kirkham, Gordon Koch, Roland Kontermann, Peter Kristensen, Arthur Lesk, Meirion Llewelyn, Benny Lo, Nigel Low, John Lund, Stefan Luzi, Stephen Mahler, Magnus Malmqvist, Roy Mariuzza, James Marks, Cara Marks, Pierre Martineau, John McCafferty, Andrew McLachlan, Cesar Milstein, Dario Neri, Michael Neuberger, Ahuva Nissim, Rosaria Orlandi, Willem Ouwehand, Didrik Paus, Olga Perisic, Terence Prospero, Terence Rabbitts, Lutz Riechmann, Stephen Russell, Trevor Rutherford, Fiona Sait, Oliver Schon, Arne Skerra, Geoffrey Smith, Sirirurg Songsivilai, Dan Teufel, Ian Tomlinson, Hiroshi Ueda, Marina Vaysburd, Dmitry Veprintsev, Martine Verhoeyen, Herman Waldmann, Edward Walker, Gerald Walter, Peter Wang, E. Sally Ward, Peter Waterhouse, Mary Waye, Roger Williams.

Harnessing Evolution to Make Medicines

Nobel Lecture, December 8, 2018 by
Gregory P. Winter
MRC Laboratory of Molecular Biology, Cambridge and Trinity College, Cambridge, UK.

INTRODUCTION

Antibodies are part of our natural defence against infectious agents such as viruses and bacteria and are raised by the immune system in response to infection or vaccination. Indeed, the immune system is a simple system for the fast evolution of antibodies against infectious agents; it generates a diverse range of antibodies and selects those that bind to the infectious agent.

Although Nature has developed antibodies to protect against infectious disease, Man has further developed and evolved antibodies for treatment of *non*-infectious disease, such as inflammatory disorders and cancer (Winter & Milstein, 1991). Man-made antibodies have been used to block the biology of protein receptors and ligands involved in inflammatory disorders, cell growth and T-cell activation; or to kill target cells by recruiting immune effector functions. Indeed, the development of antibodies for treatment of non-infectious disease has revolutionised the pharmaceutical industry, an industry previously dominated by chemical drugs, particularly for treatment of auto-immune inflammatory diseases and cancer. For example, in recent years the antibody Humira (or adalimumab), used for treatment of rheumatoid arthritis, has been the world's best-selling pharmaceutical drug. Of the top ten best selling drugs in 2016, six were antibodies (Table 1).

TRADE NAME	DISEASE	COMPANY	SALES ($bn)
1. Humira*	rheumatoid arthritis	AbbVie	16.1
2. Harvoni	hepatitis C	Gilead	9.1
3. Enbrel*	rheumatoid arthritis	Amgen/Pfizer	8.9
4. Rituxan*	NHL	Roche/Biogen	8.6
5. Remicade*	rheumatoid arthritis	J&J/Merck	7.8
6. Revlimid	multiple myeloma	Celgene	7.0
7. Avastin*	cancers	Roche	6.7
8. Herceptin*	breast cancer	Roche	6.7
9. Lantus	diabetes (insulin)	Sanofi	6.0
10. Prevnar	pneumonia (vaccine)	Pfizer	5.7

Table 1. Sales of top ten pharmaceutical drugs, 2016. Antibodies are marked*. Source: *Genetic Engineering and Biotechnology News.*

The development of such antibody pharmaceuticals has required multiple technological inventions, and a molecular understanding of the disease and of antibody structure, function and genetics.

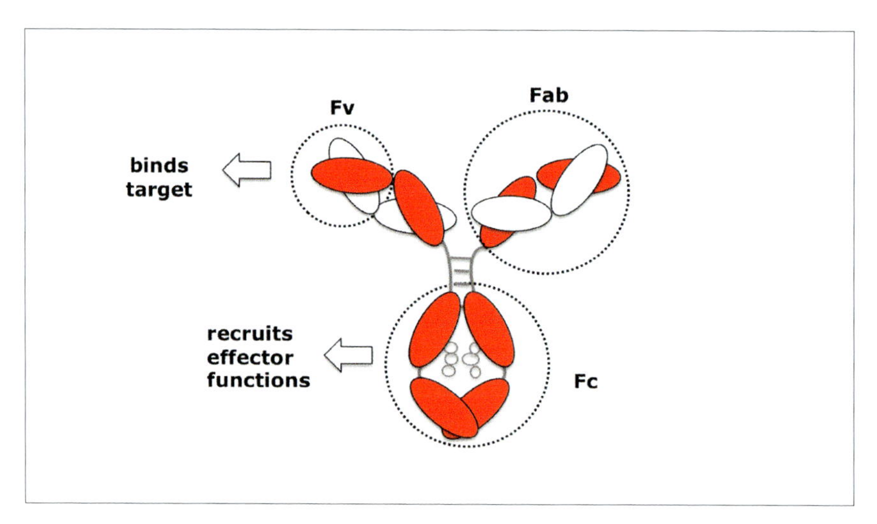

Figure 1. Structure of IgG. Heavy chains (red), light chains (white) with classical antibody fragments (Fv, Fab and Fc) marked.

ANTIBODY STRUCTURE, FUNCTION AND GENETICS

An IgG antibody (Figure 1) is a large (150,000 Da) Y-shaped molecule, two arms and a stem, comprising four chains, two heavy and two light of linked protein domains. The heavy and light chain variable domains (abbreviated VH and VL) at the end of the arms come together to form a protein scaffold of beta-sheet, surmounted by six loops of variable sequence. Antibodies protect against infectious agents by binding to the target through the variable loops and blocking the process of infection. In addition, the antibody stem can recruit immune effector functions such as complement activation, phagocytosis and antibody-dependent cellular cytotoxity (ADCC) to kill the infectious agent.

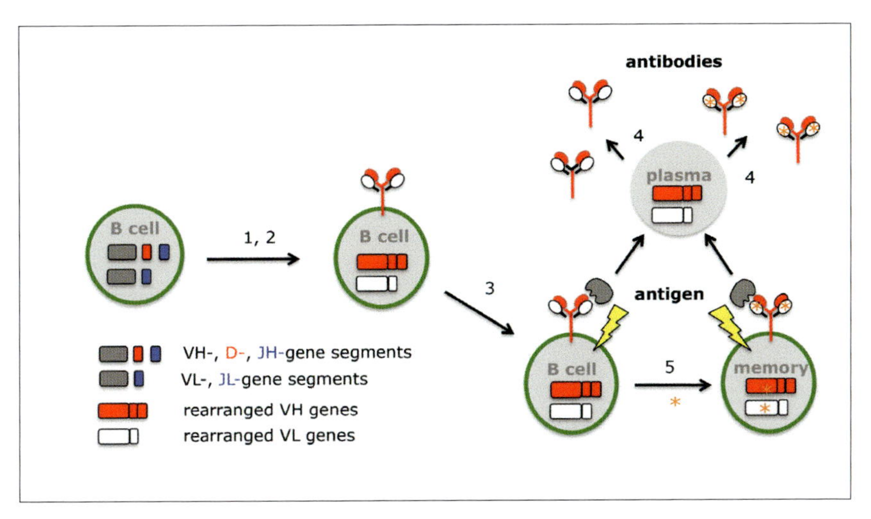

Figure 2. Strategy of the immune system for making antibodies. (1) rearrangement of V-gene segments (Hozumi 1976); (2) surface display of antibody on B-cell; (3) antigen-driven selection; (4) secretion of soluble antibody from plasma cell; (5) affinity maturation.

The diversity of the antigen-binding loops is created through rearrangement of different germ-line segments of DNA during human B-cell development (Figure 2). In heavy chains it is created through combinations of V_H, D and J_H gene segments, each of which has multiple members, to give the rearranged VH genes. In light chains it is created through combinations of two sets of gene segments (Vκ and Jκ; Vλ and Jλ) to give the rearranged VL genes. Editing at the segment junctions generates further genetic diversity, as does the random combination of heavy and light chain genes in each B-cell. After DNA rearrangement, the encoded antibody is expressed on the surface of the B-cell. As the process takes place independently in many B-cells, it generates a library of cells in which each member expresses a unique antibody on its surface.

When the displayed antibody binds to cognate antigen, the B-cell is stimulated to differentiate. It gives rise to plasma cells, which are factories for production of antibodies against the antigen, and to memory B-cells. In memory B-cells, the antibody genes are targeted for random mutation, and the mutant antibodies displayed on the cell surface. On further encounter with antigen, the mutant antibodies compete for limiting antigen; those cells displaying antibodies with the highest binding affinity are favoured for further rounds of differentiation. This leads to an antibody response that improves with repeated immunisation, a process known as affinity maturation. For general review, see Schroeder & Cavacini, 2010.

ANTIBODY ENGINEERING

The first step in the creation of antibody pharmaceuticals was the invention of hybridoma technology by Koehler and Milstein (1975) at the MRC Laboratory of Molecular Biology. Mice were immunised with the target antigens, the spleens harvested, and the responding B-cells immortalised by cell fusion. The hybrid cells (or hybridomas) were then screened to identify those making monoclonal antibodies against the target cells. Although this technology generated many useful research reagents, including against human proteins and cells, the mouse monoclonal antibodies were seen as foreign when injected into patients, compromising their use in the clinic. Attempts to make human monoclonal antibodies against human cancer cells proved impossible, not least because the human immune system has tolerance mechanisms that prevent it making antibodies against self-antigens.

By the mid 1980s, solutions began to emerge through the application of protein engineering, in which the antibody genes were altered and the altered antibodies expressed in a host cell. Attempts were made to express antibodies in bacteria, but this gave very poor yields and the antibodies

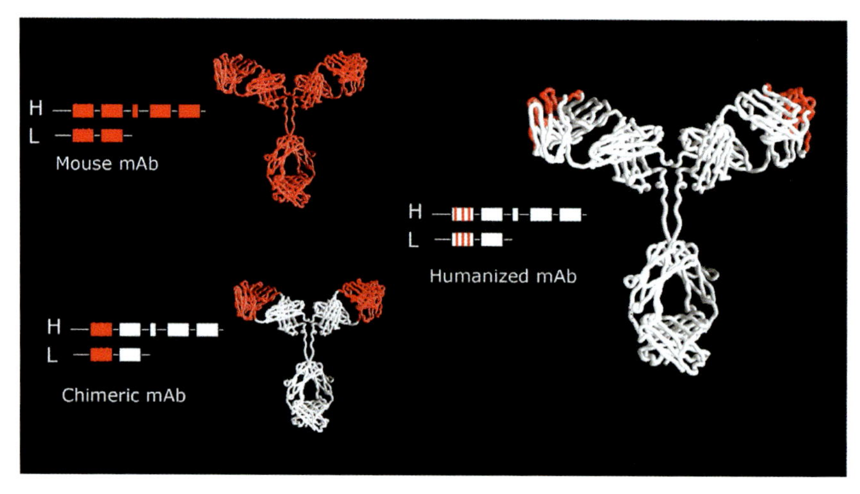

Figure 3. Humanising rodent monoclonal antibodies. IgG polypeptide backbone and gene structure, mouse origin (red) and human origin (white).

had to be refolded from intracellular inclusion bodies (Boss *et al.*, 1984; Cabilly *et al.*, 1984). Lymphoid cells proved to be more suitable hosts for the expression, secreting folded and functional antibodies into the medium (Rice & Baltimore, 1982; Neuberger, 1983; Ochi *et al.*, 1983; Oi *et al.*, 1983).

The first focus of the protein engineers was to turn mouse monoclonal antibodies into their human counterparts (Winter, 1989; Winter & Milstein, 1991) (Figure 3). First came the simple chimeric antibodies in which entire antigen-binding domains were transplanted from mouse to human antibodies. Such antibodies were one-third mouse and two-thirds human in origin, binding to the same antigen as the mouse hybridoma and triggering human effector functions (Boulianne *et al.*, 1984, Morrison *et al.*, 1984, Neuberger *et al.*, 1985).

Then came "humanised" antibodies, in which only the antigen-binding loops were grafted from the mouse antibodies into human antibodies (Jones et al., 1986; Riechmann *et al.*, 1988; Verhoeyen *et al.*, 1988).

As the inventors of humanised antibodies (which are up to 95% human in origin), we argued that as the antigen-binding loops differ between human antibodies, such humanised antibodies might be regarded as synthetic human antibodies. One of these humanised antibodies (Riechmann *et al.*, 1988), directed against the lymphocyte antigen CD52, was used to treat patients with non-Hodgkins lymphoma. It was tolerated over the 30-day course of treatment and destroyed a large mass of spleen tumour (Hale *et al.*, 1988). This appears to have been the first clinical use of an engineered antibody. Later, clinicians at the Cambridge Department of

Cinqaero (*reslizumab*) (IL5/asthma)	Lucentis (*ranibizumab*) (VEGF-A/AMD)
Nucala (*mepolizumab*) (IL5/asthma)	Avastin (*bevacizumab*) (VEGF-A/several C)
Xolair (*omalizumab*) (IgE/asthma)	Herceptin (*trastuzumab*)(HER2/HER2+ BC)
Lemtrada (*alemtuzumab*) (CD52/MS)	Perjeta (*pertuzumab*) (HER2/HER2+ BC)
Tysabri (*natalizumab*) (VLA-4/MS,CD)	Gazyvaro (*obinutuzumab*) (CD20/CLL)
Ocrevus (*ocrelizumab*) (CD20/MS)	Empliciti (*elotuzumab*) (SLAMF7/MM)
RoActemra (*tocilizumab*) (IL-6R/RA)	Tecentriq (*atezolumab*) (PD-L1/NSCLC)
Cimzia (*certolizumab pegol*)(TNF/RA, CD)	Keytruda (*pembrolizumab*) (PD-1/melanoma)
Entyvio (*vedolizumab*) (I-$\alpha 4\beta 7$/CD)	Synagis (*palivizumab*) (RSV/RSV infection)
Soliris (*eculizumab*) (C5/PNH)	

Table 2. Humanised antibodies approved by the US Food and Drug Administration. Each antibody listed in order of trade name, non-proprietary name, pharmaceutical target and disease area. MS = Multiple Sclerosis, CD = Crohn's Disease, RA = Rheumatoid Arthritis, PNH = Paroxysmal Nocturnal Haemoglobulinuria, AMD = Acute Macular Degeneration, BC = Breast Cancer, CLL = Chronic Lymphocytic Leukaemia, MM = multiple myeloma, NSCLC = Non Small Cell Lung Cancer, RSV = Respiratory Syncytial Virus.

Clinical Neurosciences developed this antibody (alemtuzumab, marketed as Lemtrada) for treatment of relapsing forms of multiple sclerosis. Many other antibodies have been humanised and approved as pharmaceutical drugs (Table 2). But in the late 1980s we didn't know that humanised (or even chimeric) antibodies would be so well tolerated in patients. We therefore began to think about ways of making fully human antibodies. A possible solution emerged from a methodological improvement in making engineered mouse antibodies.

ANTIBODY LIBRARIES

One of the rate-limiting steps in making engineered antibodies was the isolation of the rearranged VH- and VL-genes from the mouse hybridoma. For this purpose, we decided to explore the use of the polymerase chain reaction (PCR), in which target regions of DNA are amplified by repeated cycles of polymerase extension of two flanking primers (Saiki et al., 1988). By comparing the nucleotide sequences of many different antibodies, we identified regions at the ends of both heavy and light chain genes that seemed sufficiently conserved to allow the design of a simple set of PCR primers. There was no way of avoiding some primer/template mismatches; indeed, we took advantage of the mismatched regions to incorporate restriction sites for cloning the amplified DNA into expression vectors. After exploring a range of experimental conditions, we identified a set of primer sequences that allowed us to amplify the rearranged VH- and VL-genes of several mouse hybridomas (Orlandi *et al*, 1989).

This method made it so much easier to clone and express antibody genes that we speculated about making recombinant antibodies directly from libraries of V-genes from the spleens of immunised mice, thereby by-passing hybridoma technology (Orlandi *et al.*, 1989). We also realised that a similar approach might lend itself to making human antibodies, including those against human self antigens. However, we first needed to find an expression host suitable for the mass screening of libraries of recombinant antibodies. We alighted on bacteria, as it had just been shown that antibody fragments could be secreted in a folded and functional form into the bacterial periplasm by attaching a signal sequence (Better *et al.*, 1988; Skerra & Pluckthun, 1988).

We tailored a plasmid vector for the bacterial expression of antibody fragments from the VH- and VL-genes as amplified by our PCR primers. After cloning the antibody genes from the hybridoma (D1.3) (Amit *et al.*, 1986), we established that the D1.3 antibody Fv fragment (associated VH and VL domains) was secreted from the transfected bacteria into the culture supernatant, and bound to hen egg lysozyme, as in the original hybridoma. We also discovered that the secreted VH domain bound to hen egg lysozyme in the absence of the VL domain, with a loss of only about tenfold in binding affinity.

We then immunised mice with protein antigens and made a VH gene expression library from the immunised spleens. Screening of the library revealed VH domain fragments with binding activities to the immunising antigen at a frequency of about 1/100. This seemed promising, and we wondered whether such single domain antibodies (or dAbs) might provide a platform of small high affinity protein domains for a range of applications. However, the isolated VH domains were "sticky", presumably due to the exposed hydrophobic surface normally capped by the VL domains. Although we believed that the poor biophysical properties of mouse (and human) VH domains could be overcome (Ward *et al.*, 1989), we did not pursue the further development of a dAb platform for some years (Jespers *et al.*, 2004a; Jespers *et al.*, 2004b; Ueda *et al.* 1984). Indeed, camels were found to have naturally occurring antibodies devoid of light chains, and the camel VH domains to have good biophysical properties (Hamers-Casterman *et al.* 1993).

In the meantime, the group of Richard Lerner (Scripps Research Institute) and the biotechnology company Stratagene had developed an interest in antibody libraries. They took the next step and created libraries of heavy and light chain (Fab) pairs from the V-genes from spleens of lymphocytes immunised with hapten. The pairings were generated randomly, and the libraries referred to as random combinatorial libraries. After mass screening of the Fab fragments, they identified several hapten binders at a frequency of 1/10,000 (Huse *et al.*, 1989). Again, this seemed promising, but we thought it unlikely that the screening method would be sufficiently powerful to find human antibodies against human self-antigens.

We therefore looked at the more powerful strategy used by the immune system itself. Could we develop a B-cell mimic, essentially a "genetic display package", with antibody displayed on the outside of the package to encounter antigen, and the antibody genes packaged within? We considered several possibilities, including the display of antibodies on mammalian cells, bacteria and bacterial viruses. Of these the filamentous bacteriophage seemed the most attractive.

PHAGE DISPLAY

Four years earlier George Smith had shown that peptides could be displayed on filamentous bacteriophage by genetic fusion to a coat protein (p3) that mediates the bacterial infectivity of the phage. He had shown that phage displaying peptide epitopes could be selected and enriched by binding to cognate antibodies (Smith, 1985). Instead of peptides, we won-

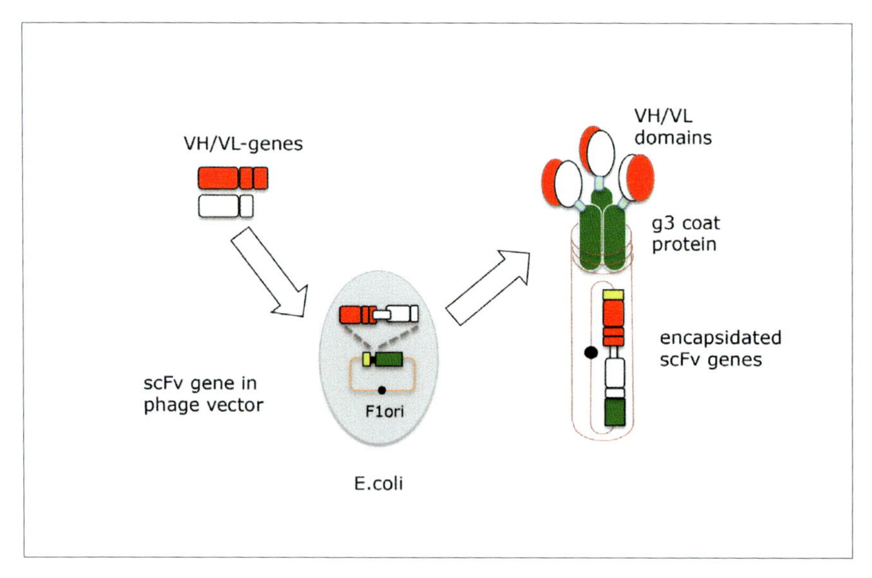

Figure 4. Cloning scFv fragment for phage display. VH genes and VH domain (red), VL genes and VL domain (white), phage 3 protein (green), leader sequence (yellow). Phage truncated and only 3 copies of p3 shown for simplicity of representation.

dered whether we could display antibody fragments on phage, and select the "phage antibodies" by binding to antigen?

We did not know whether the displayed antibodies would fold or be efficiently incorporated into the phage coat. Nor did we know whether the antibody-p3 fusions would survive proteolysis in the bacterial periplasm or mediate phage infectivity. We therefore checked with our model D1.3 antibody fragment. To avoid dissociation of the domains, we decided to link the VH and VL domains together with a short flexible peptide, so creating a single chain Fv fragment (scFv) (Bird *et al.*, 1988; Huston *et al.*, 1988). We then fused the antibody gene to the phage p3 gene in a phage vector (Figure 4). After transfection of the recombinant vector DNA into bacteria, we tested the recombinant phage for binding to hen egg lysozyme using an ELISA assay. As we had hoped, the recombinant phage bound specifically to hen egg lysozyme (McCafferty *et al.*, 1990). We now mixed the phage antibody with a large excess of wild-type phage and showed that we could enrich the phage antibody by rounds of selection on antigen-affinity columns. Indeed, the phage antibodies were enriched one thousand-fold in a single round of selection, and one million-fold over two rounds of selection.

We then used the phage to display and enrich libraries of antibody fragments from immunised mice. We decided to follow the mouse antibody response to the hapten phenyloxazolone (phOx), as this response had been characterised in molecular detail using hybridoma technology

(Berek *et al.*, 1985). We started with a random combinatorial library of about 2 x 10^5 phage antibody clones derived from spleen IgG mRNA of a mouse immunised with the hapten (Clackson *et al.*, 1991). After affinity selection, we identified multiple binders. Although we had expected to find binders, we had not anticipated how readily we would find them. On the one hand, we had expected the Ig mRNA to be enriched towards antibodies directed to the immunising hapten, particularly due to the elevated Ig mRNA in plasma cells (Schibler *et al.*, 1978; Hawkins & Winter, 1992). On the other hand, we had expected the frequency of original VH and VL pairings (and comprising those elicited by immunisation) to be low given the random combinatorial process and the number of B-cells used to make the library.

Sequence analysis of the binders revealed a further factor – "promiscuity" – in which the same VH-gene was paired with different VL-genes, and vice-versa. Indeed, after shuffling one such pairing with the original VH- gene or VL-gene libraries respectively, we found binders in which each domain had made new pairings. After recloning the antibody genes for bacterial expression as soluble antibody fragments, we identified pairings with binding affinities comparable to those of hybridomas (Kd = 10 nM) made from the same spleens (Gherardi & Milstein 1992). Together with the results from the Scripps (Huse *et al.*, 1989), this confirmed that the library technology had the potential to bypass hybridoma technology, at least for making mouse monoclonal antibodies against haptens.

HUMAN ANTIBODIES FROM PHAGE LIBRARIES

We then faced up to the next challenge – to make human antibodies and to do so without immunisation. We first made some technical improvements to make the libraries more diverse and larger. To make the libraries more diverse, we took the peripheral blood lymphocytes from unimmunised human donors and amplified the rearranged human VH- and VL-genes from naïve and primary response B-cells (IgM) cDNA (Marks *et al.,* 1991a) using a set of "family-based" PCR primers (Marks *et al.*, 1991b). To make the libraries larger, we turned to phagemid vectors in which the antibody fusion is encoded by a phagemid, and a helper phage provides other g3 subunits (Bass *et al.*, 1990). The phagemid vectors have higher bacterial transfection efficiencies than phage vectors, allowing the creation of larger libraries. The phagemid vectors also simplified the expression of soluble fragments. By interposing an amber stop codon between the antibody gene and the p3 protein, we could switch between display on phage in amber suppressor strains of bacteria (and helper phage), or secretion of soluble fragments in non-suppressor strains (Hoogenboom *et al.*, 1991).

From a library of 10^7 phagemid clones, we isolated human antibody fragments binding to a foreign antigen (turkey egg lysozyme), a hapten (phOx) (Marks *et al.*, 1991a), and to several human self-antigens (Griffiths et al., 1993). Although the binding of the soluble fragments was specific for each target, the binding affinities of the antibody monomers were in the micromolar range. We set about improving these affinities by subjecting the selected clones to further rounds of diversification and selection, as in the immune system.

For example, starting with the human phOx antibody (binding affinity Kd = 320 nM), we reshuffled the VH gene with the entire VL gene library, selecting for hapten-binding under stringent conditions (Marks *et al.*, 1992). This led to new VL partners, and an antibody fragment with a 20-fold improved binding affinity, which we improved a further 15-fold by shuffling with a library of the VH gene segments. Overall, we had achieved 300-fold improvement in binding affinity (K_d = 1 nM).

In another example, we grew the phage displaying the human phOx antibody in a bacterial mutator strain. After multiple rounds of growth and stringent selection with the hapten, we achieved a 100-fold improvement in binding affinity (Kd = 3.2 nM). This was comparable to the affinities of mouse monoclonal antibodies made by repeated immunisation to the same hapten. We could even construct a genealogical tree from the sequences of the mutants at different rounds and identify four sequentially acquired mutations which were together responsible for the improved affinity (Low *et al.*, 1996).

In the selection process, we typically used longer washes, more disruptive washing conditions or lower concentrations of antigen to distinguish between phage antibodies with different affinities, (Hawkins *et al.*, 1992). In this context the phagemid vectors had a further advantage over phage vectors – "monovalent" display (Lowman & Wells, 1991; Lowman *et al.*, 1991). With phage vectors, we would expect five antibody heads on each phage and potentially highly avid binding to solid phase antigen. In practice few phage have five antibody heads – they suffer proteolysis, and the phages bind to solid phase antigen with variable avidities, making it difficult to distinguish between those binders with high affinity and those with high avidity. The use of phagemid vectors, in which incorporation of the helper phage p3 reduces the phage valency, leads to so-called "monovalent" phage, and helps to eliminate the avidity component from affinity selections.

SYNTHETIC HUMAN ANTOBODIES

As well as antibody libraries from the rearranged VH- and VL-genes from human B-cells, we developed synthetic human antibody libraries. We created the synthetic libraries from the human gene segment building blocks (V_H, D and J_H; Vκ and Jκ; Vλ and Jλ) – such libraries have the advantage that the composition and diversity of the library can be predetermined. In 1989, the sequences of most of the human VH- and VL-genes was unknown, so we first had to clone and sequence all the human V-gene segments, starting with the V_H segments (Tomlinson *et al.*, 1992). To the ends of each of 49 human V_H gene segments (Hoogenboom & Winter, 1992; Nissim *et al.*, 1994) we introduced an artificial D-segment comprising random nucleotide sequence of variable lengths, and a J-segment. These rearranged VH gene segments were combined with a single fixed rearranged VL gene and cloned for display on phage. From a library of 10^8 phagemid clones we obtained micromolar binders to haptens and proteins, comparable in affinity to libraries of similar size made from naturally rearranged V-genes.

After cloning the Vλ- gene (Williams & Winter, 1993) and Vκ- gene (Cox *et al.*, 1994) segments, and building the corresponding synthetic rearranged VL gene repertoires, we set about creating an even larger combinatorial library. However, we were limited by the efficiencies of transformation of *E. coli* by phagemid DNA. We therefore developed a new strategy (Waterhouse *et al.*, 1993) in which we infected bacteria harbouring a heavy chain library (on a plasmid) with a light chain library (on a phage). The two chains were combined on the same (phage) replicon within the bacterium by Cre catalysed recombination at *loxP* sites, creating combinatorial libraries, potentially as large as the number of bacteria infected. In this way we created a library of $> 10^{10}$ clones (Griffiths *et al.*, 1994) including almost all the human V-gene segments. Selection of this library yielded antibodies to a wide range of antigens, against both foreign and self-antigens. The binding affinities were in the nanomolar range, and comparable to the affinities of hybridomas after affinity maturation. Indeed, using such large libraries reduced the need for affinity maturation.

THERAPEUTIC ANTIBODIES FROM PHAGE

In parallel with our work in the MRC, the start-up company Cambridge Antibody Technology (CAT) was working with commercial partners to develop human therapeutic antibodies. Their most spectacular success was the development of the phage antibody adalimumab, which is targeted against TNFalpha and used for treatment of autoimmune inflammatory diseases such as rheumatoid arthritis, psoriasis and Crohn's disease. When CAT started the work on adalimumab in the early 1990s, the random combinatorial libraries of human antibodies were not large or diverse enough. We therefore tried a different strategy, starting with a mouse monoclonal antibody against TNFα(Jespers et al., 1994), as illustrated in Figure 5. For example, the mouse heavy chain can be paired with a library of human light chains and selected for binding to TNFα. The selected human light chain can then paired with a library of human heavy chains, and selected for a fully human antibody that binds to the same antigen. This chain shuffling strategy, essentially a molecular version of Theseus's paradox (or grandfather's axe paradox), synthesises a human antibody using a mouse antibody as template. It was this strategy that gave rise to adalimumab.

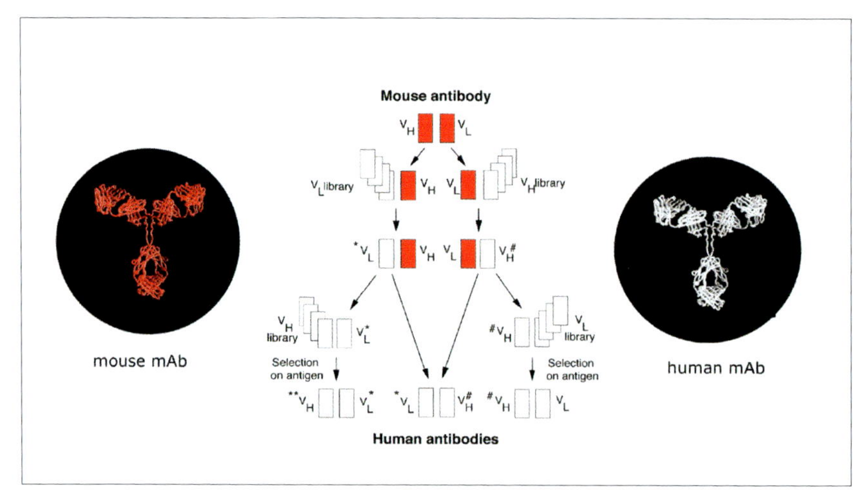

Figure 5. Humanising antibodies by chain shuffling. IgG polypeptide backbone and VH and VL gene structure, mouse origin (red) and human origin (white).

Subsequently Cambridge Antibody Technology (Medimmune/AstraZeneca) has focused on creating very large libraries (>10^{11} clones). Table 3 shows the pharmaceutical target classes to which human phage antibodies have been made using the same library (J. Osbourn, Medimmune, personal communication), and Table 4 those human therapeutic antibodies already approved by the US Food and Drug Administration.

Growth factor: PIGF, VEGF-2, GDF-8
Chemokine: CXCL13
Ion Channel: P2X4
Receptor: IL-21R, PSGL-1, TRAIL-R1, GM-CSFa2
GPCR: GLP1R, GIPR
Cytokine: IL-6, Blys, APRIL
Protease inhibitor: PAI-1
Peptide: Ghrelin, NKB, gp41

Table 3. Pharmaceutical target classes against which phage antibodies have been raised using the same large library. In order target class and molecular target.

Humira (*adalimumab*) (TNF/RA,CD)
Benlysta (*belimumab*) (BAFF/SLE)
Cablivi (*caplacizumab*) (vWF/TTP)
Tremfya (*guselkumab*) (IL23/Psoriasis)
Takhzyro (*lanadelumab*) (PK/HAE)
Portrazza (*necitumumab*) (EGFR/NSCLC)
Cyramza (*ramucirumab*) (VEGFR2/Cancer)
Lumoxiti (*moxetumumab pasudotox*)(CD22/HCL)
Bavencio (*avelumab*) (PD-L1/MCC)
ABthrax (*raxibacumab*) (Anthrax/Inhalational anthrax)

Table 4. Human antibodies made by phage display and approved by the US Food and Drug Administration. Each antibody listed in order of trade name, non-proprietary name, pharmaceutical target and disease area. SLE = Systemic Lupus Erythematosus, TTP= Thrombotic Thrombocytopenic Purpura, HAE = Hereditary Angioedema, HCL = Hairy Cell Leukaemia, MCC = Merkel Cell Carcinoma. Other abbreviations as in Table 2.

In conclusion, antibody libraries and phage display have provided the key elements for the creation of a fast evolutionary system for the generation of fully human antibody medicines.

REFERENCES

Amit A.G., Mariuzza R.A., Phillips S.E.V. & Poljak R.J. (1986) "Three-dimensional structure of an Antigen-Antibody Complex at 2.8A Resolution." *Science* **233**, 747–753

Bass S., Greene R. & Wells J.A. (1990) "Hormone phage: An enrichment method for variant proteins with altered binding properties." *Proteins* **8**, 309 –314.

Berek C., Griffiths G.M. & Milstein C. (1985) "Molecular events during maturation of the immune response to oxazolone". *Nature* **316**, 412–418.

Better M., Chang C.P., Robinson R.R. & Horwitz A.H. (1988) "Escherichia coli secretion of an active chimeric antibody fragment." *Science* **240** (4855), 1041–1043.

Bird R.E., Hardman K.D., Jacobson J.W., Johnson S., Kaufman B.M., Lee S., Lee T., Pope S.H., Riordan G.S. & Whitlow M. (1988) "Single chain antigen-binding proteins." *Science* **242**, 423–426.

Boss M.A., Kenten J.H., Wood C.R. & Emtage J.S. (1984) "Assembly of functional antibodies from immunoglobulin heavy and light chains synthesized in E. coli." *Nucl. Acids. Res.* **12**(9), 3791–3806

Boulianne G.L., Hozumi, N. & Shulman M.J. (1984) "Production of functional chimaeric mouse/human antibody." *Nature* **312**, 643–646.

Cabilly S., Riggs, A.D., Pande H., Shively J.E., Holmes W.E., Rey M., Perry J., Wetzel R. & Heynecker H.L. (1984) "Generation of antibody activity from immunoglobulin polypeptide chains produced in *Escherichia coli." Proc. Natl. Acad. Sci. USA* **81**, 3273–3277

Clackson T., Hoogenboom H.R., Griffiths A.D. & Winter G. (1991). "Making antibody fragments using phage display libraries." *Nature* **352**(6336), 624–628.

Cox J.P., Tomlinson I.M. & Winter G. (1994). "A directory of human germ-line V kappa segments reveals a strong bias in their usage." *Eur J Immunol* **24**(4), 827–836.

Gherardi E. & Milstein C. (1992) "Original and artificial antibodies" *Nature* **357**, 201–202.

Griffiths A.D., Malmqvist M., Marks J.D., Bye J.M., Embleton M.J., McCafferty J., Baier M., Holliger P., Gorick B.D., Hughes-Jones N.C., Hoogenboom H.R. & Winter G. (1993). "Human anti-self antibodies with high specificity from phage display libraries." *EMBO J* **12**, 725–734.

Griffiths A.D., Williams S.C., Hartley O., Tomlinson I.M., Waterhouse P., Crosby W.L., Kontermann R.E., Jones P.T., Low N.M., Allison T.J., Prospero T.D., Hoogenboom H.R., Nissim A., Cox J.P., Harrison J.L., Zaccolo M., Gherardi E. & Winter G. (1994). "Isolation of high affinity antibodies directly from large synthetic repertoires." *EMBO J* **13**(14), 3245–3260.

Hale G., Dyer M.J., Clark M.R., Phillips J.M., Marcus R., Riechmann L., Winter G. & Waldmann H. (1988). "Remission induction in non-Hodgkin lymphoma with reshaped human monoclonal antibody CAMPATH-1H." *Lancet* **2**(8625), 1394–1399.

Hamers-Casterman C., Atarhouch T., Muyldermans S., Robinson G., Hamers C., Bajyana Songa E., Bendahman N. & Hamers R (1993). "Naturally occurring antibodies devoid of light chains." *Nature* **363**, 446–448.

Hawkins R.E. & Winter G. (1992). "Cell selection strategies for making antibodies from variable gene libraries: trapping the memory pool." *Eur J Immunol* **22**(3), 867–870.

Hawkins R.E., Russell S.J. & Winter G. (1992). Selection of phage antibodies by binding affinity. Mimicking affinity maturation. *J Mol Biol* **226**(3), 889–896.

Hoogenboom H.R., Griffiths A.D., Johnson K.S., Chiswell D.J., Hudson P. & Winter G. (1991). "Multi-subunit proteins on the surface of filamentous phage: methodologies for displaying antibody (Fab) heavy and light chains." *Nucleic Acids Res* **19**(15), 4133–4137.
Hoogenboom H.R. & Winter G. (1992). "By-passing immunisation. Human antibodies from synthetic repertoires of germline V_H gene segments rearranged *in vitro.*" *J Mol Biol* **227**(2), 381–388.
Huse W.D., Sastry L., Iverson S.A., Kang A.S., Alting-Mees M., Burton D.R. Benkovic S.J. & Lerner R.A. (1989) "Generation of a large combinatorial library of the immunoglobulin repertoire in phage lambda" *Science* **246**, 1275–1281
Huston J.S., Levinson D., Mudgett-Hunter M., Tai M,., Novotny J., Margolies M.N., Ridge R., Bruccoleri R.E., Haber E., Crea R & Oppermann H. (1988) "Protein engineering of antibody binding sites: recovery of specific activity in an anti-digoxin single-chain Fv analogue produced in Escherichia coli." *Proc.Natl.Acad. Sci. USA* **85**, 5879–5883
Jespers L., Schon, O, James, L.C., Veprintsev D. & Winter G (2004). "Crystal structure of HEL4, a soluble, refoldable human VH single domain with germ-line scaffold." *J Mol Biol* **337**, 893–903.
Jespers L., Famm K., Schon, O. & Winter G. (2004). "Aggregation resistant proteins selected by thermal cycling." *Nature Biotech.,* **22**, 1161–1165.
Jespers L.S., Roberts A., Mahler S.M., Winter G. & Hoogenboom H.R. (1994). "Guiding the selection of human antibodies from phage display repertoires to a single epitope of an antigen." *Biotechnology (NY)* **12**(9), 899–903.
Jones P.T., Dear P.H., Foote J., Neuberger M.S. & Winter G. (1986). "Replacing the complementarity-determining regions in a human antibody with those from a mouse." *Nature* **321**(6069), 522–525.
Kohler G. & Milstein C. (1975). "Continuous cultures of fused cells secreting antibody of predefined specificity." *Nature* **256**, 495–497.
Low N.M., Holliger P. & Winter G. (1996). "Mimicking somatic hypermutation: affinity maturation of antibodies displayed on bacteriophage using a bacterial mutator strain." *J Mol Biol* **260**(3), 359–368.
Lowman H.B & Wells J.A. (1991). "Monovalent phage display: a method for selecting variant proteins from random libraries." *Methods* **3**, 205–216
Lowman H.B., Bass S.H., Simpson N. & Wells J.A. (1991). "Selecting high affinity binding proteins by monovalent phage display." *Biochemistry* **30**, 10832–10838
Marks J.D., Hoogenboom H.R., Bonnert T.P., McCafferty J., Griffiths A.D. & Winter G. (1991a). "By-passing immunization. Human antibodies from V-gene libraries displayed on phage." *J Mol Biol* **222**(3), 581–597.
Marks J.D., Tristem M., Karpas A. & Winter G. (1991b). "Oligonucleotide primers for polymerase chain reaction amplification of human immunoglobulin variable genes and design of family-specific oligonucleotide probes." *Eur J Immunol* **21**(4), 985–991.
Marks J.D., Griffiths A.D., Malmqvist M., Clackson T.P., Bye J.M. & Winter G. (1992). "By-passing immunization: building high affinity human antibodies by chain shuffling." *Biotechnology (NY)* **10**(7), 779–783.
McCafferty J., Griffiths A.D., Winter G. & Chiswell D.J. (1990). "Phage antibodies: filamentous phage displaying antibody variable domains." *Nature* **348**(6301), 552–554.
Morrison, S.L., Johnson M.J., Herzenberg L.A. & Oi V.T. (1984). "Chimeric human antibody molecules: mouse antigen-binding domains with human constant region domains." *Proc. Natl. Acad. Sci. USA* **81**(21), 6851–6855

Neuberger M.S. (1983). "Expression and regulation of immunoglobulin heavy chain gene transfected into lymphoid cells." *The EMBO Journal* **2**(8), 1373–1378

Neuberger M.S., Williams, S.G.T, Mitchell E.B, Jouhal S.S., Flanagan J.G. & Rabbitts T.H. (1985). "A hapten-specific IgE antibody with human physiological effector function." *Nature* **314** (6008) 268–270.

Nissim A., Hoogenboom H.R., Tomlinson I.M., Flynn G., Midgley C., Lane D. & Winter G. (1994). "Antibody fragments from a 'single pot' phage display library as immunochemical reagents." *EMBO J* **13**(3) 692–698.

Ochi O., Hawley R.G., Hawley T., Schulman M.J., Traunecker A, Kohler G. & Hozumi N. (1983). "Functional immunoglobulin M production after transfection of cloned immunoglobulin heavy and light chain genes into lymphoid cells." *Proc. Natl. Acad. Sci. USA* **80**, 6351–6355.

Oi V.T., Morrison S.L., Herzenberg L.A. & Berg P. (1983). "Immunoglobulin gene expression in transformed lymphoid cells." *Proc. Natl. Acad. Sci. USA* **80**, 825–829.

Orlandi R., Gussow D.H., Jones P.T. & Winter G. (1989). "Cloning immunoglobulin variable domains for expression by the polymerase chain reaction." *Proc Natl Acad Sci USA* **86**(10) 3833–3837.

Riechmann L., Clark M., Waldmann H. & Winter G. (1988). "Reshaping human antibodies for therapy." *Nature* **332**(6162) 323–327.

Rice D. & Baltimore D. (1982). "Regulated expression of an immunoglobulin k gene introduced into a mouse lymphoid line." *Proc Natl Acad Sci USA* **79**, 7862–7865

Saiki R.K., Gelfand D.H., Stoffel S., Scharf S.J., Higuchi R., Horn G.T., Mullis K.B. & Ehrlich H.A. (1988). "Primer-directed enzymatic amplification of DNA with a thermostable DNA polymerase." *Science* **239**(4839), 487–91

Schibler U., Marcu K.B. & Perry R.P. (1978). "The synthesis and processing of the messenger RNAs specifying heavy and light chain immunoglobulins in MPC-11 cells." *Cell* **15**, 1495–1509

Schroeder H.W. & Cavacini L. (2010). "Structure and Function of Immunoglobulins" *J Allergy Clin Immunol.* **125**(202):S41–S52.

Skerra A. & Pluckthun A. (1988). "Assembly of a functional immunoglobulin Fv fragment in Escherichia coli." *Science* **240**(4855), 1038–1041

Smith G.P. (1985). "Filamentous fusion phage: novel expression vectors that display cloned antigens on the virion surface." *Science* **228**(4705), 1315–7.

Hozumi N. & Tonegawa S. (1976). "Evidence for somatic rearrangement of immunoglobulin genes coding for variable and constant regions." *Proc Natl Acad Sci USA* **73**, 3628–32

Tomlinson I.M., Walter G., Marks J.D., Llewelyn M.B. & Winter G. (1992). "The repertoire of human germline VH sequences reveals about fifty groups of VH segments with different hypervariable loops." *J Mol Biol* **227**, (3) 776–798.

Ueda H., Kristensen P. & Winter G. (2004). "Stabilization of antibody VH-domains by proteolytic selection." *J Mol Catalysis B: Enzymatic*, **28**, 173–179.

Verhoeyen M., Milstein C. & Winter G. (1988). "Reshaping human antibodies: grafting an anti-lysozyme activity." *Science* **239**(4847), 1534–1536.

Ward E.S., Gussow D., Griffiths A.D., Jones P.T. & Winter G. (1989). "Binding activities of a repertoire of single immunoglobulin variable domains secreted from Escherichia coli." *Nature* **341**(6242), 544–546.

Waterhouse P., Griffiths A.D., Johnson K.S. & Winter G. (1993). "Combinatorial infection and in vivo recombination: A strategy for making large phage antibody repertoires." *Nucleic Acids Res* **21**(9), 2265–2266.

Williams S.C. & Winter G. (1993). "Cloning and sequencing of human immunoglobulin V lambda gene segments." *Eur J Immunol* **23**(7), 1456–1461.
Winter G. (1989). "Antibody engineering." In: *Phil Trans R Soc Lond B Biol Sci* **324**(1224), 537–547.
Winter G. & Milstein C. (1991). "Man-made antibodies." *Nature* **349**(6307), 293–299.

Chemistry 2019

John B. Goodenough, M. Stanley Whittingham and Akira Yoshino

"for the development of lithium-ion batteries"

The Nobel Prize in Chemistry, 2019

Presentation speech by Professor Olof Ramström, Member of the Royal Swedish Academy of Sciences; Member of the Nobel Committee for Chemistry, 10 December 2019.

Your Majesties, Your Royal Highnesses, Esteemed Nobel Laureates, Ladies and Gentlemen,

Imagine a world in which we have unlimited access to stored electric energy wherever we are. Perhaps the electricity is also coming from renewable energy sources that fluctuate over time, such as wind and sunlight, resulting in an uneven power supply. Not so long ago, this was very much a dream, and repeatedly storing large amounts of electricity for on-demand use was a major challenge. Today, however, we have come a long way to meet this goal, and the lithium-ion battery has increasingly become a viable solution.

Ever since Volta's pile, we have known that batteries can store and convert chemical energy into electricity. Still, it has proven to be tremendously arduous to develop efficient and rechargeable batteries. This is certainly true for the powerful lithium-ion battery, and many groundbreaking discoveries have been required to enable its development.

As the name implies, lithium-ion batteries are based on the element lithium in its ionic form. Around 200 years ago, this element was discovered by Arfwedson and Berzelius while analysing a mineral from the island of Utö in the Stockholm archipelago. They also named the element after the Greek word for "stone", even though lithium is our lightest metal. It soon became evident that lithium has several remarkable properties. The atoms are among the smallest we have, and their propensity for releasing an electron makes the element reactive, yet suitable for electrochemistry. Scientists gradually realised that these important properties would be useful for batteries with very high capacity.

Our Laureates took hold of these challenges and managed to tame lithium into what eventually became the lithium-ion battery. Stanley Whittingham worked with ion transport and superconductivity and discovered a titanium-containing material that could efficiently take up and release lithium ions. This revelation led to a lithium-based battery with a potential of around 2 V.

John Goodenough identified new, stable, lithium ion-binding electrode materials based on oxides and phosphates and was able to demonstrate batteries with potentials of over 4 V. Akira Yoshino developed stable carbon-based materials able to host lithium ions and replace reactive lithium metal. In combination with oxide electrodes, he was then able to show lithium-ion cells with high voltage. These discoveries laid the groundwork for the modern lithium-ion battery.

The discoveries of our Laureates have led to a dramatic change in our society. They have contributed to the so-called "mobile revolution", which has resulted in powerful, portable electronics. They have enabled the transition we are witnessing in the transport sector, with increased use of electrically-driven vehicles. They have simplified our use of reusable energy sources for the temporary storage of electrical energy and its subsequent on-demand use. This Prize is also a clear example of the fact that many crucial technical advances that transform our everyday lives have their origin in chemistry. Only with in-depth knowledge of the properties of metals and metal ions, their electrochemistry and interactions with different substances and materials could this progress be made. These chemistry discoveries have thus really paved the way for a simplified everyday life and an improvement in our environment.

John Goodenough, Stanley Whittingham and Akira Yoshino:

You have made groundbreaking discoveries in chemistry that has led to the development of the lithium-ion battery. This is a truly great achievement for the benefit of humankind. On behalf of the Royal Swedish Academy of Sciences I wish to convey to you our warmest congratulations. May I now ask you to step forward and receive your Nobel Prizes from the hands of His Majesty the King.

John B. Goodenough. © Nobel Prize Outreach AB. Photo: A. Mahmoud

John B. Goodenough

Biography

CHILDHOOD

I was born in Jena, Germany, in July 1922, to American parents, Erwin Ramsdell Goodenough and Helen Miriam (Lewis) Goodenough. My father was working on his D. Phil. dissertation on the Church Fathers at Oxford University at the time of my birth. My parents lived in Oxford, England, for three years and my father enjoyed the culture of the Weimar Republic; he spent much of his long summer vacations in Germany as well as in Rome. My grandfather was able to support my father before the market crash of 1928. After my parents returned to the U.S. from Oxford, my father became a professor of the history of religion at Yale University. My grandfather Goodenough had bought for him with a large mortgage an old house on a five-acre lot complete with an adjoining woodshed, a large barn, an ice house, and a windmill for pumping our water from a spring in the back lot. A huge wisteria vine on which we could climb was supported on the south side of the house by a giant maple tree and on the east side by a trellis. On the north side, a U-shaped driveway leading past the house to a barn encircled apple trees. The windmill, which invited young boys to climb it, was then quickly replaced by an electric pump and the coal-fired furnace by an oil furnace. However, ice was still hauled to an icebox and kerosene for the stove was fetched from a little house on the north border of the property. A large veranda in the front of the house faced west overlooking a two-bar fence before a row of elm and maple trees bordering Amity Road, the main bus route between New Haven and Waterbury. We were located seven miles north of Yale University, which is in downtown New Haven.

My older brother, Ward, and I shared the north bedroom. Ward was the leader; I was the tag-along when he would tolerate me. My world was with my dog, Mack, in the nearby meadows and woodlands where there was so much life to discover and so much wonder to experience. I liked collecting trophies, whether it was butterflies, seashells, or animal skins. A special

Figure 1. Left: My older brother Ward (right) and I (left) outside our house in Woodbridge. *Middle*: With my bicycle circa 1930. *Right*: My younger sister Hester (right side) and I (left side).

room halfway up to the hayloft in our large barn was where I housed my skin and hawk-wing collection. Despite many pleasures that I experienced throughout my childhood, my years through 7 were difficult for me. How I struggled to learn to read! I read mechanically without the ability to catch easily the meaning of a paragraph. I never was a good reader, and through my school years, I worked hard to cover my deficiency. I also had a deep sense of insecurity that only lifted slowly as I grew older.

EARLY EDUCATION

I went to a private grammar school in downtown New Haven about a mile from my father's office in Jonathan Edwards College at Yale University. My father drove me to school on his way to work. I went with a lunch pail; the meal at school was too expensive. I didn't mind eating separately. I made friends easily as I enjoyed sports on the playground and being mischievous with the other boys in the back row during French class.

I arrived with a full scholarship to Groton School on a warm September day in 1934. I was shown my cubicle in the Hundred House dormitory; each student had a bed, a dresser, and a place to hang his suits. Suits and neckties were worn at all times but on the athletic field; a stiff collar and patent-leather shoes were worn to dinner. On the first floor of the dorm, there were assigned desks in the study hall for First and Second Formers; Third Formers had a study hall in Brooks House at the other side of the chapel. Older boys had personal one- or two-boy studies.

The School day was regulated by bells. A School House bell rang out across the Circle at 7:00 a.m. After breakfast, the school day began with a

15-mintue chapel service at 8:15 before the trek to our homeroom desk. First Formers had their own homeroom; all others had desks in a large study hall where announcements were made after morning classes before being dismissed to Hundred House for lunch. Afternoons were for doing homework in a study hall for an hour followed by mandatory participation in sports until supper. The Rector kept boys occupied every hour of the day. The only exception was Sunday, a day for writing home, attending chapel morning and evening, and reading in the library.

Each class, with the exception of First Form and Sixth Form Sacred Studies, was assigned three sections: B, A, and upper A. I entered as a B student in all subjects. Even so, I was overwhelmed the first few weeks. The Rector taught First Form Sacred Studies. On our first class test, we were asked to name the 12 disciples of Jesus. As I was a poor reader with no religious training, I think I remembered no more than one. Latin class with the football coach went no better, but English grammar under Zahner turned out to be a memorable learning experience. In my Fifth Form year I had been advanced from the B through the A to the upper-A division in all my classes, and I managed to graduate magna cum laude at the head of my Form.

Little time was spent in New Haven after I left home for School. Invitations to spend days or weeks with classmates were common. One summer we were in Colorado and others were mostly relieved by working as camp counselor and lifeguard. I was 14 when I first became a counselor at a YMCA camp. The summer I turned 17, my father added $150 to my savings to allow me to go to Finland with a group from Putney School.

UNDERGRADUATE YEARS

I entered Yale University as an undergraduate student in 1940. As I left home, I determined never again to take money from my parents; I had no

Figure 2. Left: My siblings Hester (left), James (right), and I (middle) in 1933. *Right*: With my classmates on the Groton School football team in 1939; I am second from the right.

idea how I would support myself at Yale, but my focus was to do well in my College Board Exams. Fortunately, I was given a summer job tutoring a grandson of the Rector; it would give me room and board for the summer and enough money to pay for a room in a dormitory at Yale the following year.

My freshman year at Yale I was permitted to take a junior course in Ethics and Aesthetics in lieu of English; a junior course translating the Greek plays, and a sophomore second-year calculus course also gave me a good start. My freshman chemistry, Qualitative Analysis, was to satisfy my science requirement and to give me the possibility of Medical School. Freshman Psychology proved to be an intellectual insult; I found Freud totally unconvincing in many of his assertions, and Pavlov's dog seemed to be more trivial than made out. I was not drawn to the behavioral sciences even though Psychiatry was quite the in-subject at the time.

A common temptation of youth is the desire to be famous or glamorous or powerful. I realized that not everyone can be "king of the mountain" even for a short time. Can being king, therefore, be what gives meaning to life? I began to understand that any meaning to a life is not to be king of a castle, but the significance and permanence of what we serve. Is service to ourselves, our tribe (ethnic group or family), or to our country the highest service? In a time of war against evil, service to the cause of a more just world and therefore to our war effort was meaningful; but the destructive means of war is always an appalling waste on all sides. I struggled to find a meaningful calling beyond the war. Perhaps it would be in science; so, for my sophomore year, I decided to enroll in the philosophy of science and in physics.

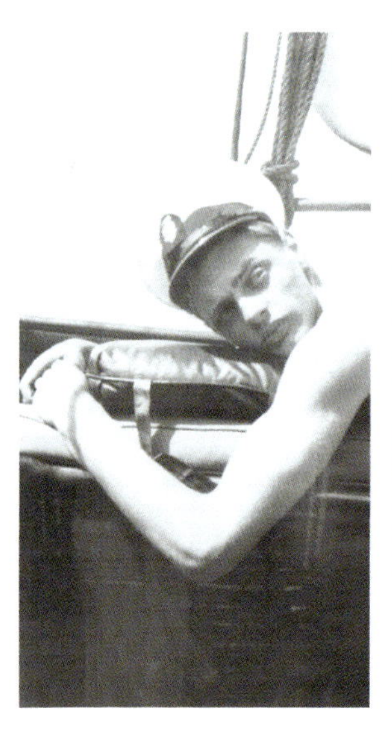

Figure 3. Left: Asleep on a sail up the New England coast. *Right*: Formal portrait of myself as an undergraduate student in 1942.

Figure 4. With the members of the secret Skull and Bones society at Yale University in 1942; I am the first person to the right of the clock in the back row.

In the summer of 1941, I tutored the son of a Chicago banker who lived in Wheaton, Illinois. Their home was a grand mansion built by the grandfather who had worked his way up to become Head of the First National Bank of Chicago. I slept on a porch with the two boys and three great Danes; I was treated as one of the family, and I earned enough money to pay for my room in Timothy Dwight College of Yale. During my latter years at Yale, a bursary job gave me 21 meal tickets during weeks when classes were in session; I was a grader of Freshman mathematics papers. But how was I to afford food during the vacations? Fortunately, the mother of my roommate, Stuart Little, would invite me to Hartford, Connecticut, to share part of my holiday there.

When I went to enlist in 1942, my mathematics professor, Egbert Miles, called me into his office and said, "John, don't sign up for the marines like all your friends. The military needs boys with backgrounds like yours to sign up for meteorology in the U.S. Army Air Corps." I had no stomach to play the hero in war, so I acted on this friendly advice. It gave me another year in college to finish my undergraduate degree, and I spent the summer of 1942 in New Haven. I was not called to active duty until February of 1943. At that time, I lacked one course for graduation, and Yale graciously gave me credit for my Army course in meteorology to grant me a summa cum laude Bachelor of Arts degree in mathematics in the spring of 1943. However, I was cognizant of the fact that, after completing successfully a second-year physics course, I had been saved from an embarrassing flunk of my first test in the Theoretical Physics course taught by Margenau by a telegram the night before calling me to active duty. I was pursuing physics after reading "Science and the Modern World" by Alfred North Whitehead. While reading that book one evening, it seemed to me that much of the intellectual ferment of my generation would be in science; and physics provides a fundamental foundation for science. If there was to be an opportunity to go to graduate school after the war, I felt that night that I should study physics.

WORLD WAR II ARMY METEOROLOGY

Figure 5. My photograph for University of Chicago International House application when serving in Army during WW II.

Entering the Army ended the pressures I had felt at Yale. The struggles to support myself, to find a calling and to finish my undergraduate requirements before being called into the Army all fell away. Most members of my Yale class were at different stages of entering the armed forces when I left, so I was not parting from a normal college experience. Adjustment to military life was not a problem for me. After a brief orientation in Boca Raton, Florida, I went by troop train to Grand Rapids, Michigan, for training to be a practicing meteorologist. The Army training, mostly by civilians, was efficient and quite professional.

Upon commission in the autumn of 1943, I was immediately posted at an air base in Houlton, Maine a few miles south of a more active air base in Presque Isle. Fighter planes were being dispatched from Presque Isle to

England. After two weeks, I found myself in charge of the weather station in Houlton. In those days we drew our own maps and made our own forecasts; there was no satellite and no computer-aided forecast from Washington.

Maine proved to be a good training experience for my next posting the following summer in Stephenville on the west coast of Newfoundland. Stephenville was the jumping-off base for the cargo B54s flying to either the Azores, the base of the northern route, or directly to England. These planes also stopped in Stephenville on their way home to Washington, D.C. The B54s had a longer range than the fighter planes. The tactical bombers were dispatched from Gander on the east side of the island.

Although almost all my forecasts were reasonably accurate, including a clearing of Eisenhower from Stephenville that landed him safely in Paris within 6 minutes of his estimated time of arrival, a forecast could be dangerously wrong.

As D-day approached, we tried to predict from the weather when the allies would storm the beaches of France. Eisenhower and our forces had bad luck with the strength of the cold front behind which they attacked. We followed closely the battle of the hedge rows and the final breakout across France.

One December day, civilian pilots flying the B54s were congratulating themselves that they were going to make it home for Christmas. When I refused to clear them for the trip to Newfoundland because a strong headwind from there to the Azores would prevent them from reaching their destination, they set out anyway. Six hours later they were back on base; the headwinds were so strong they had barely cleared the islands.

At the end of the war in Europe in 1945, my thoughts began to return to my struggle with a Christian commitment. In Sta. Maria, a Lutheran Chaplain bowed his head to give thanks before eating, and that simple act in the Mess Hall stirred up into consciousness questions I had suppressed since leaving civilian life. I decided I should read the Bible to let it speak for itself; honest dialogue surely was where I ought to begin. However, life on the base was not conducive to this discipline; the fantasies of youth and the comraderies of my fellow soldiers usurped my leisure attention. With the surrender of Japan, our thoughts turned to the question of how to return to civilian life. A letter from headquarters invited me to stay in the Army as a meteorologist; I was then a Captain. However, I thought our responsibilities in peacetime would be less than those we had assumed in war, so the invitation was declined.

Although Law did not attract me as a profession, I entertained the possibility of studying to become an international lawyer. As this idea was developing in my mind, I received a TELEX in the spring of 1946 to return from the Azores to Washington, D.C., within 48 hours. I packed up my duffle bag with great excitement; my turn to go home had come, and I was to embark on a new adventure!

In Washington, I was told that I was one of 21 returning officers who had been chosen to do graduate study in either Physics or Mathematics at the University of Chicago or at Northwestern University; we would remain in-grade, but under the command of the Quartermaster Corps. My mathematics professor at Yale, Egbert Miles, had not forgotten me! His act was unusual; except for him, I was to return as an unknown to begin life all over again. My debt to Professor Miles is profound. He it was who put my name forward when some educators became aware that a sum of unspent money was available; Egbert Miles thought it would be best spent reintegrating a few promising scholars to civilian life by giving them an opportunity to go to graduate school. From Washington I was to go immediately to Chicago as I should have been there 24 hours earlier. That night, on my way to Chicago, a vivid memory returned; I saw myself reading "Science and the Modern World" by Alfred North Whitehead the day I decided I should study Physics if I ever had the opportunity on my return from the war. This opportunity was here! I felt called to sign up for Physics at the University of Chicago the next day even though I believed I

Figure 6. From left: Ruth (Ward's wife), myself, my mother, Ward (my older brother), James (my younger brother), Hester (my younger sister), and two of my nieces (Ward's daughters).

would not qualify if they tested my aptitude for the subject. When I went to register, Professor Simpson said to me, "I don't understand you veterans. Don't you know that anyone who has ever done anything significant in physics had already done it by the time he was your age; and you want to begin?!" But my decision was made. I had decided to study physics at the University of Chicago.

GRADUATE YEARS

After serving in the US Army as a meteorologist in World War II, I went to the University of Chicago to do graduate study in Physics from 1946 to 1951. The University of Chicago is located a few blocks from Lake Michigan on the south side of the city; it borders an open strip of grass, called The Midway, that separates 55th street, which runs west from the lake. On the north side of The Midway between the campus and the park was International House; it provided rooms and meals for graduate students and visiting scholars, men in one wing and women in the other with a common room and other facilities between and beneath the dormitories. I managed to secure a room there from the autumn of 1946 until I left Chicago in 1951.

Under president Hutchinson, the University of Chicago had a two-year undergraduate program followed by entrance into an Upper Division for graduate study. Unlike in England, where the Advanced (A) Levels before university provided the grounding for a university specialty, these first two years at Chicago were designed to provide a broad liberal education for intelligent citizenship. This arrangement was a perfect match for me. I had not majored in science at Yale but had entertained many subjects in search of a general training. However, students entering the physics program in 1946 were coming from very diverse backgrounds. Almost all had been physics majors, and many came with considerable experimental experience from service assignments at Los Alamos or the MIT Radiation Laboratory. Moreover, the Department was a bit overwhelmed by the number of students they had felt obliged to admit to the program.

My first textbook at Chicago was a tome on mechanics presumably designed some years before 1939 to help students in Cambridge, England, to pass the tripods with the help of a tutor. The first 10 pages so intimidated me that the book was quickly discarded! However, I had the good fortune that first semester to have professors who endeavored to communicate the fundamentals. In the spring of 1947, an examination covering the material of the first year was introduced for the purpose of reducing the class size by 50%. I had done well enough my first year to be exempted from this exam.

Later years proved more difficult. The Physics Department had decided to adopt the Oxford-Cambridge British system of self-study, but without tutors! Lectures covered aspects of a subject that interested the professor; the student was expected to develop on his own the context for the topics covered. Class tests were not necessarily on the lecture topics. Edward Teller, for example, only appeared for three lectures the entire semester I took his class. Moreover, texts on modern physics appropriate for students were scarce or not available. In my course on electromagnetic theory, the text used one set of units, the lecturer another, and the exams were in a third. Although Quantum Mechanics was being developed in the 1920s and 1930s, the war had prevented the appearance of a good text for beginning students. It was easy, therefore, for the student to become so absorbed with the mathematical derivations that the physics was lost. Fortunately, Enrico Fermi introduced us to quantum mechanics. His class on nuclear physics covered ideas developed during the war for which there was no text. His lectures seemed clear enough, but when I attempted to solve his assigned problem for the day, I often found that he assumed we already had the background needed to appreciate fully his expositions. Three of the veteran students developed his lectures into a text that was later used widely. There was no electrical engineering at Chicago. When I asked for a course that would teach me electronics, as I needed to build experimental equipment, I was told to go read the literature. Veterans who had been trained in electronics during the war had an advantage for developing into experimentalists.

At the end of four years, we took a 32-hour written examination on four successive days. Students were only told that the examination would cover all aspects of physics and pertinent topics in mathematics and chemistry. The first eight-hour day consisted of about 32 shorter questions; the second, eight more difficult questions. On the third day there

Figure 7. Left: Myself as a graduate student. *Middle*: Irene and I on our wedding day, June 1951. *Right*: My graduation from the University of Chicago, 1951.

were only four questions, two experimental and two theoretical; the student was to answer three of them. On the fourth and final day there was only one question. The student was allowed to use the library to answer this one. For example, one such final problem was to write a proposal for funding an experiment that required design of a bathysphere for a deep-ocean study, a defense of the scientific significance of the study, and a design of the instruments to be used to accomplish the desired measurements. The first time I took this examination, I did well enough to be granted an MS degree, but I would have to take it a second time to be allowed to go on for a Ph.D. On my second try six months later, I was so discouraged after the third day that I almost didn't sit the fourth. I went out and played a game of softball that evening. More relaxed afterwards, I decided I had nothing to lose if I went for the fourth day. I have never really understood why they allowed me to go on for the Ph.D. degree. Only 10% of those with whom I entered the Department in 1946 had made it through this hurdle. In subsequent years, the Department changed this procedure; the veterans and foreign students that entered in 1946-1948 were more mature than those that followed. I knew that I had been exposed to the challenges and practice of the physics profession at its highest level; but for me it was a challenge indeed!

The Physics Department of the University of Chicago also had a policy that no professor was to attach his name to the publication of work reported in a Ph.D. dissertation. Work with the professor as a Research Assistant could not be part of a Ph.D. dissertation. The objective was to ensure that the Ph.D. dissertation represented original research developed and executed only by the student. I knew that I didn't want to go into nuclear physics, so I opted to do my research in solid-state physics. Professor Clarence Zener was the obvious choice. Zener was involved in the physics of metals, so I asked him to take me on as a student in his group. I was to come back the following Thursday to learn of his decision. That Thursday I entered his office with some trepidation. To my relief, he said "Yes, you can be my student." Then he added, "Now you have two things you must do. The first is to find your research problem and the second is to solve it. Good day!"

In fact, Zener proved to be helpful. First, he gave me a Research Assistant position measuring the internal friction of iron wires doped with carbon or nitrogen. Second, once a week he had a bag lunch with his students. One of them was to give a lecture on some topic that he assigned. After my first lecture, he called me to his office and asked whether I had found a research problem in the topic assigned. So, this was his game! In my next assignment I found a topic that proved too hard for me to solve. In my third round, the problem I found was too easy. Finally, in my fourth round I found one that would work for me. I would calculate how the interaction of the Fermi surface with the Brillouin-zone boundaries of non-cubic metal alloys would influence or change their structure. In momentum space, the position of the Fermi surface of a metal depends on the electron density in the conduction band, and the Brillouin zone is determined by the translational symmetry of the periodic potential in which the electrons move.

While I was engaged with this problem, Zener took a job as Director of the Westinghouse Research Laboratory in East Pittsburgh, PA. He invited his students to join him. My position there as a Research Engineer my final year enabled me to get married to Irene Wiseman Goodenough the Spring before we moved to Pittsburgh. When I was writing up my dissertation, Zener informed me that my employment at Westinghouse would be terminated; I was to start looking for another job.

The week before my final defense back in Chicago, I went to the American Physical Society meeting in Washington D.C. to present my work and to look for a job. After my 10-minute talk, an old man in the front row stood up and said, "That's fine, young man, but you do not have the correct Brillouin zone for the hexagonal-close-packed structure!" The old man was Brillouin. The Head of the Chicago Physics Department was in the audience and witnessed the embarrassed silence after Brillouin had spoken. I went back to my hotel room demolished! Any Ph.D. defense the following week would be lost, I thought. However, that weekend I was able to show that although Brillouin was mathematically correct, he was

Figure 8. Irene and I in New Hampshire.

physically wrong. There is no energy discontinuity across the zone face that I had omitted. The zone that I had used was the one that was physically meaningful. When I defended my thesis in Chicago the next week, another professor challenged me. He had a student working on the same problem from a different point of view. This time I was able to better the challenge, and the confrontation had pleased the examiners. I was finally awarded the Ph.D. degree!

LINCOLN LABORATORY YEARS

When I received my Ph.D., three options were offered to me: (1) to be an Assistant Professor in the Physics Department of the University of Pennsylvania, (2) to be a Research Engineer at the MIT Lincoln Laboratory, and (3) to be a Research Fellow at Harvard University. The position at the Massachusetts Institute of Technology seemed to be the best match for me, and I went to Boston with an inner assurance.

The MIT Lincoln Laboratory was supported by the Air Force to create a defense against aircraft; the ballistic missile was not yet a threat. The defense system brought together the radar installations developed at MIT during World War II, communications, and the digital computer. In 1952, the digital computer ran on vacuum tubes and filled the space of a large dance hall; but it had no memory. Jay Forrester of the Electrical Engineering Department of MIT had invented the concept of a random-access memory (RAM) storing binary numbers, 0 and 1. His memory used as the memory element a ferromagnetic transition-metal alloy with a square B-H hysteresis loop formed by rolling alloy tapes into thin sheets. Frustrated by the inability to switch the magnetization direction fast enough, Forrester had concluded that his problem was due to eddy currents in the metallic alloys. Since the entire project depended on a faster RAM memory of the digital computer, he decided to investigate the possibility of developing a square hysteresis loop in a ferrimagnetic oxide that was an insulator. Ferrimagnetic oxides had been developed secretly in France and Holland during World War II. I was assigned to a small group to develop the square hysteresis loop in a ceramic that cannot be rolled. The magneticians of the day assumed it would not be possible, but the group of ceramists and electrical engineers that I joined were empirically synthesizing and testing ferrimagnetic spinels that showed some promise; the oxospinels contained Mg, Mn, and Fe and were supplied by a small ceramics company.

Within three years, systematic empiricism and quality control enabled the experimentalists of our group to develop a recipe for reproducible fabrication of polycrystalline ceramic cores with the needed squareness of their M-H loop. This success involved optimizing the composition and specifying accurately the firing time at specified temperatures as well as the cooling rate for some unknown atomic-ordering process during synthesis. My contribution was, first, to identify the factors that controlled the shape of the M-H hysteresis loop and to show that the lower value of M in the ferrimagnetic oxides compared to the ferromagnetic alloys alleviated the requirement of aligned crystallographic axes between grains. Next, I showed that the switching speed was controlled by an intrinsic damping factor, not eddy currents, and by the magnitude of the driving field H, which was limited by the application to less than twice the critical field H_C at which the magnetization is reversed. We were fortunate to have a larger H_C in the oxides than in the rolled alloys. Although I had not yet identified the defect that triggered nucleation at H_C of a domain of reverse magnetization that would also grow at H_C to switch the total direction of M, I had shown that the atomic-ordering process occurring at the annealing temperature was associated with a critical concentration of manganese in the oxide. I had also recognized that a distortion from cubic to tetragonal symmetry occurring above this critical concentration of manganese was due to a cooperative orbital ordering at the manganese. This recognition introduced a fundamental new insight into a factor that determines crystal structure. It is now referred to as a cooperative Jahn-Teller effect since Jahn, as a student of Teller, had years before shown that where an isolated molecule has an orbital degeneracy in a state of high symmetry, the molecule is made more stable by a deformation to lower symmetry that removes the degeneracy. It was only some years later that I was able to show that the critical atomic order that we were controlling was a chemical inhomogeneity induced by a dynamic site distortion that cost less elastic energy in the crystal if it occurred cooperatively within manganese-rich regions. Nevertheless, I was able to predict that ferrospinels containing a critical concentration of Cu^{2+} ions would also yield the desired square M-H loop. I had also been able to apply the insight of a cooperative orbital ordering to articulate chemical rules for the sign, parallel or antiparallel, of the interactions between atomic magnetic moments. These rules are now known as the Goodenough-Kanamori rules; Kanamori subsequently provided a mathematical formalism justifying these rules.

One Friday afternoon after we had delivered the fast RAM with a ferromagnetic spinel memory element, Jay Forrester summoned the group to his office. I thought he might give us a raise. Instead, he thanked us for solving his problem and asked, "Now that you have worked yourselves out of a job, what are you planning to do?" The response of half of the group was to take their know-how to industry. My response was to spend the weekend thinking what we should do next. I came up with the idea of a magnetic-film memory in which all the individual atomic moments would

Figure 9. Myself (the fourth from left in the back standing) at a group meeting of Division 8 (Solid State Division back then) in Lincoln Laboratory in January 1964. (Reprinted with permission. Courtesy of MIT Lincoln Laboratory, Lexington, Massachusetts).

switch simultaneously rather than sequentially. On paper it promised to increase the switching speed one-thousand-fold. However, magnetic cores could be made smaller and realization of reliable switching of films meant slowing of their switching time. Since the film technology was more demanding, the eventual difference in switching speeds was not great enough for a move to magnetic films except for a few niche operations. With the advent of fast transistors that could be miniaturized, the Whirlwind Digital Computer and its magnetic memory became obsolete, but this technology was a critical step in the evolution to today's supercomputers and laptops.

With the exodus of half of the group and its leader, I was put in charge of the remnant and charged with realizing the magnetic-film memory and rebuilding the ceramic facility. After two years, I gave the magnetic-film project to Donald Smith who asked for it and devoted my time to devising experiments with transition-metal compounds, mostly oxides, that would reveal how competing interactions between atomic moments would give unexpected and/or complex magnetic order; I also investigated the role of cooperative orbital ordering in determining not only magnetic order, but also magnetostrictive phenomena that could be used in devices. This work resulted in my first book, published in 1961, "Magnetism and the Chemical Bond". At the same time, I realized that the transition-metal oxides and sulfides also permitted a systematic study of the transition from the localized-electron behavior responsible for atomic magnetic moments to itinerant-electron behavior as the strength of the interatomic interactions between atomic magnetic moments increases beyond a critical strength. Intraatomic interactions localize the electrons to an atomic site; interatomic interactions delocalize them. Itinerant electrons belong equally to all the like atoms of a periodic array, which allows metallic conductivity and suppression of any spontaneous magnetism. Studies of phenomena at this cross-over led to a long review, "Metallic Oxides", published in 1971; it was translated in French into a book, "Les oxydes des métaux de transition" published in 1973. Clearly, the move to Lincoln Laboratory had been a good match; it had allowed me to find my own scientific voice and to contribute at a critical point to the development of the digital computer, a revolutionizing technology for a growing, diverse global population.

In the mid-1960s, I was moved with the ceramics and magnetic measurement part of my group to the Solid State Division where I was also to have charge of the chemists investigating the growth of single crystals and the semiconductive materials finding application in the blossoming fields of microelectronics, photovoltaics, lasers, and solid-state lighting. A high-pressure facility was also available in that group; it was a tool well-suited for my studies of the transition-metal oxides. I transferred to my new group a vibrating-coil magnetometer we had built for the purpose of making magnetic measurements under pressure. Working with John Longo and James Kafalas was very productive. High-pressure synthesis and the ability to make magnetic measurements under pressure proved fruitful, but I was forced to suspend these studies in 1970 until I moved to Texas in 1986.

An amendment to a bill of Congress in about 1970 forbade research in a government Laboratory that was not targeted towards a specific application, and I was ordered to terminate my fundamental studies. At the time, I had just started a study of the copper-oxide system in which Bednorz and Muller made the discovery in 1986 of high-temperature superconductivity that won them the Nobel prize in physics; but I was not looking for superconductivity in 1970. However, I had done enough work on the transition from localized to itinerant electronic behavior to know that lattice instabilities are found at this crossover. This knowledge enabled me to identify the critical role of these instabilities not only in the surprising phenomenon of high-temperature superconductivity, but also in that of the colossal magnetoresistance discovered later in the manganese oxides. However, this insight was resisted by the orthodox theoretical physics community for nearly 20 years.

At Lincoln Laboratory in 1970, I turned my attention to the problem of renewable energy and energy conservation. It was obvious already in 1970 that our dependence on foreign oil was making the country as vulnerable as the threat of ballistic missiles from Russia. Solar energy was an obvious renewable source to be harnessed; our profligate use of energy made conservation an obvious target also. Since solar energy is variable in time and location, it was also obvious that we needed to find a way to store the solar energy that is converted into electricity. The best place to store electrical energy is to convert it to transportable chemical energy. Two clean

Figure 10. At my home office in Boston.

routes to storing electrical energy in chemicals are electrolysis, as in the electrolysis of water to produce hydrogen, and in the anode of a rechargeable battery. Both options interested me. For improved efficiency of a power plant, I proposed use of the exhaust heat in a solid oxide fuel cell. Although we had, at Lincoln Laboratory, the facilities and interested scientists and engineers to tackle research and development targeting these applications, we were told that we were an Air Force laboratory, and that energy was to be the domain of the National Energy Laboratories and of Industry. Politics, not potential productivity, is the bull in the china shop of science administration. It was a discouraging moment, and I realized it was time for me to leave Lincoln Laboratory.

Figure 11. Gordon Research Conference, New Hampton School, August 1974. I am the third person from right on the back row.

TENURED YEARS: OXFORD

During the late 1970s and early 1980s, I continued my career as head of the Inorganic Chemistry Laboratory at the University of Oxford. In 1976, Oxford had four chemistry laboratories, each headed by a Class A Professor: Organic, Physical, Theoretical, and Inorganic.

The Oxford educational system provided me a relatively smooth transition from a research laboratory to an academic post. Most of the teaching is done by the Dons in the Colleges; lectures are a supplement. The Dons act as coaches to self-teaching in preparation for the big final examination at the end of the third year. They also act as admissions officers to their college and compete with the other colleges for best scores in the finals. Laboratory instruction, lectures, and research are carried out in the Chemistry Laboratory. Competition for scholarships for D. Phil students between the three large Chemistry Laboratories – Organic, Physical, and Inorganic – sometimes required diplomacy between the professors. Fortunately, good applicants to Inorganic Chemistry always put me in a favorable position.

For my research program, I initially selected two primary targets; the direct methanol-air fuel cell and the photoelectrolysis of water. The former requires for its electrolyte a solid proton (H^+-ion) conductor that is an electronic insulator and can operate near 300°C or an anode that is catalytically active for the oxidation of methanol (CH_3OH) below 80°C and chemically stable in an acidic solution. I soon found that good proton conductivity in a solid electrolyte only occurs where the solid is wet, which means operating below 80°C. Our unsuccessful search for a sufficiently active and chemically stable anode for a direct methanol-air fuel cell introduced me to the field of electrochemistry. Our attempt to realize a practical electrode for photoelectrolysis also involved electrochemistry. I had hoped to be able to use most of the spectrum of visible light by using a filled as well as a nearly empty d-electron band of a transition-metal oxide. However, the filled d-electron band proved to be too narrow for this strategy to be practical. The alternative was to attach a dye to the surface of the oxide. This exercise provided a good D. Phil. thesis, but it did not give a practical solution. I realized it would probably be better to separate the steps in the process by coupling a photovoltaic cell to an electrolysis cell in order to store solar energy as chemical energy in hydrogen gas, a portable fuel. More successful was my effort to identify a suitable cathode material for a lithium battery, an effort that has done much to bring together the solid-state chemist and the electrochemist.

The most mobile working ion in a rechargeable battery is the H^+ ion. But these protons are only mobile in an aqueous acidic or alkaline electrolyte. To avoid electrolysis of the water, the single cell of an aqueous-electrolyte rechargeable battery is restricted to a voltage less than

Figure 12. Left: Me (front row, second from right) with my colleagues at Oxford University, 1982. *Right*: Reunion of Groton School of the class of 1940 and their wives.

1.5 V if the battery is to have a long shelf life. This restriction limits the energy density of a rechargeable battery, which is why the advent of the cellphone and the laptop computer had to await the arrival of the lithium rechargeable battery. The working Li^+ ion of a lithium battery is mobile in a nonaqueous electrolyte, which permits single-cell voltages over 4 V. However, realization of a competitive rechargeable lithium battery requires identification of electrode materials into/from which Lithium can be inserted/extracted reversibly over a large solid-solution range. Moreover, the active redox couples of the insertion-compound electrodes must have energies matched to the allowable energy of the electrolyte if larger voltages are to be achieved.

Before I left Lincoln Laboratory, I had been asked to monitor the Na-S battery project at the Ford Motor Co. This assignment introduced me to electrochemistry and the problem of designing fast cation conduction in a solid electrolyte. With Henry Hong, a crystallographer and chemist, I explored framework structures for a 3D Na^+ conductor as against the 2D Na^+ conduction in β-alumina. We came up with the $Na_{1+3x}Zr_2(Si_xP_{2-x}O_4)_3$ framework structure that was called NASICON (NA SuperIonic CONductor) by colleagues after I had left for Oxford. This structure is now being explored further for cathodes and solid electrolytes of sodium rechargeable batteries.

In about 1974, Brian Steele of Imperial College, London, was aware of the Rouxel and Schöllhorn work on the chemistry of reversible Li^+ insertion into layered MS_2 sulfides and suggested at a conference the use of TiS_2 as the cathode of a Lithium rechargeable battery. Primary Lithium batteries using a flammable organic liquid carbonate electrolyte with an ethylene-carbonate additive to passivate the Lithium anode from reducing the electrolyte had been marketed. M. Stanley Whittingham was a postdoc at Stanford with Bob Huggins and Fred Gamble was a physics student there with Ted Geballe studying intercalated TiS_2 as a 2D superconductor. They were hired by the Exxon Mobil Corporation to develop a commercial Li/TiS_2 rechargeable battery, and in 1976, Whittingham

Figure 13. Working at my home office in New Hampshire.

reported fast, reversible Li^+ into TiS_2 in a rechargeable Li/TiS_2 cell. However, on charge, the lithium anode formed whiskers (dendrites) that grew across the electrolyte on repeated charges to cause an internal short-circuit that ignited the flammable organic-liquid electrolyte. This effort was, therefore, terminated in the U.S. However, the concept of a reversible intercalation of Li^+ into layered compounds was established, and several laboratories were exploring Li^+ insertion between the layers of graphitic carbon. At the SONY Corp. of Japan, they were planning to use lithiated graphite as the anode with a TiS_2 cathode.

In 1978, an undergraduate thesis at Oxford on the structure of the $LiMO_2$ oxides reminded me of work I had done with Donald Wickham in the 1950s on $Li_xNi_{2-x}O_2$. The MO_2 oxides are not layered as is TiS_2; the electrostatic repulsive energy between the O^{2-} ions of MO_2 sheets is larger than the dipole-dipole Van der Waals binding energy. However, layered $LiMO_2$ is stabilized by the Li^+ ions between the MO_2, sheets, and the ions are well-ordered provided the sizes of the Li^+ and M^{3+} ions are sufficiently different from one another. I decided to investigate how much lithium can be extracted reversibly from a well-ordered $LiMO_2$ layered oxide. Since I wanted an M^{4+}/M^{3+} redox couple that had an energy well below the Li^+/Li^0 couple of a metallic Lithium anode, I chose to study chromium, cobalt, and nickel for the M atom. An experimental physicist, Koichi Mizushima, had just come from the University of Tokyo to work with me at Oxford. I teamed him with my chemist post-doctoral assistant, Phillip Wiseman, to work on this investigation. We found that over half of the lithium could be removed reversibly with cobalt or nickel as the M atom; each of these $Li_{1-x}MO_2/Li$ half-cells gave an output voltage near 4V.

No battery company in England, Europe, or the U.S. was interested in licensing a patent for these cathode materials; they could not imagine starting with a discharged cathode. The University of Oxford was not interested at that time in the intellectual property of its academics. As I was working with scientists of the AERE Laboratory in Harwell, a town near Oxford, to obtain joint funding for battery research from the European Economic Community (EEC), I arranged for them to apply for a patent on the understanding that, once they had retrieved their filing expenses, my two colleagues and I would share any revenue. On the day of signing, I was told that the AERE Harwell lawyers would not proceed unless we signed all our rights away. Not knowing either the full potential of our invention or any other option, we signed our rights away. Meanwhile, others were exploring the chemistry of Li insertion into layered compounds. In Switzerland, Rachid Yazami showed that reversible Li insertion/extraction into graphite occurs at only 0.2 eV below the electrochemical potential of metallic lithium without dendrite formation. In Japan, Akiro Yoshino of the Asahi Kasei Corp. then realized that he could assemble a discharged rechargeable battery using graphite as the anode and my $LiCoO_2$ as the cathode. Scientists at the SONY Corporation commercialized this Li-ion battery to market the first cell telephone that launched the wireless revolution. AERE Harwell received many millions of pounds; we received nothing. I was disappointed that not even a contribution to St. Catherine's College was forthcoming. However, the joy of having helped to enable a technology that has transformed for the better so many lives is reward enough.

Cobalt is expensive and toxic. In 1981, Michael Thackeray came from South Africa to work with me. He had been inserting lithium into magnetite, Fe_3O_4, the original ferrospinel used by the Greeks for navigation. He wanted to develop a less expensive cathode than $Li_{1-x}CoO_2$. Bruno Scrosati of Rome had reported a similar experiment in a seminar I had attended two weeks before Thackeray's arrival. I was skeptical of this report as I knew that the spinel structure cannot tolerate excess cations. Therefore, I asked Thackeray, when he arrived, to repeat his experiment in my laboratory. When he confirmed the insertion of lithium into magnetite, I realized that the insertion of lithium must be displacing the tetrahedral-site iron to the empty octahedral sites of the structure to form an ordered rock-salt phase. This insight made me realize that the spinel octahedral-site $[M_2]O_4$ array represented a three-dimensional framework into/from which lithium could be inserted/extracted reversibly. Therefore, I told Thackeray to insert lithium into the manganese spinel $Li[Mn_2]O_4$.

The $Li_{1+x}[Mn_2]O_4$/Li half-cell gave a flat 3 V open-circuit voltage. Thackeray would later extract lithium from the tetrahedral sites; the $Li_{1-x}[Mn_2]O_4$/Li half-cell gave a 4 V open-circuit voltage. A sharp shift of 1eV in the energy of the Mn^{4+}/Mn^{3+} couple occurs where the L^{i+} ions change their occupancy cooperatively from all-tetrahedral to all-octahedral sites. I gave Thackeray the patent rights to the spinel framework, but that patent was changed in South Africa to cover only the 4-V range of $Li^x[Mn_2]O_4$ ($0<x<1$). Donald Murphy of the Bell Telephone Laboratory had independently prepared the spinel framework [Ti2]S4 by extracting copper chemically from $Cu[Ti_2]S_4$. In the sulfospinel framework, lithium enters only octahedral sites and the $[Ti_2]S_4$/Li half-cell gives a voltage identical to that of the original TiS_2/Li half-cell. In Li_xTiS_2, lithium also occupies only octahedral sites. Chemical instability on cycling $Li_x[Mn_2]O_4$ over the 4-V solid-solution range $0< x <1$ prevented its commercialization as the cathode of a lithium battery. However, addition of some nickel and lithium to the $[Mn_2]O_4$ framework has provided chemical stability at the expense of discharge capacity. NISSAN has used this stabilized cathode material in their initial hybrid car.

At Oxford, in addition to the cathode materials for a Li-ion battery, I learned about oxide surface reactions with the medium in which they existed, including the zeta potential in aqueous solutions and Li^+ attraction to an oxide surface in a non-oxide solid to create Li^+ vacancies for Li^+ conduction in the non-oxide solid medium. With the realization of structural reconstructions at an oxide surface, I understood how it frustrated study of heterogeneous catalysis by oxides. Therefore, I investigated the partial oxidation of acrolein to methacrolein on a Keggin 12-molybdophosphate of known surface structure to determine the role of a stable reduced molybdenum Mov displaced from an oxygen vacancy in the catalytic process, a possibility occurring to me from my earlier studies of solid molybdenum-oxide chemistries.

At Lincoln Laboratory I had initiated work to strengthen interactions between the solid-state chemist and solid-state physicist; at Oxford, I brought together chemistry and electrochemistry to stimulate both communities. It was for work fostering these interactions that I was to be awarded the Japan Prize in 2002 and given the highest awards of the Material Research Society, the Electrochemical Society, and the 3M Society.

AUSTIN YEARS

In 1986, the politics in the laboratory at Oxford anticipating my retirement had begun, so I accepted a call to take the Virginia H. Cockrell Centennial Chair of Engineering at the University of Texas at Austin. Retirement before age 67, was no longer required in the U.S. which has enabled

Figure 14. Left: Irene and I in Japan (1985). Right: Irene and I in Russia (1991).

me to keep working for another 32 years in Texas as a Professor of Materials Science and Engineering.

In 1986, Bednorz and Mueller of IBM Zurich reported their discovery of high-temperature superconductivity below 40 K in a copper oxide while exploring whether a dynamic Jahn-Teller electron-lattice interaction at octahedral Cu ions might provide a needed electron-lattice interaction for superconductivity. Their superconductive compound was shown in Japan to be $La_{2-x}Sr_xCuO_4$. While I was setting up a chemistry laboratory in Texas with the aid of a postdoc, Arumugam Manthiram whom I brought with me from Oxford, a superconductive transition at 90

Figure 15. Drawing a diagram of a battery on the board in my office at the University of Texas in Austin, 2012.

K in $YBa_2Cu_3O_{7-\delta}$ was announced. Hugo Steinfink, a crystallographer in our ME Department who had supervised Henry Hong and had brought me to Texas, determined the structure of the 90 K superconductor the day before the structure was also announced at the "Woodstock of Physics" in New York. Subsequently, Arumugam Manthiram and I explored extensively the chemistry of the copper-oxide superconductors.

In 1987, a letter from the University of Jilin in China asked if I would take a physics student to do his Ph.D. thesis with me for graduation from the University of Jilin. The physics student is now Research Professor Jian-Shi Zhou in my group; interested in high-pressure studies of solids; he had followed the high-pressure work we did at the MIT Lincoln Laboratory. Since a discarded and broken copy of the Kafalas high-pressure cell resided in the laboratory of Hugo Steinfink, I accepted Jian-Shi Zhou, who was able to refurbish the Kefalas cell; and he has remained with me ever since. Over the last 30 years, Jian-Shi Zhou has built up a competitive high-pressure facility, a single-crystal furnace, the ability to synthesize under pressure to 26 GPa, and to measure structures as well as magnetic and transport properties under pressure and the thermal properties of his materials at ambient pressures.

At MIT, we had shown that $SrRuO_3$ is a ferromagnetic metal and that the paramagnetic susceptibility of the $Sr_{1-x}Ca_xRuO_3$ system exhibits a change to a negative Weiss constant, but with no long-range magnetic order in $CaRuO_3$. Much later, Rob Cava of Princeton University showed a peculiar transition in the paramagnetic susceptibility of $Sr_{1-x}CaxRuO_3$ near the Curie temperature of $SrRuO_3$, and Jian-Shi Zhou pointed out that the Cava data represented formation of a Griffiths phase in which the magnetic ions are diluted by nonmagnetic ions, which means there is a segregation into Ru^{IV} with magnetic 4d electrons and Ru^{IV} with nonmagnetic 4d electrons localized by spin-orbit coupling. He further found itinerant-electron ferromagnetism in the high-pressure $Sr_{1-x}BaxRuO_3$ perovskites with an abrupt transition under pressure from ferromagnetism to Pauli paramagnetism in cubic $BaRuO_3$. The structure-composition property relationships in the single-valent oxoperovskites have been shown to be rich as also have the mixed-valent manganese perovskites.

On arriving in Austin, I set up my chemistry laboratory for work on electrochemistry and ionic transport in solids with a view to continue work on energy-related materials. We finally had a chemical hood installed by Christmas! By that time, not only had the structure and composition of the original copper-oxide superconductor been identified; doping of the phase had led to the discovery of another copper-oxide phase that becomes superconductive at 90 K, well above the boiling point of liquid nitrogen! Announcement of this discovery in the New

York Times created a stampede of crystallographers in Japan and the U.S. to be the first to determine the structure of this new phase. Hugo Steinfink solved the structure and was the first to announce it at a crystallographic meeting in Austin the day before the "Woodstock" of the American Physical Society Meeting in New York City where other groups also announced the solution. Steinfink's contribution was overshadowed by the trumpets of the larger laboratories that had solved the structure independently.

Although I had not previously been interested in working on the superconductive phenomenon, this development in a transition-metal oxide captured my attention; Manthiram and I began to study the chemistry of these copper oxides. It became immediately evident to me that the extraordinary superconductivity in the copper oxides was occurring in a phase intermediate between an antiferromagnetic parent compound that was an insulator and a non-superconductive metallic phase. Superconductivity was appearing at a much higher temperature than predicted by the existing BardeenCooper-Schrieffer theory and at a transition from localized to itinerant electronic behavior in a transition-metal ceramic, a transition that I had explored in these perovskite-related oxides in the 1960s. However, in this case the crossover was occurring in a mixed-valent system whereas I had been studying it in single-valent systems. I knew that lattice instabilities were always encountered at this crossover in single-valent systems, so I suspected that a dynamic segregation of localized and itinerant electrons was

Figure 16. My wife (back row, second from right) and I (back row, third from right) at Kinkakuji Temple, Japan, April 29, 2001.

occurring in the mixed-valent copper oxides. This idea was aggressively dismissed by the leading solid-state theorists who were the opinion leaders for the majority of physicists; but when Zhou arrived, I persisted to explore this possibility with him. Manthiram became a professor in his own right and returned to the development of energy materials independently of me.

We had a first-rate high-pressure facility in Austin. Together, Zhou and I have explored the transition from localized to itinerant electronic behavior in other mixed-valent and single-valent transition-metal oxides with perovskite-related structures. We have clearly demonstrated a dynamic segregation into localized-electron and itinerant-electron domains in other mixed-valent systems as well as in the copper oxides.

The copper-oxide problem is complicated by the existence of two types of phase segregation, one a dynamic segregation into localized-electron and itinerant-electron domains and the other a static phase segregation of the superconductive phase from the antiferromagnetic parent phase on the one side and the metallic over doped phase on the other side. As the oxidation state of the superconductive CuO_2 sheets of the copper oxides increases, the charge carriers introduced by the oxidation change their character. In the antiferromagnetic phase, isolated carriers occupy a volume of about 6 copper centers. Where the oxidation is too great for the charge carriers to remain isolated from one another by electrostatic coulomb forces, they condense below room temperature into spin-paired carriers in a volume of four copper centers or into multiple spin-paired carriers in itinerant-electron chains along the Cu-O-Cu bond axes. This phenomenon represents a dynamic phase segregation in which the charge carriers are mobile. A structural distortion may trap the carriers in static itinerant-electron stripes separated by ribbons of localized electrons. These static stripes have been observed by conventional neutron-diffraction experiments, but a dynamic phase segregation requires a faster experimental probe. A pulsed neutron-diffraction experiment coupled with a pair-distribution-function analysis of the data is such a probe. This technique was developed by Takeshi Egami while he was at the University of Pennsylvania. His preliminary data show evidence of the predicted phase separation, but the complexity associated with isolated charge-carrier pairs coexisting with chain segments has made difficult a definite statement about the nature of the dynamic normal state in the superconductive phase. We have established the existence at room temperature of isolated charge carriers with a volume of 5 to 6 copper centers in the underdoped phase; in the metallic overdoped phase the localized electrons appear only as fluctuations. How the paired charge carriers become long-range ordered to give superconductivity is still to be resolved. I have pointed out that this can be done by coupling the domains of paired electrons to cooperative lattice vibrations (phonons) to give vibronic charge carriers.

In my chemistry laboratory, we also continued to develop materials for the lithium rechargeable battery and the solid oxide fuel cell. I assigned to my engineering student, Akshaya Padhi, and my post-doctoral fellow, Nanjunda Swami, the task to explore the relative energies of transition-metal redox couples in the NASICON framework $M_2(XO_4)_3$ that we had shown in Lincoln Laboratory supports fast transport of the Na^+-ion guest species. Different transition-metal atoms M and polyanions (XO_4) can be accommodated in the framework, with the charge of the framework being balanced by Li^+-ion guests over the range $0< x <5$ in $Li_xM_2(XO_4)_3$. In this framework, the energy of the redox couples appear to be essentially insensitive to the location and number of Li^+ ions in the interstitial space. I was also interested in knowing how those redox energies shifted on changing from $(SO_4)^{2-}$ to $(PO_4)^{3-}$ or $(AsO_4)^{3-}$ polyanions. During the course of this work, Padhi found that Lithium can be extracted reversibly from $LiFePO_4$, which has the olivine structure. The Li_xFePO_4/Li cell gives a constant 3.45 V open-circuit voltage over the range $0< x <1$. Made as small particles, this cathode is capable of extremely fast rates of charge and discharge. The Hydro Quebec Corporation licensed the patent granted to the University of Texas; but the A123 company in Cambridge, Massachusetts, was the first to market the $LiFePO_4$/C battery and to demonstrate its use in medium-power applications such as electric power tools and small electric vehicles.

Our work on the solid oxide fuel cell has involved the development of new oxide-ion electrolytes as well as new electrodes. Dr. Kevin Huang worked with me as a post-doctoral fellow on these problems before going to become a key player in the Siemens-Westinghouse development of a commercial hydrogen-fueled solid oxide fuel cell. After his departure, I investigated a novel class of anode materials that can operate on natural gas without becoming poisoned by sulfur impurities in the gas. The move from the internal combustion engine to batteries to power our automobiles would reduce distributed CO_2 emissions responsible for global warming; the development of electrical energy storage is needed to make viable the substitution of solar, wind, and nuclear energy sources for the fossil fuels that emit CO_2 on burning; the sequestering of CO_2 and other pollutants emitted from coal-fired power can reduce CO_2 emissions; and the introduction of more efficient energy distribution and use can reduce energy consumption. All represent urgent challenges confronting the scientific-engineering community today. The implementation of a serious national effort to meet these challenges, an effort initiated in the early 1970s, was stalled by special interests more concerned with profits than with our national vulnerability and the global environment.

Figure 17. Left: With my research group, Austin, 2019. *Right*: My 96th birthday, July 2018.

In the end, I have had an extraordinary journey, but it is the many colleagues who have worked with me over the years that I wish to thank for making it extraordinary. They are the ones who have performed the experiments and each of them kept an open dialogue with the aim to teach me as much as I tried to teach them.

REFERENCES
John B. Goodenough, *Witness to Grace*, PublishAmerica, 2018, ISBN: 1-60474-767-6.

The Pathway to Discovering Practical Cathode Materials for the Rechargeable Li^{+}-Ion Battery

Nobel Lecture, December 8, 2019 by
John B. Goodenough
The University of Texas at Austin, Austin, Texas, USA.

A BATTERY STORES electric power as chemical energy in two electrodes, the anode and the cathode, which are separated by an electrolyte. The chemical reaction between the electrodes has an ionic and an electronic component. The electrolyte transports the ionic component inside a cell and forces the electronic component to traverse an external circuit. This chemical reaction is reversible in a rechargeable battery. The following is a personal narrative of the insights that led to the development of the first practical rechargeable Li^{+}-ion battery cathode, $LiCoO_2$, and the following research that developed from this discovery.

In the 1960s, Jean Rouxel in France and Robert Schroeder in Germany were exploring the chemistry of reversible insertion of lithium into the van der Waals gap of layered transition-metal sulfides. At that time, rechargeable batteries used strongly acidic (H_2SO_4) or alkaline (KOH) aqueous electrolytes that offered fast hydrogen-ion (H^{+}) diffusion. The most stable cells used an alkaline electrolyte and a layered nickel oxyhydroxide (NiOOH) as the cathode into which H+ is inserted reversibly to form the hydroxide

Ni(OH)2. However, an aqueous electrolyte limits the voltage of the cell and, therefore, the density of electric power that a battery can deliver.

In 1967, Joseph Kummer and Neill Weber of the Ford Motor Company discovered fast sodium-ion diffusion above 300°C in a ceramic electrolyte and assembled a sodium–sulfur rechargeable battery that used this solid ceramic as the electrolyte, molten sodium as the anode and a carbon felt with molten sulfur as the cathode. Their battery was commercially impractical owing to its high operating temperature of above 300°C, but it stimulated research into solid electrolytes and alternative battery strategies. At the time, I was working on transition-metal oxides at the MIT Lincoln Laboratory, and I was asked to monitor the development of this battery. The assignment introduced me to electrochemistry and to the challenge of developing a better oxide-based sodium-ion conductor. In response, Henry Hong and I developed a $Na_{1+x}Zr_2Si_xP_{3-x}O_{12}$ electrolyte that had a framework structure and supported fast sodium ion conductivity; it was dubbed NASICON (NA SuperIonic CONductor) by colleagues after I left for the University of Oxford in 1976.[1]

In the early 1970s, an oil crisis exposed the vulnerability of US society, among others, to its dependence on imported oil and subsequently prompted investigations into solar and wind energy as potential sources for electric power. However, the intermittency of solar and wind meant that this power must be stored if these renewable sources were to be useful. Thus, there was a desire for rechargeable batteries that could serve this purpose efficiently.

At Oxford, I was free to work on problems related to energy. Initially, I selected two primary targets for my research program: the direct methanol-air fuel cell and the photoelectrolysis of water. The former requires a solid proton (H^+-ion) conductor for its electrolyte that is an electronic insulator and can operate near 300°C or an anode that is chemically stable in an acidic solution and catalytically active for the oxidation of methanol below 80°C. I soon found that good proton conductivity in a solid electrolyte can only occur when the solid is wet, which means operating below 80°C for fear of drying the solid. Our search for a sufficiently active and chemically stable anode for a direct methanol-air fuel cell proved unsuccessful, but it introduced me to the field of electrochemistry. Our attempt to realize a practical electrode for photoelectrolysis also involved electrochemistry. I had hoped to be able to use most of the spectrum of visible light by using a filled as well as a nearly empty d-electron band; however, the lower lying filled energy band proved to be too narrow for this strategy to be practical. The alternative was to attach a dye to the surface of the oxide. This exercise provided a good D. Phil. thesis, but it did not give a practical solution. I realized it would probably be better to separate the steps in the process by coupling a photovoltaic cell to an electrolysis cell to store solar energy as chemical energy in hydrogen gas as a portable fuel. More successful was my effort to identify a suitable cathode material for a lithium battery, an effort that has done much to bring together the solid-state chemist and the electrochemist.

The H^+ ion is the most mobile working ion in a rechargeable battery. However, protons are only mobile in an aqueous acidic or alkaline electrolyte. To avoid electrolysis of the water within a rechargeable battery with an aqueous electrolyte, the single cell voltage is restricted to less than 1.5 V if the battery is to have a long shelf life. A schematic of the open-circuit energy diagram of such a battery is shown in Figure 1.[2] This restriction limits the energy density of a rechargeable battery, which is why the advent of the cell telephone and the laptop computer had to await the arrival of the lithium rechargeable battery. The working Li^+ ion in a lithium battery is mobile in a nonaqueous electrolyte, which permits single-cell voltages over 4 V. Primary batteries using a lithium anode and an organic-liquid electrolyte were known at the time, so the next step was to use the chemistry demonstrated in Europe of reversible lithium-ion insertion into transition-metal layered sulfide cathodes to create a rechargeable battery. However, realization of a competitive rechargeable lithium battery required the identification of a cathode material into/from which Li^+ could be inserted/extracted reversibly over a large solid-solution range. Moreover, the active redox couples of the insertion compound electrode must have energies matched to the allowable energy of the electrolyte if larger voltages are to be achieved. Before I left Lincoln Laboratory, M. Stanley Whittingham had demonstrated intercalation of lithium into TiS_2 at the EXXON research laboratory.[3] The structure of TiS_2 consists of sheets of edge-shared $TiS_{6/3}$ octahedra held together by weak dipole-dipole (van der Waals) forces; lithium can be inserted/extracted reversibly between the layers over the solid-solution range Li_xTiS_2 ($0 \leq x \leq 1$). With a metallic Lithium anode and a TiS_2 cathode, the TiS_2/Li single cell has a voltage of about 2.4 V over most of the solid-solution range. However, it was soon discovered that a metallic Lithium anode cannot be used in a practical rechargeable battery. On repeated charge/discharge cycles, dendrites from the anode grow across the electrolyte to the cathode to short-circuit the battery explosively. Therefore, this effort to realize a lithium battery was abandoned.

I had realized that it would not be possible to obtain a layered sulfide with a higher voltage versus a Lithium anode and, therefore, that an alternative insertion-compound anode would lower the output voltage to a value where the battery would not be competitive with a nickel/metal-hydride aqueous electrolyte system. A competitive lithium battery would require an oxide into/from which lithium can be inserted/extracted reversibly at a high rate over a large solid-solution range for its cathode.

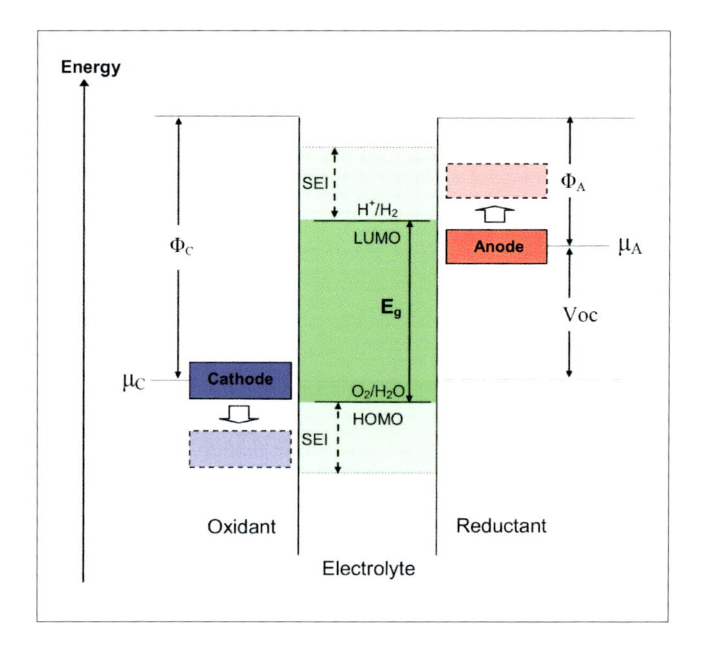

Figure 1. Schematic open-circuit energy diagram of an aqueous electrolyte. ΦA and ΦC are the anode and cathode work functions. E_g is the thermodynamic stability window of the electrolyte. A μ_A > LUMO[a] and/or a μ_C < HOMO[b] requires a kinetic stability by the formation of a solid electrolyte interphase (SEI) layer. Reprinted with permission from Goodenough, J. B.; Kim, Y. Challenges for Rechargeable Li Batteries. *Chem. Mater.* 2010, **22** (3), 587–603. https://doi.org/10.1021/cm901452z.. Copyright 2010 American Chemical Society.

In 1978, an undergraduate thesis at Oxford on the structure of the $LiMO_2$ oxides reminded me of work I had done with Donald Wickham in the 1950s on $Li_xNi_{2-x}O2$.[4] The MO_2 oxides are not layered as is TiS_2; the electrostatic energy between the O^{2-} ions of MO_2 sheets is larger than the dipole-dipole van der Waals binding energy. However, layered $LiMO_2$ is stabilized by the Li^+ ions between the MO_2 sheets, and the ions are well-ordered provided the sizes of Li^+ and M^{3+} ions are sufficiently different from one another. I decided to investigate how much lithium can be extracted reversibly from a well-ordered $LiMO_2$ layered oxide. I knew that to maximize the working voltage of the battery, we wanted to choose a material with an M^{4+}/M^{3+} redox couple that had an energy well below the Li^+/Li^0 couple of a metallic Lithium anode. My prior experience working on transition-metal oxides at Lincoln Laboratory provided me with the knowledge to know that chromium, cobalt, and nickel each had a d-band redox energy sufficiently far below the Fermi level of lithium to be inter-

[a] Lowest Unoccupied Molecular Orbital
[b] Highest Occupied Molecular Orbital

esting to study as the M atom. An experimental physicist, Koichi Mizushima, had just come from the University of Tokyo to work with me at Oxford. I teamed him with my chemist post-doctoral assistant, Phillip Wiseman, to work on this investigation. We found that over half of the lithium could be removed reversibly from the structure with cobalt or nickel as the M atom; each of these $Li_{1-x}MO_2$/Li half-cells gave an output voltage near 4 V.[5,6] $LiCoO_2$, the structure of which is provided in Figure 2, was particularly interesting owing to the fact that it could be synthesized with the lithium layers and CoO_2 layers completely separate and ordered, with little presence of cation mixing. This feature allows for lithium to be reversibly inserted/extracted without hindrance within the lithium layer. $LiNiO_2$ is difficult to make in this way directly from lithium precursors; some Ni^{3+} ions always reside in the lithium layer, disturbing the perfect ordering desired for the layered $LiMO_2$ structural motif. However, no battery company in England, Europe, or the U.S. was interested in licensing a patent for these cathode materials; they could not imagine starting with a discharged battery even though the battery was to be rechargeable.

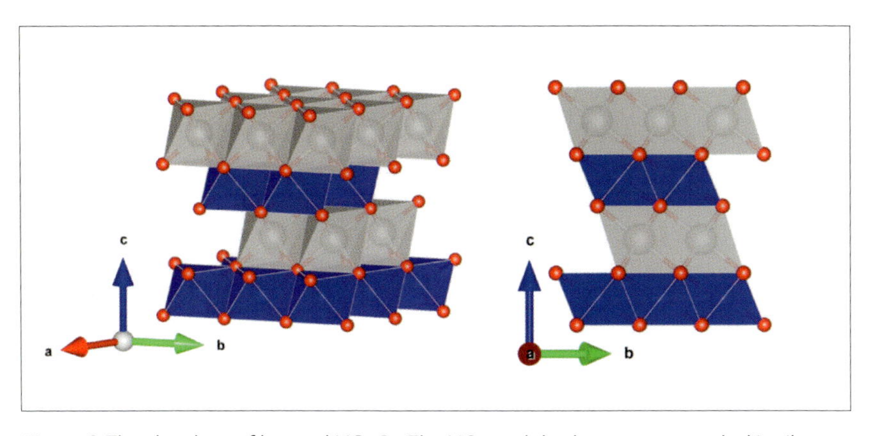

Figure 2. The structure of layered $LiCoO_2$. The $LiO_{6/3}$ octahedra are represented in silver and the CoO6/3 octahedra are represented in blue.

In 1982, Rachid Yazami and Philippe Touzain reported the dendrite-free intercalation of lithium into graphitic carbon with a potential near that of metallic lithium – see Figure 3a–b.[7] It did not take long for Akira Yoshino at Asahi Kasei Corporation to realize that this graphitic carbon electrode could be used as a discharged anode to go with our discharged $LiCoO_2$ cathode, and he assembled the first C/$LiCoO_2$ Li-ion cell. This battery cell was then licensed and developed by the SONY Corporation, who were hoping to find a rechargeable battery of high energy density so that they could market a wireless telephone. Thus, the wireless revolution was born.

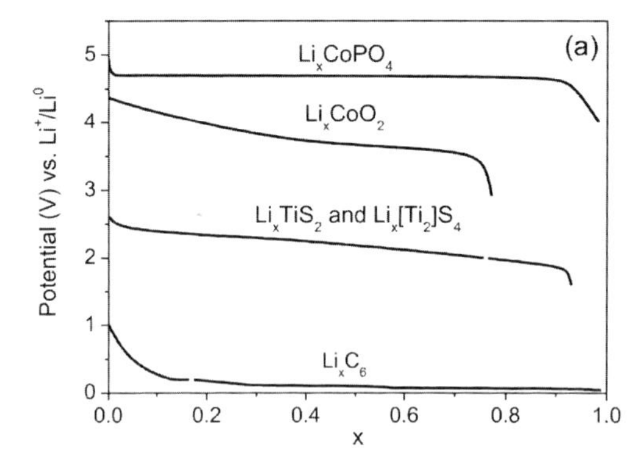

Figure 3. (a) Voltage profiles versus Li^+/Li^0 of the discharge curves of Li_xC_6, Li_xTiS_2 and $Li_x[Ti_2]S_4$, Li_xCoO_2, and $Li_xC_oPO_4$. (b) Schematic of their corresponding energy vs density of states showing the relative positions of the Fermi energy in an itinerant electron band for LixC6, the Ti^{4+}/Ti^{3+} redox couple for Li_xTiS_2 and $Li_x[Ti_2]S_4$, the Co^{4+}/Co^{3+} redox couple for Li_xCoO_2, and the Co^{3+}/Co^{2+} redox couple for $Li_xC_oPO_4$. Reprinted with permission from Goodenough, J. B.; Kim, Y. Challenges for Rechargeable Li Batteries. *Chem. Mater.* 2010, 22 (3), 587–603. https://doi.org/10.1021/cm901452z. Copyright 2010 American Chemical Society.

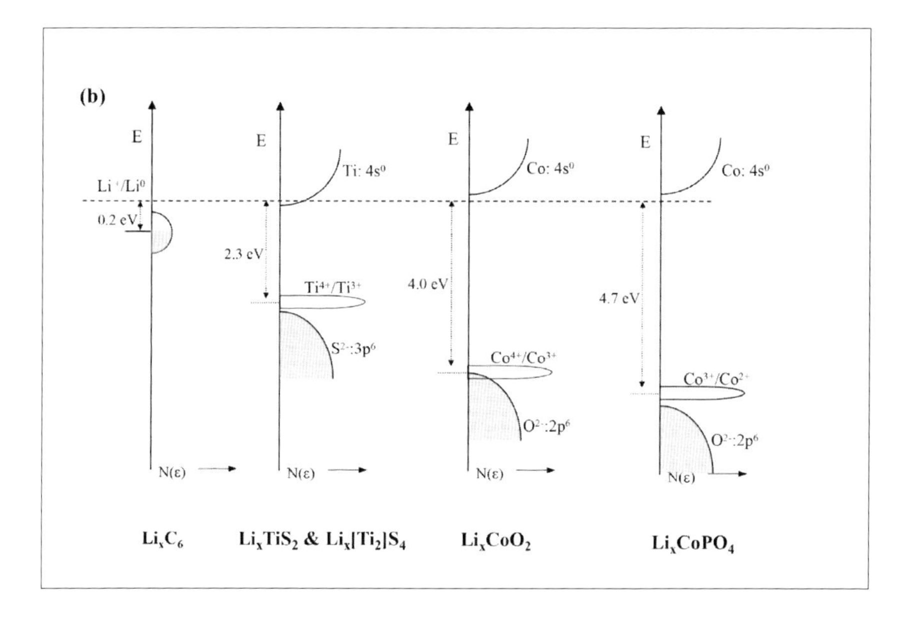

Cobalt is expensive and toxic. In 1981, Michael Thackeray came from South Africa to work with me on inserting lithium into the ferrospinel magnetite, Fe_3O_4. He wanted to develop a less expensive cathode than $Li_{1-x}CoO_2$. Bruno Scrosati of Rome had reported a similar experiment in a seminar I had learned of two weeks prior to Thackeray's arrival. I was skeptical of this result as I knew that the spinel structure cannot tolerate excess cations. Therefore, I asked Thackeray to repeat this experiment in my laboratory. When he confirmed the insertion of lithium into magnetite, I realized that the insertion of lithium must be displacing the tetrahedral-site iron to the empty octahedral sites of the structure to form an ordered rock-salt phase. This insight made me realize· that the spinel octahedral-site $[M_2]O_4$ array represented a three-dimensional framework into/from which lithium could be inserted/extracted reversibly. Therefore, I asked Thackeray to insert lithium into the manganese spinel $Li[Mn_2]O_4$.[8] The $Li_{1+x}[Mn_2]O_4$/Li half-cell gave a flat 3V open-circuit voltage. Thackeray would later extract lithium from the tetrahedral sites; the $Li_{1-x}[Mn_2]O_4$/Li half-cell gave a 4 V open-circuit voltage.[9] A sharp shift of 1 eV in the energy of the Mn^{4+} /Mn^{3+} couple occurs where the Li^+ ions change their occupancy cooperatively from all-tetrahedral to all-octahedral sites. Donald Murphy of the Bell Telephone Laboratory had independently prepared the spinel framework $[Ti_2]S_4$ by extracting copper chemically from $Cu[Ti_2]S_4$. In the sulfospinel framework, lithium enters only octahedral sites and the $[Ti_2]S_4$/Li half-cell gives a voltage identical to that of the original TiS_2/Li half-cell (Figure 3 a–b). In Li_xTiS_4, lithium occupies only octahedral sites. Chemical instability on cycling $Li[Mn_2]O_4$ over the 4-V solid-solution range $0 \leq x \leq 1$ prevented its commercial adoption as a lithium battery cathode.

Frameworks containing $(XO_4)^{n-}$ polyanions instead of oxide ions can not only open up the interstitial space for fast Li^+ -ion or Na^+-ion transport, but also provide the strong oxygen covalent bonding needed to lower the top of the O-2p or, with $(XS_4)^{n-}$, the top of the S-3p bands sufficiently to provide a $V_{oc} > 5.0$ V versus lithium. In our earlier search for a framework oxide giving fast 3D Na^+-ion transport in Lincoln Laboratory, we showed that $Na_{1+3x}Zr_2(P_{1-x}Si_xO_4)_3$ with x = 2/3 gives a superior Na^+-ion conductivity.[1] Therefore, I suggested exploring Li^+-ion insertion into $Fe_2(XO_4)_3$ frameworks with X = Mo, W, or S. Arumugam Mathiram showed that the voltage of these compounds jumps from $V_{oc} \approx 3.0$ V with X = Mo or W to 3.6 V versus lithium with X = S, each operating on the Fe^{3+}/Fe^{2+} couple.[10, 11] This experiment demonstrated the influence, through the inductive effect, of the counter cation X on the Fe^{3+}/Fe^{2+} redox energy. At the University of Texas at Austin, I assigned to my engineering student, Akshaya Padhi, and my post-doctoral fellow, Kirakodu Nanjundaswamy, the task to explore the relative energies of transition-metal redox couples in the NASICON framework: $M_2(XO_4)_3$. Different transition-metal atoms and polyanions (XO_4) can be accommodated in the framework with the charge of the framework being balanced by Li^+-ion guests over the range $0 \leq x \leq 5$ in $Li_xM_2(XO_4)_3$.[12] In this framework, the energy of the redox couples appear to be essentially insensitive to the location and number of lithium ions in the interstitial space. I was also interested in knowing how these redox energies shifted on changing from $(SO_4)^{2-}$ to $(PO_4)^{3-}$ or $(AsO_4)^{3-}$ polyanions. During the course of this work, Padhi found that Lithium can be extracted reversibly from $LiFePO_4$, which has the olivine structure.[13] The Li_xFePO_4/Li cell gives a constant 3.45 V open-circuit voltage over the range $0 \leq x \leq 1$. Made as small platelet particles, this cathode is capable of extremely fast rates of charge and discharge owing to the one-dimensional Li+-ion conduction channel being perpendicular to the face of the platelet. This discovery prompted the investigation of many other phosphate and pyrophosphate framework structures to elucidate further the effect of anion-framework on the energy of transition-metal redox couples in these materials, the results of which are summarized in Figure 4 for the Fe^{2+}/Fe^{3+} couple.[14]

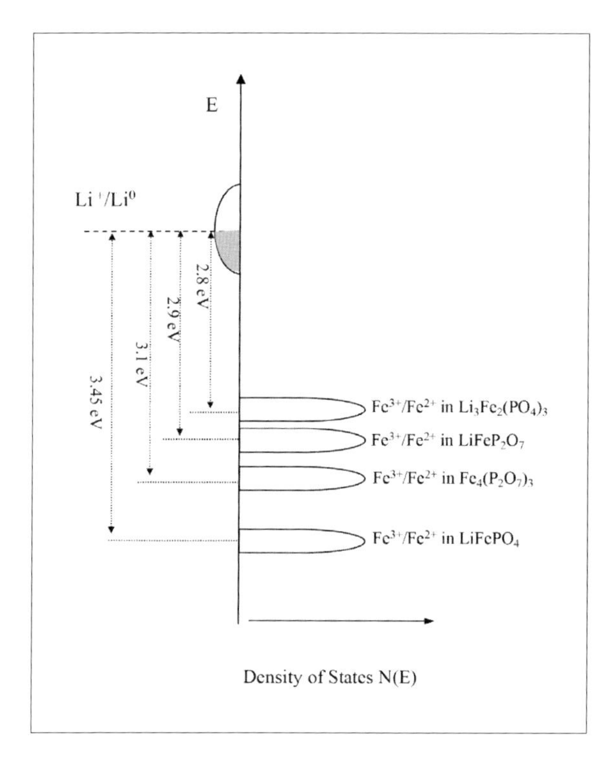

Figure 4. Relative energy levels of the Fe^{3+}/Fe^{2+} redox couple in different phosphate framework structures. Reprinted with permission from Goodenough, J. B.; Kim, Y. Challenges for Rechargeable Li Batteries. *Chem. Mater.* **2010**, 22 (3), 587–603. https://doi.org/10.1021/cm901452z. Copyright 2010 American Chemical Society.

REFERENCES

1. Goodenough, J. B.; Hong, H. Y.-P.; Kafalas, J. A. Fast Na+-Ion Transport in Skeleton Structures. *Materials Research Bulletin 1976,* **11** (2), 203–220. https://doi.org/10.1016/0025-5408(76)90077-5.
2. Goodenough, J. B.; Kim, Y. Challenges for Rechargeable Li Batteries. *Chem. Mater.* 2010, **22** (3), 587–603. https://doi.org/10.1021/cm901452z.
3. Whittingham, M. S. Electrical Energy Storage and Intercalation Chemistry. *Science* 1976, **192** (4244), 1126–1127. https://doi.org/10.1126/science.192.4244.1126.
4. Goodenough, J. B.; Wickham, D. G.; Croft, W. J. Some Magnetic and Crystallographic Properties of the System $Li^{+}_{x}Ni^{++}_{1-2x}ni^{+++}xO$. *Journal of Physics and Chemistry of Solids* 1958, **5** (1), 107–116. https://doi.org/10.1016/0022-3697(58)90136-7.

5. Mizushima, K.; Jones, P. C.; Wiseman, P. J.; Goodenough, J. B. LixCoO2 (0<x≤1): A New Cathode Material for Batteries of High Energy Density. *Materials Research Bulletin* 1980, **15** (6), 783–789. https://doi.org/10.1016/0025-5408(80)90012-4.
6. Thomas, M. G. S. R.; David, W. I. F.; Goodenough, J. B.; Groves, P. Synthesis and Structural Characterization of the Normal Spinel $Li[Ni_2]O_4$. *Materials Research Bulletin* 1985, **20** (10), 1137–1146. https://doi.org/10.1016/0025-5408(85)90087-X.
7. Yazami, R.; Touzain, P. Composes Ioniques Du Graphite Avec $NiCl_2$, BF^-_4 et K^+ Pour Le Stockage Electrochimique de l'energie. *Solid State Ionics* 1983, **9–10,** 489–494. https://doi.org/10.1016/0167-2738(83)90282-5.
8. Thackeray, M. M.; David, W. I. F.; Bruce, P. G.; Goodenough, J. B. Lithium Insertion into Manganese Spinels. *Materials Research Bulletin* 1983, **18** (4), 461–472. https://doi.org/10.1016/0025-5408(83)90138-1.
9. Thackeray, M. M.; Johnson, P. J.; de Picciotto, L. A. Electrochemical Extraction of Lithium from $LiMn_2O_4$. *Materials Research Bulletin* **19** (2), 179–187. https://doi.org/10.1016/0025-5408(84)90088-6.
10. Manthiram, A.; Goodenough, J. B. Lithium insertion into Fe2(MO4)3 frameworks: Comparison of M = W with M = Mo. *J. Solid State Chem.* 1987, **71**, 349. https://doi.org/10.1016/0022-4596(87)90242-8.
11. Manthiram, A.; Goodenough, J. B. Lithium Insertion into Fe2(SO4)3 Frameworks. *J. Power Sources* 1989, **26**, 403–408. https://doi.org/10.1016/0378-7753(89)80153-3.
12. Padhi, A. K.; Nanjundaswamy, K. S.; Masquelier, C.; Goodenough, J. B. Mapping of Transition Metal Redox Energies in Phosphates with NASICON Structure by Lithium Intercalation. *J. Electrochem. Soc.* 1997, **144** (8), 2581. https://doi.org/10.1149/1.1837868.
13. Padhi, A. K.; Nanjundaswamy, K. S.; Goodenough, J. B. Phospho-olivines as Positive-Electrode Materials for Rechargeable Lithium Batteries. *J. Electrochem. Soc.* 1997, **144** (4), 1188. https://doi.org/10.1149/1.1837571.
14. Padhi, A. K.; Nanjundaswamy, K. S.; Masquelier, C.; Okada, S.; Goodenough, J. B. Effect of Structure on the Fe^{3+}/Fe^{2+} Redox Couple in Iron Phosphates. *J. Electrochem. Soc.* **1997**, 144 (5), 1609. https://doi.org/10.1149/1.1837649.

M. Stanley Whittingham.

M. Stanley Whittingham

Biography

I WAS BORN on December 22, 1941 in the Carlton suburb of Nottingham in England in the middle of the Second World War. My father, William Stanley Whittingham, was a civil engineer and the first in the family to go to college, and my mother Dorothy Mary (nee Findley) was a chemist before marriage. My father was responsible for repairing the runways in East Anglia, so we were constantly on the move. I spent my childhood in Lincolnshire county attending primary school in Grimsby and travelled each day by train and bus from home to school. When I was around age 10, the family moved to Swallow Hill House in Thurlby, near Bourne. Looking east from our house the next higher hill was in Germany, as we were on the edge of the fens. I attended Stamford School in Stamford, about 8 miles away. Stamford is a medieval town then on the main A1 road from London to Edinburgh; it was my father who was in charge of building the by-pass around the town, that eliminated the need to go through the alternating one-way street in the center of town. It was there that I got attracted to science through the inspired teaching of Major Lamb and Squibs Bowman in chemistry and physics respectively. The school had a well-equipped new science building. The school and my teachers in 1960 are shown in Figure 1. The school dates from 1532, and its chapel (formerly known as St Paul's church and shown in Figure 1) originated in 1152. The St. Paul's Street in front of the school is named after the church. School days started with chapel at 8:45 am, and finished at 3:45 pm six days a week, with Wednesday and Saturday afternoon reserved for sports. I remember Squibs Bowman started a sailing club at a local gravel pit, and I got deeply involved in that. But my parents got totally hooked, so they took up sailing in the English Channel when they retired to Flushing in Cornwall. It was at this time that I got very interested in growing cacti, and my family built a greenhouse at our house so I could grow my hundreds of them, together with more useful tomato plants which I sold to neighbors. I still grow cacti and am an active mem-

Figure 1. Stamford School (left) and staff (right) in 1960. Courtesy of Stamford School.

ber of the Desert Botanical Garden in Phoenix, Arizona where my daughter and her family live.

In the autumn of 1959, I went for several days to Oxford to take the entrance examination, which included several languages as well as chemistry, physics and mathematics. After I was offered admission to New College in Oxford University to study chemistry, Headmaster Basil Deed tutored me in Latin to pass the then required classics examination before I could go to Oxford in October 1960.

COLLEGE

Arriving in Oxford, I took the 1st Public Examination. This was followed by the Final week-long examinations after three years of study. A fourth year was spent doing full-time research. The college rooms were somewhat rudimentary then, with no running water. A "scout" brought a bowl of hot water each morning to my room to wash and shave with. Breakfast was served in the hall until 9 am, and the lectures that also started at 9 am were in the University chemistry buildings a good 10–15 minutes' walk away. These were not mandatory but were very helpful for the weekly tutorials. Each week I prepared a lengthy paper on the weekly topic and then presented and discussed it with my tutor for an hour. I still remember going to Peter Dickens' lodgings in the College each Sunday morning for this, and his wife Mary would prepare biscuits and tea. Although life at Oxford afforded much time to watch 1st class cricket in the Parks behind the Science buildings, much time was spent in doing required laboratory experiments.

At New College, my key tutor was Peter Dickens, under whom I would do my Part II undergraduate research and my D. Phil. Peter attracted me to Solid State Chemistry. Also, at that time metallurgy was part of the Inorganic Chemistry Department, so I took metallurgy classes from William Hume-Rothery, who was deaf at that time but whose lectures were intriguing. Those were exciting times, with the US Air Force Office of Scientific Research in London supporting my undergraduate research on the recombination kinetics of oxygen atoms on the tungsten oxide bronzes. This was the sputnik era, and there was interest in possible atomic reactions on the nose cones of satellites. This work resulted in my first publication in the Transactions of the Faraday Society.

For my D. Phil. studies I won a Gas Council scholarship to study catalysts to convert coal gas to natural gas. However, just before starting this research the UK struck natural gas in the North Sea and the Gas Council gave me the freedom to do whatever research I liked. I studied the reduction by hydrogen of the same tungsten bronzes and a range of tungsten oxides and discovered how the fast ionic mobility of the alkali ions dictated the reduction reaction pathway.

POSTDOCTORAL STUDIES

In 1968, I moved to the materials center at Stanford University in California to work with Robert Huggins on advanced materials. After about three months there, Bob moved to Washington, DC for a two-year stint as Program Officer for the Materials Research Centers. I became the de-facto leader of his group for those two years, a great learning experience. There I studied the fast ion transport of alkali ions in the recently discovered beta-alumina materials, amongst other compounds, using the mixed conducting tungsten bronzes as electrodes. For this beta alumina work, I received the young author award of the Electrochemical Society. I still remember going to the Shamrock Hotel in Houston to pick up the award at a black-tie dinner. Norman Hackerman was President of the Electrochemical Society, and he brought in the Texas Rangers on their horses to the reception around the hotel pool.

At Stanford I met Georgina Andai on a trip to the San Francisco Opera in August 1968 organized by the Bechtel International Center. Georgina had just arrived from Queens College in New York to study for her graduate degree in Latin American Literature. Georgina was an immigrant from Budapest, Hungary by way of South and Central America. We were married in the Stanford Chapel on March 23rd, 1969, and spent our honeymoon travelling around four of the Hawaiian Islands. About a month later, we spent a week at a solid-state chemistry meeting in Scottsdale, Arizona at a Moorish looking hotel, organized by Leroy Eyring and Mike O'Keefe. The field was just taking-off in the USA, unlike Europe where it was strong. We had two children whilst at Stanford, Jenniffer and Michael. We were the first family living in Stanford married student housing, where the wife was the student. I still remember the looks when I went to a

Figure 2. Key advisors: Peter Dickens, Oxford University (died December 2019), and Robert Huggins, Stanford University.

Figure 3. Charging our electric vehicle in Bermuda in March 2019.

spouses meeting – what are you doing here? When our second child was born, we rented a home on Maureen Avenue in Palo Alto, a nice two-bedroom house with a garden where the children could play winter or summer. Just before we left California, our landlady offered to sell us the home for $30,000! Last year, we celebrated our 50th Anniversary in Bermuda, and rented one of the first electric vehicles on the island and learnt all about range anxiety (Figure 3).

EXXON CORPORATE RESEARCH LABORATORY

In 1972, I moved to the newly formed Corporate Research Laboratories of Esso (now ExxonMobil), who were initiating research efforts in energy beyond petroleum and chemicals. It was there, whilst working with such key scientists as Fred Gamble and Arthur Thompson in a very vibrant and intellectually stimulating interdisciplinary group, that I discovered the key role that intercalation played in the reversibility of chemical reactions. I was asked to describe this finding to a committee of the Exxon Board of Directors in New York; this would be described today as an elevator pitch. Within a week, Corporate Research was given the go-ahead to build a team to develop this invention. Exxon treated research like drilling an oil-well; not all will work, but some will but for success serious investment was needed. They did this and established a large lithium battery engineering, development and manufacturing effort. That effort resulted in the first rechargeable lithium-ion batteries, which was published in Science in 1976, some three years after the patents were filed.

The field of solid-state electrochemistry was just getting started due to materials scientists getting involved in energy research. One of the earliest international meetings in the area was a two-week NATO conference in Belgirate, Italy in 1972 which had many junior scientists present but also senior legends like Carl Wagner, the father of defects in solids. I was sitting next to him in the conference photo. Some 20 years later, when this meeting was repeated the field had dramatically changed – the lithium battery had arrived. I still remember arriving at the hotel in Belgirate and commenting to the owner that we had been there 20 years earlier and he said let me check. He went to his file cabinet, and pulled out a photograph of the earlier meeting, and saying there you are.

At that time, the journals available for publishing our work were limited. Materials Research Bulletin tended to be a favorite for both myself and John Goodenough. So North-Holland in the late 1970s proposed that I and Hans Rickert from Germany start a new journal for the field of ion transport in solids. I said there was no need. They responded by saying let's do a survey of the field. Well, the field voted over 90% in favor of a journal, and the first issue of Solid State Ionics was published in 1980. I remained Principal Editor for 20 years. At that time the journal published in English, French and German.

At Exxon, I moved from a bench scientist to group head to Director of the Solid State and Catalytic Sciences Laboratory in Corporate Research. Then in the early 1980s I became Manager of the Chemical Engineering Technology Division of Exxon Engineering, which at that time was expanding fast in order to explore synthetic fuels, such as shale oil and coal liquefaction/gasification. An interesting assignment that certainly was a broadening experience, but not really aligned with my interest in research. After a brief period in research at Schlumberger leading a high-powered group of scientists understanding rock science (most of whom are now in academia), I returned to academia in 1988.

BINGHAMTON UNIVERSITY

In the fall of 1988, I joined the Chemistry department of Binghamton University (State University of New York) with the goal of introducing materials across the curriculum, and to emphasize to students that science is interdisciplinary. I was the founding Director of the Institute for Materials Research and led it until 2018 and was the driving force behind building the graduate program in Materials Science and Engineering, which I also led for more than a decade. The move to Binghamton was great for me. I enjoyed teaching young excited students.

In 1993 I spent two months as a JSPS Fellow at the University of Tokyo in the physics department with Professor Suematsu. I arrived on April 1st in Tokyo just at the beginning of the cherry blossom season, and the students there took me to the tradition of drinking sake under the cherry trees. I managed to get back to my lodging at the Tokyo Institute of Technology, where I spent my first week until space was available at the International House of the U. Tokyo in Roppongi. I got much fitter there with the healthy food and the commute to the University each day, more than half a mile walk to the subway and another half a mile from the subway to the University. I still remember the lines of university faculty at the McDonalds at the subway exit picking up their coffee each morning. Each week I would leave Tokyo to give lectures at other universities and labs throughout Japan. Whilst there, my secretary reminded me by fax that there was a proposal call from the U.S. Department of Energy due in a few days, and was I going to apply. The next day I sent a handwritten proposal by fax to her; she typed it up and submitted it. That was the start of my DOE funding which continues to this day. Elaine still works with me and came to Stockholm for the Nobel celebrations. Half-way through my stay in Tokyo, my father passed away, and I flew back to England. At that time cash was dominant in Japan, so I had to go to the bank to get the cash for the ticket; even that was an experience as all transactions were performed by hand in a large ledger. For the last week of my stay in Tokyo, my family joined me, and we managed to squeeze into the apartment, which was little larger than our present living room.

On returning from Japan to Binghamton, based perhaps on my prior management experience at Exxon, I was invited to become the Vice-Provost for Research at Binghamton (the chief research officer on campus). I did this for five years part-time, whilst still carrying out my own research. I also took on the responsibility as Vice-Chair of the Board of Directors of the Research Foundation of SUNY; the senior faculty member on the

Figure 4. Stan Whittingham and John Goodenough working together at the Battery500 consortium in 2018 in Berkeley, California.

Board. In 2007, I took a lead role in a DOE workshop on the research needs for energy storage. A series of these workshops led to the formation of the Energy Frontier Research Centers in 2009, and I was a member of one led by Clare Grey at Stony Brook (SUNY). When Clare moved to Cambridge University in 2011, I took over as Director with an associated position at Stony Brook. On renewal in 2014, the Center moved to Binghamton. This Center has been an exciting and invigorating experience comprising some of the leading scientists around the country and enabled a fundamental understanding of the reactions occurring in lithium battery electrodes. I am also a member of a more applied battery consortium, the Battery 500 group, whose goal is to enable batteries with an energy density of 500 Wh/kg. This has allowed John Goodenough and me to work together again (Figure 4).

MY FAMILY AND FINAL NOTES

I have been very fortunate in my life by having a very supportive family, starting with my late parents who supported my goals of being a scientist, of being a cactus fanatic and of my two-year move to California, which they told me later they knew would be permanent. My wife, Georgina, has been especially supportive over the last 50+ years putting up with the long hours and all the travel. She now has her own career as a Professor

Figure 5. The Whittingham family at the Nobel Ceremony. © Nobel Media AB. Photo: Alexander Mahmoud.

of Spanish and Latin-American literature at the Oswego campus of SUNY, and is also thoroughly enjoying interacting and teaching the younger generations. Our children now have children of their own, the oldest of whom are now in college. Both of them vowed when they were old enough, they would move back to California, which they did and all four of our grandchildren were born in the San Francisco Bay area. Besides our immediate family, I was joined in Stockholm by my brother William, a PhD chemist from Cambridge and my niece Helen, a PhD Materials Scientist also from Cambridge. Both live in Britain, as do my two sisters, Anne and Susan. The family is shown at the Nobel Ceremony in Figure 5.

My life and that of my family has changed dramatically since the Nobel's recognition of our team's work on lithium batteries, that was announced whilst I was at a battery meeting in Ulm, Germany. My wife was the last of the family to hear the news as her phone was off due to the Yom Kippur holiday on October 9th. That of my University has also changed; the excitement was palpable from the virtual press conference from my room in Ulm to the reception when I returned to Binghamton (Figure 6). It is indeed humbling to join all those famous previous Nobel Laureates. I was also very pleased to join my colleagues John Goodenough and Akira Yoshino in this recognition.

Finally, I must thank all my past and present friends and collaborators at Oxford, Stanford, Exxon, Schlumberger, Binghamton and the chemistry, materials and battery communities without whom my work and this recognition would not have been possible. It is my hope that this recognition will allow all of us to achieve a cleaner and more sustainable world for our children and grandchildren.

Figure 6. The excitement at Binghamton University on October 18th. Photo: Jonathan Cohen.

REFERENCES

1. Peter G. Dickens and M. Stanley Whittingham, "The catalytic behavior of the tungsten bronzes", *Trans. Faraday Soc.* **61** (1965), 1226.
2. M. Stanley Whittingham and Robert A. Huggins, "Transport properties of silver beta alumina," *Journal of the Electrochemical Society,* **118** (1971), 1–6.
3. M. Stanley Whittingham and Robert A Huggins, "Beta alumina – Prelude to a revolution in solid state electrochemistry," *NBS Special Publication* **364**, Washington DC: U.S. Department of Commerce and National Bureau of Standards, 1972, 51–62.
4. M. Stanley Whittingham, "Electrical energy storage and intercalation chemistry," *Science,* **192** (1976), 1126–1127.
5. M. Stanley Whittingham "The role of ternary phases in cathode reactions," *Journal of the Electrochemical Society,* **123** (1976), 315–320.
6. M. Stanley Whittingham, "Chalcogenide Battery," US Patent 4009052 and UK Patent 1468416, filed 1973 issued 1977.
7. M. Stanley Whittingham, "Chemistry of intercalation compounds: metal guests in chalcogenide hosts," *Progress in Solid State Chemistry,* **12** (1978), 41–99.
8. M. Stanley Whittingham and Allan J. Jacobson, *Intercalation Chemistry,* Academic Press, 1982.
9. M. Stanley Whittingham, "Lithium Batteries and Cathode Materials", *Chemical Reviews,* **104** (2004), 4271–4301.
10. M. Stanley Whittingham, "The Ultimate Limits to Intercalation Reactions for Lithium Batteries", *Chem. Rev.*, **114** (2014), 11414–11443.
11. M. Stanley Whittingham, Jia Ding, and Carrie Siu, "Can Multielectron Intercalation Reactions Be the Basis of Next Generation Batteries?," *Accounts of Chemical Research,* **51** (2018), 258–264.

The Origins of the Lithium Battery

Nobel Lecture, December 8, 2019 by
M. Stanley Whittingham
Binghamton University, Binghamton, NY, USA.

AS NOTED BY THE ROYAL SWEDISH ACADEMY of Sciences, “Lithium-ion batteries have revolutionized our lives since they first entered the market in 1991. They have laid the foundation of a wireless, fossil fuel-free society, and are of the greatest benefit to humankind.” The idea for rechargeable lithium batteries started in 1972 in the Corporate Labs of Exxon, within a group studying the impact of intercalating electron donors on the superconductivity of the layered disulfides. This breakthrough was based on a fundamental understanding from a solid-state viewpoint of the reactions in solids and particularly of the fast motion of ions within solids. The discovery of fast ion transport in β-alumina, and the need for mixed conducting solids with a wide-stoichiometry range to measure its conductivity played a key role in developing lithium batteries. All of today’s lithium-ion batteries rely on the original intercalation concept used in titanium disulfide. There are still many scientific opportunities to improve the energy density, performance and lifetime of lithium batteries.

INTRODUCTION

Alessandro Volta (1745–1827) built the first electrochemical cell, known as the Volta Pile. Examples of the Pile may be found in the temple built in his memory at the southern end of Lake Como, as shown in Figure 1. He described these in a presentation to the Royal Society in London in French in 1800.[1] These cells are still made by almost every science student, including those grade students at the British Ambassador's house in Stockholm on December 8th, 2019. In 1859, Plante invented the lead acid battery which still has the largest share of the battery market. In 1966 Leclanche described the original C-MnO_2 cell, which was improved by Urry into today's alkaline cell. In 1899 Jungner described the Ni/Cd cell, which had an active life in portable devices, but has now been mostly phased out because of its toxicity.

The origins of the lithium-ion battery are intimately associated with the discovery and development of fast ion transport of ions in solids. Whereas, Volta originated the study of batteries, it was Michael Faraday (1791–1867) who built the foundation of the science of electrochemistry. It was also he who first reported the discovery of ionic conductors in the early 19th century in silver sulfide, and predicted that many others would be found:[2]

> "432. The effect of heat in increasing the conducting power of many substances, especially of electricity of high tension, is well known. I have lately met with an extraordinary case of this kind, for electricity of low tension, or that of the voltaic pile, and which is in direct contrast with the influence of heat upon metallic bodies, as observed and described by Sir Humphry Davy (2)."
> "433. The substance presenting this affect is sulfuret of silver."
> "434. There is no other body with which I am acquainted that, like sulphuret of silver, can compare with metals in conducting power for electricity of low tension when hot, but which unlike them, during cooling, loses in power, whilst they, on the contrary gain. Probably however, many others may, when sought be found."

It was half a century later that Nernst (1864–1941; Chemistry Nobel Prize, 1920) used the high oxygen ion conductivity of yttria stabilized zirconia, $0.85ZrO_2{:}0.15Y_2O_3$, as a light source[3]. When a current is passed through the ceramic in air, it heats up until it becomes white hot and glows. It also was the first case of potential thermal runaway. The conductivity of ionic conductors increases with increasing temperature, so a simple current cannot be used, and the control system was sufficiently complex that it was only commercialized by Westinghouse for a short period at the beginning of the 20th century, 1900–1913.

Figure 1. Volta's Temple in Como, Italy.

THE TUNGSTEN BRONZES, 1964–1972

The tungsten bronzes, M_xWO_3, where M is typically an alkali metal or protons, were first described by Wohler in 1824,[4] and reviewed by our Oxford group in 1968.[5] They have been of great interest ever since because of their brilliant colors, which are a function of the value of x[6] as shown in Figure 2; it is this change of color which allowed their use in electrochromic displays, such as in the 787 aircraft. As x can vary from 0 to 1, and the free electron concentration is directly proportional to x, they represented an excellent opportunity to study the impact of the relative importance of electronic vs geometric factors in heterogeneous catalysis.[7] The electronic properties were of greater importance, but subsequent studies of their reduction by hydrogen showed how important the structure was in determining the route by which they were reduced.[8] This reduction was enhanced by the rapid transport of the alkali ions in the lattice, and was the first indication that these materials were mixed conductors, that is showing both high electronic conductivity and high ionic conductivity. Na_xWO_3 forms a range of crystalline structures. At the highest sodium contents the structure is cubic and the same as perovskite, as shown in Figure 3 (left); the sodium ions diffuse in four sided tunnels. As the sodium content is reduced, the structure becomes more complex and tetragonal, and there are three different possible alkali sites; the sodium ions are shown in the larger five-sided channels. At the lowest sodium levels, distorted perovskite structures are observed. For the larger potassium ions, a third structure with hexagonal tunnels is found[5]. Large single crystals of these bronzes can be formed by electrolysis of molten tungstates and the thermodynamics of this process has been determined[9].

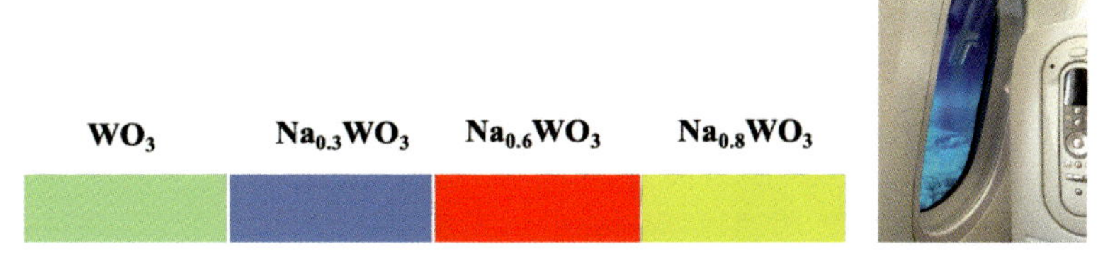

Figure 2. Schematic of the colors of the tungsten bronzes, and right an electrochromic window in the 787 aircraft.

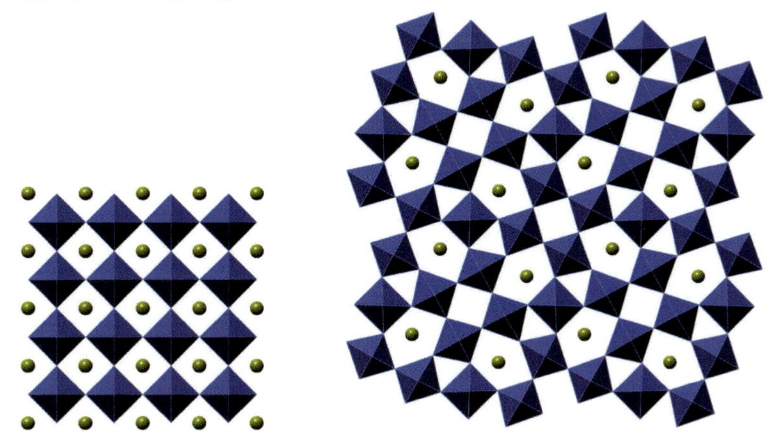

Figure 3. The cubic and tetragonal structures of Na_xWO_3. The gold circles represent the sodium ions, and the blue octahedra represent WO_6 cornered shared units.

β-ALUMINA, 1968–1972

Sodium β-alumina was initially thought to be just another form of Al_2O_3, but Pauling showed that it had a layered structure, comprising γ-Al_2O_3 blocks separated by Al-O-Al pillars, as shown in Figure 4 (top). Its nominal composition is $NaAl_{11}O_{17}$. The publication of its fast sodium ion conductivity by Yao and Kummer in 1967,[10] represented a major breakthrough and kicked-off the whole new field of Solid State Ionics. As Huggins and Whittingham predicted[11] in 1972 "β-alumina – Prelude to a Revolution in Solid State Electrochemistry", the field took off exponentially.

The mobile sodium ions reside midway between three pillars as shown in Figure 4 (bottom); there is typically a 15% excess of these ions. The excess ions reside between three other sodium ions as shown. These defects are critical to the conductivity, and the diffusion mechanism is an interstitialcy knock-on process, so no defects need to be created for a cation to diffuse through the lattice, which results in their low enthalpy of motion.

The ionic conductivity of sodium β-alumina is so high that it cannot be measured using the traditional metal electrodes, like platinum. Electrodes had to be found that were reversible to both electrons and ions. The solution

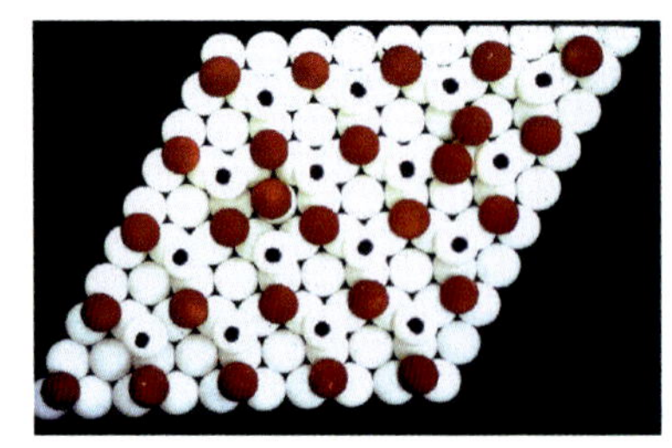

Figure 4. The structure of sodium β-alumina showing (top) the 4 layer γ-Al_2O_3 sheets separated by the Al-O-Al pillars, and (bottom) the diffusion plane of β-alumina showing the mobile cations (red), and the Al-O-Al pillars (oxygen – white spheres, aluminum – black circles). The two defect options for the excess cations are shown; the dumb-bell defect is observed for silver and the interstitial one for sodium.

was the mixed conductor Na_xWO_3 and specifically the larger tunnel structure $Na_{0.4}WO_3$, rather than the denser perovskite structure. Figure 5 (left) shows the schematic of the cell used; the sodium β-alumina single crystal was extracted from a brick used to line glass tanks. The resulting ionic conductivity is shown in Figure 5 (right). The sodium β-alumina has a single interstitialcy mechanism from 800°C to liquid nitrogen temperatures [12]. The conductivity of sodium in β-alumina is higher than that of sodium in solid sodium metal. The equivalent lithium vanadium oxides, were also used as electrodes for measuring the lithium ion conductivity of Li-β-alumina.

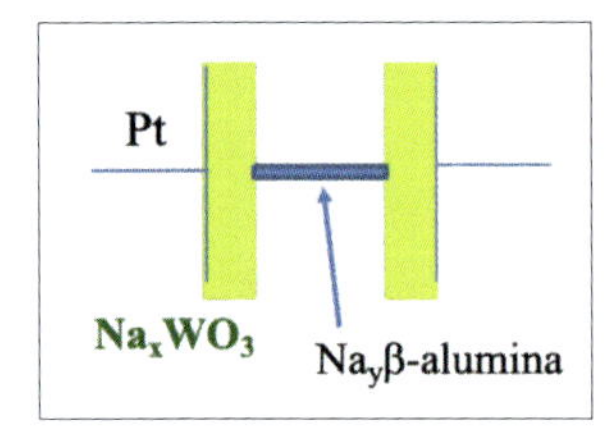

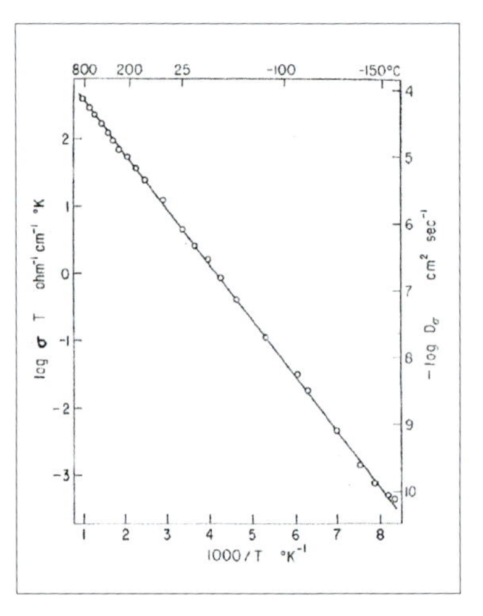

Figure 5. (Left) Schematic of the measurement cell, and (right) ionic conductivity of sodium β-alumina. Reprinted with permission from ref.[12] Copyright 1971, The American Institute of Physics.

The Ford Motor Company had proposed using β-alumina in a battery comprising a molten sodium anode and a molten sulfur cathode. It operates in the temperature range of 300–350°C. The Zebra version of the cell, which also uses a molten sodium anode but a NiCl2 cathode has also been extensively researched [13]. It has the advantage that it can be built in the discharged state, Ni metal + NaCl, and so does not require dry rooms. Essentially all work stopped in 2016 due to the much lower cost and flexibility of lithium-ion cells and their ability to operate under ambient conditions.

STATUS OF BATTERY CHEMISTRY IN 1972, TITANIUM DISULFIDE AND WHAT IS INTERCALATION

In 1972, battery scientists did not recognize ternary phases and non-stoichiometry, so that the reaction of V_2O_5 with lithium was expressed as:

V_2O_5 + 2Li reacts to V_2O_4 + Li_2O

and for the common dry cell as

$2MnO_2$ + $2H^+$ reacts to Mn_2O_3 + H_2O

However, both of these reactions are now known to occur by insertion respectively of lithium and sodium ions into the lattice to form $Li_xV_2O_5$ and MnOOH.[14]

In 1972 the layered dichalcogenides were of interest because of their superconducting behavior. They were able to intercalate a range of electron donors such as alkyl amines and other organic and organometallic species. It was found by this author that the K_xTaS_2 superconductor could be synthesized by immersion of TaS_2 crystals in an aqueous KOH solution, indicating that the free energy of this reaction was sufficiently high to prevent the evolution of hydrogen gas in the presence of water. This suggested that the layered dichalcogenides could be used to store energy. Of the dichalcogenides, titanium disulfide was found to be particularly preferred because not only is it the lightest in weight but is also a metallic conductor.[15] Titanium disulfide was found to be an almost ideal cathode because:

- There is no need for the addition of an electronic conductor, such as carbon black
- Li and TiS_2 form a single solid solution from TiS_2 to $LiTiS_2$, so no energy expended on forming new phases
- Li lithium was found to diffuse very quickly so that thick electrodes and even single crystals can be used,[16,17] and

- Other ions, such as Na and Mg can be similarly intercalated to form the basis of secondary batteries.[16]

When lithium is inserted into the van der Waals gap of the layered structure of titanium disulfide there is less than a 10% expansion of the lattice, as shown schematically in Figure 6 (left). On removal of the lithium, the lattice reverts back to its initial state. The perfect reversibility of this reaction is known as an intercalation reaction. This reaction is now often used in dictionaries as an example of intercalation. The original definition of intercalation refers to the insertion of February 29th into the calendar every leap year. Intercalation reactions are important in many areas of chemistry and medicine [18]. The layered structure of the $LiTiS_2$ is shown in Figure 6 (right), the sulfur atoms are packed in a hexagonal close-packed stacking, (hcp) and only differs from the layered oxides in the stacking sequence. The $LiMO_2$ materials are cubic close packed (ccp), but on removal of lithium can convert to hcp, as found for CoO_2. This is detrimental to the long-term reversibility of the reaction.

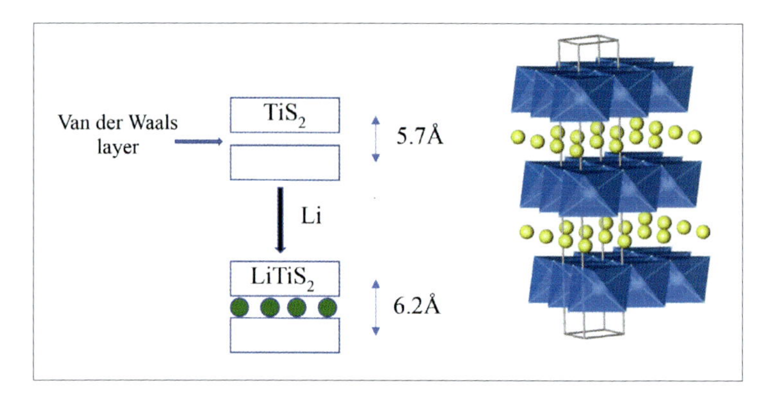

Figure 6. (Left) schematic of lithium intercalation into TiS_2, and (right) layered structure of $LiTiS_2$.

The initial studies of titanium disulfide were made using a lithium anode, and ether based electrolytes, in particular dioxolane.[16,17,19] The almost perfect characteristics of the Li/TiS2 couple allowed for high rate systems for hundreds of cycles as shown in Figure 7.[20] The difference between the charge ad discharge curves is just due to the resistance losses in the electrolyte. The reduction in the cell voltage as the lithium content increases is due to the conduction band being steadily filled as shown schematically in Figure 7 right. Although the discharge curve appears smooth, the lithium ions do tend to order as identified by Thompson using the then new electrochemical potential spectroscopy technique.[21,22] At the high rates used here the jump time is shorter than the ordering time, hence the ordering is minimized. This has been observed more recently in the Li/$FePO_4$ system, where at higher rates the system behaves as a single phase reaction, Li_xFePO_4, but at low rates phase separation occurs and a mix of $FePO_4$ and $LiFePO_4$ is observed.[23] The first operando studies of the reactions in batteries were performed on the Li/TiS_2 system by Chianelli *et al.* at Exxon in the late 1970s.[24–26]

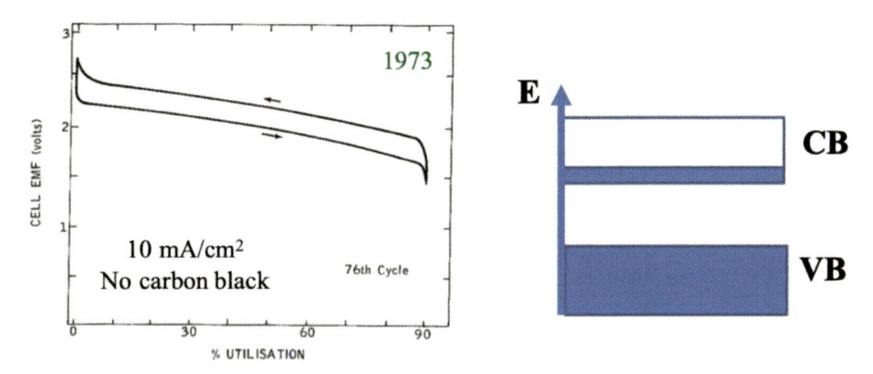

Figure 7. (Left) Discharge/charge curve of Li/TiS_2 at 10 mA/cm² (reprinted with permission from ref,20 copyright 1978 Elsevier); (Right) Schematic of electronic structure of $LiTiS_2$. (Right) Schematic of the TiS_2 band structure.

Exxon Enterprises, the new venture arm of Exxon, built a range of lithium batteries in the 1970s and early 1980s. Figure 8 (left) shows a paperweight used as a demonstration of lithium batteries; besides the battery it contains a solar cell that keeps the battery charged and a clock. This one was built in 1977 and is still operable today, indicating the long cycle life of well-constructed cells. Some of these cells are now in the Nobel Prize Museum in Stockholm. Several of these cells were tested by Rutgers University in 2015 and were found to still have more than 50% of their original capacity.[27] In addition, cells as large as 6” x 4” x 1” were constructed and exhibited at the Chicago 1977 EV show and turned a motorcycle battery on and off throughout the meeting (see Figure 8). The morphology of the TiS_2 used in the cells consisted of large platelets as shown in the Figure; these were produced by the reaction of titanium sponge with sulfur. When formed form the commercially viable route of $TiCl_4$ and H_2S, the platelet appear more star-like. Because of the tendency of lithium metal to form dendrites and/or finely divided lithium, pure lithium anode cells tend to be unstable. Thus, in the commercial cells Exxon used the safer LiAl alloy as anode.[28] This was formed in-situ when the electrolyte was added to cells containing sheets of aluminum and lithium. In addition, cells could be built lithium metal free, using a pre-intercalated cathode, i.e. $LiTiS_2$.[16]

Titanium disulfide can be formed either as the layered structure or in the spinel form. The two differ in the stacking of the sulfur layers, the for-

Figure 8. (left) Paper weight utilizing a Li/TiS_2 cell from 1977 and operating in the author's office in 2019; (middle) large TiS_2 cells exhibited at the 1977 EV show in Chicago, and (right) morphology of the TiS_2 crystals.

mer is hexagonal close packed and the spinel is cubic close packed, and as a result diffusion of the lithium ions is two dimensional and in the latter three dimensional. The spinel is formed by removal of the copper from CuTi2S4. The lithium diffusion is comparable in both phases.[29] Both titanium disulfide phases can also readily intercalate the magnesium ion, which has a comparable ionic size to lithium. Van der Ven's team showed that the potentials for the layered Mg/TiS_2 couple and the spinel Mg/Ti_2S_4 couple for magnesium contents from 0 to 0.5 Mg/Ti, are about 0.8–1.0 volts lower than those of the corresponding Li cells.[30] Nazar's team in 2016 published experimental data on the same systems with almost identical voltage characteristics.[31,32] The spinel phase showed excellent reversibility at 60°C. However, the resulting energy densities are less than half of those of the corresponding Li/TiS_2 couples, indicating that it is unlikely that Mg intercalation cells are a viable alternative to Li cells.

THE DEVELOPMENTS OF LITHIUM-ION BATTERIES POST DISULFIDES

In the late 1970s research accelerated to provide higher energy density and/or lower cost replacements for Li/TiS_2. All the subsequent rechargeable Li-ion batteries use intercalation reactions for both electrodes. The key cell components and history are shown in Figure 9 and Table 1. MoliEnergy in Vancouver used British Columbia's naturally occurring molybdenite ore to develop a Li/MoS_2 battery; Dahn had done his PhD studying the Li/TiS_2 system. Goodenough, who was studying the magnetic behavior of $LiCoO_2$, recognized that it had a similar structure to $LiTiS_2$ and studied it in lithium cells.[33] It can provide for cells around 4 volts, and thus loses a lower percentage of its energy density when coupled with a safe intercalation anode, such as those based on carbon. It was the carbon-based anode developed by Yoshino at Asahi Kasei that made the Li-ion battery safe and enabled SONY to market a viable commercial product in 1991.

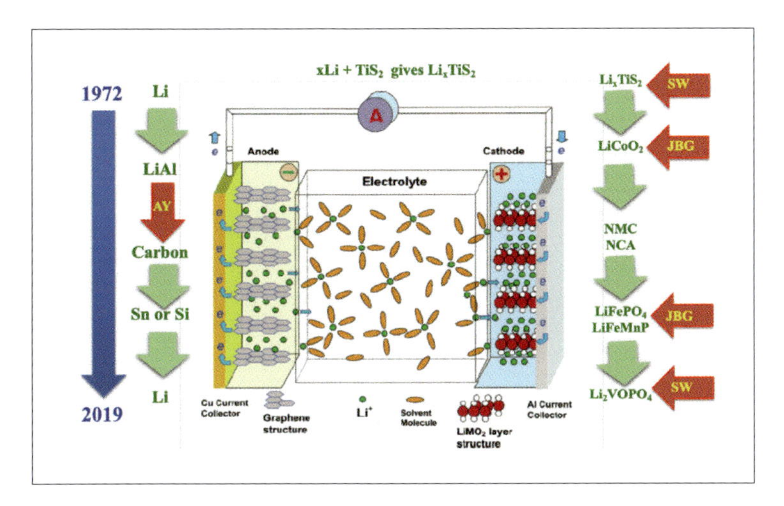

Figure 9. The components of a Li-ion battery showing the layered sulfide/oxide cathode and a layered graphite anode. These are shown as a function of time with the key contributions of the three laureates listed (AY=Yoshino; SW=Whittingham: JBG=Goodenough).

Table 1. Intercalation Reactions are the Basis of all Li-Ion Batteries

- 1970s Exxon (**Whittingham**)
    - $LiAl/TiS_2$
- 1980s Moli Energy (**Dahn**)
    - Li/MoS_2
- 1980s Oxford (**Goodenough**)
    - $LiCoO_2$
- 1980s Asahi Kasei (**Yoshino**)
    - C (coke)
- 1991 SONY
    - $C/LiCoO_2$

The C/LiCoO2 cell eventually dominated the portable battle market, displacing the Ni/Cd and Ni/MeH batteries within the last ten years. The LiCoO2 cathode still dominates applications where volumetric energy density is most important and cost is secondary such as smart phones and computers. However, the high cost of cobalt, its natural low abundance and the use of child labor in Congo demanded that its content be reduced. In addition, the use of materials is very inefficient, as seen in Table 2, which shows that today's cylindrical cells have less than 25 % of their theoretical energy density.

Chemistry	Size	Wh/L theoretical	Wh/L actual	%	Wh/kg theoretical	Wh/kg actual	%
$LiFePO_4$	54208	1980	292	14.8	587	156	26.6
$LiFePO_4$	16650	1980	223	11.3	587	113	19.3
$LiMn_2O_4$	26700	2060	296	14.4	500	109	21.8
$LiCoO_2$	18650	2950	570	19.3	1000	250	25.0
$Si/C/LiMO_2$ SONY VC7	18650	2950	725	24.6	1000	264	26.4

The theoretical values in the table assume only the active components, and no volume or weight for lithium beside that in the cathode.

Table 2. Energy densities of some commercial cells, adapted from ref.[34]

There are several ways to increase the energy density of these cells:

- The carbon is a major issue; it takes up half the volume of the cell as 72 gm of carbon are needed to hold 7 gm of lithium
- Need to use the full lithium in the cathode, not the 50–60% used today
  - o Need to find cathodes that can intercalate more than one lithium ion
- Need cathodes with higher ionic and electronic conductivities
  - o Will allow thicker electrodes, which will reduce the necessary amount of current collectors and separators

Since the 1970s the energy densities of Li-ion cells have been steadily increasing to over 250 Wh/kg.[35] There is no reason why the energy density should not exceed 350–400 Wh/kg for intercalation-based cells within the next five years, with an upper dream of 500 Wh/kg or 1 kWh/liter. This last is just 50% of the theoretical values and is the goal of the US DOE Battery 500 consortium. The first step must be modification of the graphitic carbon anode. This is an ongoing challenge, but additions of 5–10% silicon are showing promise. Below two approaches to "closing the gap" between actual and theoretical are described.

THE 1st CYCLE LOSS OF THE LAYERED OXIDE CATHODE

The $LiCoO_2$ cathode has been replaced for most applications by the NMCA class of material, where much of the cobalt has been replaced by nickel, manganese and/or aluminum. Two of particular interest are: $LiNi_{0.8}Mn_{0.1}Co_{0.1}O_2$ and $LiNi_{0.85}Co_{0.10}Al_{0.05}O_2$. The higher the nickel content the higher the energy density for any given charge voltage, but the less thermally stable the material becomes, as shown in Figure 10 (left). The instability is determined predominantly by the Ni content; the charging voltage has much less effect.

All these NMC cathodes lose 10–15% capacity on the first cycle as shown in Figure 10 (right).[36] If this could be recovered, then the cell's capacity could be increased by 10%, allowing energy densities of over 400 Wh/kg to be attained. It seems probable that a sharp drop-off in the diffusion coefficient of the lithium at lithium contents over around 0.8 in LixMO2 is the reason; this is shown in Figure 11.[36]

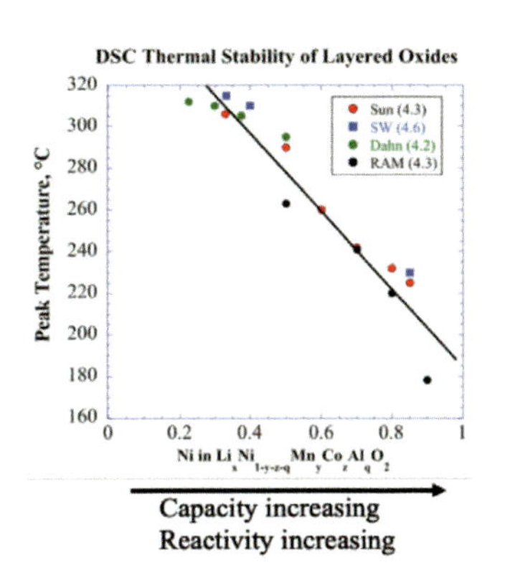

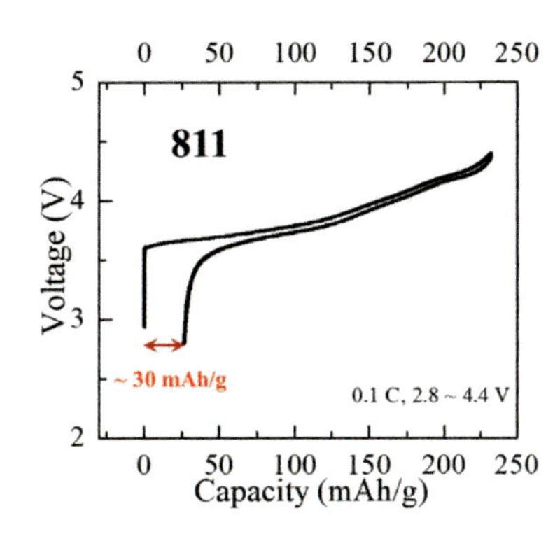

Figure 10. (Left) The thermal stability of several NMCA materials charged from 4.2 to 4.6 volts from four different research groups, showing the almost linear behavior of thermal instability to the nickel content; (right) the typical 1st cycle of an 811 NMC material charged to 4.4 volts at a C/10 rate,[36] reproduced with the permission of the American Chemical Society.

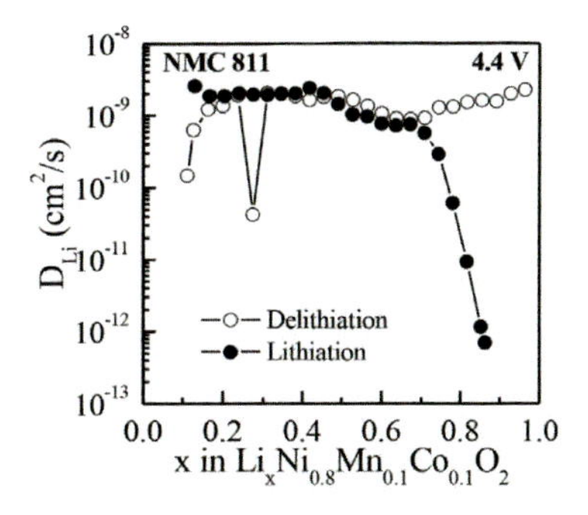

Figure 11. The lithium diffusion coefficient in $Li_xNi_{0.8}Mn0.1Co_{0.1}O_2$,[36] reproduced with the permission of the American Chemical Society.

However, it appears to be not that simple, as $LiCoO_2$ (LCO), shows essentially no 1st cycle loss as indicated in Figure 12. This Figure compares 811 with LCO, when both are charged to a depth of 120 mAh/g. The 811 shows a 20% 1st cycle loss, which is reduced to 6% on raising the temperature to 45°C, suggestive of a rate controlling reaction. In contrast, for LCO the loss is less than 3% at 21°C, and reduces to 2% at 45°C. The challenge is to make the 811 and other NMC materials behave in this respect like LCO.

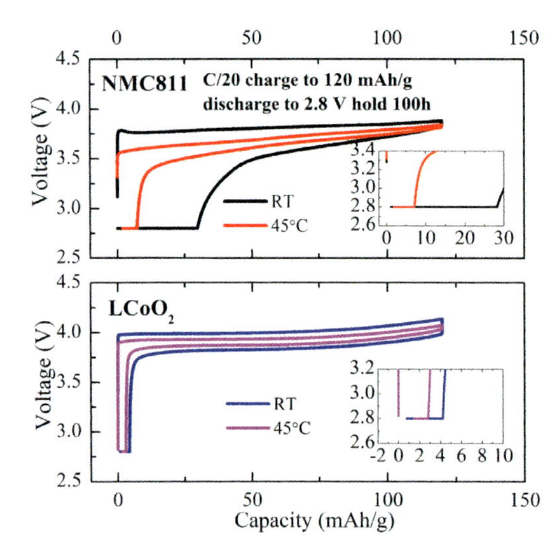

Figure 12. The 1st cycle behavior of NMC811 compared to LCO. Unpublished work at Binghamton (Hui Zhou).

TWO LITHIUM CATHODE MATERIALS, LI_xVOPO_4

An alternative to closing the gap on the layered oxides is to reversibly intercalate more than one lithium ion per redox center. Vanadium is a particularly attractive redox center because the vanadium can readily go between the +5 and +3 oxidation states, and it is the 4th most abundant transition metal. V_2O_5 itself can reversibly cycle 1.5 Li/V, but the structure is not stable and converts to disordered rock salt, $[Li_3V_2]O_5$ with a cell voltage ranging continuously from 3.5 to 1.5 volts.[37] The olivine, Li_xFePO_4, has been extensively used commercially, particularly in China, because of its low cost and enhanced stability, but its capacity is only 170 mAh/g;[23] the science of this cathode material has been recently reviewed.[23] Vanadium forms several phosphates. We have found that one of these, ε-$VOPO_4$, is particularly attractive as the cathode of a lithium battery. Its structure is shown in Figure 13. It contains VO_4 tetrahedra and highly distorted VO_6 octahedra, with chains of O-V=O—V, where the V-O bonds are alternately short and long along the chain. On lithium intercalation to Li_2VOPO_4 there are no longer any vanadyl groups. When formed by a solvothermal process, cuboid particles of 100–200 nm in size are formed and these give essentially the theoretical capacity of 305 mAh/g at low rates as shown in Figure 13.[38] There is no loss in capacity or change in shape of the cycling curves, indicating that the structure is stable to the repeated insertion and removal of two lithium ions. This opens up a new avenue for the exploration of high energy density cathodes.

Figure 13. (Left) Structure of ε-VOPO4, with the vanadyl chains coming in/out of the paper. Reprinted with permission from ref.[39] Copyright 2005 The Electrochemical Society. (Center) the solvothermally formed cuboid particles, and (right) the electrochemical behavior at a C/20 rate. Reprinted with permission from.[38] Copyright 2019 The Royal Society of Chemistry.

CONCLUSIONS

Over the last half century, lithium ion batteries have come from an idea to domination of energy storage for both portable and stationary applications ranging from milliwatt hours to proposed gigawatt grid storage. They enabled the electronics revolution and helped several 3rd world countries to bypass wired phones.

> The academy said: 'They (lithium batteries) have laid the foundation of a wireless, fossil fuel-free society, and are of the greatest benefit to humankind,'. Colleagues we hope that our discoveries will let all of us work together to build a cleaner environment, make our planet more sustainable, and help mitigate global warming, leaving a cleaner legacy to our children and grandchildren.[40]

Many countries and states have mandates for reduction of carbon dioxide emissions, which will drive the growth of electric vehicles of all sizes; some more recent examples that I have personally experienced are shown in Figure 14. There is still much to do to improve batteries for enhanced vehicle penetration into the market. These needs include:

- Higher energy density, so the batteries can be smaller and lighter
- Improved range vs cost, including operating without range penalty in more extreme weather conditions
- Sustainable systems, which must include clean recycling
- Safer batteries, which will require different electrolytes and possibly solid-state systems.

Figure 14. Three examples of electric vehicles. (Left) My wife in one of the first rental electric vehicles on Bermuda that we drove on our 50th anniversary in March 2019 [yes, range anxiety is real]. (Center and right) Two Peterbilt electric vehicles that I drove at their test track north of Seattle in Washington State, October, 2019.

Grid storage is also essential to enable renewable energy storage such as wind and solar; batteries are an essential here as there appear to be severe limits on additional pumped-hydro facilities. I am proud that my own New York State is taking the lead here – having the largest mandate, > 3 GWh, for grid storage. Numerous solar-battery facilities are being constructed around New York State as well as throughout the world. Battery storage load shifting is already economically viable today. Some examples are shown in Figure 15. Grid storage will have several advantages:

- Will allow cleaner sustainable technologies
o Will enable renewable energy storage
- Will lead to a more efficient grid
o Will help to eliminate costly peaker plants, which are often the worst polluters
- Will assist in mitigating global warming (messing up)
- Will make communities more self-contained, protecting against calamitous events, such as hurricanes.

Figure 15. (left) A load shifting battery facility at a sub-station south of Saratoga Springs, New York, 2019; (middle) a 16 MWh battery facility in West Virginia, USA to smooth wind power and to shift the load, and (right) a former battery facility in Binghamton, NY.

ACKNOWLEDGEMENTS

I thank Exxon for initiating and supporting our team's work on rechargeable batteries. Those were inspiring days with our great team, and I thank them all: Arthur Thompson, Russ Chianelli, Fred Gamble, Bernie Silbernagel, John Panella, Robert Hamlen and my boss and mentor for many years Alan Schriesheim. Much of my work at Binghamton has been supported by the U.S. Department of Energy since 1993. The NMC work was supported by the Assistant Secretary for Energy Efficiency and Renewable Energy, Office of Vehicle Technologies of the U.S. Department of Energy through the Advanced Battery Materials Research (BMR) Program under Award No. DE-EE0006852 and BMR (Battery500 Consortium) under Award No. DE-EE0007765 respectively. The $VOPO_4$ work was supported as part of NECCES, an Energy Frontier Research Center funded by the U.S. Department of Energy, Office of Science, Office of Basic Energy Sciences under Award Number DE-SC0012583. New York State and the University also provided much support for our battery facility, for which I am also very grateful. I would also like to thank the entire worldwide battery research community for their comradeship over the last fifty years; it has been an amazing journey. I would like to thank my two advisors, Peter Dickens at Oxford University and Robert Huggins at Stanford University who guided me for many years. Finally, I would like to thank my wife and family for all their support over the last half century; without them this would not have been possible.

REFERENCES

1. Volta, A. On the electricity excited by the mere contact of conducting substances of different kinds. *The Royal Society,* June 26. 1800.
2. Faraday, M. *Experimental Researches in Electricity*; Taylor and Francis: London, 1839.
3. Nernst, W. Nernst glower. *Z. Elektrochem.* 1900, **6**, 41.
4. Wohler, F. Ann. Physik. *Ann. Physik* 1824, **2**, 345.
5. Dickens, P. G.; Whittingham, M. S. The Tungsten Bronzes. *Quart. Rev. Chem. Soc.* 1968, **22**, 30.
6. Dickens, P. G.; Quilliam, R. M. P.; Whittingham, M. S. The Reflectance Spectra of the Tungsten Bronzes. *Mater. Res. Bull.* 1968, **3**, 941.
7. Dickens, P. G.; Whittingham, M. S. The catalytic behavior of the tungsten bronzes. *Trans. Faraday Society* 1965, **61**, 1226.
8. Dickens, P. G.; Whittingham, M. S. In *Reactivity of Solids*; Anderson, J. S.; Roberts, M. W.;Stone, F. S., Eds.; Chapman and Hall, 1972.
9. Whittingham, M. S.; Huggins, R. A. Electrochemical preparation and characterization of alkali metal tungsten bronzes, M_xWO_3. *NBS Special Publications* 1972, **364**, 51.
10. Yao, Y. Y.; Kummer, J. T. Ion transport in beta alumina. *J. Inorg. Nucl. Chem.* 1967, **29**, 2453.

11. Whittingham, M. S.; Huggins, R. A. Beta Alumina – Prelude to a Revolution in Solid State Electrochemistry. *NBS Special Publications* 1972, **364**, 139.
12. Whittingham, M. S.; Huggins, R. A. Measurement of Sodium Ion Transport in Beta Alumina Using Reversible Solid Electrodes. *J. Chem. Phys.* 1971, **54**, 414.
13. Galloway, R.; Haslam, S. The ZEBRA electric vehicle battery: Power and energy improvements. *J Power Sources* 1999, **80**, 164.
14. Whittingham, M. S. The Role of Ternary Phases in Cathode Reactions. *J. Electrochem. Soc.* 1976, **123**, 315.
15. Thompson, A. H. Electron-Electron Scattering in TiS_2. *Phys. Rev. Lett.* 1975, **35**, 1786.
16. Whittingham, M. S. Chalcogenide Battery. *U.S. Patent 4009052 and U.K. Patent* 1468416 1973.
17. Whittingham, M. S. Electrical Energy Storage and Intercalation Chemistry. *Science* 1976, **192**, 1126.
18. *Intercalation Chemistry*; Whittingham, M. S.; Jacobson, A. J., Eds.; Academic Press: New York, 1982.
19. Whittingham, M. S. The Role of Ternary Phases in Cathode Reactions. *Electrochemical Society Abstracts* 1975, **1975-1**, 40.
20. Whittingham, M. S. Chemistry of Intercalation Compounds: Metal Guests in Chalcogenide Hosts. *Prog. Solid State Chem.* 1978, **12**, 41.
21. Thompson, A. H. Electrochemical Potential Spectroscopy: A New Electrochemical Measurement. *J. Electrochem. Soc.* 1979, **125**, 608.
22. Thompson, A. H. Lithium Ordering in Li_xTiS_2. *Phys. Rev. Lett.* 1978, **40**, 1511.
23. Whittingham, M. S. The Ultimate Limits to Intercalation Reactions for Lithium Batteries. *Chem. Rev.* 2014, **114**, 11414.
24. Chianelli, R. R. Microscopic Studies of Transition Metal Chalcogenides. *J. Crystal Growth* 1976, **34**, 239.
25. Chianelli, R. R.; Scanlon, J. C.; Rao, B. M. L. Dynamic X-Ray Diffraction. *J. Electrochem. Soc.* 1978, **125**, 1563.
26. Chianelli, R. R.; Scanlon, J. C.; Rao, B. M. L. In situ studies of electrode reactions: The mechanism of lithium intercalation in TiS_2. *J. Solid State Chemistry* 1979, **29** (3), 323.
27. Pereira, N.; Amatucci, G. G.; Whittingham, M. S.; Hamlen, R. Lithiume titanium disulfide rechargeable cell performance after 35 years of storage. *J. Power Sources* 2015, **280**, 18.
28. Rao, B. M. L.; Francis, R. W.; Christopher, H. A. Lithium-Aluminum Electrode. *J. Electrochem. Soc.* 1977, **124**, 1490.
29. James, A. C. W. P.; Goodenough, J. B. Lithium Ion Diffusion in the Defect Thiospinel $Li_xC_{0.07}TiS_2S_4$ ($0.1 \leq x < 1.85$). *Solid State Ionics* 1988, **27**, 37.
30. Emly, A.; Van der Ven, A. Mg Intercalation in Layered and Spinel Host Crystal Structures for Mg Batteries. Inorg. Chem. 2015, 54, 4394–4402.
31. Sun, X.; Bonnick, P.; Duffort, V.; Liu, M.; Rong, Z.; Persson, K. A.; Cederd, G.; Nazar, L. F. A high capacity thiospinel cathode for Mg batteries. *Energy & Environmental Science* 2016, **9**, 2273.
32. Sun, X.; Bonnick, P.; Nazar, L. F. Layered TiS2 Positive Electrode for Mg Batteries. *ACS Energy Letters* 2016, 1, 297.
33. Mitzushima, K.; Jones, P. C.; Wiseman, P. J.; Goodenough, J. B. $L_{ix}CoO_2$ ($O < x \leq 1$): A New Cathode Material for Batteries of High Energy Density. *Mat. Res. Bull.* 1980, **15**, 783.
34. Whittingham, M. S. History, Evolution and Future Status of Energy Storage. *Proc. IEEE* 2012, **100**, 1518.

35. Crabtree, G.; Kocs, E.; Trahey, L. The energy-storage frontier: Lithium-ion batteries and beyond. *MRS Bulletin* 2015, **40** (12), 1067.
36. Zhou, H.; Xin, F.; Pei, B.; Whittingham, M. S. What limits the capacity of layered oxide cathodes in lithium batteries? *ACS Energy Letters* 2019, **4**, 1902.
37. Delmas, C.; Cognac-Auradou, H.; Cocciantelli, J. M.; Ménétrier, M.; Doumerc, J. P. The $Li_xV_2O_5$ system: An overview of the structure modifications induced by the lithium intercalation. *Solid State Ionics* 1994, **69**, 257.
38. Siu, C.; Seymour, I. D.; Britto, S.; Zhang, H.; Rana, J.; Feng, J.; Omenya, F. O.; Zhou, H.; Chernova, N. A.; Zhou, G.et al. Enabling multi-electron reaction of e-$VOPO_4$ to reach theoretical capacity for lithium-ion batteries. *Chem. Comm.* 2018, **54**, 7802.
39. Song, Y.; Zavalij, P. Y.; Whittingham, M. S. e-$VOPO_4$: Electrochemical Synthesis and Enhanced Cathode Behavior. *J. Electrochem. Soc.* 2005, **152**, A721.
40. Whittingham, M. S. Chemistry Nobel Banquet Speech. 2019. (see this volume on p. XX).

Akira Yoshino. © Nobel Prize Outreach AB. Photo: A. Mahmoud

Akira Yoshino

Biography

I WAS BORN in Suita City, Osaka Prefecture, in 1948. My father, Sojiro, was an electrical engineer who worked at a power company. My mother worked at a bank until she married, after which she became a housewife. Suita is located about 10 km north of the center of Osaka City. My home was surrounded by bamboo groves. As a child, I enjoyed playing in this nature-filled environment.

My path to chemistry began in the fourth grade of elementary school, when my teacher recommended that I read *The Chemical History of a Candle* by Michael Faraday. The book made it easy to understand why a candle burns, what happens when it burns, and why it has a wick. This stimulated my curiosity, and I became fascinated with chemistry. In 1966, I entered Kyoto University as a student of the Department of Petrochemistry, Faculty of Engineering. My major was quantum organic chemistry. At the time, the Department of Petrochemistry was a very prestigious one at Kyoto University. The faculty included Professor Kenichi Fukui, who would go on to receive the 1981 Nobel Prize in Chemistry. My university research involved quantum chemistry theory and observation of experimental results. I focused on organic photochemistry in particular. Professor Fukui was the mentor of my mentor at the university. I entered graduate school at the same university and studied photochemistry of charge transfer complexes. By irradiating ultraviolet to charge transfer complexes of 1,2,4,5-tetracyanobenzene as electron acceptor with electron donors such as toluene, I discovered that a previously unknown reaction occurred.

After receiving a master's degree from Kyoto University, I joined Asahi Kasei Corp. in 1972. I was assigned to a laboratory in Kawasaki City, Kanagawa Prefecture, and began my career as a corporate researcher. My work was basic exploratory research, which meant I was expected to find the seeds of new technology. I had to decide what subjects to research, and I performed experiments myself. My first idea was to develop a new interlayer film for the laminated safety glass of automobiles. At the time, polyvinyl butyral film was used. I worked eagerly find a new material to replace it, but without success. The performance requirements for safety glass are extremely demanding, and I was unable to obtain suitable characteristics. Polyvinyl butyral is still used as interlayer film in laminated safety glass today. I was up against a formidable opponent.

In 1974, I married my wife Kumiko. We have been blessed with two daughters, Miho and Yuko, and a son, Satoshi. My next project at work was to develop nonflammable thermal insulation material. Saving energy was a hot topic at the time, and high-performance insulation was considered important. Polyurethane foam and polystyrene foam were available, but they are both flammable. I tried to develop inorganic polymer of phosphate as a nonflammable alternative, but this too was unsuccessful. Although I was able to reach fairly high insulation performance, the material had insufficient mechanical strength. My third project was for something like what we now call photocatalysts. I wanted to develop a product with a sterilizing and deodorizing effect by activating the oxygen in the air using sunlight. The foundation of this research was the photochemistry I had studied in college, specifically photosensitized oxidation. I kept at it for four years and achieved reasonably good results, but once again I could not succeed. The market was too immature. Photocatalyst technology would not be commercialized until thirty years later.

In 1981 I began looking for my next line of research, and this would turn out to be a fateful year for me. There was a new material that everyone was excited about: electroconductive polyacetylene. Although it was a plastic, surprisingly it could conduct electricity. It was discovered by Professor Hideki Shirakawa, who would go on to receive the 2000 Nobel Prize in Chemistry along with Professor Alan Heeger and Professor Alan MacDiarmid. At the time, Professor Shirakawa was doing joint research with Kyoto University. I visited Kyoto University, the first time for me to return to my alma mater in a long time. I was amazed when he showed me a sample of polyacetylene in a test tube. It had such a metallic luster that I could scarcely believe it was plastic. I decided that this amazing material would be my next subject of research. I visited Kyoto University many times after that, and he taught me all about it, including how to polymerize it. The year 1981 turned out to have another surprise in store as well: Professor Fukui's Nobel Prize in Chemistry. The frontier molecular orbital theory for which Professor Fukui received the Nobel Prize was the very theory that explained why polyacetylene was electroconductive.

I synthesized polyacetylene at the Asahi Kasei laboratory in Kawasaki. Next, I began thinking about what kind of product to use it for. Among polyacetylene's many unique properties, I was most interested in its electrochemical properties. This meant that polyacetylene could be used as a battery material. What's more, since its electrochemical reactions were reversible, it could be used as a rechargeable battery material. I surveyed the battery research being done at the time and learned of Professor Stanley Whittingham's then novel concept of intercalation. He first applied this concept to battery cathode material in 1976, and after that there was a groundswell of work to develop a small and lightweight rechargeable battery utilizing the concept of intercalation. Successful commercialization, however, proved extremely difficult due to issues with the anode material. I was thrilled to find that polyacetylene would work as an anode material which could be doped with cations such as Li ions. I confirmed this idea and concluded that the ideal use for polyacetylene was as anode material for a rechargeable battery.

I was not an expert on battery technology, but from that point forward I became deeply involved in battery technology. I began tests to evaluate the polyacetylene I synthesized as rechargeable battery anode material. The results were encouraging. The material had large charge and discharge capacity, and it didn't deteriorate even after repeated charge/discharge cycles. The next step was to decide what cathode material to pair with it. But this turned out to be a bigger problem than I had anticipated. There weren't any cathode materials that were suitable. Since polyacetylene doesn't contain Li ions, I needed a cathode material that contained Li ions. I struggle in vain to find one until I at last encountered Professor John Goodenough's 1980 paper on lithium cobalt oxide ($LiCoO_2$) as a new cathode material. Without hesitation, I assembled a test cell using $LiCoO_2$ cathode and polyacetylene anode. This was the first instance of the new rechargeable battery system that would later become known as the lithium-ion battery. It was 1983. While the weight of the new rechargeable battery was reduced to about one-third that of a nickel-cadmium battery, the volume was unfortunately unchanged. This was a disappointment because the objective was to achieve both lighter weight and smaller size. Nevertheless, I began to show my new battery system to potential customers. I didn't know whether they considered smaller size to be more important than lighter weight or not. Unfortunately, I was told that smaller size was indeed more important.

Once again, I found myself at a crossroads. The reason I couldn't reduce the size of my new battery system was that the specific gravity of polyacetylene is 1.2, relatively low. While the low specific gravity was advantageous for achieving lighter weight, it turned out to be disadvantageous for achieving smaller size. My calculations indicated that I needed a material with a specific gravity of at least 2 in order to reach the desired size. It immediately occurred to me that carbonaceous material would have a specific gravity of 2 or more while having conjugated double bonds similarly to polyacetylene. I evaluated the many carbonaceous materials then available but could find none that would function as anode material. When I was beginning to lose hope, I learned of a new kind of carbon fiber called VGCF that was being developed at another laboratory of Asahi Kasei. I obtained a sample and found that it performed very well as an anode material. VGCF is made using a very special process that gives it a very special crystalline structure. This special crystalline structure is what makes it work well as an anode material. By replacing polyacetylene with carbonaceous material, I thus completed the basic configuration of a practical new rechargeable battery. After further development of technology required for commercialization, the first lithium-ion batteries made their debut in the world in the early 1990s.

On October 9, 2019, I received a surprising notification from Stockholm. I had been chosen for the Nobel Prize in Chemistry along with Professor Goodenough and Professor Whittingham for the development of lithium-ion batteries. Two reasons were given for the award. The first reason was the huge impact that lithium-ion batteries had for the achievement of today's mobile IT society. The second reason was the vital contribution that lithium-ion batteries are expected to make for the achievement of a sustainable society moving forward. While the latter of the two is still a work in progress, emboldened by my receipt of the Nobel Prize I am determined to do my part to make it happen.

Brief History and Future of the Lithium-Ion Battery

Nobel Lecture, December 8, 2019 by
Akira Yoshino
Honorary Fellow of Asahi Kasei Corp, Tokyo & Professor
of Meijo University, Nagoya, Japan.

1 DEVELOPMENTAL PATHWAY OF THE LIB

1.1. What is the LIB?

The lithium-ion battery (LIB) is a rechargeable battery used for a variety of electronic devices that are essential for our everyday life. Since the first commercial LIB was manufactured and sold in Japan in 1991, the LIB market has continued to grow rapidly for nearly 30 years, playing an important role in the development of portable electronic products such as video cameras, mobile phones, and laptop computers. Furthermore, the market of LIBs in electric vehicles is expanding extremely fast, as is that in applications for large-scale energy storage systems. The LIB can also facilitate the practical use of a higher proportion of renewable energy sources in smart grid systems by providing storage to balance out differences in power generation and consumption over time. With these developments, it is anticipated that the LIB market will reach the scale of US$20 billion in 2020.

Before starting my story of the development of the LIB, let me explain how the battery works and how it differs from other batteries. As shown in Table 1, batteries can be classified by two basic aspects; whether they disposable (primary) or rechargeable (secondary), and by the type of electrolyte employed, either aqueous or nonaqueous. Aqueous electrolyte is

	Aqueous electrolyte battery	**Nonaqueous electrolyte battery (high voltage/high capacity)**
Primary battery (disposable)	Manganese dry cell, Alkaline cell	Metallic lithium battery
Secondary battery (rechargeable)	Lead-acid battery, Nickel-cadmium battery, Nickel-metal hydride battery	Lithium-ion battery

Table 1. Classification of batteries.

water that contains ions, whereas nonaqueous electrolyte is an organic solvent that contains ions.

Manganese dry cells and alkaline cells are most commonly used for small electrical goods such as flashlights and clocks. They are disposable and contain aqueous electrolyte. Lead-acid batteries, nickel-cadmium batteries, and nickel-metal hydride batteries also contain aqueous electrolyte, but they are rechargeable. Lead-acid batteries are commonly used for car batteries. These widely used aqueous batteries are easily manufactured.

Generally, battery performance is evaluated in terms of electromotive force and capacity. Electromotive force refers to the voltage generated by a battery. This determines the energy density of the battery, which is the available energy of the battery in a given size. The higher the electromotive force, the smaller the battery can be to run a certain device. Battery capacity represents the maximum amount of energy that can be extracted from the battery under certain specified conditions, and this is determined by the amount of active material contained in the battery. Larger capacity is better, of course, but the capacity of an individual battery can change depending on its age and, if it is rechargeable, the number of times and conditions under which it has been charged and discharged.

Aqueous batteries have a disadvantage in the available electromotive force. In principle, it is limited to around 1.5 V, the voltage at which water of the electrolyte begins to dissociate by electrolysis. Therefore, aqueous batteries need to become larger as they become more powerful. Batteries that use aqueous electrolyte thus face a natural limit in terms of energy density, which therefore restricts the scope for reduction of size and weight for a given capacity.

On the other hand, nonaqueous electrolyte batteries can obtain an electromotive force of 3 V or more per cell, offering much greater possibilities in terms of increasing energy density. An important example is the metallic lithium battery, a primary battery which had already been commercialized when I started my research on the LIB in 1981. It uses nonaqueous electrolyte and metallic lithium as a negative electrode material.

Reviewing these batteries, it is clear that a nonaqueous secondary battery was highly desirable, and the market started to seek one in the late 1970s. Professor M. Stanley Whittingham proposed the application of electrochemical intercalation using compounds such as TiS2 as the battery cathode, and research was actively conducted.1 Demand for the new battery was, however, truly heightened by the revolutionary advancement of information technology which occurred in the early 1980s, bringing portable electronics into fashion. This led a growing need for small and lightweight rechargeable batteries, and the obvious first step was to convert the metallic lithium primary battery into a secondary battery. Unfortunately, even the best efforts could not succeed for two main reasons: 1) under charging, lithium tends to precipitate on the negative electrode in the form of dendrites, which easily cause short-circuiting, and 2) the high chemical reactivity of metallic lithium resulted in poor battery characteristics, including inadequate cycle durability due to side reactions, and moreover posed an insurmountable problem in terms of safety due to the inherent risk of a thermal runaway reaction. Please look at Figure 1. These photographs were taken at an experiment to assess the safety of the metallic lithium battery in 1986. Just 20 seconds after a battery cell was smashed by a steel weight, it started to burn intensely. This experiment strongly indicated the necessity to seek new electrode materials other than metallic lithium to ensure the safety of the battery.

Current commercial LIBs do not contain metallic lithium. They are defined as nonaqueous secondary batteries using carbonaceous material as the negative electrode, and transition metal oxides containing lithium ions (most often $LiCoO_2$) as the positive electrode. Carbonaceous material is basically charcoal, and $LiCoO_2$ is a metallic oxide, a kind of ceramic. In the completely discharged state, lithium atoms are only contained as part of the cathode. Under charg-

Figure 1. Safety test of primary battery using metallic lithium as anode.

ing, lithium ions are released from the cathode and migrate through the electrolyte into the carbonaceous material of the anode. The reverse reaction occurs during discharging, and electric energy is stored or released by repeating these reactions reversibly. Cell reaction without chemical transformation provides stable battery characteristics over a long service life, including excellent cycle durability with little degradation by side reactions, and excellent storage characteristics. The operating principle of the LIB is illustrated in Figure 2.

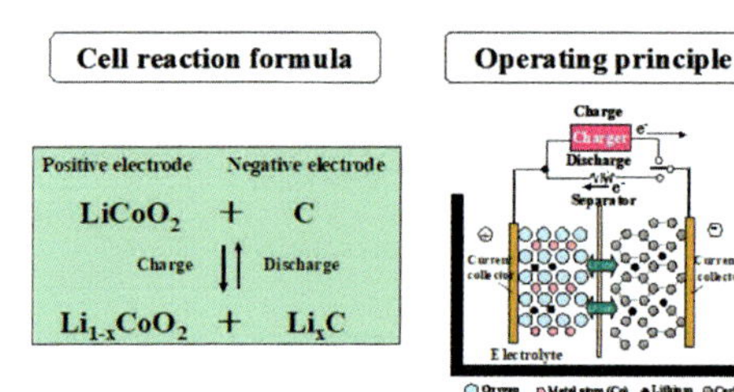

Figure 2. Electrochemical cell reaction formula and operating principle of LIB.

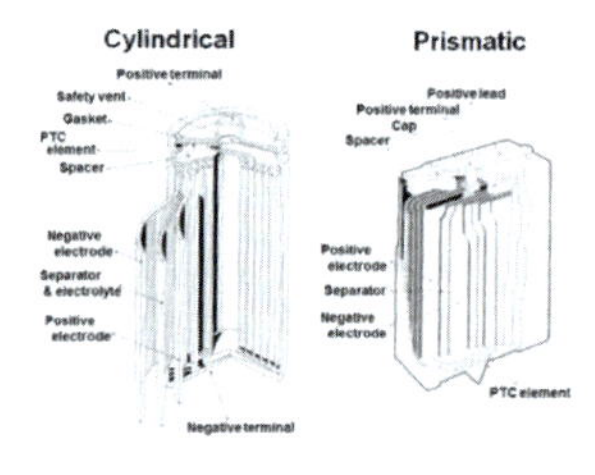

Figure 3. Cell structure of LIB.

Figure 3 shows the structure of the LIB cell. One on the left is the cylindrical type, which is commonly used for laptop computers, while the prismatic type on the right is for mobile phones. With such a structure the LIB achieves high cell voltage and high energy density, which is the key for small size and light weight. Without the invention of the LIB, we would not have been able to enjoy the information technology revolution led by the widespread use of mobile phones and portable computers. Next, I would like to describe the path I took to develop the world's first LIB.

1.2 The beginning

What happened in 1981?

My journey for the discovery of the LIB started in 1981. Technological development achieved this year was also very significant. The US space shuttle Columbia succeeded in its first flight. Also, IBM launched the world's first PC which ran a version of MS-DOS, the operating system released by Microsoft in the same year.

Among all the astonishing events that happened in that year, the most significant for me was Japan's first receipt of the Nobel Prize in Chemistry. Dr. Kenichi Fukui was awarded the prize for establishing the frontier molecular orbital theory. In simple terms, this theory claims that we are able to obtain a good approximation for chemical reactions or properties of a particular material by calculating the movement of electrons contained in the material. This implies the ability to predict chemical reactions without performing actual experiments, and therefore scientists are able to identify what they are looking for much more quickly. Recently, computer simulation based on this theory is being used to identify the effects and side effects of medicine, which is a great contribution to society.

Polyacetylene: The plastic that conducts electricity

The frontier molecular orbital theory triggered a huge advance for the creation of new materials for the LIB electrodes. In order to see how it worked, we need to consider how pi electrons behave in the case of alternating single and double bonds, which is called a conjugated system.

Importantly, the conjugated system allows the pi electrons to move across all the adjacent aligned orbitals. Here, the pi electrons do not belong to a single atom or bond but spread consistently across the whole group of atoms. Imagine that there is a material formed by carbon atoms joined together with infinite chains of conjugated double bonds. What kind of property would this material have? The pi electrons in such a material would work similarly to free electrons in metallic solids, which allow the substance to conduct an electric current. This means it is possible to create a plastic material that conducts electricity.

A number of scientists reached the conclusion that, in theory, it is possible to prepare the simplest material with structure shown in Figure 4 called polyacetylene (PA). PA is prepared by polymerizing acetylene gas, which has carbon-carbon triple bonds. Unfortunately, most attempts by scientists to synthesize PA with electric conductivity simply failed. It turned out to be rather difficult to form the infinite chains of conjugated double bonds, and so the pi electron could not move around the entire mass. Consequently, such materials did not exhibit electric conductivity.

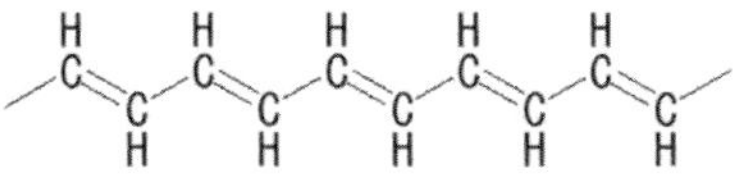

Figure 4. Chemical structure of polyacetylene.

The first successful preparation of highly conductive PA was reported by Professor Hideki Shirakawa in the 1970s. He used a method called thin-film polymerization to produce the material with metallic luster and silvery texture, which was highly conductive. Professors Shirakawa, Alan Heeger, and Alan MacDiarmid were awarded the Nobel Prize in Chemistry in 2000 for this astonishing discovery.

Batteries made of plastic

The PA discovered by Professor Shirakawa held amazing properties as a plastic. In addition to being a conductor, the material could also act as a semiconductor, meaning it could be used to make transistors, and moreover it could be used to make solar cells. These unique properties sound rather magical considering the great similarity between PA and polyethylene (PE). Shown in Figure 5, PE is a non-conductive material and the most common plastic around us.

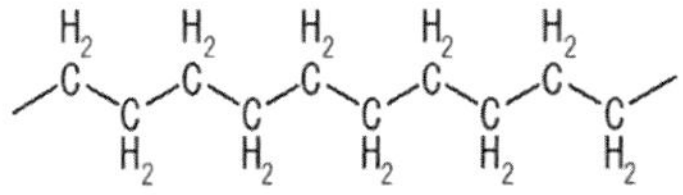

Figure 5. Chemical structure of polyethylene.

Let's look at Figure 4 again and compare the structure of PA with that of PE in Figure 5. They share the same basic structure except that the carbon atoms in PA connect to each other with alternating single and double bonds. The pi electrons in the double bonds enable the material to have these fascinating characteristics.

The property of PA that attracted me the most was its capacity to transfer electrons as an electrochemical reaction when an external voltage is applied. This characteristic is exactly what is required for the material of the battery electrodes. By applying voltage, we can eliminate negatively charged pi electrons from the bond to create positively charged PA. The removed electrons can be returned just as easily. This means that a battery with a PA electrode can charge and discharge. This positively charged PA is called p-doped PA and was expected to be cathode material. On the contrary, we can add extra electrons to the double bond structure to create negatively charged PA. This is called n-doped PA, which was thought to be suited to anode material. As you can see, PA with its pi bond structure shows extremely unique properties that allow it to be either the anode or cathode. Most researchers working at the time focused on p-doped PA because of its stability when contacting air and water, and its high electromotive force that reaches 4 V or more. I, however, decided to develop the anode with n-doped PA. As I mentioned, unstable metallic lithium anode material was preventing the commercialization of non-aqueous secondary batteries in the 1980s, so finding an alternative anode material would be a significant advancement of battery technology. I also found that n-doped PA could be highly stable as long as its contact with water is totally avoided. Being inspired by the achievements of Fukui and Shirakawa, my research to develop a new secondary battery started in 1981.

1.3 Encounter with perfect cathode material

Development of n-doped PA anode started off with enhancing the functionality of the material by improving the catalyst methods and properties of the electrolyte, etc. Usually, the quality of electrodes is first assessed separately for the anode and cathode, and then overall assessment is performed in the assembled state. As research progressed, the quality of PA anode got significantly better, and my confidence in this material was heightened. However, when I finally decided to assess the anode in the battery, I realized a serious problem. There was no suitable cathode material. Here is a list of the major cathode materials available for nonaqueous secondary batteries in the early 1980s: TiS_2, VSe_2, V_2S_5, $Fe_{0.25}V_{0.75}S_2$, $Cr_{0.75}V_{0.25}S_2$, $NiPS_3$, $FePS_3$, $CuCo_2S_4$, CuS, $NbSe_3$, MoS_3, Cr_3O_4, V_6O_{13}, V_2O_5, MoO_3. They all turned out to be unusable with our PA anode because of the lack of lithium in them. These materials were designed to be paired with metallic lithium anode, so they didn't have to contain lithium. Formula (1) shows the chemical reaction that occurs in a battery using metallic lithium as an anode and TiS2 (titanium disulfide) as a cathode.

$$Li + TiS_2 \rightarrow LiTiS_2 \qquad (1)$$

During discharging, lithium ions in the anode are transferred to the TiS_2 cathode, which forms $LiTiS_2$, and these ions go back to the anode during charging. Obviously, this ion transference does not occur once metallic lithium anode is replaced with PA. It was such a disappointment, since our PA had quite superior quality as anode.

Months passed without being able to find any suitable cathode material, but at the end of 1982 the situation suddenly changed for the better. It was the last working day of 1982, and I was clearing up my office for the New Year. I managed to dig out some research papers that I intended to read but hadn't gotten around to. Since I had some time to kill in the afternoon, I decided to read them. This was when I encountered the 1980 paper published by a research team led by Professor John B. Goodenough, who was at Oxford University at the time.[2] It was my "eureka" moment. The research reported the discovery of the new compound $LiCoO_2$, which was described as a good cathode material for secondary batteries.

This compound has a crystalline structure comprised of alternating layers of Li cations and cobalt oxide anions. This allows not only the release of Li ions from the outer layer of the crystal but also from inside the structure, making a large number of ions available for transfer. Furthermore, $LiCoO_2$ enabled the achievement of cell voltage of 4 V or more, providing high battery capacity. Importantly, the report stated that the authors hadn't found a suitable anode material to form a secondary battery. I instantly drew a formula of the chemical reaction between PA and $LiCoO_2$, which looked like Formula (2).

$$PA + LiCoO_2 \rightarrow PA\text{–} xLi^{+}{}_{x} + Li_{(1-x)}CoO_2 \qquad (2)$$

This promising-looking formula worked for real inside a test tube battery which we urgently made to try out the electrode combination at the beginning of 1983. This was such an emotional moment that marked the birth of the first LIB using a PA anode. Not only achieving superior safety than existing secondary batteries with a metallic lithium anode, the first LIB was considerably lighter than any other competitors at the time. In the same year, I successfully filed a patent for the new principle of the first LIB using PA for the anode and $LiCoO_2$ for the cathode. This is the official public acknowledgement of the first LIB.

This innovation wouldn't have happened if any of the following three events didn't occur: the formation of the frontier molecule orbital theory by Kenichi Fukui, synthesis of highly conductive PA by Hideki Shirakawa, and formulation of $LiCoO_2$ by Professor John B. Goodenough. To me, it was a miracle that these events coincidentally happened at roughly the same time, when I was exactly in the right environment to utilize them for the first LIB.

1.4 Development of anode material

After completing the principle of the first LIB using PA anode and $LiCoO_2$ cathode in 1983, we continued trying to improve battery performance in order to achieve commercialization. As the research progressed, however, I began to find several shortcomings associated with PA. One is its low chemical stability when exposed to oxygen and heat. Exposure to oxygen could be prevented by sealing the battery completely, but the deterioration that occurs under heat was much harder to avoid. Another major limitation of PA stemmed from its low real density, 1.2 g/cm^3, which results in low battery capacity for a given volume. Since the real density of lead is 13.6 g/cm3, the new battery with PA was much lighter than a lead-acid battery, but the volume did not differ greatly when their capacity was matched. This almost certainly meant that it would be impossible for us to make a small battery with PA even if the heat stability problem could be solved.

Although we hadn't totally given up on PA, we started to look for alternative materials with the conjugated molecular structure. The most favorable candidate was carbonaceous material, which has relatively high real density of above 2 g/cm3. As represented in Figure 6 below, carbonaceous material has a molecular structure with carbon atoms connected by alternating single and double bonds.

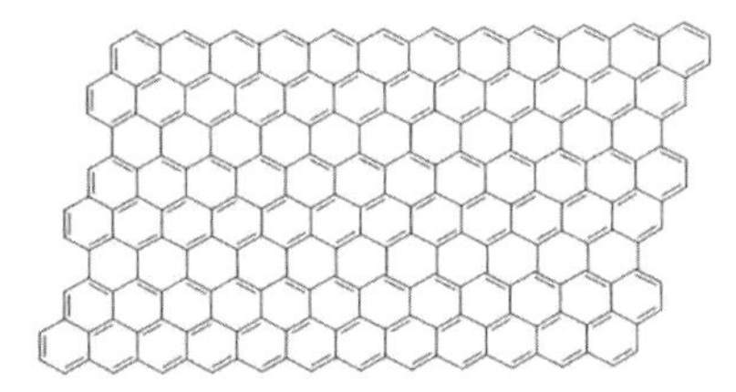

Figure 6. Chemical structure of carbonaceous material.

We gathered various carbonaceous materials from manufacturers and tested their functionality as an anode. With huge disappointment, we could not find a workable sample at all. The year was 1984, and I was very frustrated not only with our inability to find a solution for PA but also with the difficulty of obtaining a suitable carbonaceous material. Our research progress came to a halt.

Fortunately, this negative situation didn't last long. Good news was actually brought by a different business division of our company. The company has diversified business fields, and one of them is fiber production. There was a fiber research center in the south of Japan, and in the 1980s they were developing a new type of carbon fiber. Their product was called VGCF (vapor-phase grown carbon fiber), prepared by a unique method that vaporizes hydrocarbon gas in a furnace at 1 000 °C, in which carbon fiber grows. The gas used for the procedure was aromatic compounds such as benzene and toluene. Once it goes through the furnace, thin carbon fiber of several microns diameter is left on the wall, and this starts to grow just like hair. This fiber is very similar to carbon nanotubes that have a cylindrical nanostructure, but VGCF's filament diameter is 1 000 times thicker. It was later found that VGCF actually has a carbon nanotube in its core, but carbon nanotubes were not discovered until 1991. At the time, we didn't even look for such a thing; we were simply overjoyed with the superior performance of VGCF when used as an anode material for the LIB. Normally, the crystallinity of the material increases with higher carbonization temperature, and 1 000 °C is considered too low to form a highly crystalline structure. However, the VGCF has relatively high crystallinity for a material carbonized at that temperature, and this unique property provided just the right degree of crystallinity to effectively intercalate and release the lithium ions when used as a negative electrode.

With this VGCF, the carbon/$LiCoO_2$ battery configuration was completed. This LIB is the first of its kind to achieve the small size and light weight that matched what was needed in the market. In 1985, I filed a patent covering the basic configuration of the new battery,[3] and other related technologies were also filed as separate patents. From this point our research started to accelerate toward commercialization.

1.5 Safety test

About the time when we discovered the VGCF anode material, our new battery had to face another challenge; proving it was safe enough to be sold on the market. As I mentioned, the nonaqueous secondary batteries previously under development had all failed to reach the market because of their lack of safety. The reason was the hazardous metallic lithium being used for the negative electrode. Therefore, theoretically speaking, replacing the metallic lithium with carbonaceous material should remove the problem. The degree of improvement was, however, the most important aspect we had to demonstrate. The safety level required for a battery to be used by the general public is extremely high, and we had no idea if our new battery would meet the target.

The method of testing was another issue as well. No guideline or regulation on testing procedures for this kind of battery existed. Referring to safety tests for similar products, we decided to apply over 10 different assessments to prove that the new LIB was safe to be sold. Without delay another problem arose. We had no idea where to perform such a potentially dangerous experiment. The outcome of the test was unpredictable, and the situation could be hazardous. The first idea was to do it on the riverbank near the research center, but it turned out that we needed to get permission from the local authority. Since we had little hope of getting permission for this, we gave up on that idea. Changing tactics, we started to look for another business division in our company that was developing something dangerous, just to see how they had dealt with a similar situation. Our company has a diversified business portfolio, and there were lots of research and development projects going on. We finally found what we were looking for: the explosives business. Our company was dealing in explosives, so it owned an explosives testing site in the mountains of southern Japan. The personnel in the explosives business did us an enormous favor, allowing us to use their site to test the safety of our battery.

The results were astonishing. The images in Figure 7 show the first battery destruction test on an LIB, in which a steel weight was dropped on the battery.[4] To be used by the general public, the chemical reaction that occurs in the battery after the massive impact should not pose any danger to people.

The result was in stark contrast to the same test performed on the metallic lithium battery described above and shown in Figure 1, which started to burn 20 seconds after the impact due to its highly reac-

Figure 7. Safety test of LIB.

tive anode. The new carbon/$LiCoO_2$ battery, on the other hand, showed no sign of danger even one hour after being smashed. This result guaranteed that we could continue our research, and ultimately it was a very positive step toward the commercialization of our LIB.

2. COMMERCIALIZATION OF THE LIB

The LIB was thus commercialized in the early 1990s as a small and lightweight rechargeable battery, making a major contribution to our current society of mobile IT. As the LIB market in mobile IT (consumer electronics) has grown for over 25 years, battery performance and reliability have increased while the cost has come down. The LIB is now entering another period of transformation in automotive (electric vehicle) applications.

I will briefly review the current LIB technology and market. Based on its characteristics of small size and light weight, LIBs have become very widely used in small consumer electronics products for mobile IT. Some 4 billion LIB cells are currently manufactured for this application in a year. The largest share of these is for mobile phones including smartphones. Looking at the history of mobile phones, I will describe the key role LIBs played in their widespread adoption. The origin of mobile phones can be traced to the "shoulder phones" which came on the market in the mid-1980s. As the name indicates, it wasn't a handheld device, but one requiring a shoulder strap to carry. Weighing around 3 kg, it was essentially a car phone redesigned to be carried around like a shoulder bag.

It was in the early 1990s that the handheld devices known as mobile phones went on sale. The first generation, called 1G, mostly used nickel-metal hydride (NiMH) batteries, though a few models did adopt the then newly developed LIB. The 1G system was analog, and the ICs operated at 5.5 V. Therefore, using NiMH which produced 1.2 V per cell, 5 cells in series were needed, while the LIB produced 4.2 V per cell, so 2 cells in series were needed. For a few years, the mobile phone market was shared by NiMH and LIB. The advent of second-generation, or 2G, mobile phones was a major turning point. The system was changed from analog to digital, and the IC voltage was reduced from 5.5 V to 3 V. The lower voltage meant that while 3 NiMH cells would still be required, the device could operate with only a single LIB cell. This was a critical difference, as it enabled the design to be greatly simplified. The mobile phone soon completely switched to LIB, and the phones themselves became much small and lighter, in turn driving their more widespread adoption. This ushered in the IT revolution from around 1995, followed by 3G and now smartphones. As such, the LIB, being a small and lightweight battery with an output of 4.2 V, undoubtedly played a key role in enabling today's IT society.

Now the LIB is at another turning point. Electric vehicles are emerging as a major LIB application. Figure 8 shows the annual LIB market by capacity of batteries delivered in GWh for mobile IT and automotive applications. While the automotive application began around 2010, its growth remained moderate for a few years as sales of electric vehicles fell short of expectations due to high cost and short driving range. This began to change in 2015, when electric vehicle market growth accelerated as an effect of China's strict environmental regulations taking effect in 2018. When China adopted a national policy of promoting electric vehicles to deal with air pollution problems such as PM2.5, European automobile manufacturers, which had been somewhat passive about electric vehicles, swiftly changed course. In 2017 the LIB market scale for automotive applications slightly exceeded that for mobile IT, and rapid growth in the automotive sector has continued ever since. By 2025, the automotive market is forecasted to be 7–8 times larger than the mobile IT market for LIBs.

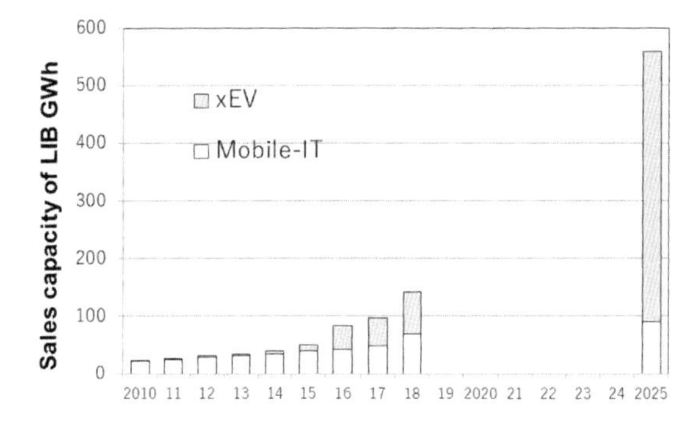

Figure 8. Capacity-based LIB market scale by category. Source: B3 Corporation Report.

3. ACHIEVING A SUSTAINABLE SOCIETY FOR THE FUTURE

Despite various discussions all around the world, we still don't have a clear solution to environmental problems. I see a trilemma between the environment, the economy, and convenience. Sustainability cannot be achieved in society if any one of these three is prioritized at the expense of the other two. How can this trilemma be overcome? I think two recent buzzwords offer a clue. We hear of CASE (connected, autonomous, shared, and electric vehicles) and MaaS (mobility as a service). Very soon, I believe these concepts will not only revolutionize transportation, but also transform our society into a truly sustainable one.

In combination with fast-evolving technologies such as AI (artificial intelligence), IoT (the internet of things), and 5G (fifth generation telecommunications networks), LIBs will play a central role. Autonomous electric vehicles with artificial intelligence will become widespread throughout society. They will be provided as a service that people subscribe to for a monthly fee. This service will not only be available in big cities, but also in rural areas. Since the vehicles will be shared by many people rather than owed individually, the cost per person will be reduced to one-seventh.

Spread throughout society, in addition to eliminating traffic accidents and congestion, such vehicles will act as a reservoir of stored energy. The electricity they store would last for ten hours if fifty power plants shut down. The vehicles charge up their batteries at fully automated charging and discharging stations when there is an ample supply of power. Conversely, they automatically discharge themselves at these same stations to supplement grid power when necessary. This facilitates greater adoption of renewable energy sources such as wind and solar whose power generation tends to be intermittent.

Such a transformation will impact every aspect of society, not just transportation. People will be able to live in comfort, with a robust economy, and minimal environmental burden. This is my vision of how the trilemma can be overcome and society can become sustainable very soon.

Current LIB performance is already approaching the level that would be required. Energy density and cost are both nearing the point where the social transformation I described would be possible. However, durability would need to improve significantly in this scenario, since the batteries would be required to undergo frequent cycles of charging and discharging. I consider that batteries would need to cycle 5,000 times, enabling vehicles to travel a total of 600,000 km, in order to enable this scenario. Technological developments to improve cycle life will thus be critical.

LIBs will play a central role in enabling a sustainable society to be achieved very soon. Battery technology linked with AI, IoT, and 5G will drive innovation throughout society. CASE and MaaS will enable an automobile-dependent society to be sustainable. The trilemma between the environment, the economy, and convenience will be overcome. This is my vision, and I will continue to devote my efforts to helping bring it about.

REFERENCES

1. M.S. Whittingham, *Journal of Electrochemical Society,* **123**, 31 (1976).
2. K. Mizushima, P.C. Jones, P.J. Wiseman and J.B. Goodenough, *Material Research Bulletin*, **15**, 783 (1980).
3. Yoshino, JP1989293, USP4668595, EP205856B2.
4. Yoshino, *Angew Chem Int Ed Engl;* 2012 Jun 11; **51**(24):5798.

Chemistry 2020

Emmanuelle Charpentier and Jennifer A. Doudna

"for the development of a method for genome editing"

The Nobel Prize in Chemistry, 2020

Presentation speech by Professor Claes Gustafsson, Member of the Royal Swedish Academy of Sciences.

Your Majesties, Your Royal Highnesses, Esteemed Nobel Laureates, Ladies and Gentlemen,

The fertilised egg cell contains all the information needed to create a human being, and this information is stored in our genetic material, our DNA. In 1953, James D. Watson and Francis H.C. Crick reported a simple, structural model for DNA, which immediately suggested how the DNA molecule can store genetic information that can be passed on to new generations of cells.

What the DNA structure did not explain was how the genetic information can be used to give rise to proteins, the main components of human tissues. During the 1950s this was an unanswered scientific question, which occupied a large number of prominent scientists. A breakthrough came on Good Friday, 1960, when a group of international scientists gathered for an informal meeting at King's College, Cambridge. The group included Sydney Brenner, Francis Crick, and François Jacob, who all later became Nobel Prize laureates. During the meeting, Jacob described the latest results from the work carried out by his colleagues and himself at the Institute Pasteur in Paris. The information provided had an almost seismic effect on Brenner and Crick, who both instantaneously realized the importance of the French findings for their own scientific work. At that very moment they understood that DNA was copied into RNA, which in turn functions as the template for protein synthesis. As so many times before and after, the free exchange of ideas and information between international scientists led to an important breakthrough.

Today we celebrate another scientific breakthrough resulting from an international collaboration, between a French scientist working in the northern part of Sweden and an American scientist based in sunny California. Ever since the early pioneers described the nature of the genetic material, scientists have tried to develop technologies that can manipulate the

DNA sequences of cells and organisms. With the discovery of the CRISPR/Cas9 genetic scissors, Emmanuelle Charpentier and Jennifer A. Doudna have provided us with a tool that can be used to change the genetic information of animals, plants and microorganisms with extremely high precision.

As so often in science, the discovery of these genetic scissors was unexpected. During Charpentier's studies of a harmful bacterium, *Streptococcus pyogenes*, she discovered a previously unknown molecule, which she called *tracrRNA*. Her work showed that tracrRNA is part of an ancient bacterial immune system, *CRISPR/Cas*, that protects the bacterium from invading viruses by cleaving their DNA. How this cleavage was accomplished was demonstrated a year later, when Charpentier and Doudna together succeeded in recreating the bacteria's genetic scissors in a test tube.

In a series of crucial experiments, the two scientists set out to simplify and reprogram the genetic scissors. In their natural form, the scissors recognise DNA from viruses, but Charpentier and Doudna proved that they could be controlled so that they can cut any DNA molecule at a predetermined site. Where the DNA is cut it is then easy to introduce new genetic information and thereby rewrite the code of life.

Since Charpentier and Doudna discovered the CRISPR/Cas9 genetic scissors in 2012 their use has increased exponentially. This tool has contributed to many important discoveries in basic research. Plant researchers have been able to develop crops with new and desired properties. In medicine, clinical trials of new cancer therapies are underway, and the dream of being able to cure inherited diseases is about to come true.

Emmanuelle Charpentier and Jennifer A. Doudna:

Your studies of the CRISPR/Cas9 genetic scissors have taken the molecular life sciences into a new epoch and are bringing a great benefit to humankind in many ways. That is a truly great achievement. On behalf of the Royal Swedish Academy of Sciences, it is my great privilege to convey to you our warmest congratulations.

Emmanuelle Charpentier. © Nobel Prize Outreach AB. Photo: Bernhard Ludewig

Emmanuelle Charpentier did not submit her autobiography and lecture. See
https://www.nobelprize.org/prizes/chemistry/2020/charpentier/facts/
https://www.nobelprize.org/prizes/chemistry/2020/charpentier/lecture/

Jennifer A. Doudna. © Nobel Prize Outreach AB. Photo: Brittany Hosea-Small

Jennifer A. Doudna

Biography

I WAS BORN ON FEBRUARY 19, 1964 in Washington, D.C., the oldest of three sisters. My father Martin K. Doudna was a speechwriter for the Department of Defense at the time and my mother Dorothy taught in community college. My family, including my siblings Ellen and Sarah, moved to Ann Arbor where my father pursued a Ph.D. in literature at the University of Michigan. After earning his degree, he received a job offer from the University of Hawaii in Hilo and moved our family there in August 1971 when I was seven.

Growing up as a "haole" (Hawaiian slang for a non-native), I felt really alone and isolated at school. This "outsider" feeling drove me to take risks and prove doubters wrong, and later influenced my choices as a scientist. In my isolation, I sought solace in books that spurred me to learn more about the world around me and how I fit in. As I made friends and expanded my social life, I fortified my reading with nature walks, hikes, bicycle riding, and explorations of lava-flow caves. With its mix of volcanoes, forests and beaches, the "Big Island" of Hawaii provided a rich palette of biological diversity that inspired my first questions as to how so much diversity came to be.

In the sixth grade, I came across a book called *The Double Helix* by James Watson that my father had laid on my bed. It told how Watson, an American biologist, and Francis Crick, an English biochemist, led the discovery in 1953 of the "spiral staircase" structure of the DNA molecule, for which they received the 1962 Nobel Prize in Physiology or Medicine with Maurice Wilkins.

DNA is formed from chains of nucleotides, each of which contains one of four nitrogen bases – adenine, thymine, guanine and cytosine. All of the information for carrying out the essential processes of life is encoded in the sequences of these AGCT base letters. A string of base letter sequences that contains instructions for making a specific protein is called a gene. *The Double Helix* explained how DNA's double-stranded helical structure –

commonly depicted as a twisted ladder that enables the molecule to unwind – exposes its sequences of AGTC letters so that the genetic instructions they carry can be copied onto messenger RNA (mRNA). In the cells of eukaryotic organisms such as humans, mRNA carries these instructions from the nucleus out into the cytoplasm where transfer RNA (tRNA) uses the information to assemble amino acids into proteins.

In reading *The Double Helix,* I was captivated by how scientists, working collaboratively, were able to meticulously piece together and solve what had been one of biology's most elusive puzzles.

Learning about the role played by Rosalind Franklin, the so-called "Dark Lady of DNA," whose X-ray crystallography images exposed DNA's helical shape, it struck me then for the first time that a woman could be a great scientist.

Despite my outstanding academic record, especially in math and science, my high school guidance counselor strongly discouraged me from pursuing a college major in chemistry, or even aspiring to be a scientist. Undeterred, I applied to and was admitted into Pomona College in California and enrolled in the fall of 1981 at the age of 17. I studied chemistry, despite briefly considering switching my major to French.

That summer I returned to Hilo and worked in the laboratory of Don Hemmes, a biology professor and longtime family friend. I was assigned to a small team investigating how a fungus, *Phytophthora palmivora*, infected papayas, which was a big problem for Hawaiian fruit growers. I learned to prepare samples for analysis in an electron microscope and follow the chemical changes that take place as the fungus advances through different stages of germination. The research revealed that calcium ions play a crucial role in the development of the fungus by signaling fungal cells to grow in response to nutrients. It was my first taste of the thrill of scientific discovery, an experience that I had read so much about, and that left me hungering for more.

In the summer of 1984, following sophomore and junior years that solidified my commitment to science, I was invited to work in the lab of my advisor, biochemistry professor Sharon Panasenko. I was tasked with growing soil-based bacteria in such a way where we could study the chemical signaling that enables the bacteria to self-organize into colonies when starved for nutrients. My method of growing the bacteria in large baking pans instead of conventional Petri dishes was acknowledged in a paper published by Sharon in the *Journal of Bacteriology,* the first time my name appeared in a scientific journal.

I graduated from Pomona College in the spring of 1985 as the top student in chemistry. I remain grateful to Pomona for a liberal arts education that exposed me to so many ideas that I otherwise might have never had come into contact with and that were key to my later success.

Upon the urging of my father, I applied to graduate school at Harvard Medical School, where I enrolled in the fall of 1985. In 1986, I elected to do my dissertation work in the lab of Jack Szostak (2009 Nobel Prize Laureate in Physiology or Medicine with Elizabeth Blackburn and Carol Greider) who discovered how telomere caps prevent chromosomes from breaking down.

Jack was in the process of switching his research focus from DNA to RNA and the role it played in the origin of life on Earth. He suspected RNA preceded DNA and encouraged me to be his first graduate student to explore this idea.

In retrospect my decision to concentrate on RNA makes perfect sense, but at the time turning away from DNA was a bold and risky move. The public spotlight then was shining on the Human Genome Project, an international effort to map and sequence all the base letters in the human genome (an organism's full complement of DNA).

Billed as "the Holy Grail of Biology," the Human Genome Project promised to revolutionize medical diagnostics and treatments with the wealth of genetic information it would provide. Mapping and sequencing DNA garnered headlines and funding, but we saw the immense value to be gained from a better understanding of the multipurpose RNA molecule in all its many forms.

Jack and I agreed that the potential for RNA to have played a starring role in the origin of life hinged upon whether RNA molecules are able to replicate themselves. To answer this question, we reengineered a self-splicing RNA intron (a segment of a DNA or RNA molecule that does not code for proteins) into an RNA enzyme (a protein catalyst) that could splice together a copy of itself. This demonstrated that RNA could function as a polymerase, an enzyme that promotes the formation of molecules such as DNA or RNA. We published the results of the study in 1989 in the journal Nature in a paper titled "RNA-catalyzed synthesis of complementary-strand RNA," marking the start of my journey in RNA research.

After receiving my Ph.D. in 1989, I continued my research on self-replicating RNA molecules in Jack's laboratory as a postdoctoral fellow, but my curiosity grew around what sort of

molecular structure would enable RNA to replicate itself. It was one of biology's toughest challenges – determining the molecular structures of RNA enzymes and other functional RNA molecules.

In the summer of 1991, I relocated to the University of Colorado in Boulder to work in the laboratory of Tom Cech, who two years earlier had won the Nobel Prize in chemistry with his collaborator Sidney Altman. In the 1980s they discovered that RNA not only serves as a genetic messenger but can also function like an enzyme. They dubbed these catalytic

RNAs "ribozymes" by combining "ribonucleic acid" with "enzyme." Their ribozyme was the same self-replicating RNA enzyme that Jack and I would later reengineer.

As a research fellow in Tom's laboratory, I used X-ray crystallography to image the structure of RNA enzymes, similar in principle to how Rosalind Franklin imaged the structure of DNA. For this work, I teamed up with a graduate student named Jamie Cate who had been using X-ray crystallography to study protein structures. Although we were able to successfully crystalize RNA enzymes for imaging, exposure to X-rays quickly destroyed the crystalline structure.

It was a fortuitous meeting with Yale University biochemists Thomas Steitz (2009 Nobel Laureate in Chemistry with Venkatraman Ramakrishnan and Ada Yonath) and Joan Steitz, a husband-and-wife team on sabbatical in Boulder for a year, where I learned of their technique for the cryogenic cooling of crystals prior to X-ray imaging. Flash-freezing crystals in liquid nitrogen preserved their crystalline integrity even when irradiated with X-rays. Steitz would share the 2009 Nobel Prize in Chemistry for his work in determining the structure of ribosomes, the macromolecules in a cell's cytoplasm that use RNA's genetic messages to synthesize proteins.

Eager to use cryo-cooling technology, I accepted an appointment at Yale in 1994 as an assistant professor of molecular biophysics and biochemistry, with Jamie Cate accompanying me as a graduate student in my new lab. With a technique devised by Jamie, we used X-ray crystallography to produce electron-density maps that enabled them to determine the location of every atom in a self-splicing RNA enzyme. With this information, we constructed structural models of the molecule, similar to what Watson and Crick did with DNA.

Just as the double-helix structure revealed how DNA is able to store and transmit the genetic code, the structural models that my team and I built showed how RNA is able to function as an enzyme capable of slicing, splicing and replicating itself, just as the double-helix structure revealed how DNA is able to store and transmit the genetic code. Tom and I were the principal investigators and Jamie Cate was the lead author on a 1996 paper published in Science titled "Crystal Structure of a Group I Ribozyme Domain: Principles of RNA Packing." The paper is considered a scientific landmark for providing the first detailed look at a large-structured ribozyme. We hoped then that our discovery would provide clues as to how scientists might be able to modify the ribozyme to repair defective genes in the future.

In 1997 I accepted an appointment as a Howard Hughes Medical Institute (HHMI) investigator. In 2000 I was named the Henry Ford II Professor of Molecular Biophysics and Biochemistry and elected into the

National Academy of Sciences. That summer I married my research partner Jamie Cate and we had a son whom we named Andrew two years later. It was also in 2002, after Andrew's birth, that Jamie and I accepted appointments as professors in the College of Chemistry at the University of California, Berkeley (UC Berkeley). We also became faculty scientists at the Lawrence Berkeley National Laboratory (Berkeley Lab), which gave us access to the Advanced Light Source, one of the world's premier sources of X-rays for crystallography research.

In the fall of 2002, soon after I started my work at UC Berkeley, there was a deadly outbreak in China of Severe Acute Respiratory Syndrome (SARS) caused by an RNA-based virus. The outbreak motivated me to pivot to RNA interference, a phenomenon that plays a fundamental role in a number of important functions, including how the human immune system fights off viral infections. RNA interference silences unwanted genetic messages, thereby blocking the production of the proteins they code for.

Using the powerful beams and sophisticated instrumentation at the Advanced Light Source, a synchrotron accelerator optimized for X-ray and ultraviolet light research, my lab and I produced crystallography images from which we determined that an enzyme known as Dicer functions as a molecular ruler to snip double-stranded RNA and initiate the process of RNA interference. We found that the molecular structure of Dicer features a "clamp" at one end to grab hold of a double-stranded RNA molecule, and a "cleaver" a set distance away at the other end to snip it.

Our findings set the stage for understanding how Dicer enzymes are involved in other phases of the RNA interference pathway serving as a guide to redesigning RNA molecules that direct specific gene-silencing pathways. Our work later led to a call from UC Berkeley microbiologist Jillian Banfield who first introduced me to the term "CRISPR."

Jill was studying the genomics of microbes that live in extreme environments as a means of finding better ways to clean up polluted sites, and repeatedly encountered a unit or length of DNA base letter sequences known as CRISPR, or Clustered Regularly Interspaced Short Palindromic Repeats, which plays a role in the defense mechanisms employed by microbes such as bacteria and archaea for protection from viruses and invading strands of nucleic acid known as plasmids. Usually located on a microbe's chromosome, a CRISPR unit is made up of base sequences called "repeats" and "spacers" that separate the repeated sequences. Interested in understanding how a combination of CRISPR and a complex of adjacent enzymes dubbed "Cas" for CRISPR-associated proteins allowed microbes to utilize small customized RNA molecules (crRNA) to protect themselves through a gene-silencing process, Jill found my lab's website from a Google search and suggested a collaboration to learn how the CRISPR-Cas immune system might work through RNA interference.

A conversation over tea at a local Berkeley café followed, where we discussed this new, mysterious biological function of CRISPR and the possibility that it was the bacterial equivalent to RNA interference. While hypotheses about CRISPR had been floated, no one had yet conducted the experiments to prove or disprove those theories.

In 2008, Jill and I, along with Mark Young at Montana State University, organized the first international conference on CRISPR research, which was held in Berkeley. That same year, research out of Northwestern University had revealed that the CRISPR-Cas system did not work through RNA interference, but instead targeted the DNA of an invader. The implication of this discovery was highly significant. If CRISPR-Cas was identifying, targeting, and cutting invasive DNA, then it had potential as a DNA editing tool. However, crucial experimental information was missing, especially as to how the CRISPR-Cas system would be able to recognize and cut out unwanted DNA.

My Doudna Lab team began filling in this missing information by first developing a purification technique that provided them with highly concentrated samples of the Cas proteins they wanted to study. With this technique, we first studied Cas1, which we discovered is able to cut up DNA in a way that helps with the insertion of new snippets of foreign DNA into a CRISPR unit during the immune system's memory-forming stage. It brought us a step closer to understanding how CRISPR steals bits of DNA from attacking phages and works that genetic information into its own, laying the groundwork for the targeting and destruction phases of the immune response.

In 2010 we used X-ray crystallography beamlines at the Advanced Light Source to produce the first atomic-scale crystal structure model of Cas6f, discovering that like Cas1, it functions as a chemical cleaver. However, we found that the job of Cas6f is to specifically and methodically slice long CRISPR RNA molecules into shorter chunks that can be used to target foreign DNA.

Our model showed that when a microbe recognizes it has been invaded by foreign DNA, it incorporates a small piece of that foreign DNA into one of its CRISPR units, which is then transcribed as a long RNA segment called the pre-crRNA. Cas6f cleaves this pre-crRNA within each repeat element to create short crRNAs containing sequences that match portions of the foreign DNA forCas proteins to use to bind the foreign DNA and silence it. These research results were reported in the journal Science in a paper titled "Sequence- and structure-specific RNA processing by a CRISPR endonuclease."

In 2011 I met Emmanuelle Charpentier at the annual American Society for Microbiology meeting in Puerto Rico. A French biochemist, microbiologist, and geneticist, and one of the world's top CRISPR-Cas researchers, Emmanuelle was at the time studying Cas9 in a CRISPR system known as type II at the Laboratory for Molecular Infection Medicine at Umeå University in Sweden. Her work had shown that in the human pathogen *Streptococcus pyogenes*, crRNAs could only be produced in the presence of a second CRISPR RNA molecule, or tracrRNA for trans-activating crRNA, and that CRISPR systems only need Cas9 to acquire immunity to viruses targeted by crRNAs.

Emmanuelle and I embarked on a collaboration to investigate how Cas9 and crRNAs function in the CRISPR microbial immune system and whether crRNA and tracrRNA could be linked into a single chimeric CRISPR RNA molecule to make the system easier to manipulate.

With Martin Jinek and Michael Hauer from my lab and Krzysztof Chylinski and Ines Fonfara from Emmanuelle's, we unraveled the components of the CRISPR-Cas9 assembly. In our discovery that Cas9 requires binding to a molecular complex of tracrRNA and crRNA which then identifies and guides the Cas9 to the invasive DNA for cleaving, we realized that the tracrRNA and crRNA complex is programmable.

The next step was to engineer a single-guide RNA (sgRNA) molecule that would have the guide information on one end and the binding handle on the other. Such a system would provide a straightforward way to cleave any desired stretch of DNA sequences in a genome. New genetic information could then be introduced into the genome using well established cellular DNA recombination technology.

The results of this momentous study were published in *Science* on August 17, 2012 (it first appeared online June 28, 2012). The paper was titled "A Programmable Dual-RNA–Guided DNA Endonuclease in Adaptive Bacterial Immunity." Emmanuelle and I were the principal investigators; the co-authors were Martin Jinek, Krzysztof Chylinski, Ines Fonfara, and Michael Hauer.

Our paper showed that an immune system evolved over eons by bacteria and other microorganisms to defend themselves against viral and other invasive DNA could be adapted into a relatively easy-to-use "CRISPR genome editing" technology for rewriting the genetic code within the cells of any organism, including humans, with unparalleled efficiency and precision.

It was an incredible, precious time of pure joy to discover that bacteria had found a way to program a warrior protein to seek and destroy viral DNA, and fortunate, even miraculous, that we could repurpose this fundamental property for an entirely different use. The joy of discovery was a feeling just like I'd felt in Don Hemmes's lab all those years before in Hawaii.

In 2014, I founded the Innovative Genomics Institute (IGI) with Jonathan Weissman to realize the potential of CRISPR genome editing in human health, climate, and agriculture. The IGI, of which I am president, is made up of researchers at UC Berkeley and the University of California, San Francisco, who are developing foundational CRISPR technologies to advance genome engineering innovations for the benefit of humanity that are accessible and affordable to all. We continue to build upon what started as our curiosity-driven, fundamental discovery project into strategies to help improve the human condition.

One of the early initiatives of the IGI centered around expanding access to treatments for sickle cell disease. By this time, I had additionally launched my first company, Caribou Biosciences, and would come to co-found companies Editas Medicine, Intellia Therapeutics, Mammoth Biosciences, and Scribe Therapeutics to bring CRISPR from the lab into the clinic in the form of CRISPR-based therapeutics and diagnostics.

I accepted the positions of Li Ka Shing Chancellor's Chair at UC Berkeley, senior investigator at the Gladstone Institutes, and adjunct professor of cellular and molecular pharmacology at UC San Francisco, persisting with my research on the exploration of delivery techniques for CRISPR-based therapies, the development of next-generation CRISPR diagnostics, and continued investigations into the structure and mechanism of CRISPR-Cas systems. Our ability to harness the immense "dual use" potential of CRISPR led to my decision to actively engage in and initiate the public discourse on the ethical use and responsible regulation of CRISPR, particularly in the case of human germline editing.

In 2015 I first called for a moratorium on the use of CRISPR in the human germline and arranged a meeting with 20 other researchers with the goal of discussing the ethics of germline gene editing. Modeled after the 1975 Asilomar conference that put forward guidelines for research on recombinant DNA and co-organized with two of its key organizers Paul Berg and David Baltimore, the discussions we conferred at the Napa Valley meeting were summarized in a report later published in *Science*. In it, we outlined the implications of human germline editing and ultimately called for a prudent path forward, including clear international guidelines, over instituting a moratorium.

I was also one of the organizers of the International Summit on Human Gene Editing in December 2015, the first international symposium at the National Academy of Sciences on the societal and ethical use of CRISPR technology. At the event, we further discussed the ethical considerations of safely applying CRISPR technology and came to a consensus similar to that which resulted from the Napa conference – that certain conditions should be met before human germline editing was permitted. As CRISPR began to enter the mainstream conversation, I detailed the ethical quandaries we faced and our rationale for calling on ongoing input from scientists and bioethicists as well as broader public discussion in my 2017 book *A Crack in Creation: Gene Editing and the Unthinkable Power to Control Evolution* co-authored with Samuel Sternberg. I had hoped that the steps Sam and I described would deter any premature attempts to perform inheritable gene editing but just one year later the

type of reckless experiment that I had feared might happen did.

At the 2018 Second International Summit of Human Genome Editing in Hong Kong, news of the world's first "CRISPR babies" born in China – a medically unnecessary, illegal human experimentation – propelled us into a new era. Along with my colleagues from the National Academies of Sciences and Medicine and the Royal Society, many of whom were participants at the 2015 Napa Valley conference, I maintained that the risks of germline editing remained too great to permit gene editing in embryos, egg cells, or sperm cells. Back in the U.S., I met with a number of senators to discuss the safest path forward for CRISPR technology – and its future impact on our health — and continued to guide conversations on the necessary regulations including the international commissions that have since been convened by the National Academies and by the World Health Organization.

In early 2020, I expanded my work with CRISPR as called for by the global COVID-19 pandemic. My colleagues at the IGI and I launched an automated coronavirus clinical testing laboratory in March 2020 built out of our research facilities over the course of just three weeks. Our group has since provided critical testing services to thousands across local and state communities. We also introduced a new initiative to develop a rapid CRISPR-based, point-of-need diagnostic test in addition to approximately two dozen additional research projects. We realized the need to further enable acts of scientific collaboration and innovation that would lead us out of the pandemic, releasing a roadmap detailing the transformation of our nonclinical labs into the testing facility and making all COVID-19 project-related intellectual property open source.

Our group continued to pursue genome engineering research that could elevate the standard of care for patients around the world and of the highest, currently unmet need. Our efforts to date, from Emmanuelle and my co-discovery of CRISPR to the genome editing revolution happening now, were chronicled by bestselling author and historian Walter Isaacson in his book *The Code Breaker: Jennifer Doudna, Gene Editing, and the Future of the Human Race* in March 2021. Shortly after, the IGI, leading a consortium of scientists and physicians across UC Berkeley, UCSF, and the University of California, Los Angeles, began the first FDA-approved clinical trial of a CRISPR-based therapy for directly correcting the genetic mutation that causes sickle cell disease. We are developing the sickle cell therapy, and future CRISPR gene therapies, to eventually treat disease from within the human body (in vivo) and extend to blood cancers, immunological conditions, and additional rare diseases that presently cannot be addressed. By improving our ability to rewrite the code of life and supporting the type of fundamental scientific research that made our discovery of CRISPR possible, I believe they soon can be.

The Chemistry of CRISPR: Editing the Code of Life/ CRISPR-Cas9: Biology and Technology of Genome Editing*

Nobel Lecture, December 8, 2020 by
Jennifer A. Doudna
University of California, Berkeley, CA, USA.

I THOUGHT I WOULD BEGIN BY TELLING YOU about the origin of the ideas around CRISPR. The work began at least two decades ago with research in microbiology laboratories that showed bacteria might have an adaptive immune system, a way to provide protection against viral infection. This adaptive immune system would protect cells from bacteriophages – viruses that inject their DNA into bacteria – through a recording system in the bacterial genome that came to be called CRISPR, which stands for Clustered Regularly Interspaced Short Palindromic Repeats.

* I'd like to begin by thanking the Royal Swedish Academy of Sciences, the Nobel Prize Committee in Chemistry; my family, including my spouse, Jamie Cate, our son, Andy, and my sisters, Ellen and Sarah Doudna; my friends, colleagues, and, of course, my former and current lab members, about whose research I will be speaking today. It's a wonderful honor to have this opportunity to share with you the science that we've done over the last few years, and to discuss the extraordinary opportunities and exciting advances that are happening right now with CRISPR-Cas9 as a genome editing technology.

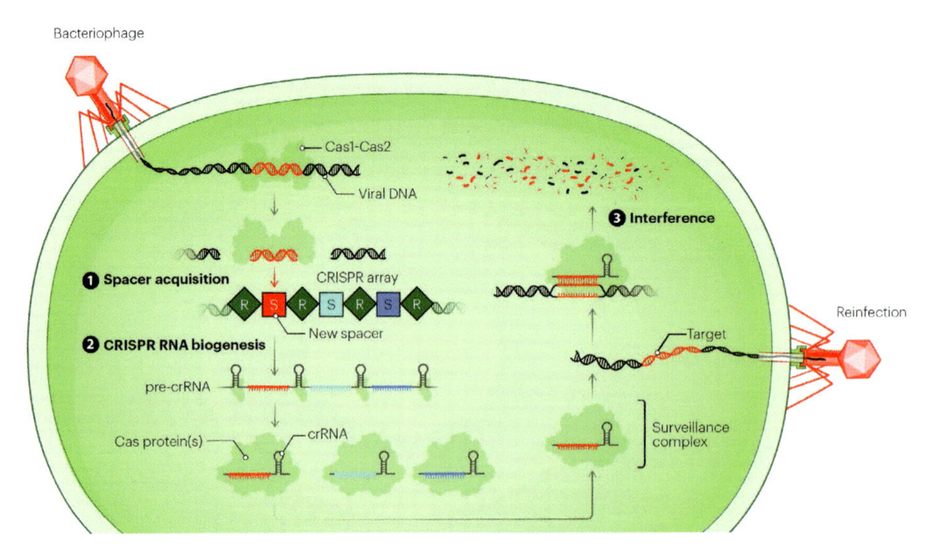

Figure 1. The illustration above shows the way the CRISPR system is thought to act in bacteria. It represents experimental results that were done in the field early on using bio-informatics and molecular genetics to understand the function of the CRISPR pathway.

We imagine that in nature, when a viral bacteriophage injects its DNA into a bacterial cell, the cell can integrate small pieces of that foreign DNA into its genome at a locus or site called CRISPR, which consists of alternating sequences called repeats and spacers. A CRISPR RNA molecule (cRNA) makes a copy of the viral DNA that is passed on to CRISPR-associated or Cas proteins, which are enzymes that use this genetic information to detect and destroy the viral DNA by cleaving it. Through the CRISPR RNA biogenesis process, the bacterial cell also acquires immunity from similar invasions in the future.

Over time, the CRISPR immune system has evolved and diversified into many different forms, but today I'm actually going to talk in particular about one type of CRISPR-Cas system that uses the protein Cas9.

Back in 2011, we had the good fortune to begin collaborating with Emmanuelle Charpentier and her student, Krzysztof Chylinski. This collaboration launched a wonderful opportunity to answer what was at the time a very interesting and intriguing question in the CRISPR field: What is the function of the Cas9 protein?

We were fascinated by this question because Cas9 had been implicated in the protection of cells against invasive DNA, particularly in *streptococcus pyogenes*, a kind of bacteria Emmanuelle's group was studying that infects humans. These bacteria have a Cas9 protein encoded in their CRISPR system that was implicated in protecting the cells from viral infection. But the question was how? In our collaboration, we addressed

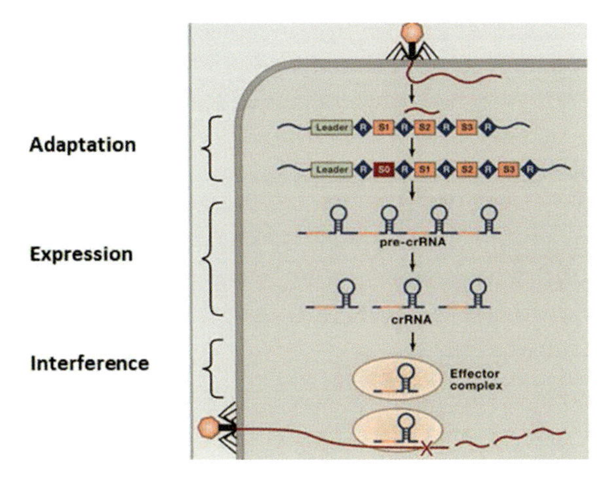

Figure 2. The illustration above shows the three steps in CRISPR acquired immunity that include adaptation, expression and interference. In the research that I did in my laboratory with a number of my former students, beginning with Blake Wiedenheft and then Rachel Haurwitz, and folks that came after them, we began investigating in particular, interference, the third step in the CRISPR pathway. Interference involves an RNA-guided detection of viral DNA.

this question by doing biochemical experiments that involved working withpurified CRISPR-Cas9 protein and the RNA that guides it to target DNA sequences and cells.

From research done by Martin Jínek, a former postdoc in my laboratory, and Krzysztof in Emmanuelle's laboratory, we found that CRISPR-Cas9 in nature is a dual RNA-guided protein.

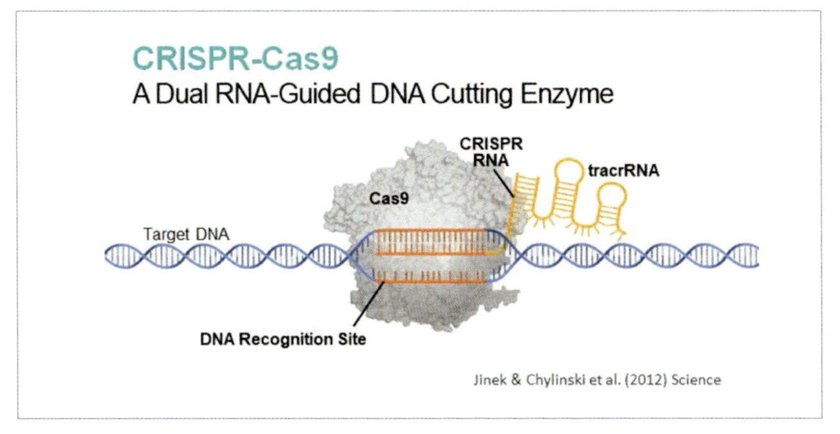

Figure 3. CRISPR-Cas9 A dual RNA-guided DNA cutting enzyme.

The Cas9 protein uses a CRISPR-RNA molecule to direct it to a target sequence of viral DNA that matches the CRISPR-RNA sequence. The process also requires a second RNA molecule called tracer – or tracrRNA – that interacts with the CRISPR RNA to guide Cas9 to target DNA sequences. Once it reaches the target, Cas9 is able to make a cut in the double helix of the viral DNA.

One of the wonderful aspects of CRISPR and this project with Emmanuelle's group is that we reached a point in our research where what began as a curiosity-driven investigation morphed into a project that had much broader implications.

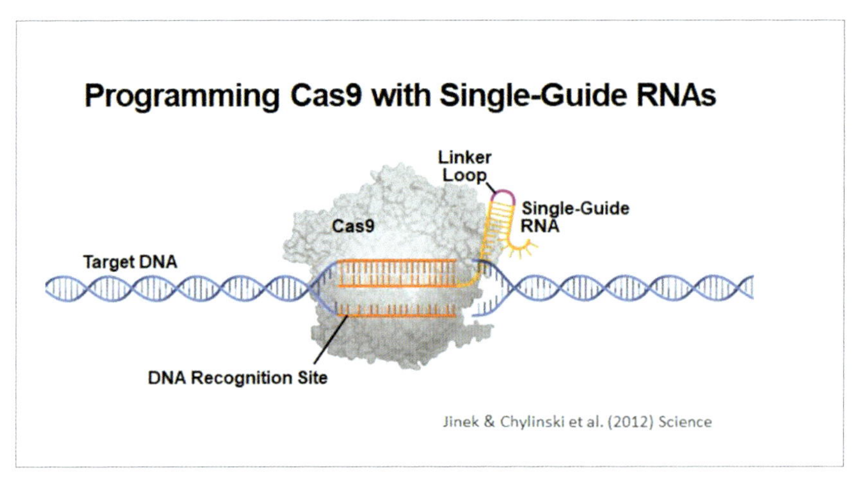

Figure 4. Programming Cas9 with Single-Guide RNAs.

Once we understood how nature uses a dual-RNA system to guide Cas9 to target DNA sequences, it became possible to engineer the dual-guide RNA as a single-guide RNA molecule that includes both the targeting information and the structural requirement for assembly with the Cas9 protein. This single-guide RNA can then be used to program Cas9, directing it to cleave double-stranded DNA at a desired sequence. To accomplish this, we take advantage of the actual targeting information in the CRISPR-RNA molecule and a short sequence of DNA in the virus called "PAM" for protospacer adjacent motif. The presence of a PAM sequence adjacent to the target sequence in the viral DNA activates the RNA-guided Cas9 to cleave the target.

I want to show you a key experiment that Martin did to test this idea of using a single-guide RNA to target Cas9 proteins to cleave specific sequences of DNA. The experiment was to design guide RNAs that would recognize several different sequences in a plasmid DNA, a circular piece of double-stranded DNA that we could purify in the laboratory.

What you're seeing in the above image are five different sites in the

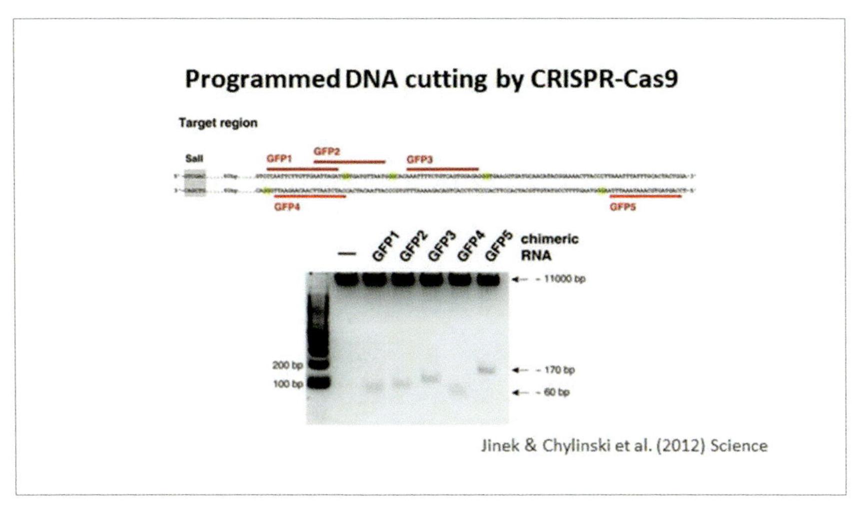

Figure 5. Programmed DNA cutting by CRISPS-Cas9.

plasmid DNA – marked in red – that were chosen as target sequences for single-guide RNAs we produced in the laboratory. Martin added those single-guide RNAs to a purified Cas9 protein and incubated them together with the plasmid DNA molecule in a laboratory test tube. To analyze the results of that experiment, he separated the different cleaved DNA products from one another in an agarose gel system shown in the image. What you can see in each lane of this gel system is that depending on where the guide-RNA was directed to interact with the plasmid DNA, Cas9 would generate a cut. By also cutting the plasmid at a separate place so the two double-stranded breaks were introduced into the plasmid at one time, we could release these cut fragments of DNA into the gel system. As you can see, each cut fragment of plasmid DNA migrated to a different position based on the size of the fragment.

I have to say that on the day Martin did this experiment and got these results, we were just incredibly excited. It was the pure joy of discovery at recognizing that we not only understood how the Cas9 bacterial protein functions, but we had also actually figured out how to engineer it as a simple two component system for directing DNA double-stranded cutting. Why was that so exciting? Well, it was, of course, interesting to know that we could harness our knowledge in this way and engineer the protein to have this desired cleavage capability. But, in addition, it also allowed us to imagine how CRISPR-Cas9 could actually be harnessed as a technology for something quite different in eukaryotic cells: cells like plant, animal and human cells, all of which treat double-strand breaks in DNA differently than the way they're treated in bacteria. I'll show this in the next couple of images.

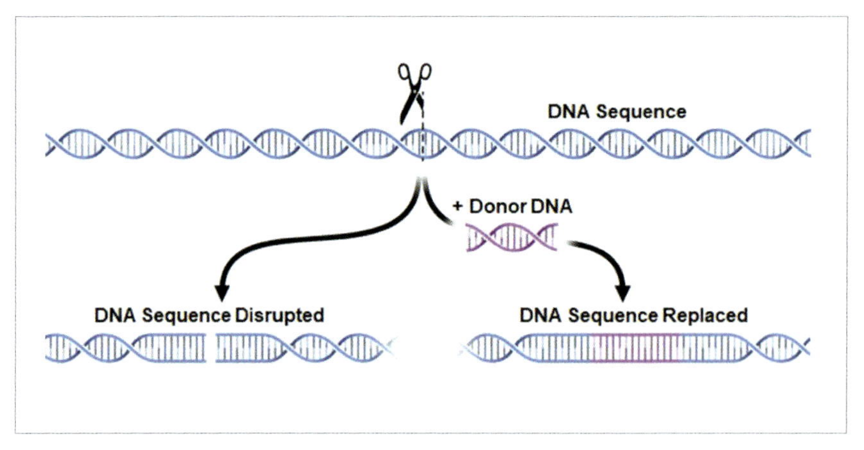

Figure 6.

Typically, in a eukaryotic cell, when the cell receives or detects a double-stranded break in its DNA, the break is detected and repaired before it can cause cell death. The repair pathways involve either a non-homologous end-joining event (left side of the image) that introduces a small disruption to the DNA sequence, or there can be an integration of DNA that has homology to the sequence of DNA flanking the double-stranded break (right side of the image). In the latter case, a new piece of genetic information is incorporated into the genome at the site of the original break.

This research had been done over the previous couple of decades before Emmanuelle and I did our work with Cas9. We recognized that the ability of Cas9 to introduce a double-stranded break into DNA at a desired position could allow scientists to introduce double-stranded breaks into a genome that could trigger the kind of repair shown in this image.

We then imagined a system where the Cas9 protein could be directed to enter the nucleus of a eukaryotic cell and – directed by its guide-RNA – search the cell's DNA for a 20-base pair sequence that matches the 20-base pair sequence of the RNA guide. When that match occurs, we now understand that Cas9 is able to unwind the DNA and generate a precise double-stranded break by cleaving each strand of DNA. The broken ends are handed off to repair enzymes in eukaryotic cells that lead to DNA repair. In the process of repairing the broken DNA, there is the opportunity to introduce a change to the genome at a precise place. This is really the definition of genome engineering.

In the next few images, I want to share a few things we've learned over the last few years about how Cas9 is able to achieve this kind of editing in the genomes of cells by triggering double-stranded breaks.

I'll start with showing a molecular model of the Cas9 protein.

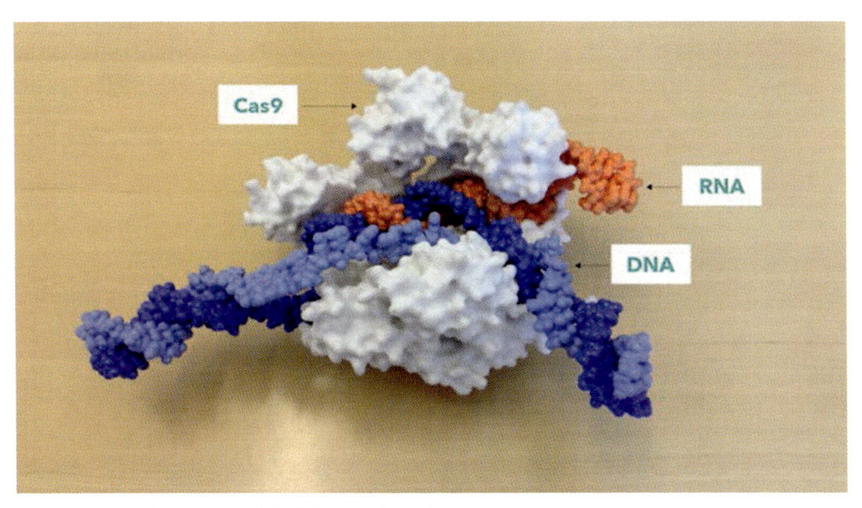

Figure 7. A molecular model of the Cas9 protein.

This model is based on a crystallographic structure that was solved originally by several different laboratories; our own, the lab of Martin Jínek, the lab of Osamu Nureki, and subsequently many others who have contributed to understanding the actual molecular basis for Cas9 functions. In this model, Cas9 is the white protein holding on to its orange guide RNA and a blue double-helical DNA molecule. I'd like to point out that in this structure, we can see the mechanism by which Cas9 uses a guide RNA molecule to interact with DNA at a precise position.

The orange and blue helix inside the Cas9 protein represents the interaction between the guide RNA and a single strand of the target DNA molecule. This is the actual way that these proteins are able to find and hold onto DNA prior to cleaving it.

The next thing we learned in studying how this cleavage event actually works is that Cas9 is a highly dynamic protein. It has to be in order to be able to handle DNA and unwind a double helix the way we know it does. It has to be able to move, and, through a series of chemical experiments that allowed us to detect motions in different parts of the protein, we discovered that Cas9 also has to be able to undergo a large conformational change as it holds on to DNA and catalyzes cutting.

The Cas9 protein morphs to the structure it forms when it binds to the guide RNA. This is the Cas9 structure that is able to search the cell, looking for a matching sequence of DNA. When that match occurs, there's an additional structural change in the protein that accommodates the DNA molecule – the RNA-DNA helix forming inside Cas9. Finally, this yellow part of Cas9 swings into position so that it can cleave the DNA strand that is attached to the guide RNA. This is a very important aspect of the chemistry by which Cas9 cuts DNA because it provides a mechanism for sensing the interaction between the guide RNA and the target DNA, and ensuring the accuracy of Cas9's cutting mechanism.

In work that was done just over the last few years, Fuguo Jiang, a former postdoc in my laboratory, and sadly now deceased, did a series of very exciting structural experiments to reveal the shape of the Cas9 protein when it's engaged on a full-length DNA molecule.

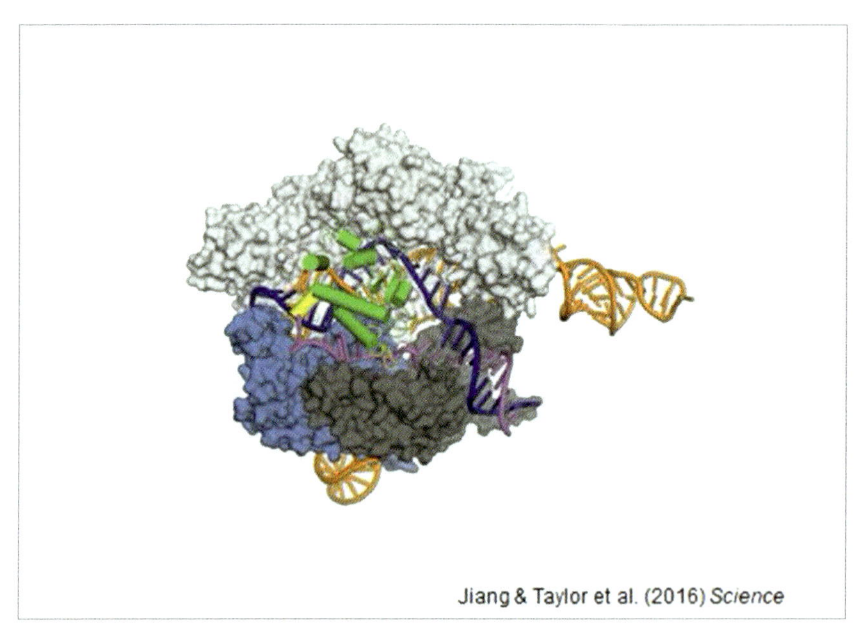

Figure 8. In the image above, the guide RNA is again shown in orange; the DNA strands are in blue and magenta. The DNA is held open by the Cas9 protein and allowed to interact with the cleavage sites in the enzyme for precise double-stranded cutting of both strands of the DNA. As shown here in green, one of the cutting parts of the enzyme swings into position to perform the catalytic chemistry required to cleave the DNA strand.

I want to also point out that in addition to these conformational changes that happen in the Cas9 protein itself, we also now understand this protein is quite dynamic in the way that it interacts with long pieces of DNA – for example, chromosomes and cells. It has to be able to move very quickly, along the length of DNA, searching for a sequence that will bear a complementary match to the guide RNA. How does that work? It seemed like an extraordinary capability to us initially.

However, in experiments that we did in the laboratory over several years with a number of former lab members and collaborators, we came up with the model shown in this image that I think is consistent with current data.

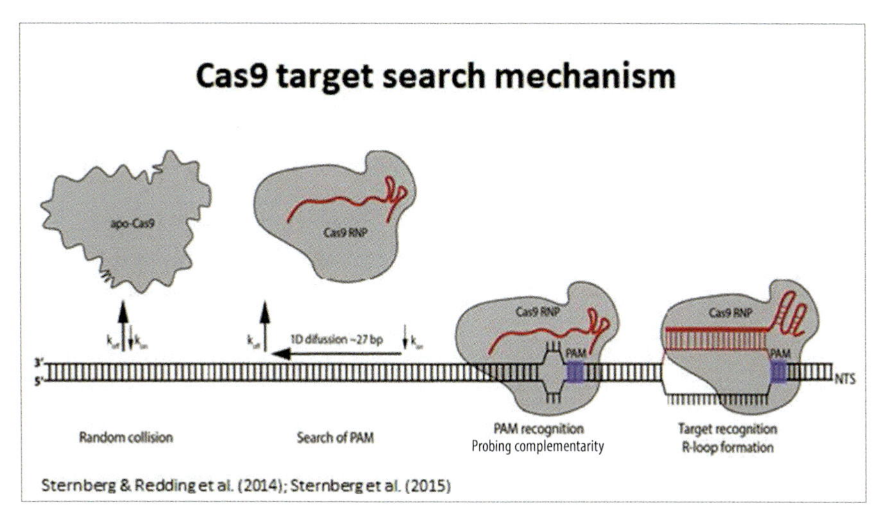

Figure 9. Cas9 target search mechanism.

The model suggests that the Cas9 protein has the ability to bind and release DNA very quickly, allowing it to search through large, really vast stretches of DNA quite fast. This model also allows interrogation of the sequence that's being searched for a match with the guide RNA, which is shown here in red. You can see that our data suggests that when Cas9 protein with its guide RNA interacts with DNA it begins to pry apart the two strands of the double helix. This allows the protein to determine if there is a complementary sequence for binding to the CRISPR-RNA. If there is, then our data suggests that the strands of the DNA continue to melt apart, allowing the RNA-DNA helix to form inside the Cas9 protein. If that helix is perfect or close to perfect, then the enzyme is triggered to cleave DNA.

This is quite an amazing mechanism that clearly allows the CRISPR immune system to search the cell very quickly, looking for viral DNAs to destroy. But in eukaryotic cells, this mechanism is equally effective at triggering double-stranded breaks that can be repaired and triggering changes in the format of genome editing that we now understand can be effectively catalyzed by Cas9.

In the next part of the talk, I want to turn my attention to where CRISPR technology is going. There's a lot that one could say here so I'm just going to hit on a few of the highlights.

CRISPR Opportunities and Challenges

Research
Public health
Agriculture
Biomedicine

Figure 10. CRISPR opportunities and challenges.

First of all, genome editing extends across all of biology. It can be used for fundamental research as well as for exciting applications in public health, agriculture, and biomedicine.

I think it's also very important to point out that genome editing can be conducted in many different kinds of cells, and fundamentally in the two kinds of cells cited, somatic cells and germ cells.

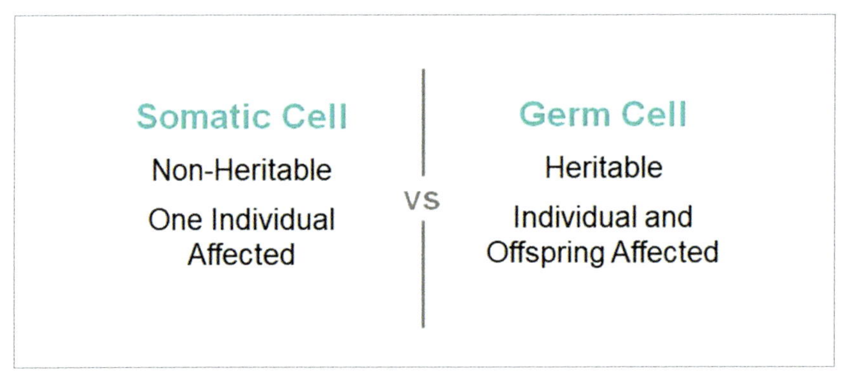

Figure 11. Somatic cell vs. germ cell.

A somatic cell is fully differentiated and does not have the ability to create a new organism. A germ cell, such as a sperm cell, an egg cell, or cells in an early embryo, is pluripotent, which means it is able to differentiate into many different cell types as an organism is forming.

If genome edits are introduced into a somatic cell, those DNA changes are not heritable; they affect only one cell or one tissue type, or one individual organism. But if genome edits are introduced into a germ cell, the DNA changes have the potential to be heritable and to introduce DNA changes that become part of not only an individual but all of that individual's progeny.

Heritability makes genome editing of germ cells a very powerful tool when we think about using it in plants or using it to create better animal models of human diseases, for example, as has been done using CRISPR-Cas9 in mice and rats. It's very different when we think about the enormous ethical and societal issues raised by the possibility of using germline editing in humans. I won't say too much more about that, but a very active area of my own work over the last few years is to think about the responsible use of CRISPR-Cas9. In particular, I want to ensure that transparency and careful thought goes into work in which CRISPR-Cas9 is applied to the human germline.

In somatic cell genome editing, however, I think there are extraordinary and exciting opportunities that we will be developing in the near future. One of these opportunities is no longer a "potential," but has actually been realized.

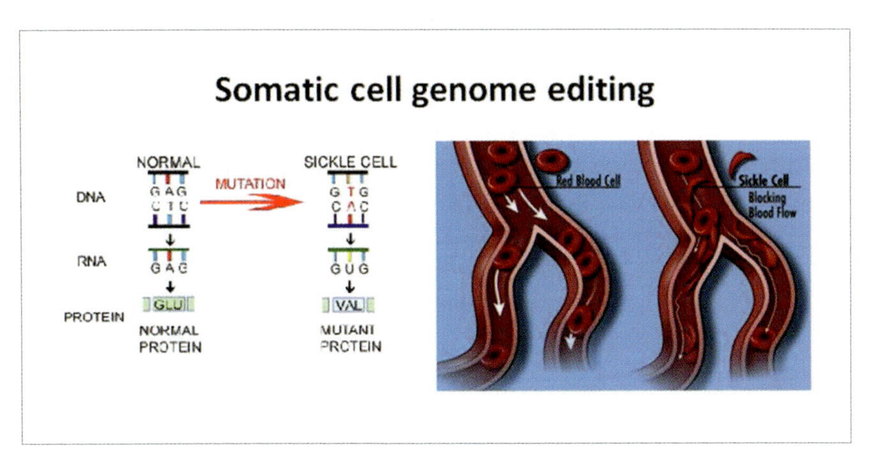

Figure 12. Somatic cell genome editing.

CRISPR-Cas9 technology has been shown to be a safe and effective way to treat sickle cell anemia in a patient by correcting the well-defined mutation that causes the disease. I think this is a very exciting way to imagine how CRISPR-Cas9 technology will impact human health in the future.

With regards to germline editing, I certainly imagine that we will see increasing applications in germ cells, including in the human germline. That has to be managed very carefully. I'm pleased that there's been an active international effort to control the use of CRISPR-Cas9 and certainly to encourage transparency.

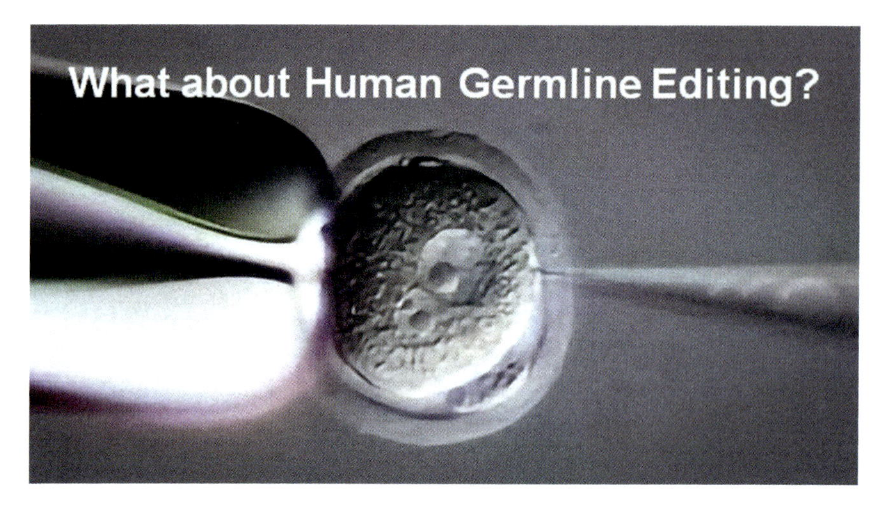

Figure 13. What about human germline editing?

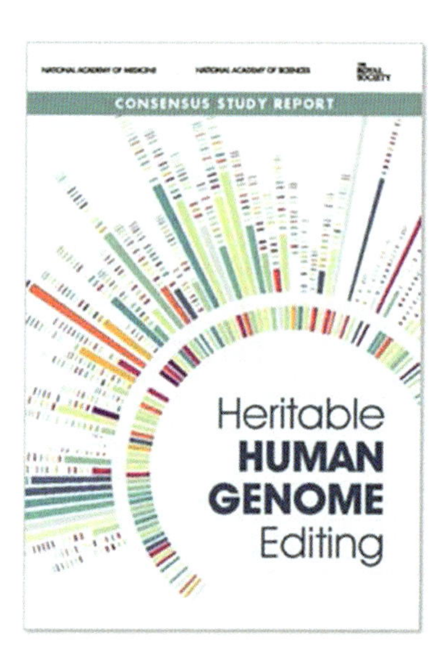

Figure 14. HHGE consensus study report.

A recent joint report from the National Academy of Medicine, the National Academy of Sciences and the Royal Society discusses the science and the technology around human germline editing and establishes criteria for using CRISPR-Cas9 in the human germline in the future.

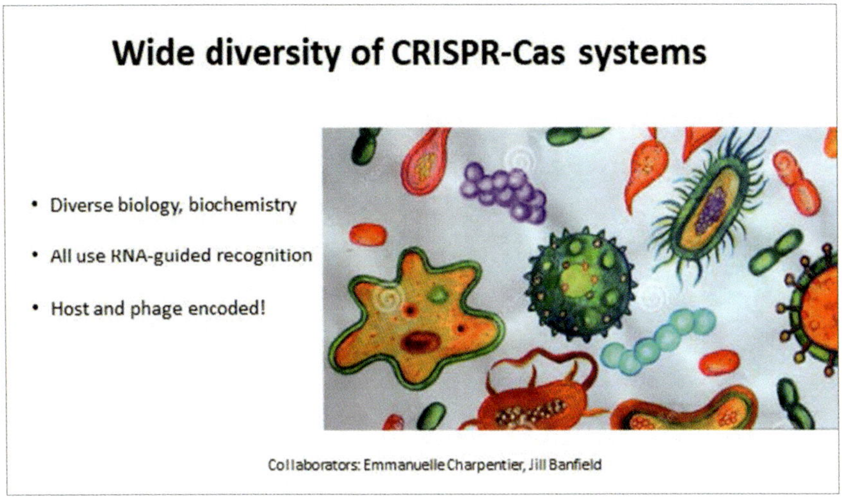

Wide diversity of CRISPR-Cas systems

- Diverse biology, biochemistry
- All use RNA-guided recognition
- Host and phage encoded!

Collaborators: Emmanuelle Charpentier, Jill Banfield

Figure 15. Wide diversity of CRISPR-Cas systems.

In the last part of my talk, I'd like to turn to where CRISPR technology is headed in the future.

One of the aspects of this work that's fascinating to me is the incredible diversity of CRISPR systems in nature. This diversity continues to drive the field in terms of fundamental biology; understanding what these systems do in their natural settings and microbes; and how they may be harnessed as technologies and other organisms, as was the case for CRISPR-Cas9.

I also wanted to mention briefly the effort to investigate new CRISPR-Cas systems. One of the recent findings we've had with our collaborator, Jillian Banfield at the University of California, Berkeley, is the discovery that bacteriophages, the viruses that bacteria use CRISPR to protect themselves from, in fact can also carry around their own CRISPR-Cas systems.

As shown in the image next page, one example, a protein we named CRISPR-CasØ, is entirely phage-encoded. It's a tiny protein but nonetheless has the RNA-guided, DNA-cutting capabilities that we discovered originally in CRISPR-Cas9. This is a fascinating example of nature's diversity and provides opportunities for future applications of CRISPR-CasØ. For example, there are cases in which one could benefit from having a very tiny protein with a small gene that could be more easily delivered into eukaryotic cells than CRISPR-Cas9.

We've also been very interested in the biochemical activities of other CRISPR-Cas proteins. I'll mention here very briefly research done originally by Alexandra East-Seletsky, a former graduate student in my laboratory, and Mitch O'Connell working in partnership with her. Alexandra and Mitch discovered that Cas13, a class of CRISPR proteins that are naturally occurring RNA-targeting enzymes, has a biochemical activity that could be harnessed as an RNA detection mechanism.

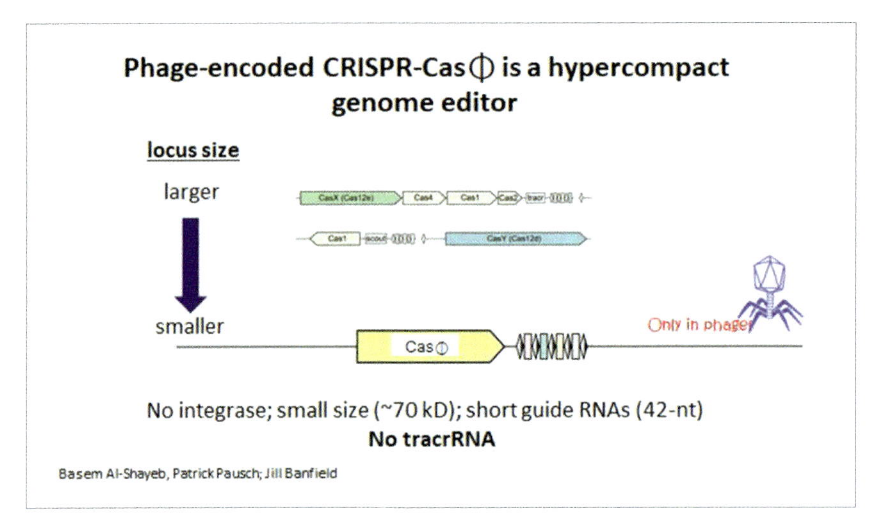

Figure 16. Phage-encoded CRISPR-CasØ is a hypercompact genome editor.

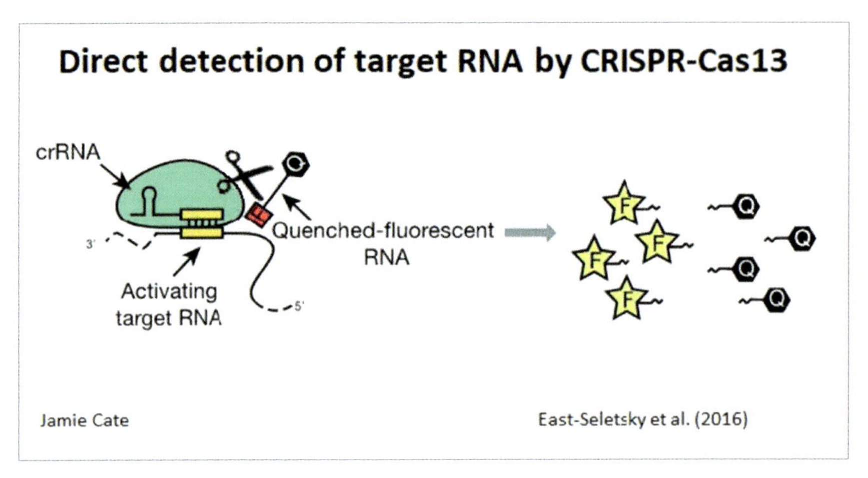

Figure 17. Direct detection of target RNA by CRIPR-Cas13.

The experiment originally performed by Alexandra and suggested by Jamie Cate is shown in the above image. The idea was to put fluorophores on to small pieces of RNA that were cleaved upon Cas13's detection of an RNA sequence using its guide-RNA.

The CRISPR-Cas13 system turned out to be highly effective at detecting RNA molecules all the way down to about picomolar levels.

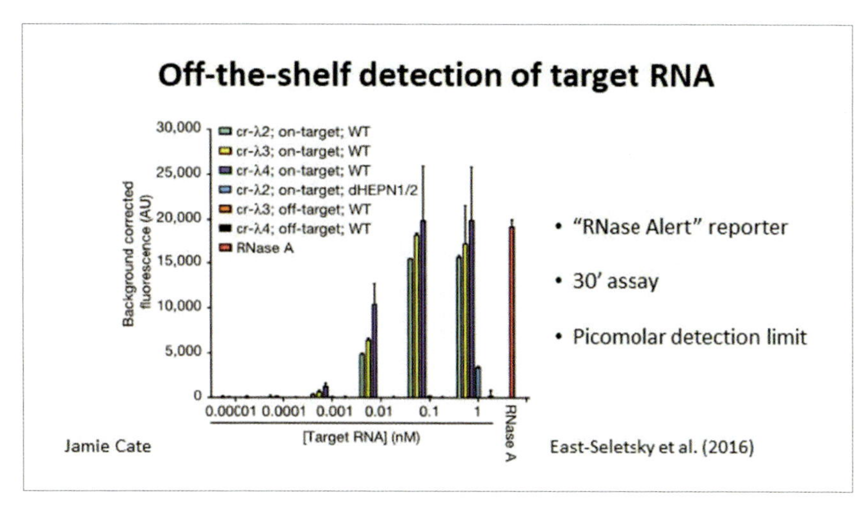

Figure 18. Off-the-shelf detection of target RNA.

This CRISPR-Cas13 experiment was carried out with an off-the-shelf kit called RNase Alert. The results, shown in the image above, triggered our interest in the biochemical activities of other families of CRISPR-Cas proteins in addition to Cas9.

In research done a couple of years ago, Janice Chen, a student in my lab at the time, showed that the family of Cas proteins called Cas12 also has the ability to cleave single-stranded molecules. In this case, Cas12 cleaves single-stranded DNA upon recognition of a double-stranded DNA target.

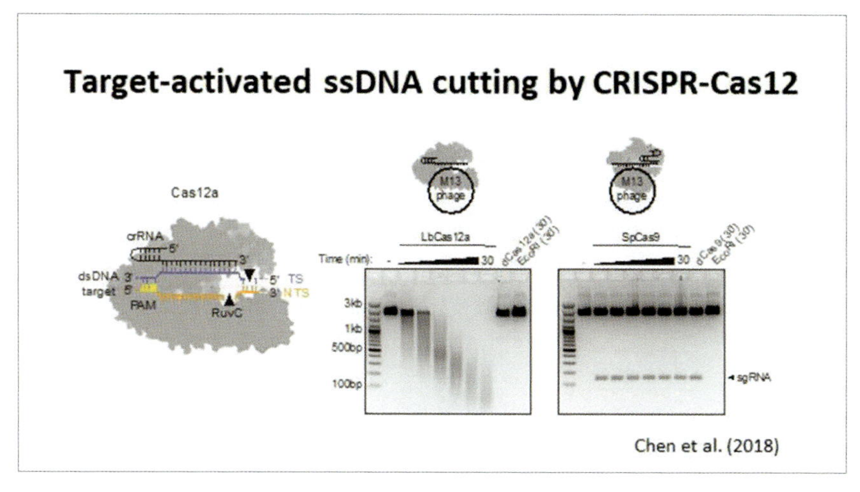

Figure 19. Target-activated ssDNA cutting by CRISPR-Cas12.

In the experiment illustrated above, Janice showed how the Cas12a protein, upon recognition of a double-stranded DNA, is able to cleave single-stranded molecules of DNA. In the cEnter panel of this image, we see very rapid degradation of a circular single-stranded DNA molecule, which is a biochemical activity we do not detect for Cas9 as shown in the panel on the right.

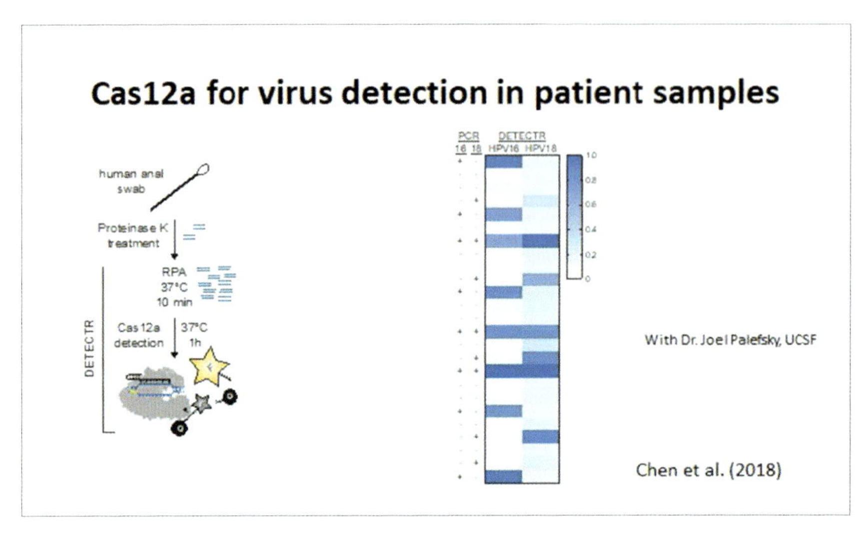

Figure 20. Cas12a for virus detection in patient samples.

Working with a colleague, Joel Palefsky at University of California, San Francisco, we were able to use Cas12a activity to detect the human papillomavirus in patient samples, and even to distinguish between two different strains of HPV as shown in the image above.

This study told us that not only can CRISPR-Cas proteins be useful for detection; they can also be useful for specificity in figuring out what type of viral signal might be present in a patient's sample.

In the current SARS-CoV-2 pandemic that has become a worldwide human health issue, we and others have been using a detection technique based on CRISPR-Cas13a to very quickly identify SARS-CoV-2 infections in patient saliva or in patient nasal swabs. Ultimately, as shown in the image below, we hope to develop a rapid Covid diagnostic technique based on CRISPR-Cas13a that would allow people to test themselves and record the results on their cell phones. This would enable people everywhere to screen themselves against the Covid virus. The programmability of CRISPR-Cas proteins, which is a fundamental property of their biology, and their ability to be harnessed as technology should also help us prepare for future pandemics.

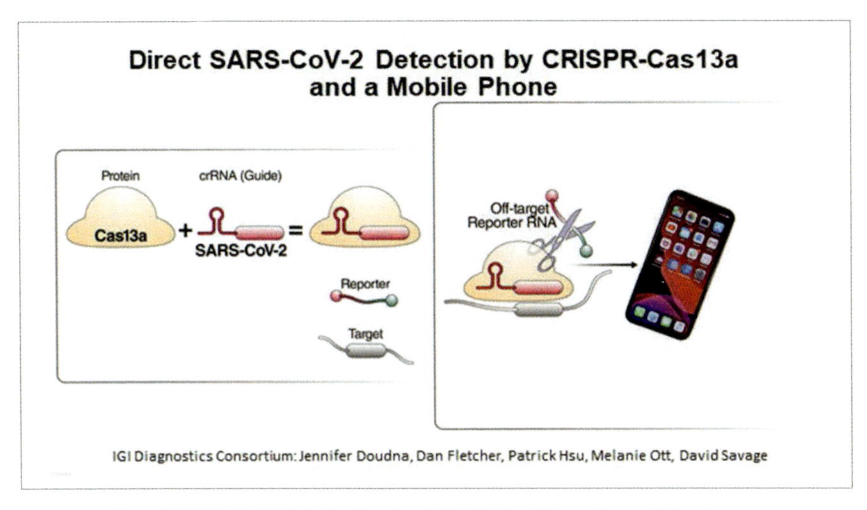

Figure 21. Direct SARS-CoV-2 detection by CRISPR-Cas13a and a mobile phone.

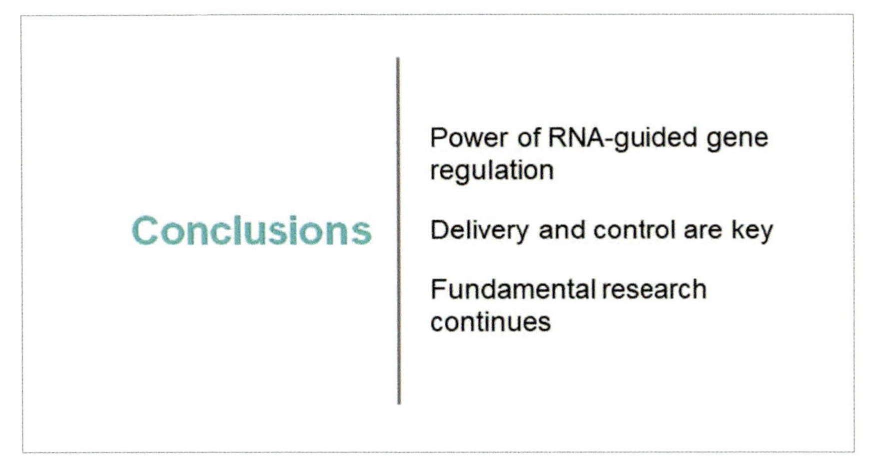

Figure 22. Conclusions.

I'd like to close by pointing out that the RNA-guided gene regulation we observe in CRISPR-Cas systems is fundamental to the way these systems work in bacteria, but also to the way they operate as technologies for genome editing and beyond.

Delivery and control are key to the future of genome editing. We need to have better ways to deliver CRISPR-Cas9 and related proteins into cell types of interest for genome editing, including into human patients. Fundamental research will continue to drive the field forward as it has in the past.

Figure 23. The possibilities are endless.

I'm excited about what will happen in the future both with fundamental research, and with the applications that will solve real world problems in human health and the environment.

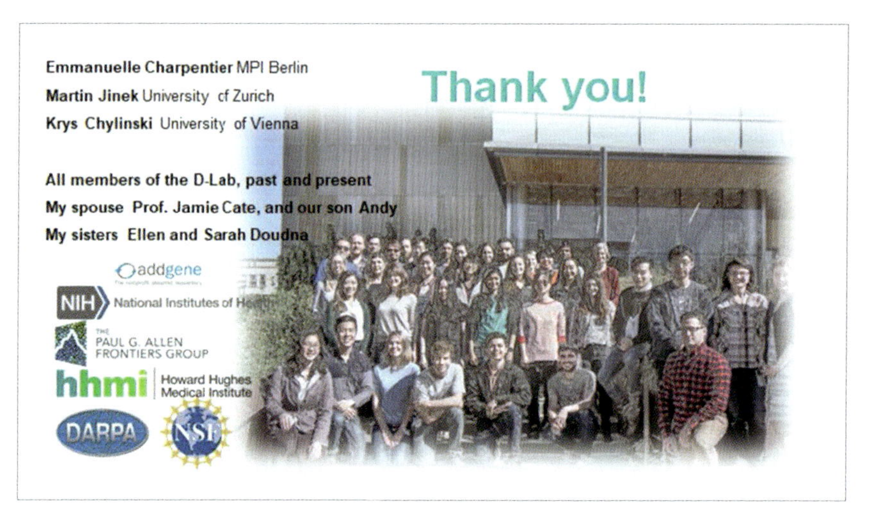

Figure 24. Thanks.

I'd like to thank my past and current lab members with whom I've had the joy of doing science together over many years. I want to thank the many colleagues who have been involved in the fields of genome engineering, in DNA repair, and in the applications of genome editing that have made the field so exciting over the last few years. And finally, I want to thank my colleagues at the University of California, Berkeley, who I've had the joy of working with, and where we share together the dedication to public education and research that makes our work so rewarding. Thank you very much.